Biotechnology in Agriculture and Forestry

Springer-Verlag Berlin Heidelberg GmbH

Biotechnology in Agriculture and Forestry 45

Transgenic Medicinal Plants

Edited by Y.P.S. Bajaj

With 130 Figures, 6 in Color, and 79 Tables

 Springer

Professor Dr. Y.P.S. BAJAJ
A-137
New Friends Colony
New Delhi 110065, India

ISSN 0934-943-X
ISBN 978-3-642-63595-3   ISBN 978-3-642-58439-8 (eBook)
DOI 10.1007/978-3-642-58439-8

Library of Congress Cataloging-in-Publication Data.

Transgenic medicinal plants/edited by Y.P.S. Bajaj. p. cm. – (Biotechnology in agriculture and forestry; 45) Includes bibliographical references and index. ISBN 3-540-65120-9 (alk. paper: hardcover) 1. Medicinal plants – Genetic engineering. 2. Transgenic plants. I. Bajaj, Y. P. S., 1936– II. Series. SB293.T735 1999 633.8'8 – ddc21 98-37572 CIP

© Springer-Verlag Berlin Heidelberg 1999
Originally published by Springer-Verlag Berlin Heidelberg New York in 1999
Softcover reprint of the hardcover 1st edition 1999

Cover design: *design & production* GmbH, Heidelberg

Typesetting: Best-set Typesetter Ltd., Hong Kong

SPIN: 10865931   31/3111 – 5 4 3 2 1 – Printed on acid-free paper

Dedicated to
Brijinder and Chetana Maya Nagpal

Preface

Over the last decade there has been tremendous progress in the genetic transformation of plants, which has now become an established tool for the insertion of specific genes. Work has been conducted on more than 200 plant species of trees, cereals, legumes and oilseed crops, fruits and vegetables, medicinal, aromatic and ornamental plants etc. Transgenic plants have been field-tested in a number of countries, and some released to the farmers, and patented.

Taking the above-mentioned points into consideration, it appeared necessary to review the literature and state of the art on genetic transformation of plants. Thus 120 chapters contributed by experts from 31 countries (USA, Russia, Canada, France, Germany, England, The Netherlands, Belgium, Switzerland, Italy, Spain, Bulgaria, Yugoslavia, Denmark, Poland, Finland, Australia, New Zealand, South Africa, China, Japan, Korea, Singapore, Indonesia, India, Israel, Mexico, Brazil, Moroco, Senegal, Cuba, etc.) have been compiled in a series composed of the following five books:

1. *Transgenic Trees* comprises 22 chapters on forest, fruit, and ornamental species such as *Allocasuarina verticillata*, *Casuarina glauca*, *Cerasus vulgaris*, *Citrus* spp., *Coffea* species, *Diospyros kaki*, *Eucalyptus* spp., *Fagara zanthoxyloides*, *Larix* spp., *Lawsonia inermis*, *Malus* × *domestica*, *Picea mariana*, *Pinus palustris*, *Pinus radiata*, *Poncirus trifoliata*, *Populus* spp., *Prunus* species, *Rhododendron*, *Robinia pseudoacacia*, *Solanum mauritianum*, *Taxus* spp., and *Verticordia grandis*.
2. *Transgenic Medicinal Plants* comprises 26 chapters on *Ajuga reptans*, *Anthemis nobilis*, *Astragalus* species, *Atropa belladonna*, *Catharanthus roseus*, *Datura* species, *Duboisia* species, *Fagopyrum* species, *Glycyrrhiza uralensis*, *Lobelia* species, *Papaver somniferum*, *Panax ginseng*, *Peganum harmala*, *Perezia* species, *Pimpinella anisum*, *Phyllanthus niruri*, *Salvia miltiorrhiza*, *Scoporia dulcis*, *Scutellaria baicalensis*, *Serratula tinctoria*, *Solanum aculeatissimum*, *S. commersonii*, *Swainsona galegifolia*, tobacco, and *Vinca minor*.
3. *Transgenic Crops I* comprises 25 chapters divided into 2 sections:
   Section I. *Cereals and grasses*, such as wheat, rice, maize, barley, sorghum, pearl millet, triticale, *Agrostis*, *Cenchrus*, *Dactylis*, *Festuca*, *Lolium*, and sugarcane.
   Section II. *Legumes and Oilseed Crops. Arachis hypogaea*, *Brassica juncea*, *Brassica napus*, *Cicer arietinum*, *Glycine max*, *Gossypium hirsutum*,

Helianthus annuus, Lens culinaris, Linum usitatissimum, Sinapis alba, Trifolium repens, and *Vicia narbonensis.*

4. *Transgenic Crops II* comprises 23 chapters on fruits and vegetables, such as banana, grapes, strawberry, kiwi, watermelon, cucumber, tomato, asparagus, carrot, cabbage, kale, turnip, rutabaga, Brussel sprouts, broccoli, sweet pea, common bean, *Luffa, Amaranthus*, horseradish, sugarbeet, chicory, cassava, sweet potato, potato, etc.

5. *Transgenic Crops III* comprises 26 chapters arranged in 2 sections: *Section I. Ornamental, Aromatic and Medicinal Plants. Anthurium, Antirrhinum, Artemisia absinthium, Begonia, Campanula*, Carnation, *Chrysanthemum, Dendrobium, Eustoma, Gentiana, Gerbera, Gladiolus, Hyoscyamus muticus, Hyssopus officinalis, Ipomoea, Leontopodium alpinum, Nierembergia, Phalaenopsis, Rudbeckia, Tagetes* and *Torenia.*

   Section II. Miscellaneous Plants. Craterostigma plantaginsum, Flaveria spp., *Moricandia arvensis, Solanum brevidens*, and freshwater wetland monocols.

These books will be of special interest to advanced students, teachers, and research workers in the field of molecular biology, genetic manipulation, tissue culture, and plant biotechnology in general.

New Delhi, April 1999

Professor Dr. Y.P.S. BAJAJ
Series Editor

Contents

XI Genetic Transformation in *Lobelia* Species
K. ISHIMARU and K. SHIMOMURA (With 16 Figures)

XII Genetic Transformation of *Papaver somniferum* L. (Opium
Poppy) for Production of Isoquinoline Alkaloids
K. YOSHIMATSU and K. SHIMOMURA (With 3 Figures)

XIII Genetic Transformation of *Panax ginseng* (Ginseng)
J.R. LIU, H.S. LEE, and S.W. KIM (With 3 Figures)

XIV Genetic Transformation of *Peganum harmala*
J. BERLIN (With 4 Figures)

XV Genetic Transformation of *Perezia* Species
J. ARELLANO and G. HERNÁNDEZ (With 5 Figures)

XXII Genetic Transformation of *Solanum aculeatissimum*
T. IKENAGA and T. MURANAKA (With 9 Figures)

XXIII Genetic Transformation of *Solanum commersonii* Dun.
T. CARDI (With 3 Figures)

XXIV Genetic Transformation of *Swainsona galegifolia*
(Darling Pea)
T.M. ERMAYANTI

XXV Transgenic Tobacco: Gene Expression and Applications
B.L.A. MIKI, S.G. MCHUGH, H. LABBE, T. OUELLET, J.H. TOLMAN,
and J.E. BRANDLE (With 5 Figures)

XXVI Genetic Transformation of *Vinca minor* L.
N. TANAKA (With 6 Figures)

List of Contributors

ALFERMANN, A.W., Institut für Entwicklungs- und Molecularbiologie der Pflanzen, Heinrich-Heine-Universität Düsseldorf, Universitätstr. 1, 40225 Düsseldorf, Germany

ARELLANO, G.J.J., Centro de Investigación sobre Fijación de Nitrógeno, Universidad Nacional Autónoma de México, Av. Universidad s/n Col. Chamilpa CP 62210 Apdo. Postal 565-A, Cuernavaca, Morelos, México

BAJAJ, Y.P.S., Biotechnology in Agriculture and Forestry, A-135 New Friends Colony, New Delhi 110065, India

BERLIN, J., GBF – Gesellschaft für Biotechnologische Forschung m.b.H., Mascheroder Weg 1, 38124 Braunschweig, Germany

BRANDLE, J.E., Southern Crop Protection and Food Research Centre, Agriculture and Agri-Food Canada, London, Ontario N5V 4T3, Canada

CARDI, T., CNR-IMOF, Research Institute for Vegetable and Ornamental Plant Breeding, via Universita' 133, 80055 Portici, Italy

CHAPUIS, L., INRA-Bordeaux, Unité de Recherches Intégrées sur la Vigne, Phytopharmacie, BP 81, Rue E Bourleaux, 33883 Villenave d'Ornon, France

CHARLWOOD, B.V., Division of Life Sciences, King's College London, Campden Hill Road, London W8 7AH, UK and Department of Chemistry, Universidade Federal de Alagoas, Campus Universitário, 57072-970 Maceió, Brazil

CORIO-COSTET, M.-F., INRA-Bordeaux, Unité de Recherches Intégrées sur la Vigne, Phytopharmacie, BP 81, Rue E Bourleaux, 33883 Villenave d'Ornon, France

DELBECQUE, J.-P., Laboratoire de Neuroendocrinologie, Université de Bordeaux I, Avenue des facultés, 33405 Talence, France

ERMAYANTI, T.M., Research and Development Centre for Biotechnology, Indonesian Institute of Sciences (LIPI), Jalan Raya Bogor Km. 46, Cibinong, 16911, P.O. Box 422, Bogor, Indonesia

FAUCONNIER, M.-L., Faculté des Sciences Agronomiques, UER Chimie Générale et Organique, Passage des Déportés 2, 5030 Gembloux, Belgium

GARNIER, F., Laboratory of Plant Molecular Biology, EA2106, Faculty of Pharmacy, 31 avenue Monge, 37200 Tours, France

GUO, Y.-W., Laboratory of Plant Biotechnology, Université Libre de Bruxelles, Chaussée de Wavre 1850, 1160 Bussels, Belgium and Laboratory of Pharmacognosy and Bromatology, Université Libre de Bruxelles, Campus plaine CP205/4, Bld. Triomphe, 1050 Brussels, Belgium

HAMDI, S., Laboratory of Plant Molecular Biology, EA2106, Faculty of Pharmacy, 31 avenue Monge, 37200 Tours, France

HAYASHI, T., Facuty of Pharmaceutical Sciences, Toyama Medical and Pharmaceutical University, 2630 Sugitani, Toyama 930-0194, Japan

HERNÁNDEZ, D.G., Centro de Investigación sobre Fijación de Nitrógeno, Universidad Nacional Autónoma de México, Av. Universidad s/n Col. Chamilpa CP 62210 Apdo. Postal 565-A, Cuernavaca, Morelos, México

HIROTANI, M., School of Pharmaceutical Sciences, Kitasato University, Shirokane 5-9-1, Minato-ku, Tokyo 108-8641, Japan

HU, Z.B., Laboratory of Biotechnology of Chinese Materia Medica, Shanghai University of Traditional Chinese Medicine, 530 Ling Ling Road, Shanghai 200032, P.R. China

IKENAGA, T., School of Environmental Sciences, Nagasaki University 1-14, Bunkyo-machi, Nagasaki 852-8521, Japan

IONKOVA, I., Faculty of Pharmacy, Department of Pharmacognosy, Medical University-Sofia, Dunav 2, 1000 Sofia, Bulgaria

ISHIMARU, K., Department of Applied Biological Sciences, Faculty of Agriculture, Saga University, 1 Honjo, Saga 840-8502, Japan

JAZIRI, M., Laboratory of Plant Biotechnology, Université Libre de Bruxelles, Chaussée de Wavre 1850, 1160 Brussels, Belgium

KAMADA, H., Gene Experiment Center, University of Tsukuba, Tsukuba, Ibaraki 305-8571, Japan

KIM, S.W., Genetic Resources Center, Korea Research Institute of Bioscience and Biotechnology, P.O. Box 115, Yusong, Taejon 305-333, Korea

KITAMURA, Y., School of Pharmaceutical Sciences, Nagasaki University, Nagasaki, 852-8521 Japan

KOBAYASHI, T., Department of Biotechnology, Faculty of Engineering, Nagoya University, Furo-cho, Chikusa-ku, Nagoya 464-8603, Japan

LABBE, H., Eastern Cereal and Oilseed Research Centre, Agriculture and Agri-Food Canada, Ottawa, Ontario K1A 0C6, Canada

LABEL, P., INRA, Research Centre of Orléans, Department of Genetic Engineering of Forest Trees, Ardon, 45160 Orléans, France

LEE, H.S., Plant Biochemistry Research Unit, Korea Research Institute of Bioscience and Biotechnology, P.O. Box 115, Yusong, Taejon 305-333, Korea

LIU, D., Laboratory of Biotechnology of Chinese Materia Medica, Shanghai University of Traditional Chinese Medicine, 530 Ling Ling Road, Shanghai 200032, P.R. China

LIU, J.R., Plant Cell and Molecular Biology Research Unit, Korea Research Institute of Bioscience and Biotechnology, P.O. Box 115, Yusong, Taejon 305-333, Korea

LOYOLA-VARGAS, V.M., Unidad de Biología Experimental, Centro de Investigación Cientifica de Yucatán, Apdo. Postal 87, CP 97310, Cordemex, Yucatán, México

MARLIER, M., Faculté des Sciences Agronomiques, UER Chimie Générale et Organique, Passage des Déportés 2, 5030 Gembloux, Blegium

McHUGH, S.G., Eastern Cereal and Oilseed Research Centre, Agriculture and Agri-Food Canada, Ottawa, Ontario K1A 0C6, Canada

MIKI, B.L.A., Eastern Cereal and Oilseed Research Centre, Agriculture and Agri-Food Canada, Ottawa, Ontario K1A 0C6, Canada

MURANAKA, T., Biotechnology Laboratory, Sumitomo Chemical Co., Ltd., 2-1, 4-Chome Takatsuka, Takatsukasa, Takarazuka, Hugo 665-8555, Japan

OUELLET, T., Eastern Cereal and Oilseed Research Centre, Agriculture and Agri-Food Canada, Ottawa, Ontario K1A 0C6, Canada

QUETTIER-DELEU, C., Laboratoire de Physiologie Cellulaire et Morphogénèse Végétales, Université des Sciences & Technologies de Lille (USTL), 59655 Villeneuve d'Ascq Cedex, France

RIDEAU, M., Laboratory of Plant Molecular Biology, EA2106, Faculty of Pharmacy, 31 avenue Monge, 37200 Tours, France

SAITO, K., Research Center of Medicinal Resources, Faculty of Pharmaceutical Sciences, Chiba University, Yayoi-cho 1-33, Inage-ku, Chiba 263-8522, Japan

SALEM, K.M.S.A., Division of Life Sciences, King's College London, Campden Hill Road, London W8 7AH, UK

SHIMOMURA, K., Tsukuba Medicinal Plant Research Station, National Institute of Health Sciences, 1 Hachimandai, Tsukuba, Ibaraki 305-0843, Japan

TANAKA, N., Center for Gene Science, Hiroshima University, Kagamiyama 1-4-2, Higashi-Hiroshima 739-8527, Japan

TOLMAN, J.H., Southern Crop Protection and Food Research Centre, Agriculture and Agri-Food Canada, London, Ontario N5V 4T3, Canada

TROTIN, F., Laboratoire de Physiologie Cellulaire et Morphogénèse Végétales, Université des Sciences & Technologies de Lille (USTL), 59655 Villeneuve d'Ascq Cedex, France

UOZUMI, N., Bioscience Center, Nagoya University, Furo-cho, Chikusa-ku, Nagoya 464-8601, Japan

USTACHE, K., INRA-Bordeaux, Unité de Recherches Intégrées sur la Vigne, Phytopharmacie, BP 81, Rue E Bourleaux, 33883 Villenave d'Ornon, France

VANHAELEN, M., Laboratory of Pharmacognosy and Bromatology, Université Libre de Bruxelles, Campus plaine CP205/4, Bld. Triomphe, 1050 Brussels, Belgium

VASSEUR, J., Laboratoire de Physiologie Cellulaire et Morphogénèse Végétales, Université des Sciences & Technologies de Lille (USTL), 59655 Villeneuve d'Ascq Cedex, France

YAMAKAWA, T., Department of Global Agricultural Sciences, The University of Tokyo, Yayoi 1-chome, Bunkyo-ku, Tokyo 113-8654, Japan

YAMAZAKI, M., Research Center of Medicinal Resources, Faculty of Pharmaceutical Sciences, Chiba University, Yayoi-cho 1-33, Inage-ku, Chiba 263-8522, Japan

YOSHIMATSU, K., Tsukuba Medicinal Plant Research Station, National Institute of Health Sciences, 1 Hachimandai, Tsukuba Ibaraki, 305 Japan

I Genetic Transformation of Medicinal Plants

Y.P.S. Bajaj[1] and K. Ishimaru[2]

1 Introduction

In spite of the great progress and developments in the synthesis of chemical compounds, plants still remain the main source of medicinal compounds flavors, dyes, pesticides, and other pharmaceutics. However, there are many species of medicinal plants which are rare, endangered, or threatened with extinction; their germplasm needs to be conserved, multiplied, and improved for various traits, especially for the production of secondary metabolites. Moreover, new genetic variability also needs to be generated and incorporated into the medicinal plants, and here biotechnology plays an important role (see Bajaj 1998). These studies have been conducted on numerous species of medicinal plants; however, *Atropa belladonna* and *Nicotiana tabacum* are the two model systems on which numerous basic research studies have been conducted. The regeneration of complete plants from isolated cells and tissues is of great significance in obtaining transgenic plants. The author's work primarily deals with the regeneration of plants from isolated protoplasts (Bajaj 1972; Bajaj et al. 1978), isolated pollen (Bajaj 1978), from cryopreserved cell suspensions (Bajaj 1976), pollen embryos (Bajaj 1977), and protoplasts (Bajaj 1988). *Atropa* has also been subjected to a number of studies on the in-vitro production of alkaloids through cell suspensions and hairy roots (Bajaj and Simola 1991).

Various in-vitro methods for gene transfer, such as embryo rescue (Bajaj 1990) and somatic hybridization (see Bajaj 1994) have been employed for wide hybridization, but each has its own limitations. For instance, transfer of genes through protoplast fusion is more or less "juggling"; there is not much specificity, the outcome may be unpredictable, and much selection is required. On the other hand, through recombinant DNA technology, a specific desirable gene-determining single trait can be incorporated into the system without concomitant transfer of undesirable traits.

The studies on various aspects of tissue culture/biotechnology have been recently reviewed (Bajaj 1998). During the past 10 years, tremendous progress

[1] Biotechnology in Agriculture and Forestry, A-137 New Friends Colony, New Delhi 110065, India
[2] Department of Applied Biological Sciences, Faculty of Agriculture, Saga University, 1 Honjo, Saga 840, Japan

Biotechnology in Agriculture and Forestry, Vol. 45
Transgenic Medicinal Plants (ed. by Y.P.S. Bajaj)

has been made in the genetic transformation of plants. At present there are about 200 plant species, including agricultural crops, trees, ornamentals, fruits, vegetables, etc., on which such studies have been conducted. Of these, about 70 species belong to medicinal plants (Table 1). The work on medicinal plants is being done primarily with a view to enhance/increase the production of various pharmaceutics, drugs, flavors, colors, and medicinal compounds under controlled conditions. In the process, sometimes new compounds, normally not found in the whole plant, are also produced in the transformed cultures (see Bajaj 1988–1999).

Of the various methods used for the transformation of medicinal plants, *Agrobacterium rhizogenes*-mediated gene transfer is preferred because of the induction of hairy roots, and increased production of secondary metabolites. Most of the studies conducted so far rely on this method (Tables 1, 2).

Studies with genetic transformation of medicinal plants have opened up new avenues, and significant progress has already been achieved, for example: (1) production of alkaloids, pharmaceutics, polyphenols, terpenoids, nema-tocidal compounds, and also some novel compounds not found in the whole plant, (2) regeneration of plants resistant to herbicides, diseases, and pests, (3) scaleup of cultures in bioreactors, (4) plants with different morphological traits, and (5) transgenic plants for the production of vaccines, enzymes etc. These developments have far-reaching in implications the pharmaceutical industry.

In this chapter various methods of genetic transformation of medicinal plants, induction and culture of hairy roots, results achieved, and an up-to-date literature on the subject are complied.

2 Methods of Genetic Transformation

Two methods have generally been used to bring about transformation in medicinal plants, i.e., direct gene transfer/uptake of DNA, and *Agrobacterium*-mediated gene transfer. However, the latter is the method of choice for most studies.

2.1 Direct Gene Transfer

In early studies, tobacco cells and protoplasts were used as a system for direct gene transfer/DNA uptake without involving any vector as a carrier. The literature on various techniques for uptake and incorporation of exogenous DNA has been reviewed (Larquin 1989). The following methods were used.

1. *Polyethylene Glycol Method.* Krens et al. (1982) incubated tobacco protoplasts with DNA in the presence of PEG, then plated them on a selective medium. The integration of purified Ti was expressed.

2. *Electroporation.* This is an efficient method for the incorporation of foreign DNA into protoplasts. It refers to the process by which macromolecules

Table 1. Summary of genetic transformation studies on medicinal plants

Plant species	Vector/method used (Agrobacterium rhizogenes A. r A. tumefaciens A. t)	Secondary metabolites/ compound studied/ results	Reference
Ajuga reptans	A. r MAFF 03-01724	β-Ecdyson	Matsumoto and Tanaka (1991), Uozumi et al. (1995)
Ambrosia trifida	A. r	Sesquiterpene, thiarubrine	Lu et al. (1993)
Amsonia elliptica	A. r A4	Indole alkaloid	Sauerwein and Shimomura (1990), Sauerwein et al. (1991a)
Anthemis nobilis	A. t C-58	Essential oil	Fauconnier et al. (1996)
Artemisia absinthium	A. r LBA9402	Volatile oil	Kennedy et al. (1993), Nin et al. (1997)
Astragalus membranaceus	A. r pRi 15834	Astragaloside, saponins	Hirotani et al. (1994), Zhou et al. (1995), Ionkova and Alfermann (1990)
Atropa belladonna	A. r pRi 15834, A4, 8196	Tropane alkaloid	Kamada et al. (1986), Knopp et al. (1988), Yun et al (1992), Yoshimatsu et al. (1997)
A. caucasica	A. r A4, 8196	Tropane alkaloid	Knopp et al. (1988)
Begonia tuberhybrida	A. t LBA4404	GUS expression	Kiyokawa et al. (1996)
Brugmansia candida	A. r LBA9402	Tropane alkaloid	Giulietti et al. (1993)
Calystegia sepium	A. r A4, 8196	Tropane alkaloid	Jung and Tepfer (1987)
Campanula medium	A. r A13	Lobetyol, lobetyolin, lobetyolinin	Tada et al. (1996c)
C. glomerata	A. r MAFF 03-01724	Lobetyol, lobetyolin, lobetyolinin	Tanaka et al. (1996b)
Cassia obtusifolia	A. r A4, 15834	Anthraquinone	Ko et al. (1988)
C. occidentalis	A. r A4, 15834	Anthraquinone	Ko et al. (1988)
C. torosa	A. r A4, 15834	Anthraquinone	Ko et al. (1988)
Catharanthus roseus	A. r LBA9402, 15834, A4	Vinca alkaloid, indole alkaloid	Parr et al. (1988), Ciau-Uitz et al. (1994), Jung et al. (1995), Garnier et al. (1996b)
C. trichophyllus	A. r 15834	Epialloyohimbine	Davioud et al. (1989)
Cinchona ledgeriana	A. r LBA9402	Cinchona alkaloid	Hamill et al. (1989)
Coreopsis tinctoria	A. r 15834	Phenylpropanoids	Thron et al. (1989). Reichling and Thron (1990)
Datura innoxia	A. r pRi 15834	Tropane alkaloid	Shimomura et al. (1991a, 1995), Boitel Couti et al. (1995)
D. candida	A. r pRi 15834	Tropane alkaloid	Shimomura et al. (1991a, 1995)
D. stramonium	A. r TR105, 15834, A4	Tropane alkaloid	Maldonado-Mendoza et al. (1993)

Table 1. *Cont.*

Plant species	Vector/method used (*Agrobacterium rhizogenes A. r / A. tumefaciens A. t*)	Secondary metabolites/ compound studied/ results	Reference
Datura sp.	*A. r* A4, 8196	Tropane alkaloid	Knopp et al. (1988)
Digitalis purpurea	*A. r* pRi 15834	Digitoxin	Saito et al. (1990, 1993)
Dioscorea bulbifera	*A. r, A. t* C58	Tumor production	Schafer et al. (1987), Conner and Dommisse (1992)
D. opposita	*A. t* A 208	Transformation	Feng et al. (1986)
Duboisia hybrid	*A. r* 15834	Tropane alkaloid	Ishimaru and Shimomura (1989), Muranaka et al. (1992, 1993a,b)
Echinacea purpurea	*A. r* 15834	Alkamides	Trypsteen et al. (1991)
Fagopyrum esculentum	*A. r* MAFF 03-01724	Procyanidin, rutin	Trotin et al. (1993), Nešković et al. (1995), Tanaka et al. (1996a)
Geranium thunbergii	*A. r* A4	Geraniin, corilagin	Ishimaru and Shimomura (1991, 1995a,b)
Glycyrrhiza uralensis	*A. r* pRi 15834 (pGSGlucl)	Glycyrrhizin	Saito (1995)
Hyoscyamus albus	*A. r* pRi 15834, A4	Tropane alkaloid	Sauerwein and Shimomura (1991), Shimomura et al. (1991a, 1995)
H. albus	*A. r* A4	Tropane alkaloid	Christen et al. (1992)
H. albus	*A. r* MAFF 03-01724, A4, 8196	Hyalbidone, tropane alkaloid	Knopp et al. (1988), Sauerwein et al. (1991b)
H. muticus	*A. r* LBA9402, A4, 15834	Root formation, tropane alkaloid	Knopp et al. (1988), Jaziri et al. (1994), Vanhala et al. (1995)
H. muticus	*A. t* C58, LBA9402	Root formation	Vanhala et al. (1995)
H. niger	*A. r* pRi 15834	Tropane alkaloid	Sauerwein and Shimomura (1991), Shimomura et al. (1991a, 1995)
H. niger	*A. r* A4, 8196	Tropane alkaloid	Knopp et al. (1988), Uchida et al. (1993)
H. reticulatus	*A. r* 15834	Tropane alkaloid	Ionkova (1992)
Lactuca virosa	*A. r* LBA9402	Sesquiterpene lactone	Kisiel et al. (1995)
Lawsonia inermis	*A. r* NCIB 8196	Lawsone	Bakkali et al. 1997
Leontopodium alpinum	*A. r* 9402	Anthocyanin, essential oil	Hook (1993)
Linum flavum	*A. r* LBA9402, 8490, 9365	Lignan	Oostdam et al. (1993)
Lippia dulcis	*A. r* A4	Hernandulcin	Sauerwein et al. (1991c)
Lithospermum erythrorhizon	*A. r* 15834	Shikonin	Shimomura et al. (1991b)
Lobelia cardinalis	*A. r* 15834, MAFF 03-01724	Lobetyol, lobetyolin, lobetyolinin	Yamanaka et al. (1996)

Species	Strain	Compound	Reference
L. chinensis	A. r 15834	Anthocyanin	Tada et al. (1996a)
L. chinensis	A. r 15834	Lobetyol, lobetyolin, lobetyolinin	Tada et al. (1995b)
L. inflata	A. r 15834	Lobeline	Yonemitsu et al. (1990), Ishimaru et al. (1992a), Tanaka et al. (1993)
L. inflata	A. r 15834	Lobetyol, lobetyolin, lobetyolinin	Ishimaru et al. (1991b, 1992c, 1993, 1995b), Tanaka et al. (1993)
L. sessilifolia	A. r 15834, MAFF 03-01724	Lobetyol, lobetyolin, lobetyolinin	Ishimaru et al. (1994)
Lotus corniculatus	A. r C58C1	Procyanidin	Morris and Robbins (1992)
Mentha citrata	A. t C58, T37	Sugar, nitrogen content, monoterpene	Spencer et al. (1993), Subroto et al. (1995)
M. piperita	A. t T37	Monoterpene	Spencer et al. (1993)
Nicandra physaloides	A. r LBA9402	Hygrine	Parr (1992)
Nicotiana tabacum	A. t pGV3845, 3304, pRi 8196	Nicotine alkaloid	Saito et al. (1989), Fecker et al. (1993)
N. rustica	A. r LBA9402	Nicotine	Rhodes et al. (1994)
Ocimum basilicum	A. r 15834, MAFF 03-01724	Rosmarinic acid	Tada et al. (1996a)
Panax ginseng	A. r 15834, A4	Ginsenoside	Yoshikawa and Furuya (1987), Ko et al. (1989), Lee et al. (1995)
Papaver somniferum	A. r MAFF 03-01724	Morphinan alkaloid	Yoshimatsu and Shimomura (1992, 1996), Williams and Eppis (1993)
Peganum harmala	A. t LBA4404, C58CI	Serotonin	Berlin et al. (1993)
Perezia cuernavacana	A. r AR12	Perezone	Arellano et al. (1996)
Phyllanthus niruri	A. r A4, A. t R-1000+121	Procyanidin	Ishimaru et al. (1992b)
Pimpinella anisum	A. t T37	Essential oils	Salem and Charlwood (1995)
Platycodon grandiflorum	A. r 15834, MAFF 03-01724	Lobetyol, lobetyolin, lobetyolinin	Tada et al. (1995a), Ahn et al. (1996)
Rauwolfia serpentina	A. r 15834	Ajmaline, serpentine	Benjamin et al. (1994)
Rubia tinctorum	A. r 15834	Anthraquinone	Sato et al. (1991)
R. peregrina	A. r LBA9402	Anthraquinone	Lodhi and Charlwood (1996), Lodhi et al. (1996)
Ruta graveolens	A. r	Anthranilic acid N-methy ltransferase	Eilert and Wolters (1989)
Salvia miltiorrhiza	A. r LBA9402, A4. 15834	Diterpenes, tanshinones	Hu and Alfermann (1993)
Sanguisorba officinalis	A. r A4	Sanguiin, hydrolysable tannin	Ishimaru et al. (1990a, 1991a, 1995a)
Scoparia dulcis	A. r pRi 15834, pARK5	Scopadulcic acid B	Yamazaki et al. (1996)
Scopolia tangutica	A. r pRi 15834	Tropane alkaloid	Shimomura et al. (1991a, 1995), Zhang (1993)
Scutellaria baicalensis	A. r pRi 15834	Flavonoids, phenylethanoids	Zhou et al. (1997)
Sesamum indicum	A. r 15834	Naphthoquinone	Ogasawara et al. (1993)

Table 1. *Cont.*

Plant species	Vector/method used (*Agrobacterium rhizogenes* A. r A. tumefaciens A. t)	Secondary metabolites/ compound studied/ results	Reference
Solanum aculeatissimum	A. r 15834	Steroidal saponin	Ikenaga et al. (1995)
S. aviculare	A. r A4	Steroidal alkaloids	Subroto and Doran (1994)
S. brevidens	A. t LBA4404	GUS expression, *uid* A gene	Liu et al. (1995)
S. cammersonii	A. t C58 (GV2260)	Shoot regeneration	Cardi et al. (1992)
S. demissum	A. t C58	Shoot regeneration	Kumar et al. (1995)
S. hjertingii	A. t C58	Shoot regeneration	Kumar et al. (1995)
S. mauritanium	A. r LBA9402	Solasodine	Drewes and Van Staden (1995)
S. papita	A. t C58	Shoot regeneration	Kumar et al. (1995)
S. stoloniferum	A. t C58	Shoot regeneration	Kumar et al. (1995)
S. verrucosum	A. t C58	Shoot regeneration	Kumar et al. (1995)
Swertia japonica	A. r 15834	Xanthone	Ishimaru et al. (1990b)
S. japonica	A. r 15834	Amarogentin, amaroswerin	Ishimaru et al. (1990b), Ishimaru and Shimomura (1996)
Swainsona galegifolia	A. r	Swainsonine	Ermayanti et al. (1994)
Tagetes patula	A. r 43057	Bithiophene, α-terthienyl	Parodi et al. (1988), Kyo et al. (1990), Hjortso and Mukundan (1994)
Valeriana officinalis	A. r R1602	Valepotriates, iridoid diester	Gränicher et al. (1992, 1994, 1995)
Vinca minor	A. r DC-AR2	Vincamine	Tanaka et al. (1995)

Table 2. Secondary products in transformed root cultures of medicinal plants (see also Tables 1, 3)

Products	Material plant	Reference
Phenolics		
Geraniin, corilagin	*Geranium thunbergii*	Ishimaru and Shimomura (1991, 1995a,b)
Sanguiin, hydrolyzable tannin	*Sanguisorba officinalis*	Ishimaru et al. (1990a, 1991a, 1995a)
Procyanidin	*Phyllanthus niruri*	Ishimaru et al. (1992b)
	Lotus corniculatus	Morris and Robbins (1992)
	Fagopyrum esculentum	Tanaka et al. (1996a)
Anthraquinone	*Rubia tinctorum*	Sato et al. (1991)
	Cassia sp.	Ko et al. (1988)
Xanthone	*Swertia japonica*	Ishimaru et al. (1990b)
Rutin	*Fagopyrum esculentum*	Tanaka et al. (1996a)
Rosmarinic acid	*Ocimum basilicum*	Tada et al. (1996a)
Anthocyanin	*Lobelia chinensis*	Tada et al. (1996a)
Lignan	*Linum flavum*	Oostdam et al. (1993)
Terpenes		
Amarogentin, amaroswerin	*Swertia japonica*	Ishimaru et al. (1990b), Ishimaru and Shimomura (1996)
Hernandulcin	*Lippia dulcis*	Sauerwein et al. (1991c)
Shikonin	*Lithospermum erythrorhizon*	Shimomura et al. (1991b)
Digitoxin	*Digitalis purpurea*	Saito et al. (1990)
β-Ecdyson	*Ajuga reptans*	Matsumoto and Tanaka (1991)
Ginsenoside	*Panax ginseng*	Ko et al. (1989), Lee et al. (1995)
Astragaloside	*Astragalus membranaceus*	Hirotani et al. (1994), Zhou et al. (1995)
Sesquiterpene lactone	*Lactuca virosa*	Kisiel et al. (1995)
Perezone	*Perezia cuernavacana*	Arellano et al. (1996)
Alkaloids		
Tropane alkaloid	*Atropa belladonna*	Kamada et al. (1986), Jaziri et al. (1994b)
	Hyoscyamus albus, H. niger	Sauerwein and Shimomura (1991), Shimomura et al. (1991a, 1995)
	Scopolia tangutica	Shimomura et al. (1991a, 1995)
	Datura innoxia, D. candida	Shimomura et al. (1991a, 1995)
	D. stramonium	Maldonado-Mendoza et al. (1993)
	Duboisia hybrid	Ishimaru and Shimomura (1989)
Indole alkaloid	*Amsonia elliptica*	Sauerwein and Shimomura (1990), Sauerwein et al. (1991a)
	Catharanthus roseus	Islas et al. (1994), Jung et al. (1995)
Morphinan alkaloid	*Papaver somniferum*	Yoshimatsu and Shimomura (1992)
Lobeline	*Lobelia inflata*	Yonemitsu et al. (1990), Ishimaru et al. (1992a), Tanaka et al. (1993)
Hyalbidone	*Hyoscyamus albus*	Sauerwein et al. (1991b)
Cinchona alkaloid	*Cinchona ledgeriana*	Hamill et al. (1989)
Vinca alkaloid	*Catharanthus roseus*	Parr et al. (1988)

Table 2. *Continued*

Products	Material plant	Reference
Polyacetylene		
Bithiophene	*Tagetes patula*	Parodi et al. (1988), Mukundan and Hjortso (1991)
Thiophene	*Rudbeckia hirta*	Daimon et al. (1993)
Lobetyol, lobetyolin, lobetyolinin	*Lobelia sessilifolia*	Ishimaru et al. (1994)
	L. inflata	Ishimaru et al. (1991b, 1992c, 1993, 1995b), Tanaka et al. (1993)
	L. chinensis	Tada et al. (1995b)
	L. cardinalis	Yamanaka et al. (1996)
	Campanula medium	Tada et al. (1996c)
	C. glomerata	Tanaka et al. (1996b)
	Platycodon grandiflorum	Tada et al. (1995a), Ahn et al. (1996)
α-Terthienyl	*Tagetes patula*	Kyo et al. (1990)

present in the extracellular medium are internalized by living cells following a brief electric pulse (Neumann et al. 1982). Riggs and Bates (1986) used this successfully for tobacco protoplasts.

3. *Microinjection.* Through this technique it is possible to transfer subcellular organelles and isolated DNA. Crossway et al. (1986) injected tobacco mesophyll protoplasts with DNA. The transformation frequency was about 14%, and no selection was required to identify transformed cell lines by Southern blot hybridization. Later, Latt and Blass (1987) injected metaphase chromosomes isolated from kanamycin-resistant cell suspension of *Nicotiana plumbaginifolia* into recipient wide-type protoplasts and obtained visual evidence.

4. *Particle Bombardment/Balistic Gun.* Klein et al. (1987) developed this sophisticated method for direct transfer of DNA, which consists in bombarding the cells with DNA-coated tungsten or gold particles. It requires the use of a high-velocity gun to accelerate the metal particles. In addition to tobacco, this method has been successfully applied to cotton (McCabe and Martinell 1993) and a number of agricultural crops.

2.2 *Agrobacterium*-Mediated Gene Transfer

Although various methods of direct gene transfer have provided significant information in basic research, they remain of purely academic interest. However, in medicinal plants, most of the transformation studies have been conducted with two soil-borne bacteria, i.e., *Agrobacterium tumefaciens* and *A. rhizogenes*, as the vectors. *Agrobacterium tumefaciens* introduces the genes for the tumor induction (Ti) plasmid of the bacterium into the plant genome,

Table 3. Increased production of secondary metabolites through transformed hairy roots

Plant species	Compound	Reference
Ajuga reptans	20-Hydroxyecdyson	Tanaka and Matsumoto (1993)
Amsonia elliptica	Indole alkaloid	Sauerwein and Shimomura (1990)
Atropa belladonna	Tropane alkaloids	Sharp and Doran (1990), Hashimoto et al. (1993)
Catharanthus roseus	Indole alkaloids	Toivonen et al. (1990), Ciau-Uitz et al. (1994), Vazquez-Flota et al. (1994)
Coreopsis tinctoria	Phenylpropanoids	Reichling and Thron (1990)
Datura candida	Tropane alkaloid	Christen et al. (1989)
D. stramonium	Scopolamine	Maldonado-Mendoza et al. (1993, 1995)
Duboisia leichhardtii	Scopolamine	Muranaka et al. (1992)
Hyoscyamus muticus	Solaretivove	Corry et al. (1993)
Lithospermum spp.	Shikonin	Chang and Sim (1996)
Lobelia inflata	Polyacetylene	Ishimaru et al. (1992c)
Nicotiana tabacum	Cadaverine, anabasine	Fecker et al. (1993)
N. rustica	Nicotine	Rhodes et al. (1994)
Panax ginseng	Ginsenosides	Inomata and Yokoyama (1996)
Papaver somniferum	Morphinan	Yoshimatsu and Shimomura (1996)
Peganum harmala	Serotonin	Berlin et al. (1993)
Sanguisorba officinalis	Sanguiin	Ishimaru et al. (1995a)
Solanum aviculare	Steroidal alkaloids	Subroto and Doran (1994)
Vinca minor	Vincamine	Tanaka et al. (1995)

causing crown gall formation in the infected plant. On the other hand, *A. rhizogens* has root-inducing (Ri) plasmid which forms hairy roots (Chilton et al. 1982).

The main advantages of transformed hairy root cultures are (1) they grow in the absence of growth regulators, (2) they have a higher rate of growth than the untransformed ones, (3) they can be accumulated in good quantities, (4) they are more stable than the cell suspensions, (5) there is an increased production of secondary metabolites (Table 3), and (6) in many cases entire plants can be easily regenerated from such roots.

3 Culture of Hairy Roots

Profuse proliferation of untransformed excised roots in cultures has been observed for a long time (Bajaj and Dionne 1968). The cultures are initiated from a variety of tissues and organs, sterilized segments of stem, leaf disks, etc. They are inoculated with a suspension of bacteria. Roots usually appear at the site of inoculation. These roots are then excised and cultured on a medium containing an antibiotic. In the course of time, in most cases in liquid media without growth regulators, profuse growth of transformed hairy roots is observed.

The growth, development, and production of secondary metabolites in hairy root cultures is influenced by a number of factors (Hilton and Rhodes

1993), such as media composition, liquid/agar medium, pH, growth regulators, light/dark, temperature, oxygen supply, etc. These aspects have recently been discussed in detail by Loyola-Vargas and Miranda-Ham (1995), and are summarized here.

The composition of the culture medium along with the growth regulators and sucrose concentration have a profound effect on the growth, development, and production of secondary metabolites. In general, simple media have proved to be more productive (Parr et al. 1988; Ishimaru and Shimomura 1991). Likewise, liquid media, in contrast to agar-solidified media, cause faster growth, resulting in an increased root biomass. Supplementing the medium with a mixture of growth regulators, IBA and kinetin increased saponin accumulation in *Panax* hairy roots (Yoshikawa and Furuya 1987). Sauerwein et al. (1991a) observed that NAA, along with light, had a positive effect on growth in *Amsonia*, whereas it was reduced in the dark. They (Sauerwein et al. 1991c) also reported 100% increase in hernandulcin production by NAA in *Lippia dulcis* hairy roots. Gibberellic acid increased scopolamine fourfold in *D. innoxia* (Ohkawa et al. 1989). Accumulation of steroidal alkaloids in *Solanum aviculare* hairy roots was improved by about 40% by GA (Subroto and Doran 1994).

In general, increase in sucrose concentration caused higher production of alkaloids. For instance, increase of sucrose from 3–5% resulted in 32% increase in hyoscyamine contents in *Datura stramonium* (Payne et al. 1987); later an eightfold increase was also reported (Hilton and Rhodes 1993). In contrast, tenfold decrease in scopolamine content was noticed when sucrose was reduced from 3 to 2% (Jaziri et al. 1988). In *Rubia tinctoria* 12% sucrose caused the highest anthraquinone production (Sato et al. 1991).

The root culture of *D. stramonium* treated with copper at 1 mM induced rapid accumulation of high levels of sesquiterpenoid defensive compounds, such as lubimin (Furze et al. 1991).

Although pH is one of the most important factors in determining growth of hairy roots, results vary in different plant species. For instance in *Scopolia japonica* (Mano et al. 1986) and *N. glauca* (Green et al. 1992), when pH was raised to 7 growth was stimulated, whereas in *D. stramonium* (Payne et al. 1987) the lag phase was longer when pH was raised from 5.8 to 7. In *Catharanthus roseus*, however, the growth of hairy roots was not affected by the initial pH of the medium (HO and Shanks 1992).

Oxygen supply to the hairy roots grown in fermenters/bioreactors can be a limiting factor; however, there are very few studies on this aspect. Aeration had a strong influence on growth and digitoxin contents in submerged shoot cultures of *Digitalis* (Hagimori et al. 1984).

Recently, Yu and Doran (1994) studied oxygen requirements and mass transfer of hairy roots of *Atropa belladonna*. Specific growth rate increased with oxygen tension up to 100% air saturation; this result demonstrates that hairy roots aerated without oxygen supplementation are likely to be oxygen-limited.

4 Achievements in the Genetic Transformation of Medicinal Plants

4.1 Production of Pharmaceutics in Hairy Root Cultures

There are numerous examples of the production of secondary metabolities in transformed cultures of medicinal plants. These studies are cited in Tables 1–6, and summarized below.

4.1.1 Alkaloids

Solanaceous medicinal plants such as *Atropa, Datura, Duboisia, Hyoscyamus, Scopolia,* etc. are major sources of supply of tropane alkaloids. Amongst numerous studies on secondary metabolism in transformed cultures, those concerning tropane alkaloid biosynthesis have been most actively and intensively performed, resulting in several useful products (see Petri and Bajaj 1989, Bajaj and Simola 1991).

Kamada et al. (1986) first reported production of hyoscyamine and scopolamine in hairy root cultures of *Atropa belladonna* at levels the same or higher than in normal plants grown in the field. The axenic cultures proliferated 60-fold based on the initial weight in 1-month culture. Jung and Tepfer (1987) also observed considerable increase in biomass and tropane alkaloid accumulation in *A. belladonna*-transformed root cultures.

The hyoscyamine 6β-hydroxylase gene of *Hyoscyamus niger* was induced to hyoscyamine-rich *A. belladonna* by a binary vector system using *A. rhizogenes* (Hashimoto et al. 1993). The engineered *A. belladonna* hairy roots showed increased amounts and enzyme activities of the hydroxylase and contained up to fivefold higher concentrations of scopolamine than the wild type.

Hairy root cultures of *H. albus, H. niger, Scopolia tangutica, Datura innoxia, D. candida, Duboisia* hybrid (M-II-8-6: *D. myoporoides* × *D. leichhardtii*), etc. mainly produced four tropane alkaloids hyoscyamine, 6β-hydroxyhyoscyamine, 7β-hydroxyhyoscyamine, and scopolamine, supplying important medicines (Sauerwein and Shimomura 1991; Shimomura et al. 1991a, 1995). Particularly, successful isolation and chemical structural characterization of 7β-hydroxyhyoscyamine, which is a biosynthetically interesting intermediate in the conversion of hyoscyamine to scopolamine and occurs in small amounts in natural plant tissues, becomes practicable by utilization of *H. albus* and *Duboisia* hybrid hairy roots (Ishimaru and Shimomura 1989).

In addition to the tropane alkaloids, various other alkaloids have also been reported in transformed cultures of various plants. Root cultures of *Amsonia elliptica* (Apocynaceae) transformed by *A. rhizogenes* A4 produced a new yohimbane derivative, 17α-O-methylyohimbine, together with the previously known indole alkaloids vallesiachotamine and pleiocarpamine (Sauerwein and Shimomura 1990). The addition of 0.5 mg/l NAA to the culture medium remarkably enhanced growth and alkaloid production (to ca. ten times that of the parent tissues in vivo) of the hairy roots. In the light condition, the hairy roots accumulated chlorophylls as green hairy roots and the production of vallesiachotamine and pleiocarpamine was promoted (Sauerwein et al. 1991a).

In *Papaver somniferum*, gene transfer was performed by *A. rhizogenes* MAFF 03-01724 (Yoshimatsu and Shimomura 1992), and tumerous trans-formed shoots were obtained which yielded morphinan alkaloids at a level comparable to that in non-transformed shoots.

Other alkaloidal constituents such as a piperidine-type lobeline in *Lobelia inflata* (Yonemitsu et al. 1990; Ishimaru et al. 1992a), a piperidone-type hyal-bidone which is a new alkaloid in *H. albus* (Sauerwein et al. 1991b), etc. were also synthesized in each hairy root culture system.

4.1.2 Polyphenols

In the hairy root cultures of *Geranium thunbergii* (Geraniaceae), the tradi-tional medicinal plant mainly used for diarrhea, the metabolism for the tannin constituent, geraniin, which is the major pharmaceutical in the aerial parts of the plant, was regulated by NH_4^+ content in the culture medium (Ishimaru and Shimomura 1991, 1995a,b). Biosynthesis of geraniin in hairy roots was activated in B5 medium (containing $2\,mM\ NH_4^+$) and not in NH_4^+-rich medium (1/2 MS; Murashige and Skoog 1962) containing $10.3\,mM\ NH_4^+$.

The rosaceous medicinal plant *Sanguisorba officinalis* also contains high molecular polyphenol compounds (particularly hydrolyzable type tannins) such as sanguiin H-6 and sanguiin H-11, which are presumed to be active constituents when treated for hemostatic, anti-phlogistic, and astringent prop-erties. Among the six clones of the hairy roots, cultured in hormone-free 1/2 MS liquid medium, five clones produced mainly sanguiin H-6 whereas the remaining clone produced an especially high level of 1,2,3,6-tetra-O-galloyl-β-D-glucose and sanguiin H-11, whose levels were more than double of those in the parent plant (Ishimaru et al. 1990a, 1991a, 1995a). The root cultures of *S. officinalis*, by careful selection of the clone and the determination of the effects of some additives (medium constituents, auxin, etc.), showed a stable supply of these high molecular tannins, coincidentally affording a useful sys-tem for biosynthetic studies of hydrolyzable tannins.

The hairy roots of *Swertia japonica* (Gentianaceae), produced a great amount of xanthones, i.e., bellidifolin, methylbellidifolin, swertianolin, and a new glycoside, 8-O-primeverosylbellidifolin, which gave a bright yellow col-oration to the tissues (Ishimaru et al. 1990b; Ishimaru and Shimomura 1996). It was biosynthetically noteworthy that the hairy roots produced only 1,3,5,8-oxygenated xanthones although in-vivo plants produce both 1,3,5,8- and 1,3,7,8-oxygenated derivatives. By comparison of enzymes and/or gene activity between transformed (hairy root) and non-transformed tissues of this plant, it may be possible to obtain enzymes and genes related to the biosynthesis (oxidation) of xanthones. *S. japonica* hairy roots also yielded two new phenyl glucosides, 5-(3'-glucosyl)benzoyloxygentistic acid and 2,6-dimethoxy-4-

hydroxyphenol 1-glucoside, together with 1-sinapoylglucoside (Ishimaru et al. 1990c; Ishimaru and Shimomura 1996). In transformed cultures the biosynthetic capability for glycosylation was activated, which made it possible to produce some new phenylglycosides.

Eight clones of *Phyllanthus niruri* (used as a euphorbiaceous folk medicine) hairy root produced seven phenolics: gallic acid, (+)-catechin, (−)-epicatechin, (−)-epigallocatechin, (−)-epicatechin 3-O-gallate, and (−)-epigallocatechin 3-O-gallate (Ishimaru et al. 1992b). Although phenolic constituents in the aerial parts of *P. niruri* (both in-vitro and in-vivo plants) were mainly hydrolyzable tannins such as geraniin, corilagin, and galloylglucose, the hairy root cultures contained flavan-3-ols whose constitutional pattern was similar to that observed in leaves of *Thea sinensis* (green tea). These cultures were expected to provide new medicinal material as antiviral (hepatitis, flu, anti-HIV, etc.) and antitumor drugs.

Ocimum basilicum (Lamiaceae) hairy roots grew well in hormone-free MS, B5, and Woody Plant (Lloyd and McCown 1980), and produced a large amount of rosmarinic acid (over 14% dry weight) together with related phenolics, lithospermic acid (over 1.7% dry weight), and lithospermic acid B (over 0.17% dry weight) (Tada et al. 1996a). These transformed tissues present a new source of antioxidant polyphenols which may be used as food additives.

Fagopyrum esculentum (Polygonaceae) hairy roots induced by *A. rhizogens* strain 15834 showed fourfold faster growth and synthesis of five flavanols (Trotin et al. 1993). Recently, flavonoid rutin has also been reported (Tanaka et al. 1996a).

4.1.3 Terpenoids

Genetic transformation of *Digitalis purpurea* (foxglove), was performed by an *Agrobacterium*-mediated system (Saito et al. 1990), and the green hairy roots obtained produced cardioactive glycosides.

Other secondary metabolites, sometimes used as natural additives for drugs and/or food chemicals, such as the sweet sesterterpenoid hernandulcin in *Lippia dulcis* (Sauerwein et al. 1991c), the bitter glycosides amarogentin and amaroswerin in *Swertia japonica* (Ishimaru et al. 1990b; Ishimaru and Shimomura 1996), shikonin derivatives in *Lithospermum erythrorhizon* (Shimomura et al. 1991b), etc. also yielded hairy roots in *Agrobacterium*-transformed tissues of each species.

4.1.4 Other Pharmaceutics

Polyacetylene compounds occur in several varieties in the Asteraceae, Umbelliferae, Araliaceae, etc. Particularly the important biological activity (cytotoxicity for tumor cells) of *Panax* sp. (Araliaceae), whose alcoholic extract (of roots) has been widely used as an anticancer drug in home treatment, is presumed to have originated in their polyacetylene constituents. Recently, in some campanulaceous plants such as *Platycodon grandiflorum* (Tada et al. 1995a; Ahn et al. 1996), *Lobelia inflata* (Tanaka et al. 1993; Ishimaru et al. 1993, 1995b), *L. chinensis* (Tada et al. 1995b), *L. cardinalis* (Yamanaka et al. 1996), *L. sessilifolia* (Ishimaru et al. 1994), etc, the production of new polyacetylene derivatives, lobetyol, lobetylolin, and lobetyolinin (Ishimaru et al. 1991b, 1992c), which have mild cytotoxic activity, were obtained in hairy roots (Table 2). Another group of compounds, such as thiophene and α-terthienyl, which have nematocidal properties, have been extracted from transgenic plants of *Tagetes* (Kyo et al. 1990, Mukundan and Hjortso 1991), and *Rudbeckia* (Daimon et al. 1993).

4.2 Transgenic Plants for the Production of New Compounds

Biotransformation attempts of some chemicals using plant cell cultures (with active enzymes) offer a good system for yielding new useful metabolites normally only poorly present in nature. Particularly for synthesis needing positional selectivity in the reactions, plant cells cultured under suitable conditions are very profitable when used as bioreactors.

Recently, *Lobelia sessilifolia* hairy roots were used for biotransformation of some phenolics (flavan-3-ols and C6-C1 phenols) which successfully produced new glucosylated compounds (Yamanaka et al. 1995). In addition, the metabolism in biotransformation of five phenolics (as substrates), protocatechuic acid, gallic acid, *trans*-cinnamic acid, *p*-coumaric acid, and caffeic acid in hairy root cultures of five plant species, i.e., *L. sessififolia*, *L. cardinalis*, *Campanula medium* (Campanulaceae), *Ocimum basilicum* (Lamiaceae), and *Fragaria* x *ananassa* (Rosaceae) was determined. In this system, the site-specific glycosylation of phenolics was successful, which is difficult when using organic chemical synthesis (Ishimaru et al. 1996).

Other examples of new compounds are piperidone-type hyalbidone in *H. albus* (Sauerwein et al. 1991b), and two new glucosides in *Swertia japonica* (Ishimaru and Shimomura 1996).

4.3 New Morphological Traits/Plant Types

A. rhizogenes-mediated transformation integrates some genes on the Ri plasmid of the bacterium, which codes enzymes for phytohormone biosynthesis, into plant genome (e.g., rol genes). Therefore, by endogeneous modification of phytohormone levels, the morphological traits and secondary metabolism of the regenerants coming from the hairy roots are sometimes modified. When the regenerants (transgenic plant) of medicinal plants have good traits for propagation, fast-growing, early-flowering and/or well-established secondary metabolism, the plants can prove to be new pharmaceutical resources. Here, some examples of transformed medicinal plantlets which showed interesting traits of morphology and biosynthesis are described.

4.3.1 Atropa belladonna

A. belladonna has a high potential for shoot formation; the transgenic entire plants could be obtained without difficulty by regeneration from hairy root cultures. The morphological changes, the so-called hairy-root syndrome, such as high root growth rate and wrinkled leaves and tropane alkaloid accumulation in transgenic A. belladonna plantlets, have been evaluated (Kamada et al. 1986; Jung and Tepfer 1987). Saito et al. (1991) introduced shooty teratomas which failed to produce tropane alkaloids at a high level, but had the ability to store and metabolize the alkaloids. Recently, Subroto et al. (1996) also established shooty teratomas (by A. tumefaciens nopaline strain T37) and used them for an experiment on biotransformation of tropane alkaloids by coculture system with A. belladonna hairy roots (by A. rhizogenes agropine strain A4). Transgonic A. belladonna regenerants are a good materials for the determination of correlationship between tissue differentiation and the secondary metabolism of the plant.

When a rolC gene was co-introduced with a kanamycin resistant (NPT II) gene under control of a cauliflower mosaic virus 35S promoter, the rolC gene was expressed strongly in leaves, flowers, stems and roots (Kurioka et al. 1992; Suzuki et al. 1993). The transformed A. belladonna regenerated from callus, which was derived by the infection with A. tumefaciens strain LBA4404 (pSK223), showed extensive flowering, extremely deceased apical dominance, pale and lanceolated leaves and smaller flowers. The morphological changes in transgenic plants were related to the degree of expression of the rolC gene. Jaziri et al. (1994b) reported that transgenic placts had thick and wrinkled leaves and short internodes.

4.3.2 Other Examples – Campanula, Wahlenbergia, Rudbeckia

The hairy root cultures of two campanulaceous plants, Campanula lactiflora and Wahlenbergia marginata, produce high content of polyacetylenes (lobetyol, lobetyolin and lobetyolinin) and are expected to be one of the new pharmaceutical resources. Adventitious shoot formation from the hairy roots and the regenerants showed some morphologically interesting traits i.e. flowering in short-time, frequent occurrence of variation of petai number etc. These regenerants, after acclimatization, could be easily cultivated in soil condition. Transgenic Rudbeckia hirta plants regenerated from hairy roots showed morphological alterations, such as short internodes, wrinkled leases, and small flowers (Daimon et al. 1993).

4.4 Herbicide-Resistant Medicinal Plants

Herbicide-tolerant plants have been obtained by selection of mutants and by genetic transformation with resistant genes. In an earlier study, the introduction of the *bar* gene in tobacco (De Block et al. 1987) resulted in herbicide-resistant plants. Also the 2,4-D monooxygenase gene *tfdA* from *Alcaligenes eutrophus* was isolated, modified, and expressed in transgenic tobacco and cotton (Bayley et al. 1992) which showed resistance to the he bicide 2,4-D. Transgenic herbicide-resistant plants of *Atropa belladonna* (Saito et al. 1992) and *Scoparia dulcis* (Yamazaki et al. 1996) were obtained by using the Ri binary vector system. Hairy roots resistant to Bialaphos were selected and plantlets were regenerated. The transgenic plants and progenies showed a resistant trait towards Bialaphos and Phosphinothricin.

4.5 Disease-, Insect-, and Nematode-Resistant Plants

The first transgenic tobacco plants with increased resistance to tomato mosaic virus (TMV) resulted from the expression of a coat protein (Abel et al. 1986). In China, almost 1 million ha of transgenic tobacco was under cultivation in 1994; such plants showed resistance to TMV, and yielded 5–7% more leaves (Dale 1995).

The bacterium *Bacillus thuringiensis* (*B.t.*)-mediated transformed tobacco (Barton et al. 1987) and cotton (a gossypol-producing plant) have shown resistance to various lepidopteran insects (Perlak et al. 1990). *B.t.* produces proteins that are toxic to insects; the larvae fed on leaves containing such proteins die. Another transgenic approach sucessfully applied to tobacco (Masoud et al. 1993) and cotton (Thomas et al. 1995) is through proteinase inhibitors. The insect-resistant plants have been field-tested (Wilson et al. 1994).

Studies have also been conducted on hairy roots of plants antagonistic to nematodes. Kyo et al. (1990) obtained one line of *Tagetes patula* hairy root culture which showed 80-fold higher α-terthienyl. *Rudbeckia hirta* is another plant that is antagonistic to nematodes in nature. Its transgenic hairy roots regenerated plants that contained α-terthienyl (Daimon et al. 1993), but the authors did not report on the nature of their resistance to nematodes. Such studies would be highly rewarding.

4.6 Transgenic Plants as Bioreactors (Goddijn and Pen 1995)

Transgenic plants are an attractive and cost-effective alternative to microbial systems for the production of biomolecules. A number of compounds have been produced by using the molecular farming approach, which has already proved feasible to produce carbohydrates, fatty acids, pharmaceutical polypeptides, industrial enzymes, and biodegradable plastic (see Goddijn and Pen 1995).

4.7 Transgenic Plants for Production of Vaccines
(Mason and Arntzen 1995)

The concept of vaccine production in transgenic plants that express foreign proteins of pharmaceutical value (Mason et al. 1992) represents an economical alternative to fermentation-based production. Transgenic plants that express antigens in their edible tissue might be used in production and delivery of inexpensive oral vaccine. For instance, potato tubers expressing a bacterial antigen stimulated humoral and mucosal immune responses when they were provided as food. Such studies are being conducted on banana, and are likely to be extended to other plants. In tobacco, the expression of hepatitis B surface antigen Ag has been demonstrated (Mason et al. 1992).

4.8 Bioreactor Production of Hairy Roots (Table 4)

Various types of bioreactors/fermenters of varying capacities have been used for the large-scale culture of cell suspensions (Hashimoto and Azechi 1988), somatic embryos (Stuart et al. 1987; Graidziak et al. 1990), and hairy root cultures (Table 4). Cell suspensions are comparatively easy, whereas roots pose certain problems as they become entangled and can be damaged by impellers (Hilton et al. 1988). However, in view of the obvious advantages of hairy roots over cell suspensions, efforts are continuing to develop/improve bioreactors that ensure large-scale production with minimum damage to the roots. Some examples of bioreactors used for hairy root culture are given in Table 4.

4.9 Release of Secondary Metabolites from Hairy Roots into the Culture Medium (Table 5)

The release of secondary metabolites into the medium is of great significance for harvesting the required product from the culture. In *Nicotiana rustica* hairy roots, up to 76% nicotine is released into the medium (Rhodes et al. 1986). In *Datura stramonium* hairy roots almost 100% scopolamine and 70% hyoscyamine was released at low pH (Saenz-Carbonell et al. 1993). Temperature variation has also been reported to affect the release of secondary metabolites.

Table 4. Production of hairy roots in bioreactors

Plant species	Bioreactor type/ capacity	Reference
Ajuga reptans	Air lift	Matsumoto and Tanaka (1991)
Amoracia rusticana	Air lift column reactor	Taya et al. (1989)
Artemisia absinthium	Column bioreactor (1l)	Kennedy et al. (1993)
Atropa belladonna	Air lift	Sharp and Doran (1990)
Catharanthus roseus	Air lift	Toivonen et al. (1990)
Datura stramonium	Column fermenter (1.5l)	Wilson et al. (1987)
	Stirred tank reactor	Hilton et al. (1988)
	Stirred tank reactor (14l)	Hilton and Rhodes (1990)
Duboisia leichhardtii	Two-stage culture, bioreactor system	Muranaka et al. (1993b)
Nicotiana rustica	Two-stem system	Rhodes et al. (1986)
Panax ginseng	Turbine-blade reactor (2l)	Inomata and Yokoyama (1996)
Phyllanthus niruri	Air lift (6l)	Ishimaru et al. (1992b)
Trigonella foenum-graecum	Air lift (9l)	Rodriguez-Mendiola et al. (1991)

Table 5. Release of some secondary metabolites from hairy roots into the culture medium

Compounds	Plant species	References
Betanin	*Beta vulgaris*	Kino-oka et al. (1992)
Hyoscyamine	*Datura stramonium*	Hilton and Rhodes (1990)
Scopolamine	*D. stramonium*	Saenz-Carbonell et al. (1993)
Hyoscyamine, scopolamine	*Duboisia leichhardtii*	Mano et al. (1989), Muranaka et al. (1992)
Shikonin	*Lithospermum erythrorhizon*	Shimomura et al. (1991b)
Nicotine	*Nicotiana rustica*	Rhodes et al. (1986), Wilson et al. (1987), Whitney (1992)
Tanshinones	*Salvia miltiorrhiza*	Hu and Alfermann (1993)

Table 6. Some examples of shooty teratoma/transformed shoot cultures

Plant species	Observations/results	Reference
Atropa belladonna alkaloids	Stored and metabolized	Saito et al. (1991), Subroto et al. (1993, 1996)
Begonia tuberhybrida	GUS expression	Kiyokawa et al. (1996)
Brassica napus	Shoot regeneration	Damgaard and Rasmussen (1997)
B. rapa	Shoot regeneration	Radke et al. (1992)
Citrus sp.	Regeneration	Moore et al. (1992)
Duboisia hybrid	Alkaloid production	Subroto et al. (1993)
Eucalyptus grandis × E. *urophylla*	Root and tumor development	Machado et al. (1997)
Mentha citrata, M. piperata	Oil glands developed, produced monoterpenes	Spencer et al. (1990, 1993)
M. citrata	Growth kinetics and stoichiometry	Subroto et al. (1995)
Moricandia arvensis	Plant formation	Rashid et al. (1996)
Nicotiana tabacum	Nicotine-fed teratomas converted nicotine to nornicotine	Saito et al. (1989, 1991)
Rhododendron cv.	Shoot regeneration	Ueno et al. (1996)
Sinapis alba	Shoot regeneration	Hadfi and Batschauer (1994)
Solanum aviculare	Biotransformation of tropane alkaloids	Subroto et al. (1993)
Solanum sp.	Shoot regeneration	Kumar et al. (1995)

For instance, in *D. stramonium* at 30 °C there was a seven-fold increase in hyoscyamine liberation as compared to 25 °C (Hilton and Rhodes 1990).

Efforts are being made to recover the released product to the maximum: up to 97% of scopolamine released into the medium by hairy roots of *Duboisia* could be recovered by a column containing the adsorbant (Muranaka et al. 1992). Some more examples of release are given in Table 5.

4.10 Culture of Shoot Teratomas (Table 6)

The aerial parts of plants, especially in-vitro-differentiated shoots, are known to produce a wide range of compounds. The in-vitro culture of shoots gener-ally requires the presence of plant hormones; however, *Agrobacterium tumefaciens*-induced shoot teratomas have the advantage of growing independent of growth regulators. A number of examples (Table 6) of shooty teratomas capable of forming secondary metabolites are now available. Shooty teratomas would be useful in cases where hairy roots do not form the secondary metabolites found in the aerial parts of the plants.

5 Summary and Conclusions

Genetic transformation has now become an established method for gene transfer in medicinal plants; more than 60 genera of such plants have been transformed. Secondary metabolite studies have been conducted on cell suspensions, hairy roots, and shooty teratomas. Of these three systems used, most of the work has been done on *Agrobacterium rhizogenes*-induced hairy roots. They have proved to be quite efficient, and have the advantage of high growth rate, being genetically more stable than cell suspensions, and growing on hormone-free media.

Some achievements in the genetic transformation of medicinal plants have been the (1) production of pharmaceutics, alkaloids, polyphenols, terpenoids, nematocidal compounds, and also some novel compounds not found in the whole plant, (2) regeneration of plants resistant to herbicides, diseases, and pests, (3) scaleup of culture in bioreactors, (4) plants with different morphological traits, and (5) transgenic plants for the production of vaccines, enzymes, etc. These developments have far-reaching implications for the pharmaceutical industry.

References and Further Reading

(also included are references not cited in the text)

Abegaz BM (1991) Polyacetylenic thiophenes and terpenoids from the roots of *Echinops pappii*. Phytochemistry 30:879–881

Abel PP, Nelson RS, De B, Hoffmann N, Rogers SG, Fraley RT, Beachy RN (1986) Delay of disease development in transgenic plants that express the tobacco mosaic virus coat protein gene. Science 232:738–743

Ahn JC, Hwang B, Tada H, Ishimaru K, Sasaki K, Shimomura K (1996) Polyacetylenes in hairy roots of *Platycodon grandiflorum*. Phytochemistry 42:69–72

Anzai H, Yoneyama K, Yamaguchi I (1989) Transgenic tobacco resistant to a bacterial disease by the detoxification of a pathogenic toxin. Mol Gen Genet 219:492–494

Arellano J, Vazquez F, Villegas T, Hernandez G (1996) Establishment of transformed root cultures of *Perezia cuernavacana* producing the sesquiterpene quinone perezone. Plant Cell Rep 15:455–458

Bajaj YPS (1972) Protoplast culture and regeneration of haploid tobacco plants. Am J Bot 59:647

Bajaj YPS (1976) Regeneration of plants from cell suspensions frozen at −20, −70, and −196°C. Physiol Plant 37:263–268

Bajaj YPS (1977) Survival of *Atropa* and *Nicotiana* pollen embryos frozen at −196°C. Curr Sci 46:305–307

Bajaj YPS (1978) Regeneration of haploid tobacco plants from isolated pollen grown in drop culture. Indian J Exp Biol 16:407–409

Bajaj YPS (1988) Regeneration of plants from frozen (−196 °C) protoplasts of *Atropa belladonna* L., *Datura innoxia* Mill. and *Nicotiana tabacum* L. Indian J Exp Biol 26:289–292

Bajaj YPS (1988–1999) Medicinal and aromatic plants I–XI. Biotechnology in agriculture and forestry. Springer, Berlin Heidelberg New York

Bajaj YPS (1989) Genetic engineering and in vitro manipulation of plant cells – technical advances. In: Bajaj YPS (ed) Plant protoplasts and genetic engineering II. Biotechnology in agriculture and forestry, vol 9. Springer, Berlin Heidelberg New York, pp 1–25

Bajaj YPS (1990) Wide hybridization in legumes and oilseed crops through embryo, ovule, and ovary culture. In: Bajaj YPS (ed) Legumes and oilseed crops I. Biotechnology in agriculture and forestry, vol 10. Springer, Berlin Heidelberg New York, pp 3–37

Bajaj YPS (1991) Automated micropropagation for en masse production of plants. In: Bajaj YPS (ed) High tech and micropropagation I. Biotechnology in agriculture and forestry, vol 17. Springer, Berlin Heidelberg New York, pp 3–16

Bajaj YPS (1994) Somatic hybridization – a rich source of genetic variability. In: Bajaj YPS (ed) Somatic hybridization in crop improvement I. Biotechnology in agriculture and forestry, vol 27. Springer, Berlin Heidelberg New York, pp 3–32

Bajaj YPS (1995) Cryopreservation of plant cell, tissue, and organ culture for the conservation of germplasm and biodiversity. In: Bajaj YPS (ed) Cryopreservation of plant germplasm I. Biotechnology in agriculture and forestry, vol 32. Springer, Berlin Heidelberg New York, pp 3–28

Bajaj YPS (1998) Biotechnology for the improvement of medicinal plants. In: S Scannerini et al. (eds) Proc Int Cong Plant biotechnology as a tool for the exploitation of mountain lands. Fondazione per le Biotecnologie, Turin May 25–27, 1997 Acta Hortic no 457, pp 37–45

Bajaj YPS, Dionne LA (1968) The continuous culture of excised potato roots. NZ J Bot 6:386–394

Bajaj YPS, Simola LK (1991) *Atropa belladonna* L.: In vitro culture, regeneration of plants, cryopreservation, and the production of tropane alkaloids. In: Bajaj YPS (ed) Medicinal and aromatic plants III. Biotechnology in agriculture and forestry, vol 15. Springer, Berlin Heidelberg New York, pp 1–23

Bajaj YPS, Gosch G, Ottma M, Weber A, Gröbler A (1978) Production of polyploid and aneuploid plants from anthers and mesophyll protoplasts of *Atropa belladonna* and *Nicotiana tabacum*. Indian J Exp Biol 16:947–953

Bajaj YPS, Furmanowa M, Olszowska O (1988) Biotechnology of the micropropagation of medicinal and aromatic plants. In: Bajaj YPS (ed) Medicinal and aromatic plants I. Biotechnology in agriculture and forestry, vol 4. Springer, Berlin Heidelberg New York, pp 60–103

Baker CM, Dyer WE (1996) Genetic transformation of *Carthamus tinctorius* L. (safflower). In: Bajaj YPS (ed) Plant protoplasts and genetic engineering VII. Biotechnology in agriculture and forestry, vol 38. Springer, Berlin Heidelberg New York, pp 201–210

Bakkali AT, Jaziri M, Foriers A, Vander Heyden Y, Vanhaelen M, Homes J (1997) Lawsone accumulation in normal and transformed cultures of henna, *Lawsonia inermis*. Plant Cell Tiss Org Cult 51:83–87

Barton KA, Whitely HR, Yang KS (1987) *Bacillus thuringiensis* endotoxins expressed in transgenic *Nicotiana tabacum* provides resistance to lepidopteran insects. Plant Physiol 85:1103–1109

Bayley C, Trolinder N, Ray C, Morgan M, Quisenberry JE, Ow DW (1992) Engineering 2,4-D resistance into cotton. Theor Appl Genet 83:645–649

Benjamin BD, Roja G, Heble MR (1994) Alkaloid synthesis by root cultures of *Rauwolfia serpentina* transformed by *Agrobacterium rhizogenes*. Phytochemistry 35:381–383

Berlin J, Rugenhagen C, Dietze P, Fecker LF, Goddijn OJM, Hoge JHC (1993) Increased production of serotonin by suspension and root cultures of *Peganum harmala* transformed with a tryptophan decarboxylase cDNA clone from *Catharanthus roseus*. Transgenic Res 2:336–344

Boitel Conti M, Gontier E, Laberche JC, Ducrocq C, Sangwan Norrel BS (1995) Permeabilization of *Datura innoxia* hairy roots for release of stored tropane alkaloids. Planta Med 61:287–290

Cardi T, Iannamico V, D'Ambrosio F, Filippone E, Larquin PF (1992) *Agrobacterium*-mediated genetic transformation of *Solanum commersonii* Dun. Plant Sci 87:179–189

Chang HN, Sim SJ (1996) Genetic transformation of *Lithospermum erythrorhizon* for increased production of shikonin. In: Bajaj YPS (ed) Plant protoplasts and genetic engineering VII. Biotechnology in agriculture and forestry, vol 38. Springer, Berlin Heidelberg New York, pp 233–242

Chilton MD, Tepfer DA, Petit A, David C, Casse-Delbart F, Tempé J (1982) *Agrobacterium rhizogenes* inserts T-DNA into the genomes of the host plant root cells. Nature 295:432–434

Christen P, Roberts MF, Phillipson JD, Evans WC (1989) High-yield production of tropane alkaloids by hairy-root cultures of a *Datura candida* hybrid. Plant Cell Rep 8:75–77

Christen P, Aoki T, Shimomura K (1992) Characteristics of growth and tropane alkaloid production in *Hyoscyamus albus* hairy roots transformed with *Agrobacterium rhizogenes* A4. Plant Cell Rep 11:597–600

Giau-Uitz R, Miranda-Ham ML, Coello-Coello J, Chi B, Pacheco LM, Lyola-Vargas VM (1994) Indole alkaloid production by transformed and non-transformed root cultures of *Catharanthus roseus*. In Vitro Cell Dev Biol 30P:84–88

Conner AJ, Dommisse EM (1992) Monocotyledonous plants as hosts for *Agrobacterium*. Int J Plant Sci 153(4):550–555

Constabel CP, Towers GHN (1989) Incorporation of 35S into dithia-cyclohexadiene and thiophene polyines in hairy root cultures of *Chaenectis douglasii*. Phytochemistry 28:93–95

Corry JP, Reed WL, Curtis WR (1993) Enhanced recovery of solavetivone from *Agrobacterium*-transformed root cultures of *Hyoscyamus muticus* using integrated product extraction. Biotechnol Bioeng 42:503–508

Couillerot E, Caron C, Audran JC, Zeches M, Jardillier JC, Chenieux JC (1997) Establishment of transformed roots of *Fagara zanthoxyloides* and analysis of their alkaloid production. Plant Cell Tissue Organ Cult (in press)

Crossway A, Oakes JV, Irvine JM, Ward B, Knauf VC, Sholmaker CK (1986) Integration of foreign DNA following microinjection of tobacco mesophyll protoplasts. Mol Gen Genet 202:179–185

Daimon H, Ito Y, Ohara A, Mii M (1993) Plant regeneration from hairy roots of antagonistic plants to nematodes induced by wild strains of *Agrobacterium rhizogenes*. Crop Prod and Improv Technol in Asia, KSCS, Korea, pp 529–535

Damgaard O, Jensen LH, Rasmussen OS (1997) *Agrobacterium tumefaciens*-mediated transformation of *Brassica napus* winter cultivars. Transgenic Res 6:279–288

Dale PJ (1995) R & D regulations and field trialing of transgenic crops. TIBTECH 13:398–403

Davioud E, Kan C, Quirion JC, Das BC, Husson HP (1989) Epiallo-yohimbine derivatives isolated from in vitro hairy-root cultures of *Catharanthus trichophyllus*. Phytochemistry 28:1383–1387

De Block M, Botterman J, Vandewiele M, Dockx J, Thoen C, Gossele V, Rao MN, Thompson C, Van Montagu M, Leemans J (1987) Engineering herbicide resistance in plants by expression of a detoxifying enzyme. EMBO J 6:2513–2518

de Laat AMM, Blaas J (1987) An improved method for protoplast micro-injection suitable for transfer of entire plant chromosomes. Plant Sci 50:161–169

Doran PM (1995) Production of chemicals using genetically transformed plant organs. Ann NY Acad Sci 426–441

Drewes FE, Van Staden J (1995) Initiation of and solasodine production in hairy root cultures of *Solanum mauritanium* Scop. Plant Growth Regul 17:27–31

Eilert U, Wolters B (1989) Elicitor induction of S-adenosyl-L-methionine, anthranilic acid N-methyltransferase activity in cell suspension and organ cultures of *Ruta graveolens* L. Plant Cell Tissue Org Cult 18:1–18

Ermayanti TM, JA McComb, O'Brien PA (1994) Growth and swainsonine production of *Swainsona galegifolia* (Andr.) R. Br. untransformed and transformed root cultures. J Exp Bot 45:633–638

Fauconnier M-L, Jaziri M, Homès J, Shimomura K, Marlier M (1996) *Anthemis nobilis* L. (Roman Chamomile): In vitro culture, micropropagation, and the production of essential oils. In: Bajaj YPS (ed) Biotechnology in agriculture and forestry Vol. 37 Medicinal and aromatic plants IX. Springer Berlin Heidelberg pp 16–37

Fecker LF, Rügenhagen C, Berlin J (1993) Increased production of cadaverine and anabasine in hairy root cultures of *Nicotiana tabacum* expressing a bacterial lysine decarboxylase gene. Plant Mol Biol 23:11–21

Feng Xinhua, Shao Q, Jiang X. (1986) Transformation of the monocot *Dioscorea opposita* using *Agrobacterium tumefaciens*. GMICN 2(1):52–59

Flores HE, Curtis WR (1992) Approaches to understanding and manipulating the biosynthetic potential of plant roots. Ann NY Acad Sci 665:188–209

Furze JM, Rhodes MJC, Parr AJ, Robins RJ, Whitehead IM, Threlfall DR (1991) Abiotic factors elicit sesquiterpenoid phytoalexin production but not alkaloid production in transformed root culture of *Datura stramonium*. Plant Cell Rep 10:111–114

Gamborg OL, Miller RA, Ojima K (1968) Nutrient requirements of suspension cultures of soybean root cells. Exp Cell Res 50:151–158

Garnier F, Carpin S, Label P, Creche J, Rideau M, Hamdi S (1996a) Effect of cytokinin on alkaloid accumulation in periwinkle callus cultures transformed with a light-inducible *ipt* gene. Plant Sci 120:47–55

Garnier F, Label P, Hallard D, Chenieux JD, Rideau M, Hamdi S (1996b) Transgenic periwinkle tissues overproducing cytokinins do not accumulate enhanced levels of indole alkaloids. Plant Cell Tissue Organ Cult 45:223–230

Giulietti AM, Parr AJ, Rhodes MJC (1993) Tropane alkaloid production in transformed root cultures of *Brugmansia candida*. Planta Med 59:428–431

Goddijn OJM, Pen J (1995) Plants as bioreactors. TIBTECH 13:379–387

Gränicher F, Christen P, Kapetanidis I (1992) High-yield production of valeoptriates by hairy root cultures of *Valeriana officinalis* L. var. *sambucifolia* Milkan. Plant Cell Rep 11:339–342

Gränicher F, Christen P, Vuagnat P (1994) Rapid high performance liquid chromatographic quantification of valepotriates in hairy root cultures of *Valeriana officinalis* L. var. *sambucifolia* Mikan. Phytochem Anal 5:297–301

Gränicher F, Christen P, Kamalaprija P, Burger U (1995) An iridoid diester from *Valeriana officinalis* var. *sambucifolia* hairy roots. Phytochemistry 38:103–105

Green KD, Thomas NH, Callow JA (1992) Product enhancement and recovery from transformed root cultures of *Nicotiana glauca*, Biotechnol Bioeng 39:195–202

Greidziak N, Diettrich B, Luckner M (1990) Batch cultures of somatic embryos of *Digitalis lanata* in gas-lift fermenters. Development and cardenolide accumulation. Planta Med 56:175–178

Hadfi K, Batschauer A (1994) *Agrobacterium*-mediated transformation of white mustard (*Sinapis alba* L.) and regeneration of transgenic plants. Plant Cell Rep 13:130–134

Hagimori M, Matsumoto T, Mikami Y (1984) Jar fermenter culture of shoot-forming cultures of *Digitalis purpurea* L. using a revised medium. Agric Biol Chem 48: 965–970

Hamill JD, Robins RJ, Rhodes MJC (1989) Alkaloid production by transformed root cultures of *Cinchona ledgeriana*. Planta Med 55:354–357

Hashimoto T, Azechi S (1988) Bioreactors for the large-scale culture of plant cells. In: Bajaj YPS (ed) Biotechnology in agriculture and forestry, vol 4. Medicinal and aromatic plants I, Springer, Berlin Heidelberg New York, pp 104–122

Hashimoto T, Yun D-Jin, Yamada Y (1993) Production of tropane alkaloids in genetically engineered root cultures. Phytochemistry 32:713–718

Hilton MG, Rhodes MJC (1990) Growth and hyoscyamine production of hairy root cultures of *Datura stramonium* in a modified stirred tank reactor. Appl Microbiol Biotechnol 33:132–138

Hilton MG, Rhodes MJC (1993) Factors affecting the growth and hyoscyamine production during batch culture of transformed roots of *Datura stramonium*. Planta Med 59:340–344

Hilton MG, Wilson PDG, Robins RJ, Rhodes MJC (1988) Transformed root cultures – fermentation aspects. In: Robins RJ, Rhodes MJC (eds) Manipulating secondary metabolism in culture. University Press, Cambridge, pp 239–245

Hirotani M, Zhou Y, Lui H, Furuya T (1994) Astragalosides from hairy root cultures of *Astragalus membranaceus*. Phytochemistry 36:665–670

Hjortso M, Mukundan U (1994) Genetic transformation in *Tagetes* species (Marigolds) for thiophene contents. In: Bajaj YPS (ed) Biotechnology in agriculture and forestry, vol 29. Plant protoplasts and genetic engineering V. Springer, Berlin Heidelberg New York, pp 365–382

Ho CH, Shanks JV (1992) Effects of initial medium pH on growth and metabolism of *Catharanthus roseus* hairy root cultures. A study with 31p and 13c NMR spectroscopy. Biotechnol Lett 14:959–964

Hook ILI (1993) *Leontopodium alpinum* Cass. (Edelweiss): In vitro culture, micropropagation, and the production of secondary metabolites. In: Bajaj YPS (ed) Biotechnology in agriculture and forestry, vol 21. Medicinal and aromatic plants IV. Springer, Berlin Heidelberg New York, pp 217–232

Hu Zhili, Alfermann AW (1993) Diterpenoid production in hairy root cultures of *Salvia miltiorrhiza*. Phytochemistry 32:699–703

Ikenaga T, Oyama T, Muranaka T (1995) Growth and steroidal saponin production in hairy root cultures of *Solanum aculeatissimum*. Plant Cell Rep 14:413–417

Inomata S, Yokoyama M (1996) Genetic transformation of *Panax ginseng* (C.A. Meyer) for increased production of ginsenosides. In: Bajaj YPS (ed) Plant protoplasts and genetic engineering VII. Biotechnology in agriculture and forestry, vol 38. Springer, Berlin Heidelberg New York, pp 253–269

Ionkova I (1992) Alkaloid production of *Hyoscyamus reticulatus* plant and transformed root culture clone. Biotech Bioeng 6:50–52

Ionkova I, Alfermann AW (1990) Transformation of *Astragalus* species by *Agrobacterium rhizogenes* and their saponin production. Planta Med 56:634

Ishimaru K, Shimomura K (1989) 7β-hydroxyhyoscyamine from *Duboisia myoporoides-D. leichhardtii* hybrid and *Hyoscyamus albus*. Phytochemistry 28:3507–3509

Ishimaru K, Shimomura K (1991) Tannin production in hairy root culture of *Geranium thunbergii*. Phytochemistry 30:825–828

Ishimaru K, Shimomura K (1995a) *Geranium thunbergii*: in vitro culture and the production of geraniin and other tannins. In: Bajaj YPS (ed) Medicinal and aromatic plants VIII. Biotechnology in agriculture and forestry vol 33. Springer, Berlin Heidelberg New York, pp 232–247

Ishimaru K, Shimomura K (1995b) Phenolics root cultures of medicinal plants. In: Atta-ur-Rahman (ed) Studies in natural products chemistry, vol 17. Elsevier Amsterdam, pp 421–449

Ishimaru K, Shimomura K (1996) Genetic transformation in *Swertia japonica*. In: Bajaj YPS (ed) Plant protoplasts and genetic engineering VII. Biotechnology in agriculture and forestry, vol 38. Springer, Berlin Heidelberg New York, pp 308–317

Ishimaru K, Hirose M, Takahashi K, Koyama K, Shimomura K (1990a) Tannin production in root cultures of *Sanguisorba afficinalis*, Phytochemistry 29:3827–3830

Ishimaru K, Sudo H, Satake M, Matsunaga Y, Hasegawa Y, Takemoto S, Shimomura K (1990b) Amarogentin, amaroswerin and four xanthones from hairy root cultures of *Swertia japonica*. Phytochemistry 29:1563–1565

Ishimaru K, Sudo H, Satake M,Shimomura K (1990c) Phenyl glucosides from a hairy root culture of *Swertia japonica*. Phytochemistry 29:3823–3825

Ishimaru K, Hirose M, Takahashi K, Koyama K, Shimomura K (1991a) Tannin production in hairy root cultures of *Sanguisorba officinalis* L. Plant Tissue Cult Lett 8:114–117

Ishimaru K, Yonemitsu H, Shimomura K (1991b) Lobetyolin and lobetyol from hairy root culture of *Lobelia inflata*. Phytochemistry 30:2255–2257

Ishimaru K, Ikeda Y, Kuranari Y, Shimomura K (1992a) Growth and lobeline production of *Lobelia inflata* hairy roots. Shoyakugaku Zasshi 46:265–267

Ishimaru K, Yoshimatsu K, Yamakawa T, Kamada H, Shimomura K (1992b) Phenolic constituents in tissue cultures of *Phyllanthus niruri*. Phytochemistry 31:2015–2018

Ishimaru K, Sadoshima S, Neera S, Koyama K, Takahashi K, Shimomura K (1992c) A polyacetylene gentiobioside from hairy roots of *Lobelia inflata*. Phytochemistry 31:1577–1579

Ishimaru K, Arakawa H, Sadoshima S, Yamaguchi Y (1993) Effects of basal media on growth and polyacetylene production of *Lobelia inflata* hairy roots. Plant Tissue Cult Lett 10:191–193

Ishimaru K, Arakawa H, Yamanaka M, Shimomura K (1994) Polyacetylenes in *Lobelia sessilifolia* hairy roots. Phytochemistry 35:365–369

Ishimaru K, Hirose M, Takahashi K, Koyama K, Shimomura K (1995a) *Sanguisorba officinalis* L. (Great Burnet): In vitro culture and production of sanguiin, tannins, and other secondary metabolites. In: Bajaj YPS (ed) Medicinal and aromatic plants VIII. Biotechnology in agriculture and forestry, vol 33. Springer, Berlin Heidelberg New York, pp 427–441

Ishimaru K, Yamaguchi Y, Shimomura K, Yoshihira K (1995b) Polyacetylene production in hairy root cultures of *Lobelia inflata*. Jpn J Food Chem 2:80–84

Ishimaru K, Yamanaka M, Terahara N, Shimomura K, Okamoto D, Yoshihira K (1996) Biotransformation of phenolics by hairy root cultures of five herbal plants. Jpn J Food Chem 3:38–42

Islas I, Loyola-Vargas VM, Miranda-Ham ML (1994) Tryptophan decarboxylase activity in transformed roots from *Catharanthus roseus* and its relationship to tryptamine, ajmalicine, and catharanthine accumulation during the culture cycle. In Vitro Cell Dev Biol 30P:81–83

Jarl CI, Rietveld EM (1996) Transformation efficiencies and progeny analysis after varying different parameters of direct gene transfer of *Nicotiana tabacum* protoplasts. Physiol Plant 98:550–556

Jaziri M, Legros M, Homes J, Van Haelen M (1988) Tropine alkaloids production by hairy root cultures of *Datura stramonium* and *Hyoscyamus niger*. Phytochemistry 27:419–420

Jaziri M, Homes J, Shimomura K (1994a) An unusual root tip formation in hairy root culture of *Hyoscyamus muticus*. Plant Cell Rep 13:349–352

Jaziri M, Yoshimatsu K, Homes J, Shimomura K (1994b) Traits of transgenic *Atropa belladonna* doubly transformed with different *Agrobacterium rhizogenes* strains. Plant Cell Tissue Organ Cult 38:257–262

Jaziri M, Shimomura K, Yoshimatsu K, Fauconnier M-L, Marlier, Homes J (1995) Establishment of normal and transformed cultures of *Artemisia annua* L. for artemisinin production. J Plant Physiol 145:175–177

Jung G, Tepfer D (1987) Use of genetic transformation by the Ri T-DNA of *Agrobacterium rhizogenes* to stimulate biomass and tropane alkaloid production in *Atropa belladonna* and *Calystegia sepium* roots grown in vitro. Plant Sci 50:145–151

Jung KH, Kwak SS, Choi CY, Liu JR (1995) An interchangeable system of hairy root and cell suspension cultures of *Catharanthus roseus* for indole alkaloid production. Plant Cell Rep 15:51–54

Kamada H, Okamura N, Satake M, Harada H, Shimomura K (1986) Alkaloid production by hairy root cultures in *Atropa belladonna*. Plant Cell Rep 5:239–242

Kennedy AI, Deans SG, Svoboda KP, Gray AI, Waterman PG (1993) Volatile oils from normal and transformed roots of *Artemisia absinthium*. Phytochemistry 32:1449–1451

Kino-oka M, Hongo Y, Taya M, Tone S (1992) Culture of red beet hairy root in bioreactor and recovery of pigment released from the cells by repeated treatment of oxygen starvation. J Chem Eng Jpn 25:490–495

Kisiel W, Stojakowska A, Malarz J, Kohlmunzer S (1995) Sesquiterpene lactones in *Agrobacterium rhizogenes*-transformed hairy root culture of *Lactuca virosa*. Phytochemistry 40:1139–1140

Kiyokawa S, Kikuchi Y, Kamada H, Harada H (1996) Genetic transformation of *Begonia tuberhybrida* by Ri *rol* genes. Plant Cell Rep 15:606–609

Klein TM, Wolf ED, Wu R, Sanford JC (1987) High velocity microprojectiles for delivering nucleic acids into living cells. Nature 327:70–73

Knopp E, Strauss A, Wehrli W (1988) Root induction on several solanaceae species by *Agrobacterium rhizogenes* and the determination of root tropane alkaloid content. Plant Cell Rep 7:590–593

Ko KS, Ebizuka Y, Noguchi H, Sankawa U (1988) Production of secondary metabolites by hairy roots and regenerated plants transferred with R_i plasmids. Chem Pharm Bull 36(10):4217–4220

Ko KS, Noguchi H, Ebizuka Y, Sankawa U (1989) Oligoside production by hairy root cultures transformed by R_i plasmids. Chem Pharm Bull 37:245–248

Komari T (1989) Transformation of callus cultures of nine plant species mediated by *Agrobacterium*. Plant Sci 60:223–229

Krens FA, Molendijk L, Wullems GJ, Schilperoort RA (1982) In vitro transformation of plant protoplasts with T_i-plasmid DNA. Nature 296:72–74

Kumar A, Miller P, Lyon J, Davie P. (1995) *Agrobacterium*-mediated transformation of five wild *Solanum* species using in vitro microtubers. Plant Cell Rep 14:324–328

Kurioka Y, Suzuki Y, Kamada H, Harada H (1992) Promotion of flowering and morphological alterations in *Atropa belladonna* transformed with a CaMV 35S-rolC chimeric gene of the R_i plasmid. Plant Cell Rep 12:1–6

Kyo M, Miyauchi Y, Fujimoto T, Mayama S (1990) Production of nematocidal compounds by hairy root cultures of *Tagetes patula* L. Plant Cell Rep 9:393–397

Larquin PF (1989) Uptake and integration of exogenous DNA in plants. In: Bajaj YPS (ed) Plant protoplasts and genetic engineering II. Biotechnology in agriculture and forestry, vol 9. Springer, Berlin Heidelberg New York, pp 54–74

Lee HS, Kim SW, Lee KW, Eriksson T, Liu JR (1995) *Agrobacterium*-mediated transformation of ginseng (*Panax ginseng*) and mitotic stability of the inserted β-glucuronidase gene in regenerants from isolated protoplasts. Plant Cell Rep 14:545–549

Leuschner C, Walton JW, Herzog G, Schultz G (1995) Chorismate mutase isoenzymesi in transformed roots of *Datura stramonium*. Plant Physiol Biochem 33:367–371

Liu THA, Stephens LC, Hannapel DJ (1995) Transformation of *Solanum brevidens* using *Agrobacterium tumefaciens*. Plant Cell Rep 15:196–199

Lloyd G, McCown B (1980) Commercially feasible micropropagation of mountain laurel, *Kalmia latifolia*, by use of shoot-tip culture. Int Plant Prop Soc 30:421–427

Lodhi AH, Charlwood BV (1996) *Agrobacterium rhizogenes*-mediated transformation of *Rubia peregrina* L.: in vitro accumulation of anthraquinones. Plant Cell Tissue Organ Cult 46:103–108

Lodhi AH, Bongaerts RJM, Verpoorte R, Coomber SA, Charlwood BV (1996) Expression of bacterial isochorismate synthase (EC 5.4.99.6) in transgenic root cultures of *Rubia peregrina*. Plant Cell Rep 16:54–57

Loyola-Vargas VM, Miranda-Ham Maria de Lourdes (1995) Root culture as a source of secondary metabolites of economic importance. In: Armason JT et al. (eds) Phytochemistry of medicinal plants. Plenum, New York, pp 217–248

Lu T, Parodi FJ, Vargas D, Quijano L, Mertooetomo ER, Hjortso MA, Fischer NH (1993) Sesquiterpenes and thiarubrines from *Ambrosia trifida* and its transformed roots. Phytochemistry 33:113–116

Machado LOR, de Andrade GM, Cid LPB, Penchel RM, Brasileiro ACM (1997) *Agrobacterium* strain specificity and shooty tumour formation in eucalypt (*Eucalyptus grandis* × E. urophylla). Plant Cell Rep 16:299–303

Maldonado-Mendoza IE, Ayora-Talavera T, Loyola-Vargas VM (1993) Establishment of hairy root cultures of *Datura stramonium*. Characterization and stability of tropane alkaloid production during long periods of subculturing. Plant Cell Tissue Organ Cult 33:321–329

Maldonado-Mendoza IE, Loyola-Vargas VM (1995) Establishment and characterization of photosynthetic hairy root cultures of *Datura stramonium*. Plant Cell Tissue Organ Cult 40:197–208

Mano Y, Nabeshima S, Matsui C, Ohkawa H (1986) Production of tropane alkaloids by hairy root cultures of *Scopolia japonica*. Agric Biol Chem 50:2715–2722

Mano Y, Ohkawa H, Yamada Y (1989) Production of tropane alkaloids by hairy root cultures of *Duboisia leichhardtii* transformed by *Agrobacterium rhizogenes*. Plant Sci 59:191–201

Mason HS, Arntzen CJ (1995) Transgenic plants as vaccine production systems. TIBTECH 13:388–392

Mason HS, Lam DMK, Arntzen CJ (1992) Expression of hepatitis B surface antigen in transgenic plants. Proc Natl Acad Sci 89:11745–11749

Masoud SA, Johnson LB, White FF, Reeck GR (1993) Expression of a cysteine proteinase inhibitor (oryzacystatin-I) in transgenic plants. Plant Mol Biol 21:655–663

Matsumoto T, Tanaka N (1991) Production of phytoecdysteroids by hairy root cultures of *Ajuga reptans* var. *atropurpurea*. Agric Biol Chem 55:1019–1025

McCabe DE, Martinell BJ (1993) Transformation of elite cotton cultivars via particle bombardment of meristems. Bio/Technology 11:596–598

Moore GA, Jacono CC, Neidigh JL, Lawrence SD, Cline K (1992) *Agrobacterium*-mediated transformation of *Citrus* stem segments and regeneration of transgenic plants. Plant Cell Rep 11:238–242

Morris P, Robbins MP (1992) Condensed tannin formation by *Agrobacterium rhizogenes*-transformed root and shoot organ cultures of *Lotus corniculatus*, J Exp Bot 43:221–231

Mukundan U, Hjortso MA (1991) Growth and thiophene accumulation by hairy root cultures of *Tagetes patula* in media of varying initial pH. Plant Cell Rep 9:927–930

Muranaka T, Ohkawa H, Yamada Y (1992) Scopolamine release into media by *Duboisia leichhardtii* hairy root clones. Appl Microbiol Biotechnol 37:554–559

Muranaka T, Kazuoka T, Ohkawa H, Yamada Y (1993a) Characteristics of scopolamine-releasing hairy root clones of *Duboisia leichhardtii*. Biosci Biotech Biochem 57:1398–1399

Muranaka T, Ohkawa H, Yamada Y (1993b) Continuous production of scopolamine by a culture of *Duboisia leichhardtii* hairy root clone in a bioreactor system. Appl Microbiol Biotechnol 40:219–223

Murashige T, Skoog F (1962) A revised medium for rapid growth and bio-assays with tobacco tissue cultures. Physiol Plant 15:473–497

Nagakai M, Kushiro T, Matsumoto T, Tanaka N, Karinuma K, Fujimoto Y (1994) Incorporation of acetate and cholesterol into 20-hydroxyecdysone by hairy root clone of *Ajuga reptans* var. *Atropurpurea*. Phytochemistry 36:907–910

Neškovic M, Miljuš-Djukić J, Ninković S (1995) Genetic transformation in *Fagopyrum esculentum* (Buckwheat). In: Bajaj YPS (ed) Plant protoplasts and genetic engineering VI. Biotechnology in agriculture and forestry, vol 34. Springer, Berlin Heidelberg New York, pp 171–182

Neumann E, Schaefer-Ridder M, Wang Y, Hofschneider PH (1982) Gene transfer into mouse lyoma cells by electroporation in high electric fields. EMBO J 1:841–845

Nin S, Bennici A, Roselli G, Marioti D, Schiff S, Magherini R (1997) *Agrobacterium*-mediated transformation of *Artemisia absinthium* L. (Wormwood) and production of secondary metabolites. Plant Cell Rep 16:725–730

O'Dowd NA, Richardson DHS (1994) Production of tumours and roots by *Ephedra* following *Agrobacterium rhizogenes* infection. Can J Bot 72:203–207

Ogasawara T, Chiba K, Tada M (1993) Production in high yield of a naphthoquinone by a hairy root culture of *Sesamum indicum*. Phytochemistry 33:1095–1098

Ohkawa H, Kamada H, Sudo H, Harada H (1989) Effects of gibberellic acid on hairy root growth in *Datura innoxia*. J Plant Physiol 134:633–636

Oostdam A, Mol JNM, van der Plas LHW (1993) Establishment of hairy root cultures of *Linum flavum* producing the lignan 5-methoxypodophyllotoxin. Plant Cell Rep 12:474–477

Parodi FJ, Fischer NH, Flores HE (1988) Benzofuran and bithiophenes from root cultures of *Tagetes patula*. J Nat Prod 51:594–595

Parr AJ (1992) Alternative metabolic fates of hygrine in transformed root cultures of *Nicandra physaloides*. Plant Cell Rep 11:270–273

Parr AJ, Peerless ACJ, Hamill JD, Walton NJ, Robins RJ, Rhodes MJC (1988) Alkaloid production by transformed root cultures of *Catharanthus roseus*. Plant Cell Rep 7:309–312

Payne J, Hamill JD, Robins RJ, Rhodes MJC (1987) Production of hyoscyamine by hairy root cultures of *Datura stramonium*. Planta Med 53:474–478

Perlak FJ, Deaton RW, Armstrong TA, Fuchs RL, Sims SR, Greenplate JT, Fischoff DA (1990) Insect-resistant cotton. Bio/Technology 8:939–943

Petri G, Bajaj YPS (1989) *Datura* spp: In vitro regeneration and the production of tropanes. In: Bajaj YPS (ed) Medicinal and aromatic plants II. Biotechnology in agriculture and forestry, vol 7. Springer, Berlin Heidelberg New York, pp 135–161

Radke SE, Turner JC, Facciotti D (1992) Transformation and regeneration of *Brassica rapa* using *Agrobacterium tumefaciens*. Plant Cell Rep 11:499–505

Reichling J, Thron U (1990) Accumulation of rare phenylpropanoids in *Agrobacterium rhizogenes*-transformed root cultures of *Coreopsis tinctoria*. Planta Med 56:488–490

Rhodes MJC, Hilton MG, Parr AJ, Hamill JD, Robins RJ (1986) Nicotine production by hairy root cultures of *Nicotiana rustica*: fermentation and product recovery. Biotechnol Lett 8:415–420

Rhodes MJC, Parr AJ, Giulietti A, Aird ELH (1994) Influence of exogenous hormones on the growth and secondary metabolite formation in transformed root cultures. Plant Cell Tiss Org Cult 38:143–151

Riggs CD, Bates GW (1986) Stable transformation of tobacco by electroporation: evidence for plasmid concentration. Proc Natl Acad Sci USA 83:5602–5606

Rodriguez-Mendiola MA, Stafford A, Cresswell R, Arias-Castro C (1991) Bioreactors for growth of plant roots. Enzyme Microb Technol 13:697–702

Sáenz-Carbonell L, Maldonado-Mendoza IE, Moreno V, Ciau-Uitz R, López-Meyer M, Oropeza C, Loyola-Vargas VM (1993) Effect of the medium pH on the release of secondary metabolites from roots of *Datura stramonium*, *Catharanthus roseus* and *Tagetes patula* cultured in vitro. Appl Biochem Biotechnol 38:257–267

Saito K (1995) Genetic transformation in *Glycrrhiza uralensis*. In: Bajaj YPS (ed) Plant protoplasts and genetic engineering VI. Biotechnology in agriculture and forestry, vol 34. Springer, Berlin Heidelberg New York, pp 204–213

Saito K, Murakoshi I, Inze D, van Montagu M (1989) Biotransformation of nicotine alkaloids by tobacco shooty teratomas induced by a T$_i$ plasmid mutually. Plant Cell Rep 7:607–610

Saito K, Yamazaki M, Shimomura K, Yoshimatsu K, Murakoshi I (1990) Genetic transformation of foxglove (Digitalis purpurea) by chimeric foreign genes and production of cardioactive glycosides. Plant Cell Rep 9:121–124

Saito K, Yamazaki M, Kawaguchi A, Murakoshi I (1991) Metabolism of solanaceous alkaloids in transgenic plant teratomas integrated with genetically engineered genes. Tetrahedron 47:5955–5968

Saito K, Yamazaki M, Anzai H, Yoneyama K, Murakoshi I (1992) Transgenic herbicide-resistant Atropa belladonna using an R$_i$ binary vector and inheritance of the transgenic trait. Plant Cell Rep 11:219–224

Saito K, Yamazaki M, Shimomura K, Yoshimatsu K, Murakoshi I (1993) Transformation in Digitalis purpurea L. (Foxglove). In: Bajaj YPS (ed) Biotechnology in agriculture and forestry, vol 22. Plant protoplasts and genetic engineering III. Springer, Berlin Heidelberg New York, pp 182–189

Salem KMSA, Charlwood BV (1995) Accumulation of essential oils by Agrobacterium tumefaciens-transformed shoot cultures of Pimpinella anisum. Plant Cell Tissue Organ Cult 40:209–215

Sato K, Yamazaki T, Okuyama E, Yoshihira K, Shimomura K (1991) Anthraquinone production by transformed root cultures of Rubia tinctorum: influence of phytohormones and sucrose concentration. Phytochemistry 30:1507–1509

Sauerwein M, Shimomura K (1990) 17α-O-methylyohimbine and vallesiachotamine from roots of Amsonia elliptica. Phytochemistry 29:3377–3379

Sauerwein M, Shimomura K (1991) Alkaloid production in hairy roots of Hyoscyamus albus transformed with Agrobacterium rhizogenes. Phytochemistry 30:3277–3280

Sauerwein M, Ishimaru K, Shimomura K (1991a) Indole alkaloids in hairy roots of Amsonia elliptica. Phytochemistry 30:1153–1155

Sauerwein M, Ishimaru K, Shimomura K (1991b) A piperidone alkaloid from Hyoscyamus albus roots transformed with Agrobacterium rhizogenes. Phytochemistry 30:2977–2978

Sauerwein M, Yamazaki T, Shimomura K (1991c) Hernandulain in hairy root cultures of Lippia dulcis. Plant Cell Rep 9:579–581

Schäfer W, Görz A, Kahl G (1987) T-DNA integration and expression in a monocot crop plant after induction of Agrobacterium. Nature 327:529–532

Sharp JM, Doran PM (1990) Characteristics of growth and tropane alkaloid synthesis in Atropa belladonna roots transformed by Agrobacterium rhizogenes. J Biotechnol 16:171–186

Shimomura K, Sauerwein M, Ishimaru K (1991a) Tropane alkaloids in the adventitious and hairy root cultures of solanaceous plants. Phytochemistry 30:2275–2278

Shimomura K, Sudo H, Saga H, Kamada H (1991b) Shikonin production and secretion by hairy root cultures of Lithospermum erythrorhizon. Plant Cell Rep 10:282–285

Shimomura K, Yoshimatsu K, Ishimaru K, Sauerwein M (1995) Tropane alkaloids in root cultures of solanaceous plants. In: Atta-ur-Rahman (ed) Studies in natural products chemistry, vol 17. Elsevier Amsterdam, pp 395–419

Sim SJ, Chang HN (1993) Increased shikonin production by hairy roots of Lithospermum erythrorhizon in a two-phase bubble column reactor. Biotechnol Lett 15:45–150

Spencer A, Hamill JD, Rhodes MJC (1990) Production of terpenes by differentiated shoot cultures of Mentha citrata transformed with Agrobacterium tumefaciens T37. Plant Cell Rep 8:601–604

Spencer A, Hamill, Rhodes MJC (1993) In vitro biosyntheis of monoterpenes by Agrobacterium-transformed shoot cultures of two Mentha species. Phytochemistry 32:911–919

Stuart DA, Strickland SG, Walker KA (1987) Bioreactor production of alfalfa somatic embryos. Hort Sci 22:800–803

Subroto MA, Doran PM (1994) Production of steroidal alkaloids by hairy roots of Solanum aviculare and the effect of gibberellic acid. Plant Cell Tissue Organ Cult 38:93–102

Subroto MA, Hamill JD, Doran PM (1993) Alkaloids from transformed shoot cultures of several solanaceous plants. In: Sargeant J et al. (eds) Proc 11th Australian Biotech Conf. Perth, pp 255–256

Subroto MA, Hamill JD, Doran PM (1995) Growth kinetics and stichiometry of Mentha citrata shooty teratomas transformed by Agrobacterium tumefaciens. Biotechnol Lett 17:427–432

Subroto MA, Kwok KH, Hamill JD, Doran PM (1996) Coculture of genetically transformed roots and shoots for synthesis, translocation, and biotransformation of secondary metabolites. Biotechnol Bioeng 49:481–494

Suzuki Y, Kurioka Y, Ogasawara T, Kamada H (1993) Transformation in *Atropa belladonna*. In: Bajaj YPS (ed) Plant protoplasts and genetic engineering III. Biotechnology in agriculture and forestry, vol 22 Spring Berlin Heidelberg New York pp 135–143

Tada H, Shimomura K, Ishimaru K (1995a) Polyacetylenes in *Platycodon grandiflorum* hairy root and campanulaceous plants. J Plant Physiol 145:7–10

Tada H, Shimomura K, Ishimaru K (1995b) Polyacetylenes in hairy root cultures of *Lobelia chinensis* Lour. J Plant Physiol 146:199–202

Tada H, Murakami Y, Omoto T, Shimomura K, Ishimaru K (1996a) Rosmarinic acid and related phenolics in hairy root cultures of *Ocimum basilicum*. Phytochemistry 42:431–434

Tada H, Terahara N, Motoyama E, Shimomura K, Ishimaru K (1996b) Anthocyanins in *Lobelia chinensis* hairy roots. Plant Tissue Cult Lett 13:85–86

Tada H, Nakashima T, Kunitake H, Mori K, Tanaka M, Ishimaru K (1996c) Polyacetylenes in hairy root cultures of *Campanula medium* L. J Plant Physiol 147:617–619

Tanaka M, Yonemitsu H, Shimomura K, Ishimaru K, Mochida S, Endo T, Kaji A (1993) Transformation in *Lobelia inflata*. In: Bajaj YPS (ed) Plant protoplasts and genetic engineering III. Biotechnology in agriculture and forestry vol 22. Springer Berlin Heidelberg New York, pp 253–264

Tanaka N, Matsumoto T (1993) Regeneration of *Ajuga* hairy roots with high productivity of 20-hydroxyecdysone. Plant Cell Rep 13:87–90

Tanaka N, Takao M, Matsumoto T (1995) Vincamine production in multiple shoot culture derived from hairy roots of *Vinca minor*. Plant Cell Tissue Organ Cult 41:61–64

Tanaka N, Yoshimatsu K, Shimomura K, Ishimaru K (1996a) Rutin and other polyphenols in *Fagopyrum esculentum* hairy roots. Nat Med 50:269–272

Tanaka N, Yamada Y, Shimomura K, Ishimaru K (1996b) Polyacetylenes in tissue cultures of *Campanula glomerata*. Plant Tissue Cult Lett 13:215–217

Taya M, Yoyama A, Kondo O, Kobayashi T, Mitsui T (1989) Growth characteristics of plant hairy roots and their culture in bioreactors. J Chem Eng Jpn 22:89–94

Thomas JC, Adams DG, Keppenne VD, Wasmann CC, Brown JK, Kanost MR, Bohnert HJ (1995) Protease inhibitors of *Manduca sexta* expressed in transgenic cotton. Plant Cell Rep 14:758–762

Thron U, Maresch L, Beiderbeck R, Reichling J (1989) Accumulation of unusual phenylpropanoids in transformed and non-transformed root cultures of *Coreopsis tinctotia*. Z Naturforsch 44c:573–577

Toivonen L, Ojala M, Kauppinen V (1990) Indole alkaloid production by hairy root cultures of *Catharanthus roseus*: growth kinetics and fermentation. Biotechnol Lett 12:519–524

Trotin F, Moumou Y, Vasseur J (1993) Flavanol production by *Fagopyrum esculentum* hairy and normal root cultures. Phytochemistry 32:929–931

Trypsteen M, Van Lijsebettens M, Van Severen R, Van Montague M (1991) *Agrobacterium rhizogenes*-mediated transformation of *Echinacea purpurea*. Plant Cell Rep 10:85–89

Uchida K, Kuroyanagi M, Ueno A (1993) Tropane alkaloid production in hairy roots of *Hyoscyamus niger* transformed with *Agrobacterium rhizogenes*. Plant Tissue Cult Lett 10:223–228

Ueno K, Fukunaga Y, Arisumi K (1996) Genetic transformation of *Rhododendron* by *Agrobacterium tumefaciens*. Plant Cell Rep 16:38–41

Uozumi N, Makino S, Kobayashi T (1995) 20-hydroxyecdysone production in *Ajuga* hairy root controlling intracellular phosphate based on kinetic model. J Ferment Bioeng 80:362–368

Uozumi N, Ohtake Y, Nakashimada Y, Morikawa Y, Tanaka N, Kobayashi T (1996) Efficient regeneration of GUS-transformed *Ajuga* hairy root. J Ferment Bioeng 81:374–378

Vanhala L, Hiltunen R, Oksman-Caldentey KM (1995) Virulence of different *Agrobacterium* strains on hairy root formation of *Hyoscyamus muticus*. Plant Cell Rep 14:236–240

Vazquez-Flota F, Moreno-Valenzuela O, Miranda-Ham ML, Coello-Coello J, Loyola-Vargas VM (1994) Catharanthine and ajmalicine synthesis in *Catharanthus roseus* hairy root cultures. Plant Cell Tissue Organ Cult 38:273–279

Whitney PJ (1992) Novel bioreactors for the growth of roots transformed by *Agrobacterium rhizogenes*. Enzyme Microb Technol 14:13–17

Williams RD, Ellis BE (1993) Alkaloids from *Agrobacterium rhizogenes* transformed *Papaver somniferum* cultures. Phytochemistry 32:719–723

Wilson AD, Flint HM, Deaton WR, Duehler RE (1994) Yield, yield components, and fiber properties of insect-resistant cotton lines containing a *Bacillus thuringiensis* toxin gene. Crop Sci 34:38–41

Wilson PDG, Hilton MG, Robins RJ, Rhodes MJC (1987) Fermentation studies of transformed root cultures. In: Moody GW, Baker PB (eds) Bioreactors and biotransformations. Elsevier, Amsterdam, pp 38–51

Yamanaka M, Shimomura K, Sasaki K, Yoshihira K, Ishimaru K (1995) Glucosylation of phenolics by hairy root cultures of *Lobelia sessilifolia*. Phytochemistry 40:1149–1150

Yamanaka M, Ishibashi K, Shimomura K, Ishimaru K (1996) Polyacetylene glucosides in hairy root cultures of *Lobelia cardinalis*. Phytochemistry 41:183–185

Yamazaki M, Son L, Hayashi T, Morita N, Asamizu T, Mourakoshi I, Saito K (1996) Transgenic fertile *Scoparia dulcis* L., a folk medicinal plant, conferred with a herbicide-resistant trait using an R_i binary vector. Plant Cell Rep 15:317–321

Yonemitsu H, Shimomura K, Satake M, Mochida S, Tanaka M, Endo T, Kaji A (1990) Lobeline production by hairy root culture of *Lobelia inflata* L. Plant Cell Rep 9:307–310

Yoneyama K, Anzai H (1993) Transgenic plants resistant to diseases by the detoxification of toxins. Biotech Plant Dis Control. Wiley-Liss, New York pp 115–117

Yoshikawa T, Furuya T (1987) Saponin production by cultures of *Panax ginseng* transformed with *Agrobacterium rhizogenes*. Plant Cell Rep 6:449–453

Yoshimatsu K, Shimomura K (1992) Transformation of opium poppy (*Papaver somniferum* L.) with *Agrobacterium rhizogenes* MAFF 03-01724. Plant Cell Rep 11:132–136

Yoshimatsu K, Shimomura K (1996) Genetic transformation in *Papaver somniferum* (opium poppy) for enhanced production of morphinan. In: Bajaj YPS (ed) Plant protoplasts and genetic engineering VII. Biotechnology in agriculture and forestry, vol 38. Springer, Berlin Heidelberg New York, pp 243–252

Yoshimatsu K, Shimomura K (1998) Isoquinoline alkaloid production by transformed cultures of *Papaver somniferum*. Phytochemistry (in press)

Yoshimatus K, Jaziri M, Kamada H, Shimomura K (1997) Production of diploid and haploid transgenic *Atropa belladonna* plants: morphological traits and tropane alkaloid production. Belg J Bot (in press)

Yu S, Doran PM (1994) Oxygen requirements and mass transfer in hairy root culture. Biotechnol Bioeng 44:880–887

Yun DJ, Hashimoto T, Yamada Y (1992) Metabolic engineering of medicinal plants: transgenic *Atropa belladonna* with an improved alkaloid composition. Proc Natl Acad Sci USA 89:11799–11803

Zhang YL (1993) Transformation in *Scopolia*. In: Bajaj YPS (ed) Biotechnology in agriculture and forestry, vol 22. Plant protoplasts and genetic engineering III. Springer, Berlin Heidelberg New York, pp 314–328

Zhou Y, Hirotani M, Rui H, Furuya T (1995) Two triglycosidic triterpene astragalosides from hairy root cultures of *Astragalus membranaceus*. Phytochemistry 38:1407–1410

Zhou Y, Hirotani M, Yoshikawa T, Furuya T (1997) Flavonoids an phenylethanoids from hairy root cultures of *Scutellaria baicalensis*. Phytochemistry 44:83–87

II Genetic Transformation of *Ajuga reptans*

N. Tanaka[1], N. Uozumi[2], and T. Kobayashi[3]

1 Introduction

Ajuga reptans, a member of the Labiatae, is a horticultural plant that produces blue flowers in spring and is used as ground cover. The plant originated in Europe and has been utilized as a medicinal herb for treatment of jaundice and rheumatism in European countries for centuries. The roots of *A. reptans* contain some steroids, such as 20-hydroxyecdysone (20-HE), ajugasterone, cyasterone, and norcyasterone, which are phytoecdysteroids (Fig. 1B,D; Tomas et al. 1992; Camps and Coll 1993). 20-HE, is identical to β-ecdysone, is a principal physiological inducer of molting and metamorphosis in arthropoda. The physiological properties of 20-HE in insects and in mammals have been investigated in many laboratories, and the possibility of their practical use in pest control (Kubo et al. 1983), in chemotherapy (Yoshida et al. 1971), and in silk production (Kozakai et al. 1990) has been demonstrated. Indeed, a Japanese company has recently succeeded in developing and selling 20-HE as a control chemical for the spinning of cocoons by silkworms.

It has been suggested that *A. reptans* might be a source of anthocyanin for the industrial production of food colorants (Callebaut et al. 1990, 1993, 1996). Furthermore, since the plant produces and stores large amounts of raffinose in its intracellular vacuoles, this plant has been utilized in a study of the mechanism of frost-hardiness (Bachmann and Keller 1995).

The productivity of secondary metabolites such as phytoecdysteroids, anthocyanin, and raffinose in *A. reptans* will be elevated by genetic engineering, in particular, gene manipulation technique. Even though the importance of *A. reptans* is becoming apparent, to our knowledge, no efforts apart from ours have been made to establish a system for the transformation, culture, and regeneration of this plant to date. The in-vitro culture, secondary metabolite, and transformation studies on *Ajuga* are summarized in Table 1.

[1] Center for Gene Science, Hiroshima University, 1-4-2 Kagamiyama, Higashi-Hiroshima, Hiroshima 739, Japan
[2] Bioscience Center, Nagoya University, Furo-cho, Chikusa-ku, Nagoya 464-01, Japan
[3] Department of Biotechnology, Faculty of Engineering, Nagoya University, Furo-cho, Chikusa-ku, Nagoya 464-01, Japan

Biotechnology in Agriculture and Forestry, Vol. 45
Transgenic Medicinal Plants (ed. by Y.P.S. Bajaj)
© Springer-Verlag Berlin Heidelberg 1999

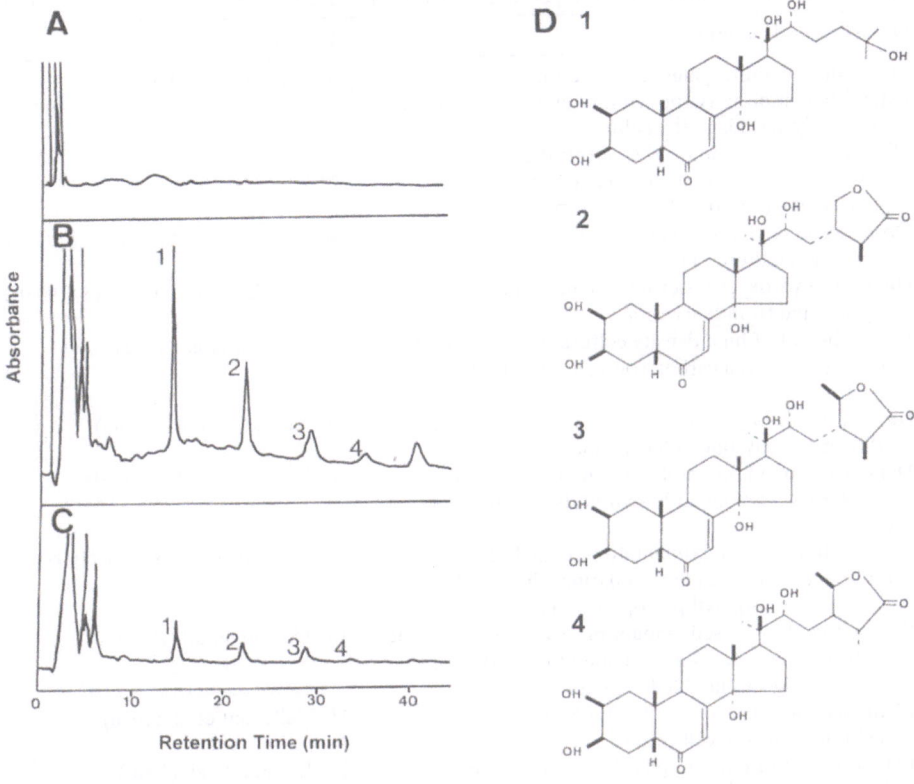

Fig. 1A–D. Profiles of their elution after HPLC of fractions from callus (**A**), hairy roots (**B**), and original roots (**C**) of *A. reptans*. Column, RP-C18; mobile phase, CH₃CN, methanol and H₂O (2:18:80, v/v); flow rate, 1 ml/min; detection, 242 nm. n-Butanol extracts dissolved in methanol (10 μl) was used. Structures of phytoecdysteroids (**D**). *1* 20-HE; *2* norcyasterone; *3* cyasterone; *4* isocyasterone. (Matsumoto and Tanaka 1991)

2 Genetic Transformation

2.1 Initiation, and Culture of Hairy Roots

A modified leaf-disk inoculation method (Noda et al. 1987) was employed for transformation of *Ajuga reptans* var. *atropurpurea*. Segments (1 × 1 cm) cut from surface-sterilized leaves of greenhouse-grown plants or of leaves of plants cultured in vitro were used as material for bacterial inoculation. Cells of *Agrolacteriun rhizogenes* strain MAFF301724 (Shiomi et al. 1987), isolated in Japan, or the derivative strain DC-AR2 (Tanaka et al. 1993b), which is kanamycin-sensitive, were grown in LB medium for 2 days, and used for infection. Both these strains harbored a mikimopine-type Ri plasmid, pRi1724, which was characterized by Tanaka and Oka (1994). After co-cultivation for 2 or 3 days on water that had been solidified with 1% agar

Table 1. Summary of in vitro culture, secondary metabolites, and transformation studies on *Ajuga*

Observations/Remarks	Reference
Production of anthocyanin in cell culture	1. Callebaut et al. (1990)
Establishment of hairy root culture and production of phytoecdysteroids in the culture	2. Matsumoto and Tanaka (1991)
Observation of location and concentration of phytoecdysteroids in in vitro micropropagated plants and cultured cells.	3. Tomas et al. (1992)
Observation of characteristics of pRi-transformed regenerant cultured in vitro	4. Tanaka and Matsumoto (1993a)
Highly 20-hydroxyecdysteroid-producing plant regenerated from hairy root	5. Tanaka and Matsumoto (1993b)
Establishment of high density culture of hairy root by using glucose as a carbon source in turbine-blade reactor	6. Uozumi et al. (1993)
Demonstration of converting cholesterol into 20-hydroxyecdysone in hairy root	7. Nagakari et al. (1994a)
Demonstration of 5β-ketol as an intermediate of 20-hydroxyecdysone in biosynthesis in hairy root culture	8. Nagakari et al. (1994b)
Investigation of compartmentation of anabolic raffinose metabolism by comparing whole-leaf tissue with mesophyll protoplasts and vacuoles	9. Bachmann and Keller (1995)
Production of increased amount of 20-hydroxyecdysone controlling intracellar PO_4^{3-} content in hairy root culture using turbine-blade reactor	10. Uozumi et al. (1995)
Identification of two types of anthocyanin acyltransferases in cell culture	11. Callebaut et al. (1996)
Establishment of efficient plantlet-regeneration system from GUS- transformed hairy roots and the possibility as a source of artificial seed production	12. Uozumi et al. (1996)
Demonstration of mevalonate as a biosynthetic origin of cyasterone and norcyasterone in hairy roots	13. Fujimoto et al. (1996)
Proposal of a mechanism of sterol biosynthesis in hairy root culture	14. Yagi et al. (1996)

in the light, the inoculated leaf segments were cultured on MS medium supplemented with 500 µg/ml vancomycin and 500 µg/ml carbenicillin, solidified with 0.2% Gellan gum (Wako Pure Chemical Industries Ltd., Japan). This medium was shown previously to be optimal for the initiation and growth of adventitious roots (Matsumoto and Tanaka 1991). Many adventitious roots emerged directly from the veins of inoculated leaves 10 to 14 days after inoculation (Fig. 2A). After removal and transfer to fresh MS medium, these roots displayed vigorous growth, with rapid elongation and conspicuous branching (Fig. 2B).

2.2 Selection of a Superior Clone of Hairy Roots

Most of the adventitious roots were confirmed to be genuine hairy roots by monitoring the accumulation of mikimopine (Isogai et al. 1990) by a rapid

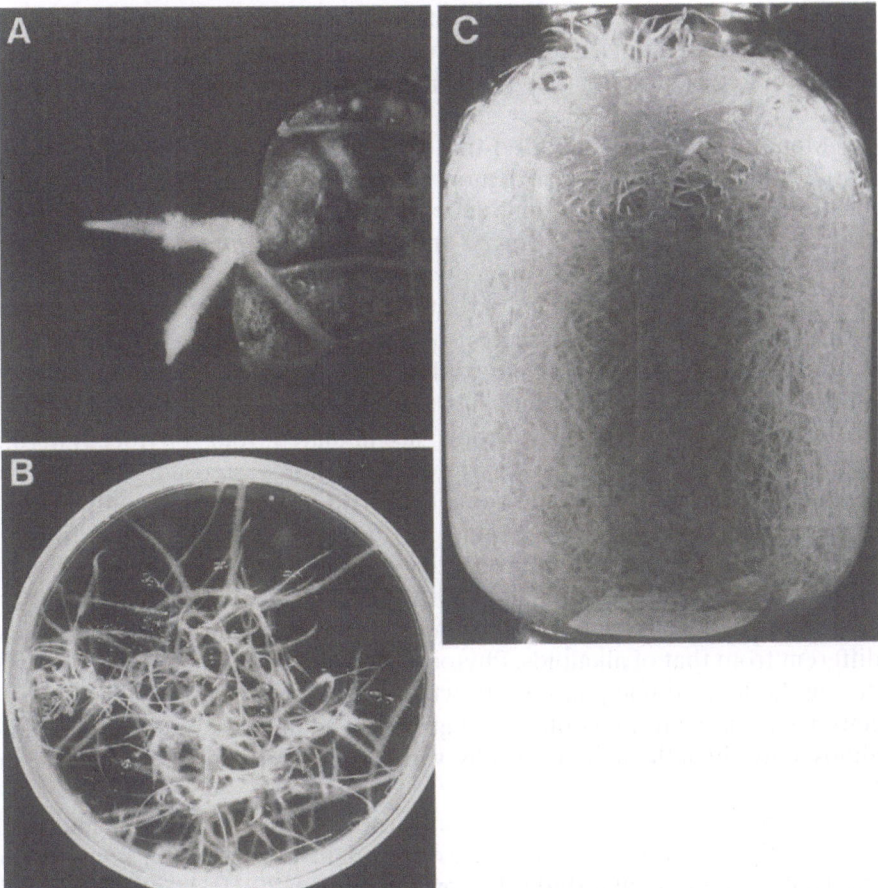

Fig. 2A–C. Induction and culture of hairy roots of *A. reptans*. **A** Emergence of hairy roots from veins of a leaf disk 2 weeks after inoculation with *Agrobacterium rhizogenes*. **B** Results of culture of hairy roots on MS medium solidified with Gellan gum for 30 days. **C** Results of culture of hairy roots in liquid MS medium in a air-lift-type fermenter for 45 days. (Matsumoto and Tanaka 1991)

paper-electrophoretic method (Tanaka 1990). In brief, hairy roots of 5–10 mm in length were crushed directly on a piece of Whatman 3-MM filter paper with a glass rod. The extract on the paper was subjected to electrophoresis at 500 V for 5 to 30 min in 5 g/l ammonium bicarbonate buffer (pH 9.8). For paper electrophoresis, a paper electrophoresis apparatus was optimal, but a vertical slab-type electrophoretic system for analysis of protein could also be used (N. Tanaka, unpubl.). In this case, the upper edge of the paper, which had been wetted with the buffer, was bent so that it hung in the upper tank of the electrophoresis apparatus. After electrophoresis, spots of mikimopine were detected by dipping or spraying with Pauly reagent (Tanaka 1990).

The transformation of adventitious roots was also confirmed by Southern blotting analysis. T-DNA fragments of pRi1724 were detected in the cellular DNA isolated as described above from almost all the adventitious roots generated (Tanaka et al. 1993a).

Mano et al. (1986) reported that a superior hairy root line should have strong proliferative ability, as demonstrated by elongation and branching. To illustrate this, they selected some superior lines of hairy roots with high proliferative ability in *Scopolia japonica* and *Duboisia leichhardtii*. Lines of hairy roots from *A. reptans* also displayed various phenotypes with respect to elongation and branching. Some hairy roots with rapid elongation grew 5 mm/day, with extensive branching. This trait of hairy roots was also evident in liquid culture. The hairy root lines that exhibited rapid elongation and considerable branching grew at much higher rates than the lines without these traits.

Almost all lines of hairy roots accumulated four phytoecdysteroids, namely, 20-HE, ajugasterone, cyasterone, and norcyasterone, in proportions similar to those in the original roots (Fig. 1B,C). By contrast, only traces of these phytoecdysteroids were found in the callus tissue derived from the original plant (Fig. 1A). High productivity of phytoecdysteroids was observed in the hairy root lines with high growth rates as compared to those with low growth rates. It is likely that the pattern of production of phytoecdysteroids is different from that of alkaloids. Phytoecdysteroids in hairy roots are produced during the logarithmic phase of growth, while alkaloids are produced during stationary phase after completion of growth. The phytoecdysteroids remained almost entirely in the hairy root cells, with less than 1 µg/ml being released into the medium.

Among more than 20 lines of hairy root cells, Ar-4 was selected as the superior line, with a 250-fold increase in weight per month in liquid culture (Matsumoto and Tanaka 1991). The productivity of 20-HE was 0.15% on a dry weight basis (Table 2).

Table 2. Levels of 20-HE in pRi-transformed regenerants of *Ajuga reptans*. (Tanaka and Matsumoto 1993a, 1993b)

	Level of 20-HE (% on a dry weight basis)				
	Cultured hairy roots	Cultured plant[a]		Cultivated plant[b]	
		Leaf	Root	Leaf	Root
Original plant	–	–	–	ND	0.030
Control plant[c]	–	0.014	0.074	ND	0.050
Hairy root line					
Ar-4	0.150	0.017	0.070	ND	0.102 (0.148[d])
Ar-8	0.130	0.017	0.067	–	–
Ar-19	0.085	0.019	0.077	ND	0.074
Ar-22	0.066	0.009	0.088	–	–

[a] Cultured for 1 month.
[b] Cultivated for 1 month after 1-month adaptation.
[c] Regenerated from an untransformed root.
[d] Cultivated for 2 months after 1-month adaptation.

2.3 Hairy Root Culture as a Material for Study of
Phytoecdysteroid Biosynthesis

Exploiting the high productivity of 20-HE by hairy roots of *A. reptans*, Nagakari et al. (1994a,b) investigated the biosynthetic pathway to phyto-ecdysteroids in plant cells, as compared to that in insect cells. Earlier feeding experiments for investigation of ecdysteroid biosynthesis in insects and plants suffered from low rates of incorporation and yield. Tritium has been used to examine the metabolic fate of certain hydrogen atoms in cholesterol. Such tracer experiments have required chemical transformations for location of the isotope after the isolation of metabolites. Cultured callus cells derived from phytoecdysteroid-producing plants, such as *A. reptans*, failed to solve this problem because of the low productivity of phytoecdysteroids (Hikino et al. 1971, 1975; Ravishankar and Mehta 1979; Matsumoto and Tanaka 1991). By contrast, since most hairy root cultures have high rates of production of secondary metabolites, Nagakari et al. studied the biosynthesis of phytoecdysteroid in the hairy root line Ar-4, feeding compounds that had been labeled with stable isotopes such as ^{2}H and ^{13}C. They demonstrated that the hairy roots of the Ar-4 line transformed ^{13}C-labeled cholesterol into 20-HE with an appreciable yield and, further, that 5β-ketol was an intermediate in this pathway. In insects, 5β-ketol is not regarded as an intermediate in the biosynthesis of 20-HE and, thus, this result clearly shows that the biosynthetic pathway to 20-HE in plants, or at least in *A. reptans*, is different from that in insect cells. Furthermore, since ^{13}C-labeled cholesterol was not incorporated into cyasterone and norcyasterone, Fujimoto et al. (1996) studied the biosynthesis pathway of these compounds by similar feeding experiments and clarified that clerosterol is converted into cyasterone. They also reported that clerosterol is derived from desmosterol in hairy roots of *A. reptans* (Yagi et al. 1996).

2.4 High-Density Culture and Production of 20-Hydroxyecdysone

Once established, the hairy root lines were genetically stable in terms of growth and product formation, but enhancement of the productivity of valu-able compounds has proved possible by implementation of various fermenta-tion techniques. Matsumoto and Tanaka (1991) reported a scaleup experiment with hairy root cultures using several air-lift-type fermenters. In particular, a hairy root culture, supported by polyurethane foam in a 1.5-l air-lift-type fermenter, which had proved effective in increasing the density of hairy root cultures from plants such as carrot and horseradish (Taya et al. 1989), gave good results (Fig. 2C).

Fed-batch culture is also effective for enhancement of production of the secondary metabolite 20-HE by hairy roots of *A. reptans*. The optimization of fermenter-scale hairy root culture requires a knowledge of substrate require-ments and rates of substrate utilization. In plant cell cultures, sucrose has generally been used as a carbon source. However, sucrose is hydrolyzed to

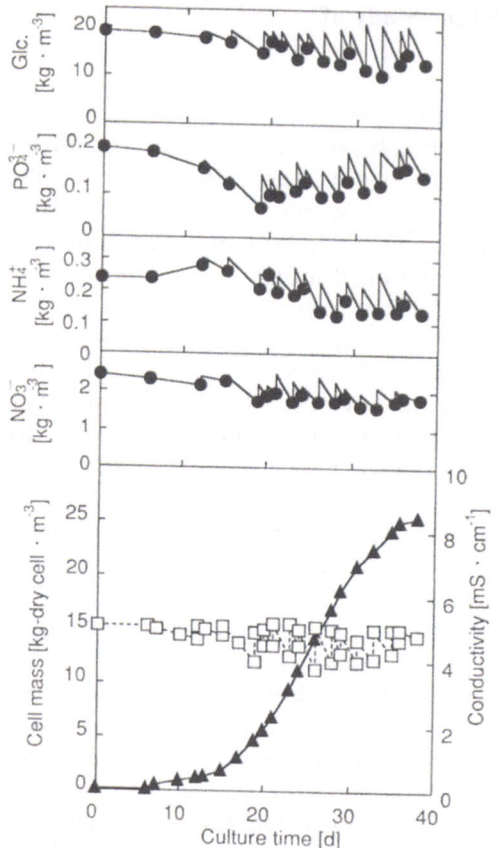

Fig. 3. Fed-batch culture of *A. reptans* hairy root using glucose as a carbon source: ▲ cell concentration estimated from conductivity decrease of the medium; □ medium conductivity. Concentration of the main components were measured and their time courses are also shown. (Uozumi et al. 1993).

glucose and fructose, which increase the osmotic pressure of the medium as cells grow. For hairy root cultures, the monosaccharide glucose, which is less expensive than sucrose, fructose, and galactose, was the best source of carbon for growth and for control of the concentration of monosaccharides during culture (Uozumi et al. 1991). A fed-batch culture using MS medium with some modification of the inorganic components, based on the biomass yield, provided the optimal condition for hairy root growth, with monitoring of the conductivity of the medium (Uozumi et al. 1993; Uozumi and Kobayashi 1994). Uozumi et al. reported that the final cell mass of *A. reptans* hairy roots cultured in a turbine-blade reactor was 27.2 kg dry cells m^{-3} at day 39, by keeping at almost constant levels of concentration of glucose, phosphate, ammonium, and nitrate ions (Fig. 3). The characteristic growth profile of the hairy root culture represents a unique growth curve (Fig. 3), different from those of bacteria and calli. After the lag phase, branching occurred vigorously, and then the individual roots elongated with low branching frequency. During the growth phase, the cell mass increased markedly and the growth pattern was linear. In the total growth profile of *A. reptans* hairy root in fed-batch culture, the linear growth phase is predominant. This growth pattern is closely associated with the characteristic of hairy root that the cell division occurs at the apical meristem.

As callus cells took up inorganic ions in culture medium, the cell mass (kg dry cells/m^3) obtained during culture (t, days) was accurately proportional to the cell yield coefficient of each inorganic ion in the medium. By contrast, the carbon source was not supplied in the same manner, since carbon is utilized to gain biomass and to maintain cell viability. According to the following balance equation, metabolic energy costs can be assessed from carbon (S, kg/m^3) utilized for growth (cell yield, $Y_{X/S}$, g dry cell/g carbon source) and carbon used to maintain existing biomass (maintenance coefficient, m, g carbon source/g dry cells/day):

$$q = -(dS/dt)/X = (1/X)(dX/dt)Y_{X/S} + m = \mu/Y_{X/S} + m \tag{1}$$

The specific growth rate (μ/day) and specific net uptake rate of carbon substrate into the cells (q, g carbon source/g dry cells/day) can be estimated under conditions where the growth rate is carbon-limited. Thus, continuous culture of callus can offer the proper conditions for estimating the cell yield and maintenance coefficient. Hairy roots cannot, however, be carried out in the continuous culture due to their morphology and the domination of the linear growth phase. For evaluation of kinetic parameters of hairy roots in linear growth phase, the relationship between hairy root growth and carbon source consumption should be rewritten in the following balance equation:

$$-dS/dt = (dX/dt)/Y_{X/S} + mX \tag{2}$$

The estimated values are summarized in Table 3 and compared to values obtained from the literature. *A. reptans* and carrot hairy roots had $Y_{X/S}$ values comparable with those of callus.

The addition of indole acetic acid (IAA) increased the growth rate of hairy roots as a result of an increase in the number of root apical meristems. Thus, the addition of IAA reduced the required duration of culture of hairy roots. However, IAA should be eliminated from the culture medium at the growth phase, because it inhibits 20-HE production. Furthermore, the 20-HE content of hairy roots increased when the intracellular level of inorganic monophosphate (PO_4^{3-}) was kept below 5 mg/g dry weight (Fig. 4). Although the intracellular phosphate content cannot be measured rapidly, a kinetic model was developed to simulate the changes in phosphate content. The results for fed-batch culture of hairy roots in a turbine-blade reactor are shown in Fig. 5. During the growth phase, phosphate was present above the critical level. IAA (1 mg/l) was added only in the initial medium and no IAA was added during the culture. During the production of the secondary metabolite (20-HE), the phosphate level was kept below 5 mg/g dry weight from 10 to 20 days after inoculation, and then phosphate was eliminated from the medium.

Table 3. Comparison of biomass yield from saccharide and maintenance coefficients for saccharide with published values. (Uozumi et al. 1993)

Organism	Substrate	Cell yield $Y_{X/S}$ (g/g)	Maintenance coefficient, m (g/g/day)	Reference
Ajuga hairy root	Glucose	0.77	0.105	Uozumi et al. (1993)
Carrot hairy root	Fructose	0.60	0.085	Uozumi et al. (1991, 1993)
Eschscholtzia californica callus	Sucrose	0.71	0.074	Taticek et al. (1990)
Apple callus	Sucrose	0.57	0.012	Pareilleux and Chaubet (1980)
Nicotiana tabacum callus	Sucrose	0.60	0.086	Stouthamer and Bettenhausen (1973)
Nicotiana tabacum callus	Glucose	0.52	0.106	Schnapp et al. (1991)
Medicago sativa callus	Lactose	0.77	0.113	Pareilleux and Chaubet (1981)

The value of $Y_{X/S}$ and im of *Ajuga* and carrot hairy roots were evaluated according to Eq. (2) in the text. The value for calli were evaluated according to Eq. (1) in the text.

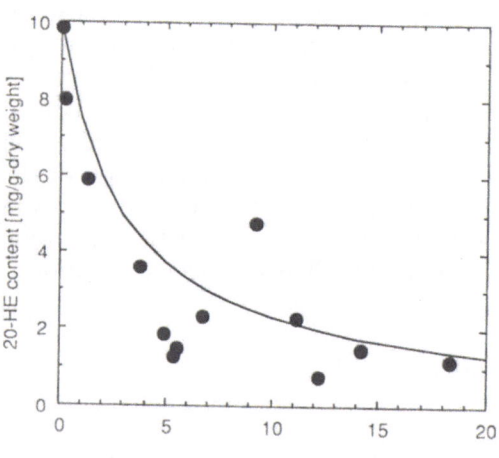

Intracellular PO_4^{3-} content [mg-PO_4^{3-}/g-dry weight]

Fig. 4. The relationship between intracellular PO_4^{3-} content and 20-HE content. *Points* represent experimental data, and the *solid line* depicts the calculated result based on the kinetic model. Medium without IAA was used for the experiment. (Uozumi et al. 1995).

Culture was carried out in a turbine-blade reactor with control of the culture conditions by reference to the conductivity of the culture medium and the simulation data (Uozumi et al. 1995). Such culture resulted in a three fold increase in 20-HE content, as compared to the fed-batch culture without phosphate control. The final cell mass and 20-HE content reached 22.2 g dry weight/l and 6.1 mg/g dry weight at 27 days, respectively, as shown in Fig. 5.

Thus, for a high-performance culture of *A. reptans* hairy roots and the production of 20-HE, it is necessary (1) to maintain a constant level of

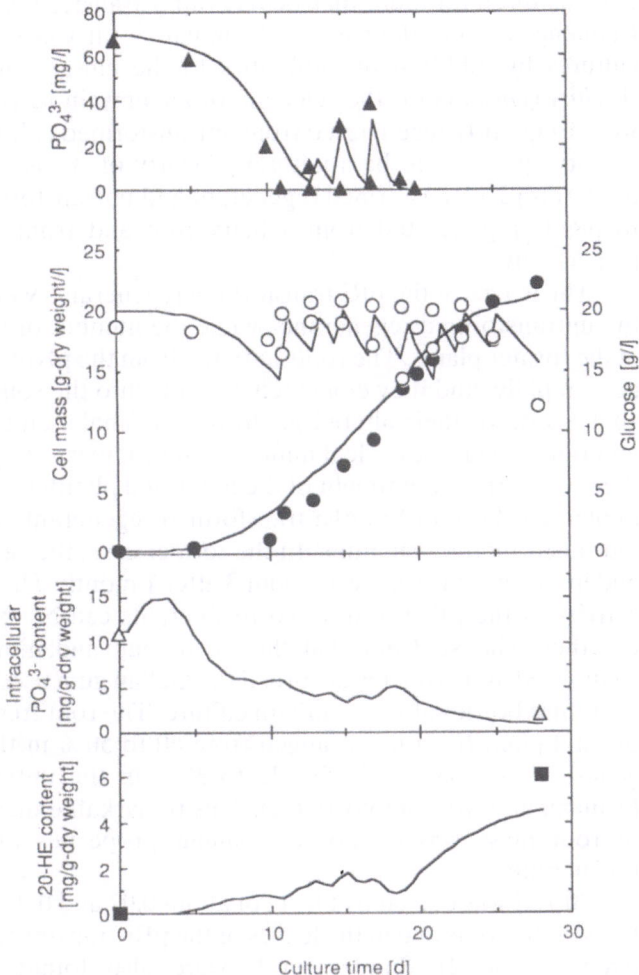

Fig. 5. Fed-batch culture of *A. reptans* hairy roots. All *points* represent experimental data. *Symbols* indicate the concentration of phosphate in the medium (▲), dry cell mass (●), concentration of glucose in the medium (○), the intracellular phosphate content (△) and the 20-HE content (■). All *solid lines* depict calculated results based on a kinetic model. (Uozumi et al. 1995).

monosaccharide and a source of nitrogen, (2) to add IAA to increase the growth rate, and (3) to reduce the concentration of phosphate to increase the 20-HE content.

2.5 Characteristics of pRi-Transformed Regenerants Cultured in Vitro

Most of the hairy roots of *A. reptans* spontaneously regenerated shoots in darkness (Tanaka and Matsumoto 1993a). In particular, hairy roots that had been cultured for more than 30 days after transfer regenerated shoots at high frequency around their bases. A similar result was obtained with hairy root cultures by addition of antibiotics to the culture medium for removal of *A. rhizogenes* and/or the selection of kanamycin-resistant lines. Since shoots were frequently regenerated from untransformed cultured roots, regenerative activity appeared to be a natural property of *A. reptans* plants. We will use the terms pRi-transformed regenerant and untransformed regenerant to refer to plants regenerated from a hairy root and from an original plant root, respectively.

The leaves of the pRi-transformed regenerants were smaller than those of the untransformed regenerants, while the number of leaves was much higher in the former plants. The roots initiated from the pRi-transformed regenerants grew rapidly, and they elongated not only into the solid medium but also into air because of their altered geotropism (Tanaka and Matsumoto 1993a). In addition to increases in leaf number and root mass, the growth rate (defined as the ratio of the fresh weight of the harvested plantlet to the fresh weight of the implanted shoot tip) of pRi-transformed regenerants was higher than that of untransformed regenerants. In the former case, the ratio ranged from 8 to 20 and in the latter case, it was about 3 after 1 month. The increased proliferative activity of the pRi-transformed regenerants can be easily related to the fact that they exhausted almost all the MS medium and, moreover, they completely solubilized and probably absorbed the Gellan gum used to solidify the medium in culture bottles after a 1-month culture. The root fresh mass as a percentage of total plant fresh mass ranged from 40 to 50% in the pRi-transformed regenerants as compared with 25 to 30% in the untransformed regenerants (Tanaka and Matsumoto 1993a). This remarkable increase in the percentage of root mass was one of the salient properties of the pRi-transformed regenerants.

20-HE was detected at level of about 0.07 and 0.01–0.02% on a dry weight basis in the roots and in the leaves of the pRi-transformed regenerants, respectively (Table 2). Similar levels were also found in the untransformed regenerants. These plants had more than twice the 20-HE content of roots of the original *A. reptans* plants that had been grown continuously in a greenhouse (about 0.03% on a dry weight basis; Table 2). However, the level in the roots of pRi-transformed regenerants was lower than that in the hairy roots. Nonetheless, 20-HE was also detected in the leaves and stems of both types of regenerant, but not in those of the original *A. reptans* plants (Table 2). Even though the culture conditions, such as the aseptic environment, the composition of the medium, the illumination, and/or the temperature, must have affected the production of 20-HE in the regenerants, we do not yet have a precise explanation for the presence of 20-HE in the leaves and stems.

2.6 Efficient Production of Transformed Hairy Roots as Artificial Seeds

The exploitation of cultured hairy roots is mainly directed towards the large-scale production of useful products or secondary metabolites in fermenters.

However, hairy root-mediated micropropagation is a promising system for delivery of plantlets and production of useful compounds. Such a system, using artificial seeds, has been examined in view of the efficient induction of roots and regeneration of shoots from hairy roots (Uozumi et al. 1992, 1994a, 1996; Uozumi and Kobayashi 1995; Nakashimada et al. 1995, 1996).

The procedure for systematic regeneration from hairy roots of *A. reptans* is shown schematically in Fig. 6. Addition of an auxin, naphthaleneacetic acid (0.1 mg/l), to the medium for hairy root culture enhanced the rate of root cell growth (Nakashimada et al. 1994, 1996). Mechanical fragmentation for 10 s in a commercial blender allowed a large number of root fragments to be produced efficiently (Nakashimada et al. 1995, 1996). When cultured in the light with a cytokinin, namely, benzyladenine (10 mg/l), shoots developed from the root fragments. Plantlets of more than 4 mm in length were able to grow into whole plants on a modified MS medium that had been solidified with agar.

This procedure was applied to hairy roots of *A. reptans* in which a gene for β-glucuronidase (GUS) had been introduced under control of the promoter of the gene for the small subunit of tomato ribulose-1,5-biphosphate carboxylase/oxygenase (*rbcS3B*) by the *A. rhizogenes*-mediated cotransformation (Uozumi et al. 1994b; Nakashimada et al. 1996). Hairy roots transformed with this construct were used for the efficient regeneration of transformed plantlets. One of the GUS-transformed regenerants, designated PRT-4, could develop to the plantlets (Fig. 7). Although the growth rate of PRT-4 in the hairy root growth phase with NAA was lower than the untransformed hairy root, the number of PRT-4 plantlets reached approximately 27 500, which was 6.4-fold higher than in the case of non-GUS-transformed hairy root. PRT-4 were developed into plants on the solid medium. GUS expression was detected in leaf and green root tissue of the regenerated plant under the light condition, but not in the white portion of the root (Table 4). These results seem to be correlated with function of the *rbcS3B* promoter.

The ability of hairy roots to regenerate did not decline during a lengthy period of subculture, while that of embryogenic callus tends to decline under similar conditions. Improvement of plants by genetic manipulation may even-

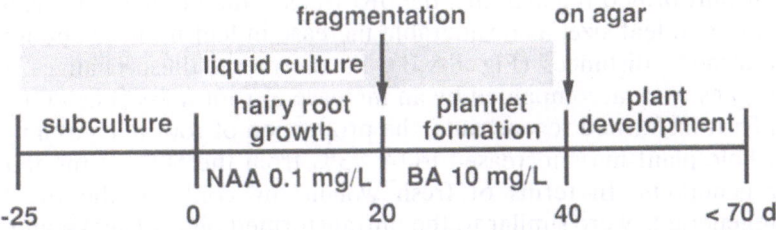

Fig. 6. Schematic representation of the procedure for regeneration of *A. reptans*. Benzyladenine and naphthaleneacetic acid are abbreviated as *BA* and *NAA*, respectively. (Uozumi et al. 1996)

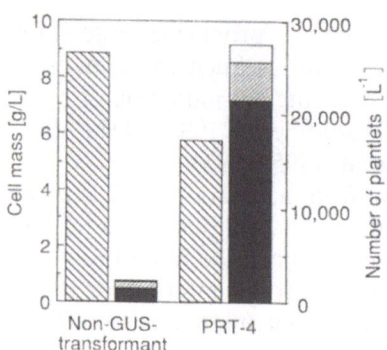

Fig. 7. Cell growth and plantlet formation of non-GUS transformant and PTR-4. ▨ indicates the final cell mass of hairy root cultured in the medium with NAA 0.1 mg/l at 20 days in the hairy root growth phase. Plantlet sizes: ■ 1–2 mm; ▨ 2.1–4.0 mm; □ more than 4 mm. (Uozumi et al. 1996)

Table 4. GUS activity (1/mnol/min^{-1} mg^{-1} protein) of PRT-4. (Uozumi et al. 1996)

Root of PTR-4[a]	Green root of PTR-4[b]	Leaf (3–5 mm) of PTR-4	Leaf (5–10 mm) of PTR-4	Leaf (5–10 mm) of control[c]
ND[d]	2.74	0.63	2.29	0.06

The size of leaf in parentheses.
[a] Non-green root of PTR-4.
[b] Green portion of PTR-4.
[c] Leaf (5–10 mm) of non-GUS-transformed hairy root grown in light.
[d] ND not detected.

tually replace conventional breeding and selection methods. From the example described above, we can see the possibility of improving hairy roots in terms of proliferative ability and the production of secondary metabolites by genetic manipulation.

2.7 Greenhouse Studies on Transformed Regenerants

During cultivation in a greenhouse, the dwarf phenotype was prominent in the aerial organs of pRi-transformed regenerants of *A. reptans*. In comparison to untransformed regenerants, the pRi-transformed regenerants exhibited a decrease in leaf size, a considerable increase in leaf number and a reduction in internode distances (Fig. 8A,B). Furthermore, these changes in the aerial organs were accompanied by an increase in root mass (Fig. 8C). Thus, in the pRi-transformed regenerants, the proportion of root mass as a percentage of whole plant mass increased to 68–73% from the 50% of the untransformed regenerants. In terms of fresh weight, by contrast, the pRi-transformed regenerants were similar to the untransformed ones (Tanaka and Matsumoto 1993b). It is likely that the decrease in weight of aerial organs due to dwarfing and the increase in root weight caused by vigorous proliferation balanced each

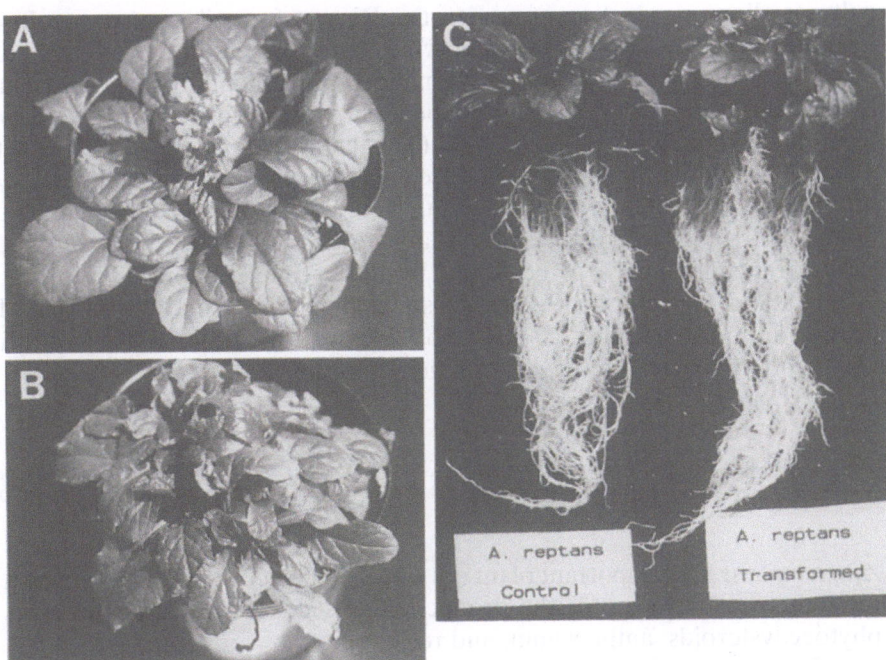

Fig. 8A–C. General view of untransformed and pRi-transformed regenerants of *A. reptans* after cultivation for 1 month in a greenhouse. **A** An untransformed regenerants. **B** A regenerant from hairy root line Ar-4. **C** Entire untransformed (*left*) and pRi-transformed (*right*) regenerants. (Tanaka and Matsumoto 1993b)

other out in the transformed regenerants. Since the hairy roots of *A. reptans* do not exhibit geotropism, aerial roots were discernible in the soil in which the transformed regenerants were cultivated.

The pRi-transformed regenerants failed completely to differentiate flowers (Fig. 8A,B). Accordingly, sexual transmission of the transformed phenotypes could not be examined. Phenotypes were stably maintained in the progeny derived from runners of parental plants. The mechanism of inhibition of floral differentiation in the pRi-transformed regenerants remains unknown.

The 20-HE content of untransformed roots was 0.03% on a dry weight basis, whereas the 20-HE content of hairy root lines Ar-4 and Ar-19 reached 0.15 and 0.08%, respectively (Table 2). The 20-HE content of the pRi-transformed regenerants derived from the Ar-4 and Ar-19 hairy root lines (0.10 and 0.07%) also revealed high productivity of 20-HE as compared to that of the untransformed regenerants (0.05%; Table 2). Thus, the productivity of the original root line was also expressed in the regenerants.

Depending on the method of root excision, numerous cultures of *A. reptans* hairy root can be established from the same hairy root line. Furthermore, the vigorous regenerative ability associated with individual hairy root cultures allows production of numerous pRi-transformed regenerants. In order to determine whether the phenotypic characteristics of the original hairy root line are expressed in the same manner in its clonal regenerants, many regenerants were induced from independent cultures of hairy roots derived from the original hairy root line Ar-4 (Tanaka and Matsumoto 1993b). After 2 months of cultivation in a greenhouse, increases in root mass and the 20-HE productivity of the various regenerants were compared. No obvious fluctuations could be detected in either characteristics of these regenerants; the former ranged from 54 to 66% and the latter ranged from 0.11 to 0.21% (Tanaka and Matsumoto 1993b). These results indicate that the capacity for rooting and production of 20-HE associated with the original hairy root line was stably transferred to its clonal regenerants.

3 Summary and Conclusions

Ajuga reptans is an important plant not only for horticultural purposes, such as ground cover, but also for the production of secondary metabolites such as phytoecdysteroids, anthocyanin, and raffinose.

Hairy roots of *A. reptans* obtained by infection with *Agrobacterium rhizogenes* grew efficiently in bench-scale fed-batch culture. They contained high level of 20-hydroxyecdysone (20-HE) when grown in modified MS medium with a controlled concentration of phosphate.

Since plants of *A. reptans* are good material not only for induction of hairy roots but also for generation of pRi-transformed regenerants, the expression of foreign genes in plant organs and transgenic plants should be quite straightforward. Exploiting the hairy root-mediated system for transformation of *A. reptans* using *A. rhizogenes* that harbors a binary vector, transgenic plants can be obtained within a mere 2 months. Moreover, *A. reptans* hairy roots, grown in large-scale culture in a fermenter, can be cut with a blender and embedded in gel beads to provide a good model for artificial seeds that allow propagation of pRi-transformed regenerants.

The pRi-transformed regenerants contained levels of 20-HE similar to those in the original hairy roots.

References

Bachmann M, Keller F (1995) Metabolism of the raffinose family oligosaccharides in leaves of *Ajuga reptans* L. Inter- and intracellular compartmentation. Plant Physiol 109:991–998

Callebaut A, Hendrickx G, Voets AM, Motte JC (1990) Anthocyanins in cell cultures of *Ajuga reptans*. Phytochemistry 29:2153–2158

Callebaut A, Decleire M, Vandermeiren K (1993) *Ajuga repens* (Bugle): In vitro production of anthocyanins. In: Bajaj YPS (ed) Biotechnology in agriculture and forestry, vol 24. Medicinal and aromatic plants V. Springer, Berlin Heidelberg New York, pp 1–22

Callebaut A, Terahara N, Decleire M (1996) Anthocyanin acyltransferases in cell cultures of *Ajuga reptans*. Plant Sci 118:109–118

Camps F, Coll J (1993) Insect allelochemicals from *Ajuga* plants. Phytochemistry 32:1361–1370

Fujimoto Y, Nakagawa T, Yamada J, Morisaki M (1996) Biosynthetic origin of C-26 and C-27 of the phytoecdysteroids cyasterone and 29-norcyasterone in *Ajuga* hairy roots. Chem Commun 2063–2064

Hikino H, Jin H, Takemoto T (1971) Occurrence of insect-moulting substances ecdysone and inokosterone in callus tissue of *Achyranthes*. Chem Pharm Bull 19:438–439

Hikino H, Jin H, Takemoto T (1975) Tissue culture of *Achyranthes* and formation of phytocedysteroids in cultured tissues. Yakugaku Zasshi 95:581–589

Isogai A, Fukuchi N, Hayashi M, Kamada H, Harada H, Suzuki A (1990) Mikimopine, an opine in hairy roots of tobacco induced by *Agrobacterium rhizogenes*. Phytochemistry 29:3131–3134

Kozakai Y, Mizusawa H, Sudo M, Matsumoto T (1990) β-ecdysone ni yoru kaiko no jukukasokusin to jouzoku no saiituka. Sanshi Kagaku to Gijyutu 19:44–47 (in Japanese).

Kubo I, Klocke JA, Asano A (1983) Effect of ingested phytoecdysteroids on the growth and development of two lepidopterous larvae. J Insect Physiol 29:307–316

Mano Y, Nabeshima S, Matsui C, Ohkawa H (1986) Production of tropane alkaloids by hairy root cultures of *Scopolia japonica*. Agric Biol Chem 50:2715–2722.

Matsumoto T, Tanaka N (1991) Production of phytoecdysteroids by hairy root cultures of *Ajuga reptans* var. *atropurpurea*. Agric Biol Chem 55:1019–1025

Nagakari M, Kushiro T, Matsumoto T, Tanaka N, Kakinuma K, Fujimoto Y (1994a) Incorporation of acetate and cholesterol into 20-hydroxyecdysone with hairy root clone of *Ajuga reptans* var. *atropurpurea*. Phytochemistry 36:907–910

Nagakari M, Kushiro T, Yagi N, Tanaka N, Matsumoto T, Kakinuma K, Fujimoto Y (1994b) 4β-hydroxy-5β-cholest-7-en-6-one as intermediate of 20-hydroxyecdysone biosynthesis in a hairy root culture of *Ajuga reptans* var. *atropurpurea*. J Chem Soc Chem Commun 15:1761–1762

Nakashimada Y, Uozumi N, Kobayashi T (1994) Simulation of emergence of root apical meristems in horseradish hairy root by auxin supplementation and its kinetic model. J Ferment Bioeng 77:178–182

Nakashimada Y, Uozumi N, Kobayashi T (1995) Plantlet production for use as artificial seed from horseradish hairy root fragmented by blender. J Ferment Bioeng 79:458–464

Nakashimada Y, Uozumi N, Kobayashi T (1996) Efficient culture method for production of plantlets from mechanically cut horseradish hairy roots. J Ferment Bioeng 81:87–89

Noda T, Tanaka N, Mano Y, Nabeshima S, Ohkawa H, Matui C (1987) Regeneration of horseradish hairy roots incited by *Agrobacterium rhizogenes* infection. Plant Cell Rep 6:283–286

Pareilleux A, Chaubet N (1980) Growth kinetics of apple plant cell cultures. Biotechnol Lett 2:291–296

Pareilleux A, Chaubet N (1981) Mass cultivation of *Medicago sativa* growing on lactose: Kinetic aspects. Eur J Appl Microbiol Biotechnol 11:222–225

Ravishanker GA, Mehta AR (1979) Control of ecdysterone biogenesis in tissue cultures of *Trianthema portulacastrum*. J Natl Prod 42:152–153

Schnapp SR, Curtis WR, Bressan RA, Hasegawa PM (1991) Estimation of growth yield and maintenance coefficient of plant cell suspensions. Biotechnol Bioeng 38:1131–1136

Shiomi T, Shirakata T, Takeuchi A, Oizumi T, Uematsu S (1987) Hairy root of melon caused by *Agrobacterium rhizogenes* biovar 1. Ann Phytopathol Soc Jpn 53:454–459

Stouthamer AH, Bettenhausen C (1973) Utilization of energy for growth and maintenance in continuous and batch cultures of microorganisms. Biochem Biophys Acta 301:53–70

Tanaka N, Matsui C (1993) Transformation in horseradish (*Armoracia rusticana*): hairy roots incited by *Agrobacterium rhizogenes* infection. In: Bajaj YPS (ed) Biotechnology in agriculture and forestry, vol 23. Plant protoplasts and genetic engineering V. Springer, Berlin Heidelberg New York, pp 135–146

Tanaka N (1990) Detection of opines by paper electrophoresis. Plant Tissue Cult Lett 7:45–47

Tanaka N, Matsumoto T (1993a) Characterization of *Ajuga* plant regenerated from hairy roots. Plant Tissue Cult Lett 10:78–83

Tanaka N, Matsumoto T (1993b) Regeneration from *Ajuga* hairy roots with high productivity of 20-hydroxyecdysone. Plant Cell Rep 13:87–90

Tanaka N, Oka A (1994) Restriction endonuclease map of the root-inducing plasmid (pRi1724) of *Agrobacterium rhizogenes* strain MAFF03-01724. Biosci Biotech Biochem 58:297–299

Tanaka N, Matsumoto T, Oka A (1993a) Molecular analysis of T-DNA region on the root-inducing plasmid (Ri) in a mikimopine-type *Agrobacterium rhizogenes* strain 1724. Ann Phytopathol Soc Jpn 59:155–162

Tanaka N, Takao M, Matsumoto T, Machida Y (1993b) Transformation of *Agrobacterium rhizogenes* strain MAFF03-01724 by electroporation. Ann Phytopathol Soc Jpn 59:587–593

Taticek RA, Moo-Young M, Legge RL (1990) Effect of bioreactor configuration on substrate uptake by cell suspension cultures of the plant *Eschschotzia californica*. Appl Microbiol Biotechnol 33:280–286.

Taya M, Yoyama A, Kondo T, Kobayashi T, Matsui C (1989) Growth characteristics of plant hairy roots and their cultures in bioreactors. J Chem Eng Jpn 22:84–89

Tomas J, Camps F, Claveria E, Coll J, Mele E (1992) Composition and location of phytoecdysteroids in *Ajuga reptans* in vivo and in vitro cultures. Phytochemistry 31:1585–1591

Uozumi N, Kobayashi T (1994) Application of hairy root and bioreactors. In: Ryu DDY, Furusaki S (eds) Advances in plant biotechnology, Elsevier Amsterdam, pp 307–338

Uozumi N, Kobayashi T (1995) Artificial seed production through encapsulation of hairy root and shoot tips. In: Bajaj YPS (ed) Biotechnology in agriculture and forestry, vol 30. Somatic embryogenesis and synthetic seed I. Springer, Berlin, Heidelberg New York, pp 170–180

Uozumi N, Kohketsu K, Kondo O, Honda H, Kobayashi T (1991) Fed-batch culture of hairy root using fructose as a carbon source. J Ferment Bioeng 72:457–460

Uozumi N, Kato Y, Nakashimada Y, Kobayashi T (1992) Production of artificial seed from horseradish hairy root. J Ferment Bioeng 74:21–26

Uozumi N, Kohketsu K, Kobayashi T (1993) Growth and kinetic parameters of *Ajuga* hairy root in fed-batch culture on monosaccharide medium. J Chem Technol Biotechnol 57:155–161

Uozumi N, Asano Y, Kobayashi T (1994a) Micropropagation of horseradish hairy root by means of adventitious shoot primordia. Plant Cell Tissue Org Cult 36:187–190

Uozumi N, Inoue Y, Yamazaki K, Kobayashi T (1994b) Light activation of expression associated with the tomato *rbcS* promoter in transformed tobacco cell line BY-2. J Biotechnol 36:55–62

Uozumi N, Makino S, Kobayashi T (1995) 20-hydroxyecdysone production in *Ajuga* hairy root controlling intracellular phosphate based on kinetic model. J Ferment Bioeng 80:362–368

Uozumi N, Ohtake Y, Nakashimada Y, Morikawa Y, Tanaka N, Kobayashi T (1996) Efficient regeneration from GUS-transformed *Ajuga* hairy root. J Ferment Bioeng 81:374–378

Yagi T, Morisaki M, Kushiro T, Yoshida H, Fujimoto Y (1996) Biosynthesis of 24β-alkyl-Δ^{25}-sterols in hairy roots of *Ajuga reptans* var. *atropurpurea*. Phytochemistry 41:1057–1064

Yoshida T, Otaka T, Uchiyama M, Ogawa S (1971) Effect of ecdysterone on hyperglycemia in experimental animals. Biochem Pharmacol 20:3263–3268

III Genetic Transformation of *Anthemis nobilis* L. (Roman Chamomile)

M. Jaziri[1], M.-L. Fauconnier[2], Y.-W. Guo[1,3], M. Marlier[2], and M. Vanhaelen[3]

1 Introduction

General account concerning the distribution, importance, and biologically active constituents of *Anthemis nobilis* has been reported by Fauconnier et al. (1996). *A. nobilis*, the so-called Roman chamomile, is a perennial herb belonging to the Asteraceae family. It is native to southwest Europe and has spread all over Europe. It grows also in southwest Asia. The flower heads, the commercial drug, are used in traditional medicine for their pharmacological properties related to several groups of secondary metabolites: a sedative property has been attributed to the presence of esters (Melegari et al. 1988), antiinflammatory activity to flavonoids (Achterrath-Tuckermann et al. 1980; Della Loggia 1986), sesquiterpene lactones (Hall et al. 1979), and azulenes (Issac and Kristen 1980), antispasmodic activity to apiin and luteoside (Duke 1987), antibacterial activity to sesquiterpenic lactones and essential oil constituents (Rodriguez et al. 1979), and cytotoxicity to the presence of nobilin, and derivatives such as 1,10-epoxynobilin and 3-dehydronobilin (Leung 1980). Extracts from *A. nobilis* are also added to cosmetics, shampoos, and bath preparations, hair dye formulas, preparations for mouth washes, creams, and gels to treat cracked nipples (Proserpio et al. 1983; Rovesti et al. 1983).

A. nobilis plants are cultivated in many European countries, Egypt, and Argentina. In France (Anjou) the production yield reaches about 1 ton of dry flowers per ha, more than 160 ha being cultivated (Bezanger-Beauquesne et al. 1986).

The dried flower head contains 0.4 to 1.5% (on a dry weight basis) of a volatile oil. The most important characteristics of this essential oil are the presence of chamazulene and a high content of low molecular weight esters (angelates as the greatest part of the esters, 65%). The accumulation of essential oil in *A. nobilis* is strictly restricted to the aerial part of the plant (Fauconnier et al. 1993, 1996).

[1] Laboratory of Plant Biotechnology, Université Libre de Bruxelles, Chaussée de Wavre 1850, 1160 Brussels, Belgium
[2] Faculté des Sciences Agronomiques, UER Chimie Générale et Organique, Passage des Déportés 2, 5030 Gembloux, Belgium
[3] Laboratory of Pharmacognosy and Bromatology, Université Libre de Bruxelles, Campus plaine CP205/4, Bld. Triomphe, 1050 Brussels, Belgium

Biotechnology in Agriculture and Forestry, Vol. 45
Transgenic Medicinal Plants (ed. by Y.P.S. Bajaj)
© Springer-Verlag Berlin Heidelberg 1999

In this chapter, *Agrobacterium*-mediated genetic transformation of *A. nobilis* is discussed. The transgenic tissue (crown gall tumors) and organs (hairy roots and shooty teratomas) cultures were established and evaluated for essential oil production.

2 Genetic Transformation

2.1 Bacterial Strains and Infection Procedures

Four *Agrobacterium* strains were used: *A. tumefaciens* (strains C58 and T37) and *A. rhizogenes* (stains A4 and ATCC 15834). The bacteria were cultured on solid YEB medium (Vervliet et al. 1975) and maintained at room temperature by subculturing every month on the same medium until required for infection.

Axenic *A. nobilis* shoot cultures (Fauconnier et al. 1996) were used for direct infection by the bacteria. For the infection by *A. tumefaciens* strain T37, the coculture method was used. The infected shoots were incubated on hormone-free MS solid or liquid medium (Murashige and Skoog 1962) at 23 °C under a 16-h photoperiod ($6\,\mu E/m^2/s$).

2.2 Establishment of Transformed Tissue and Organ Cultures

After 3–4 weeks of incubation, crown gall tumors appeared at the inoculated sites of *A. nobilis* shoot cultures infected with *A. tumefaciens* (strain C58), and hairy roots appeared from the inoculated sites of shoots with *A. rhizogenes*. From the shoots cocultured with *A. tumefaciens* (strain T37), the development of numerous adventitious shoots (shooty teratomas) was observed at the surface of the explants.

The genetic transformation of the different cultures was confirmed by opine detection (Petit et al. 1983), indicating the integration of the bacterial T-DNA into the plant genome. Transformed tissues and organs were then transferred for two subcultures to fresh hormone-free MS solid (for crown gall tumors) or liquid medium containing Claforan (500 mg/l). The antibiotic was omitted after the two subcultures. The axenic transformed cultures obtained under these conditions grew vigorously in a hormone-free medium, showing the characteristic features of the transformed organs: for the hairy roots, a high degree of lateral branching and for the shooty teratoma, new shoots radiating from stolon-like structures.

2.3 Morphological Traits and Essential Oil Content in Normal and Transformed Cultures

Transgenic tissue and organ cultures were derived from *A. nobilis* tissue after infection with *Agrobacterium* species (wild-type strains) (Fig. 1). These cul-

Fig. 1A–D. Transformation in *Anthemis nobilis* L. **A** Normal shoot cultures were used for infection with *Agrobacterium* sp. **B** Three-week-old crown gall tumors (*arrow*) induced after direct infection of *A. nobilis* shoot cultures with *A. tumefaciens* (strain C58). **C** Transformed *A. nobilis* shoot cultures induced by *A. tumefaciens* (strain T37). **D** Hairy root (*arrow*) emerging from petiole segments infected with *A. rhizogenes* (strain A4). (Fauconnier et al. 1996)

tures were established in order to study the capability of the transformed cultures to accumulate the essential oil. Table 1 summarizes the characteristics of the established transformed tissue and organ cultures of *A. nobilis*. This study was conducted in order to obtain variant cultures with modified metabolic characters and to evaluate the potential for genetic manipulation of the plant secondary metabolism.

Gas chromatography-mass spectrometry (GC-MS) (Fauconnier et al. 1993) was used for the analysis of the essential oil constituents in various plant samples (Table 2). Although excretory reservoirs have never been found, the essential oil content of the *A. nobilis* crown gall tumors was 0.25% of the dry weight, and the composition of the essential oil was comparable to that of the flowers. The analysis shows that tumor extracts contain numerous esters such as isobutyl and isoamyl angelates (16.6 and 6.5%, respectively), isobutyl and 2-methylbutyl isobutyrates (6.4 and 9%, respectively). These observations are similar to those of Reichling and Beiderbeck (1991), who showed that the essential oil fraction of 4-week-old tumors of *Chamomilla recutita* was to a large extent identical with that of the shoots; *trans*-β-farnesene, *trans*-α-farnesene, and the *cis/trans*-en-in-dicycloethers were identified.

Table 1. Transformation in *A. nobilis*: morphological and molecular characteristics of the transgenic cultures

Bacteria	Infection procedure	Family of opines	Culture condition	Morphology of the transformed material	Essential oil composition
A. tumefaciens					
C58	Direct	Nopaline & agrocinopine	Solid HF-MS	Crown gall tumors	Similar to that of the flowers
T37	Coculture	Nopaline and agrocinopine	Liquid HF-MS	Shooty teratomas	Comparable to that of the normal shoot cultures
A. rhizogenes					
A4	Direct	Agropine	Solid or liquid HF-MS	Hairy roots	No characteristic esters were detected in
ATCC15834	Direct	Agropine	Solid or liquid HF-MS	Hairy roots	either hairy root extracts

The analysis of the regenerated plantlets showed that the essential oil content was four times higher (0.30% dry weight) than in the shoot cultures (0.08% dry weight); in addition, tiglate and angelate esters were not detected. α-Farnesene represented 91% of the total essential oil extracted from the shoot cultures. From these observations, it was concluded that the establishment of the root system seems to be important for essential oil biosynthesis in *A. nobilis* plant. Moreover, the shoot cultures need an appropriate stimuli for the biosynthesis and accumulation of an essential oil similar to that of rooted plant. The root system might be a source for appropriate precursors and/or specific enzymes for essential oil biosynthesis. These observations are in total agreement with the recent report of Ferreira and Janick (1996) concerning the root system as an enhancing factor for the production of the sesquiterpene lactone, artemisinin, in shoot cultures of *Artemisia annua* (Asteraceae). It was indeed demonstrated that the presence of roots enhances the formation of artemisinin in shoot cultures, despite the fact that roots were essentially artemisinin-free under both in vivo and in vitro conditions.

In the plant, the accumulation of volatile compounds is restricted to particular morphological structures, the glandular hair. This tissue-specific expression and regulation is consequently correlated with a low yield of the essential oil content in the undifferentiated cultures. We therefore examined the possibility of producing essential oil in transformed shoot cultures (shooty teratomas) and in hairy root cultures. Though it is well demonstrated that shoot cultures maintained on cytokinin-rich medium have the potential for synthesis of shoot-derived secondary metabolites, the advantage of transformed shoot cultures is that no exogenous phytohormones are required. In the case of *A. nobilis*, transformation with nopaline strains of *A. tumefaciens* (strain T37) resulted in the induction and formation of shooty teratomas. The

Table 2. Essential oil composition in normal and transformed tissue and organ cultures of *A. nobilis*. The composition is given in % of the total essential oil content

	Normal plant material			Transformed plant material	
	Flowers	Aerial parts in vitro plantlets	Shoot cultures or teratomas shooty teratomas	Crown gall tumors	Hairy root cultures
Essential oil constituents					
Linolyl 2-methylbutanoate	–	–	–	–	68.1
Geranyl 2-methylbutanoate	–	–	–	–	6.4
α-Pinene	5.9	3.3	1.6	6.6	–
β-Pinene	1.1	0.1	–	1.0	–
Geraniol	–	–	–	–	0.5
Limonene	0.1	0.9	1.6	0.8	–
α-Farnesene	–	22.2	91.1	–	0.6
trans-Pinocarveol	3.0	1.5	–	1.6	–
1-Hexanol	–	2.8	–	–	0.3
3-Hexenol	–	3.0	–	–	–
2-Hexanol	–	8.5	–	–	–
2-Methyl-2 buten 1-ol	8.4	4.9	–	–	–
Benzyl alcohol	–	–	–	1.2	–
Nerolidol	–	–	–	–	–
3-Hydroxy-2-butanone	–	–	–:	0.6	–
trans-2-Hexenal	–	2.2	0.6	–	0.4
Neryl propionate	–	5.5	3.0	–	–
Isobutylisobutyrate	6.6	2.2	–	6.4	–
2-Butylbutenoate	1.3	–	–	1.4	–
Isobutyl 2-methylbutyrate	2.5	0.7	–	3.8	–
2-Methyl butylisobutyrate	4.6	–	–	9.0	–
2-Methyl butenylacetate	3.1	1.6	–	–	–
2-Methyl butyl 2-methylbutyrate	–	–	–	2.5	–
Isobutyl angelate	14.3	7.8	–	6.5	–
Methallyl angelate	10.8	3.5	–	–	–
Isoamyl angelate	12.8	1.3	–	16.6	–
Total essential oil content (% dry wt.)	0.50	0.30	0.08	0.25	nd

nd: not determined.

essential oil content of the transformed shoot cultures was similar to that of normal shoot cultures, indicating that a high level of endogenous cytokinin is not a prerequisite for the induction of essential oil biosynthesis. The effect of cytokinins on essential oil production may be an indirect effect of root inhibition rather than a direct effect of essential oil biosynthesis. The GC-MS analysis revealed no significant modifications in the essential oil composition of both tissues.

Since shooty teratomas are not able to root, it was therefore not possible to examine if the essential oil yield in the aerial part (shoot culture) was influenced by the establishment of the root system. The question arising from these observations concerned the precise role of root metabolism in the essential oil composition.

The transformed roots cultured under dark conditions are unable to accumulate the characteristic esters found in the essential oil. However, GC-MS analysis shows numerous mono- and sesquiterpenes, among which β-caryophylene and α-farnesene were the most important. Linalol and geraniol 2-methylbutanoate esters were also identified. Linalyl 2-methylbutanoate represents more than 60% of the total essential oil extracted from the *A. nobilis* hairy root cultures. Many secondary metabolites were reported to be inducible by greening of the tissues (Saito et al. 1989a,b, 1993). This was correlated with the fact that these compounds are normally produced in the leaves of the mother plants. It has been reported that green normal or transformed roots from some plant species grown in the light produced certain levels of useful secondary metabolites characteristic of the aerial parts of the plant (Jaziri et al. 1995).

3 Summary and Conclusions

The essential oil content of the crown gall tissue of *Anthemis nobilis* was 0.25% of the dry weight, and the composition of the essential oil was comparable to that of the flowers. From the data related to the essential oil content and composition of the normal and transformed shoot cultures as well as from the regenerated plantlets, it was concluded that the establishment of the root system affects the essential oil composition of the regenerated plantlets. The analysis of rooted plantlets showed that their essential oil content was four times higher than in the shoot cultures without roots. Finally, transformed root cultures were unable to accumulate typical esters found in the essential oil.

From our perspective, advances in understanding the induction and the regulation of essential oil biosynthesis would be better achieved by investigating experiments using differentiated cultures rather than undifferentiated tissue culture system. In addition, it was clearly demonstrated that the genetic transformation of plant cells with *Agrobacterium* sp., which can be considered a facet of plant-microbe interaction, resulted in a consequent modification of the metabolic activity of the plant cells, and particularly in those metabolic activities that are related to the biosynthesis, accumulation, and turnover of secondary metabolites.

Acknowledgments. Mondher Jaziri is an associate researcher of the Fonds National de la Recherche Scientifique Belgium.

References

Achterrath-Tuckermann U, Kunde R, Flaskamp E, Issac O, Thiemer K (1980) Pharmacological investigations with compounds of chamomile. Planta Med 39:38–50

Bezanger-Beauquesne L, Pinkas M, Torck M (1986) Les plantes dans la thérapie moderne. Maloine SA, Paris, pp 11–283

Della Loggia R (1986) The role of flavonoids in the antiinflamatory activity of *Chamomilla recutita*. In: Cody V, Middleton E Jr, Harborne JB (eds) Plant flavonoids in biology and medicine: biological pharmacological, and structure-activity relationships. Alan R Liss, New York, pp 481–484

Duke A (1987) Handbook of medicinal herbs. CRC Press, Boca Raton, pp 111

Fauconnier M-L, Jaziri M, Marlier M, Roggemans J, Wathelet JP, Lognay G, Severin M, Homès J, Shimomura K (1993) Essential oil production by *Anthemis nobilis* L. tissue culture. J Plant Physiol 141:759–761

Fauconnier M-L, Jaziri M, Homès J, Shimomura K, Marlier M (1996) *Anthemis nobilis* L. (Roman Chamomile): In vitro culture, micropropagation, and the production of essential oils. In: Bajaj YPS (ed) Biotechnology in agriculture and forestry vol 37. Medicinal and aromatic plants IX. Springer, Berlin Heidelberg New York, pp 16–37

Ferreira JFS, Janick J (1996) Roots as an enhancing factor for the production of artemisinin in shoot cultures of *Artemisia annua*. Plant Cell Tissue Organ Cult 44:211–217

Hall IH, Lee KH, Starnes CO (1979) Antiinflamatory activity of sesquiterpene lactones and related compounds. J Pharm Sci 68:537–542

Issac O, Kristen G (1980) Alte und neue Wege der Kamillentherapie. Die Kamille als Beispiel für moderne Arzneipflanzen-Forschung. Med Welt 31:1145–1149

Jaziri M, Shimomura K, Yoshimatsu K, Fauconnier M-L, Marlier M, Homès J (1995) Establishment of normal and transformed cultures of *Artemisia annua* L. for artemisinin production. J Plant Physiol 145:175–177

Leung AY (1980) Encyclopedia of common natural ingredients used in foods, drugs and cosmetics. John Wiley, New York, pp 110–112

Melegari M, Albasini A, Pecorari P, Vampa G, Rinaldi M, Rossi T, Bianchi A (1988) Chemical characteristics and pharmacological properties of the essential oil of *Anthemis nobilis*. Fitoterapia 59:445–449

Murashige T, Skoog F (1962) A revised medium for rapid growth and bio-assays with tobacco tissue cultures. Physiol Plant 15:473–497

Petit A, David C, Dahl GA, Ellis JG, Guyon P, Casse-Delbart F, Tempé J (1983) Further extention of the opine concept: plasmids in *Agrobacterium rhizogenes* cooperate for opine degradation. Mol Gen Genet 190:204–214

Proserpio G, Martemlli A, Petri GF (1983) Elementi di Fitocosmesi. SEPEM, Milan, pp 649–650

Reichling J, Beiderbeck R (1991) *Chamomilla recutita* (L.) Rauschert (Camomile): in vitro culture and the production of secondary metabolites. In: Bajaj YPS (ed) Biotechnology in agriculture and forestry vol 15. Medicinal and aromatic plants III. Springer, Berlin Heidelberg New York, pp 156–175

Rodriguez E, Towers GHN, Mitchell JC (1979) Biological activities of sesquiterpene lactones. Phytochemistry 15:1573–1580

Rovesti P, Boni U, Patri G (1983) Le Erbe. Fabbri SpA, Milan, pp 142–145

Saito K, Yamasaki M, Yamakawa K, Fujisawa S, Takamatsu S, Kawaguchi A, Murakoshi I (1989a) Lupin alkaloids in tissue culture of *Sophora flavescens* var. angustifolia: greening-induced production of matriin. Chem Pharm Bull 37:3001–3004

Saito K, Yamazaki M, Takamatsu S, Kawaguchi A, Murakoshi I (1989b) Greening-induced production of (+)-lupanine in tissue culture of *Thermopsis lupinoides*. Phytochemistry 28:2341–2344

Saito K, Yamazaki M, Shimomura K, Yoshimatsu K, Murakoshi I (1993) Transformation in *Digitalis purpurea* L. (Foxglove). In: Bajaj YPS (ed) Biotechnology in agriculture and forestry vol 22. Plant Protoplasts and genetic engineering III. Springer, Berlin Heidelbery New York, pp 182–189

Vervliet G, Holsters M, Teuchy H, Van Montagu M, Schell J (1975) Characterization of different plaqueforming and defective temperate phages in *Agrobacterium* strains. J Gen Virol 26:33–48

IV Genetic Transformation of *Astragalus* Species

I. IONKOVA

1 Introduction

1.1 Importance of *Astragalus*

The genus *Astragalus* (Fabaceae family) comprises ca. 2000 species and has a wide distribution. Various aspects of chemical constituents, pharmacology, and in-vitro culture studies have been recently reviewed by the author (see Ionkova 1995).

The medicinal use of *Astragalus* species dates back over 1000 years, This plant was well known to Theophrast and Dioscurides (Trease 1983). Radix Astragali (*A. membranaceus, A. mongholicus*) is a very old and well-known drug in traditional Chinese medicine. The different species are used in traditional medicine in Bulgaria, Turkey, Rumania, Yugoslavia, Greece, Russia, China, Japan, Mongolia, Korea, and other European and Asiatic contries (Stepashkina 1959; Stojanov 1972; Ionkova 1983) Some extracts or isolated compounds are also being clinically tested (Alania 1985; Zhon 1985; Chu et al. 1988; Isaev et al. 1989; Ionkova 1995). Some medicines have been introduced in clinical practice (Turova 1974; Gubanev et al. 1976; Isaev et al. 1989). In recent years, research on the *Astragalus* species has been undertaken, due to their pharmacological constituents and medicinal application as an immunostimulant or as a substitute for *Panax ginseng*.

The biologically active constituents of *Astragalus* species represent mainly three classes of chemical compounds: polysaccharides, saponins, and flavonoids, and have been extensively studied. The indolizidine alkaloids swansonine and swansonin N-oxide have also been isolated from spotted lockoweed (*A. lentiginosus*, Rizk 1986) and *A. oxyphysus*. Their occurrence in a number of *Astragalus* species has been investigated (Molyneux et al. 1991). Two new arylindolizidine alkaloids have now been obtained from the aerial parts of Indian *Astragalus polycanthus* – polycanthine (Michael 1995). Polyhydroxyindolizidines are a recently discovered class of alkaloids, occurring primarily in the family Fabaceae. The group as a whole has remained undiscovered until recently, probably due to their failure to react with alkaloid-specific chromogenicreagents such as iodoplatinate and Dragendorff's

Department of Pharmacognosy, Faculty of Pharmacy, Medical University – Sofia, 2 Dunav Str., 1000 Sofia, Bulgaria

Biotechnology in Agriculture and Forestry, Vol. 45
Transgenic Medicinal Plants (ed. by Y.P.S. Bajaj)
© Springer-Verlag Berlin Heidelberg 1999

reagent (Molyneux et al. 1991). The toxicity and biological activity of these alkaloids resides in their general properties as inhibitors of glycosidases (Elbein and Molyneux 1987) Swainsonin is a potent and specific inhibitor of α-mannosidase (Dorling et al. 1989). Its potential as an antimetastatic immunomodulator has proved highly topical.

1.2 Need for Genetic Transformation

Due to their complex structures, alkaloids, saponins, polysaccharides, and flavonoids are still most efficiently produced by plants. Within the Fabaceae, production of these compounds has been reported for several genera (Ionkova 1983). However, there are several problems associated with this production method (Verpoorte et al. 1993) Variable quantities and qualities of the plant material, plants that need to grow several years before they are ready for harvesting (*Astragalus* roots), and overcollecting of endangered species (*A. membranaceus, A. mongholicus*) are just a few of the problems connected with the production of these natural products, Therefore, cultured cells rather than plants are a possible alternative production method.

Studies on cell culture in *Astragalus* began with the report on callus cultures of *A. hamosus* and *A. boeticus* (Ionkova and Alfermann 1990). To achieve industrial production, a stable high-producing cell lines must be obtained. For this, two approaches are being used (1) screening and selection for high-producing cell lines, and (2) optimization of growth and production medium and conditions (Ionkova 1991, 1992; Ionkova et al. 1993). Although the growth of the cell cultures of *Astragalus* was fairly good, the amount of saponins and flavonoids produced in the cells was lower than in intact plants (Ionkova 1995).

It has been shown in many plant cell cultures that growth without differentiation is incompatible with the expression of a secondary metabolic pathway (Cristen and Roberts 1993) and that the use of differentiated cultures results in considerably higher and stable production of the compounds of interest (Ionkova and Alfermann 1994). Since the root is mostly the site of saponin and polysaccharide biosynthesis, in vitro root cultures from different *Astragalus* species have been able to produce large amounts of secondary metabolites (Ionkova 1995). A summary of studies on hairy root cultures of *Astragalus* is given in Table 1.

2 Genetic Transformation Studies (Table 1)

2.1 Bacterial Strains and Transformation Methodology

In order to produce highly productive transformed root cultures, it is essential to use plant material selected for its interesting metabolite spectrum and with a high content of biologically active substances relative to other plants of the

Table 1. Summary of studies on hairy root cultures of *Astragalus* species

Astragalus spp.	Explant used for raising HR	Remarks	Reference
A. membranaceus	Seedlings	*Agrobacterium rhizogenes* ATCC 15834, aminobutyric acid	Asamizu et al. (1990)
A. hamosus,	Leaves/stems A. boeticus	*A. rhizogenes* R1604, saponins	Ionkova and Alfermann (1990)
A. hamosus	Leaves/stems	*A. rhizogenes* R 1604, LBA 9402, soyasapogenol B, sterols	Ionkova (1991)
A. gummifer	Leaves	*A. rhizogenes* ATCC 15834, NIAES 1724, mucilage	Isa (1991)
A. membranaceus	Leaves	*A. rhizogenes* TR 105, LBA 9402, cycloartane saponins	Ionkova (1993)
A. membranaceus, A. mongholicus, A. gummifer	Leaves, stems	*A. rhizogenes* LBA 9402, ATCC 15834, TR 105, polysaccharides	Ionkova et al. (1993)
A. mongholicus	Leaves/stems	*A. rhizogenes* LBA 9402, astragalosides I, II, III	Tappe et al. (1994)
A. membranaceus	Leaves	*A. rhizogenes* ATCC 15834, cycloartane saponins, agroastragalosid II	Hirotani et al. (1994a,b)
A. englerianus, A. glycyphyllos, A. monspessulanus	Leaves/stems	*A. rhizogenes* TR 105, R 1604, LBA 9402, ATCC 15834, saponins, polysaccharides	Ionkova (1995)
A. membranaceus	Leaves	*A. rhizogenes*, two triglkycosidic triterpene astra galosides	Zhou et al. (1995)

same *Astragalus* genus. The selection of plants (Ionkova 1983, 1995) showing desirable properties was the first step to obtain hairy root cultures of *Astragalus*.

Transformation Procedure. Hairy root cultures have been established from a variety of *Astragalus* species. The bacteria were maintained on solid YMB media as described earlier (Ionkova 1995). Before infection, the bacteria were grown for 48h at 26°C, 80ppm in liquid YMB. The bacterial concentration may also affect the infection and thus the number of roots formed. Concentration below or much above the optimum (0.9×10^8 cells/ml) may result in lower availability of bacteria for transforming the plant cells. It was observed that infection should be carried out immediately after wounding in order to obtain the maximal number of roots as reported for other plants (Vanhala et al. 1995).

Young leaves or stems of sterile *Astragalus* plants were wounded with a sterile needle and the different bacteria from the liquid media were spread onto the midrib in order to generate transformed root cultures. The hairy root cultures were maintained on MS medium without phytohormones, containing

2–6% (w/v) sucrose in 300-ml flasks on a gyratory shaker at 25 °C in darkness, as already described (Ionkova 1995). The elimination of *Agrobacterium* was sometimes difficult. Generally, it was possible to establish transformed roots in axenic culture after subculturing a few times at 4-week intervals in liquid media containing the antibiotic Claforan.

The experiment was repeated three times. Statistical calculation (regression analysis and one factor analysis of variance) were carried out on a Hewlett Packard computer using a statistical program.

Effect of Bacterial Strains. Four different bacterial strains, TR 105, R 1601, ATCC 15834, and LBA 9402, were used (Table 2). The phenotypic response results from the insertion into the plant genome of T-DNA (transfer DNA) was different (Fig. 1). In some of the more recalcitrant *Astragalus* species, successful transformation was achieved with a variety of *Agrobacterium rhizogenes* strains showing different host characteristics. However, different strains of bacteria showed various abilities to induce hairy roots on the leaf explants of the same *Astragalus* species. The difference in virulence could be explained by the plasmids harbored by bacterial strains (Cardelli et al. 1985).

Morphology of Hairy Roots. The morphology of the hairy roots depended not only on the inorganic elements in the culture media, but also on the *Astragalus* species. The roots were formed mainly on the midrib of the leaf and stem, and they differed morphologically, depending on the species. *A. glycyphyllos*-HR

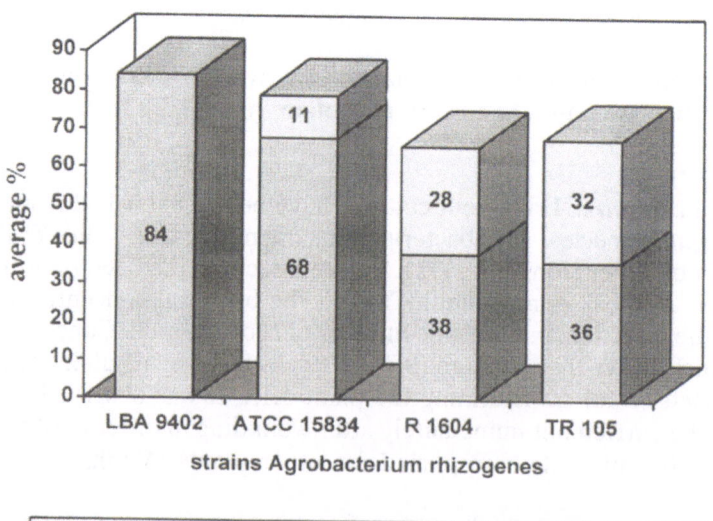

Fig. 1. Phenotypic response of *Astragalus* spp. with different bacterial strains

transformed with ATCC were brownish, thick and long, while the *A. membranaceus* ATCC-roots were orange, thin, with branching, and *A. gummifer* ATCC-roots were white, thin, and long. Our investigations have shown that *Astragalus* hairy roots could be divided morphologically into three groups even when the same *Agrobacterium* strain (ATCC 15834) was used. The same results were found by Nguyen et al. (1992, cited in Vanhala et al. 1995) working with *Psoralea* species transformed with *Agrobacterium* strain LBA 9402.

Initial experiments showed that transformation by TR 105 and R 1601 occurs at a low frequency compared to results observed using strains ATCC 15834 and LBA 9402. The last two show good virulence characteristics. The bacterial strain LBA 9402, which contains the plasmid pRi 1855, shows less dependence on the internal auxin levels in the *Astragalus* plant, with regard to its ability to produce roots, than several others (Cardarelli et al. 1985).

Effect of Acetosyringon. In order to improve the frequency of transformation, two methods were attempted. It has been demonstrated that Agrobacteria are attracted to the host across chemical gradients of phenolic compounds released by injured plant cells. One specific highly active compound in this respect has been identified as acetosyringone, which has been reported to increase *Agrobacterium*-mediated transformation frequencies in a number of plants (Vanhala et al. 1995). In some of the more recalcitrant species, successful transformation was achieved by including 10 µM acetylsyringone in the medium in which the bacteria are suspended. The addition of acetylsyringone to the cocultivation medium changed the frequency of transformation and resulted in a substantial increase in the number of transformation events (Table 2). This compound, produced during the wounding response of plants, activates the *vir* genes of *Agrobacterium*, aiding plasmid T-DNA transfer (Rhodes et al. 1987). The number of individual roots formed has been used as a measure of virulence. The results show that acetylsyringone can be used to increase *Agrobacterium*-mediated transformation frequencies in *Astragalus* species as well.

The susceptibility of *Astragalus* species to infection with the same *Agrobacterium* strain was very variable (Table 2). Some species (*A. membranaceus, A. mongholicus, A. monspessulanus*) have more difficulty in establishing transformed roots. In some species (*A. glycyphyllos, A. hamosus, A. boeticus*), a profusion of roots appears directly at the site of inoculation, but in others (*A. englerianus, A. mongholicus, A. missouriensis, A. sulcatus*) callus was formed initially and transformed roots emerged subsequently from it. Depending on the species and the *Agrobacterium* strains, a profusion of roots appears within 1–4 weeks.

2.2 Stability, Growth Kinetics, and Confirmation of Transformation of Hairy Roots

The hairy root cultures (HR) from different *Astragalus* species grown rapidly in simple media without phytohormones (Ionkova 1991, 1992) were stable in

Table 2. Response of *Astragalus* species (leaf explants) to produce hairy roots within 28 days after infection with various strains of *Agrobacterium rhizogenes*. The bacteria were perculated overnight without (a) or with (b) acetosyringone (10 µM)

Astragalus sp.	Treatment of bacteria	Producing roots (%)			
		TR 105	R 1601	ATCC1584	LBA 9402
A. boeticus	a	5	10	40	85
	b	10	50	70	100
A. brachycera	a	50	*	80	90
	b	70	*	100	100
A. canadensis	a	50	*	90	100
	b	90	*	90	100
A. englerianus	a	0(cal)	0(cal)	50(cal)	80
	b	20	30(cal)	100	100
A. falcatus	a	70	*	100	100
	b	100	*	90	100
A. glycyphyllos	a	80	70	100	100
	b	90	100	100	100
A. gummifer	a	30	50	70	100
	b	30	80	100	100
A. hamosus	a	30	100	80	100
	b	50	100	90	100
A. membranaceus	a	0(cal)	0(cal)	25	50
	b	5	10	33	70
A. missouriensis	a	20(cal)	20(cal)	50	60
	b	20(cal)	50	50	80
A. mongholicus	a	0(cal)	0	33(cal)	20
	b	5	15	50	85
A. monspessulanus	a	0(cal)	0(cal)	20(cal)	20
	b	0(cal)	10	50	80
A. oxyglotis	a	60	*	60	90
	b	80	*	80	100
A. sulcatus	a	30(cal)	*	50	60
	b	30(cal)	*	60	90

Abbreviation: (cal) callus formation. * no examined.

their growth rate and polysaccharide (Fig. 2) and saponin (Fig. 3) production over a period of more than 5 years in culture.

To prove the genetic transformation, the opines manopine/agropine were identified in several HR clones of *Astragalus* species. Preparation of extracts from plant material and HR clone, opine analysis by high voltage paper electrophoresis and TLC, and detection with alkaline silver nitrate reagent were carried out as described earlier (Ionkova and Alfermann 1990, 1994; Ionkova 1992). HR of all *Astragalus* species were tested positive for opines, showing that the HR contained the appropriate enzymes from the plasmids responsible for opine synthesis. The roots of parent plant species did not synthesize opines.

HR of *Astragalus* spp. have a profusion of root hairs, a high degree of lateral branching, and absence of geotropism, resulting in high growth rates. After a short lag phase, the HR started to grow rapidly and the fresh weight

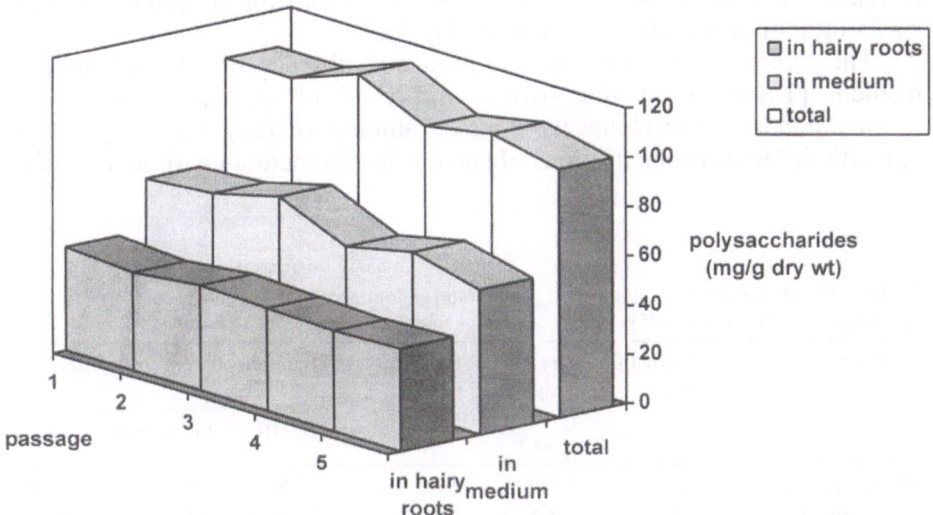

Fig. 2. Stability of polysaccharide production in *A. mongholicus* HR-A5 over a period of more than 5 years in culture

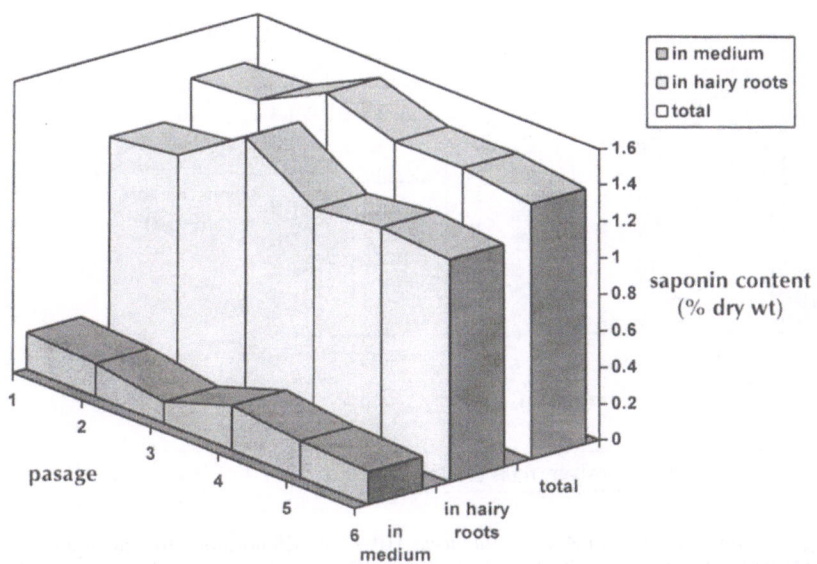

Fig. 3. Stability of saponin production in *A. membranaceus* hairy roots over a period of more than 5 years in culture

increased 35–40-fold for *A. mongholicus*, 14–15-fold for *A. gummifer*, and
20–25-fold for *A. membranaceus*, until day 30.

The effects of sucrose concentration and absence of NH_4NO_3 in the
medium (Table 3) on the growth and yield of saponins (Fig. 4) in
A. membranaceus HR clones were also examined. Investigation was carried
out with different concentrations of sucrose in MS medium with ammonium

Table 3. Effect of medium condition on growth and saponin content in hairy roots of *Astragalus membranaceus*. (Ionkova 1995)

Medium	Growth rate[a]	Dry weight (g/100 g fr wt.)	Production index[b]	Saponin content[c] (mg/g dry wt.)		
				In HR	In medium	Total
MS sucrose 2%	15	3.742	210	13.66	0.35	14.01
MS sucrose 4%	24	3.928	260	9.86	0.99	10.85
MS sucrose 6%	21	6.348	63	2.87	0.14	3.01
MS without NH$_4$NO$_3$ sucrose 2%	14	5.964	175	11.27	1.22	12.49

[a] Growth rate was determined by increase of fresh weight after 30 days of culture. The values are the quotient of the fresh weight after 30 days of culture and the fresh weight of the inoculum.
[b] Production index = growth × total saponin content.
[c] The amount of the saponin was determinated by densitometry as described in Ionkova (1995).

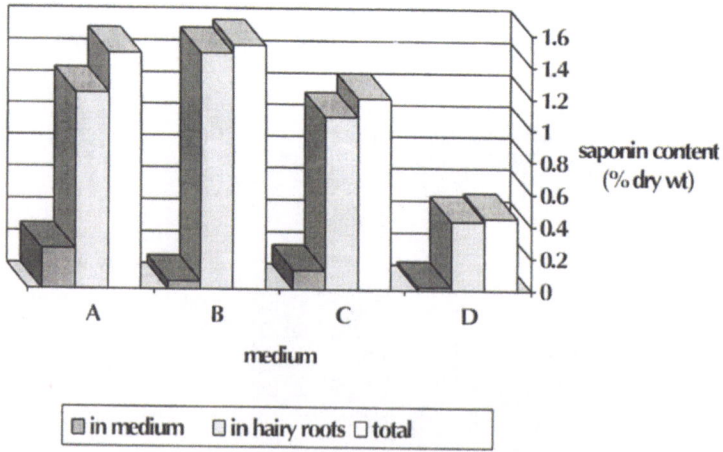

Fig. 4. Saponin content of *A. membranaceus* HR-L3 in MS hormone-free medium.
A MS without ammonium nitrate and 2% sucrose; *B* MS with ammonium nitrate and 2% sucrose;
C MS with ammonium nitrate and 4% sucrose; *D* MS with ammonium nitrate and 6% sucrose;
The saponin content was measured on day 30 of cultivation

nitrate (Fig. 5); 4% sucrose is most suitable for both biomass yield and saponin content. Low (2%) concentration promoted saponin biosynthesis, but inhibited growth of HR. High (6%) concentration promoted growth of *Astragalus* HR but yield of saponins was very low.

Using an HR line of *A. membranaceus*, the effect of the nitrogen source on growth and saponin production was studied (Fig. 6). They were investigated in

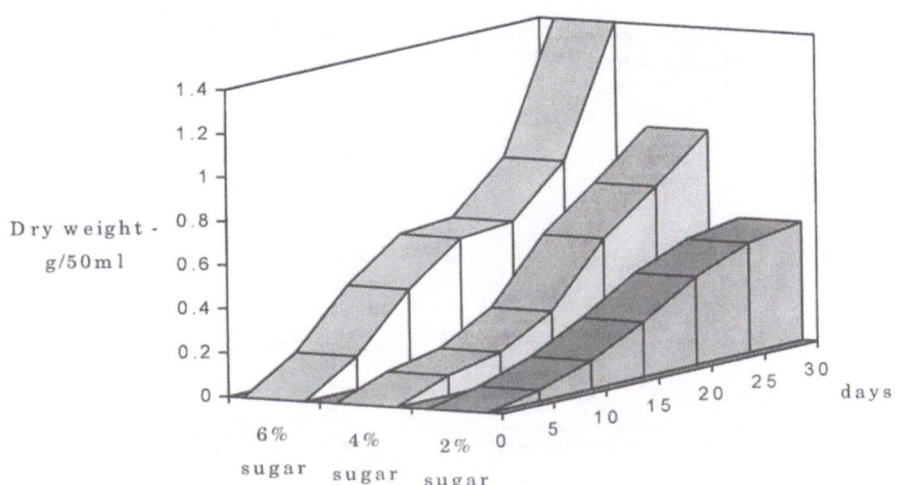

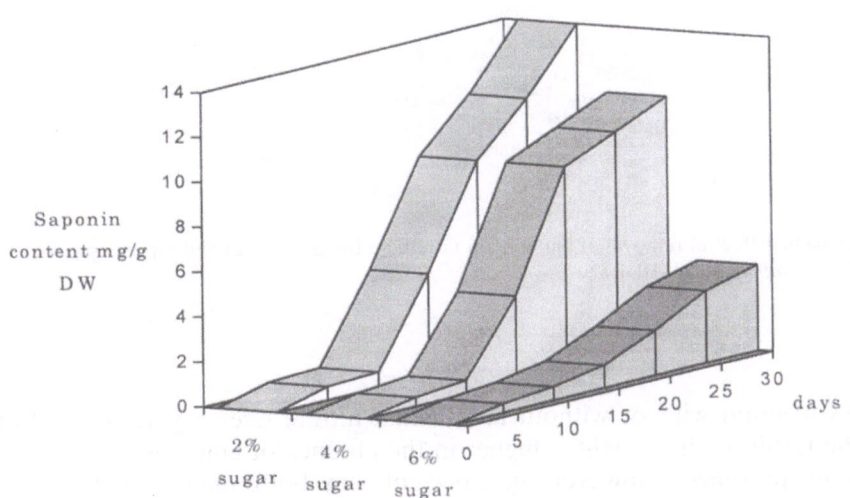

Fig. 5. Growth and saponin production of the *A. membranaceus* hairy roots on MS medium with different concentrations of sucrose

a

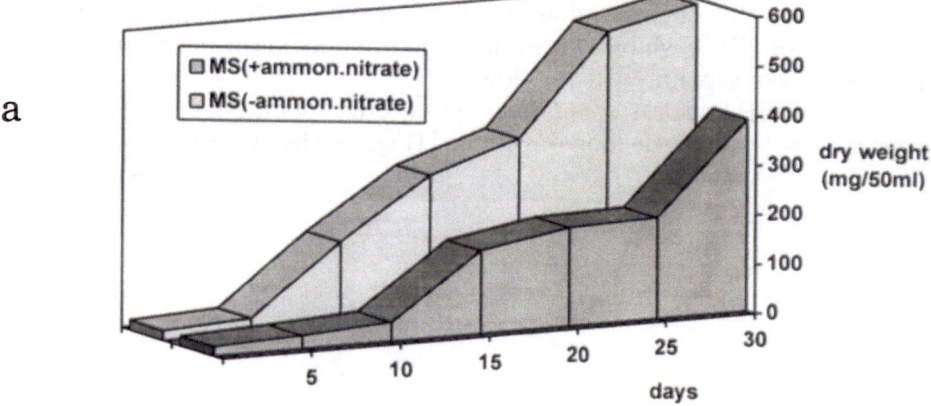

b

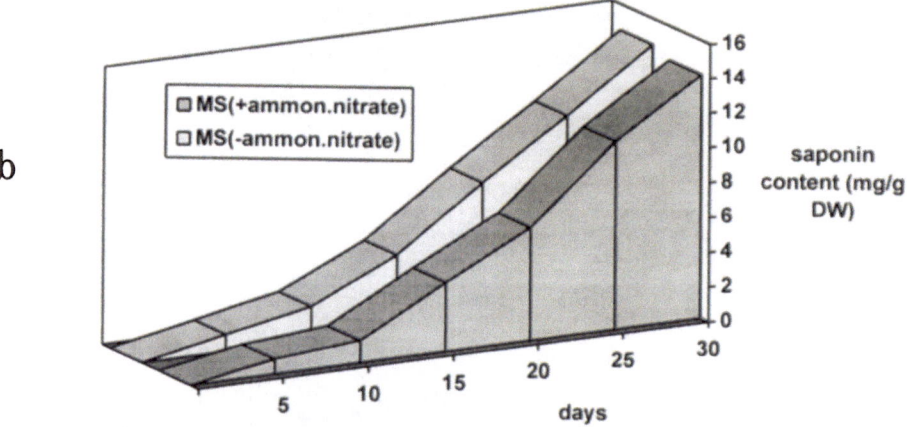

Fig. 6a.b. Effect of removal of ammonium nitrate on the growth (**a**) and saponin production (**b**) of *A. membranaceus* HR-L7

MS medium with or without ammonium nitrate over a period of 30 days. The resulting dry weight is higher in the absence of ammonium nitrate than in its presence. However, unexpectedly, in both media similar saponin contents were observed. Saponin production in HR cultures of *A. membranaceus* was not impaired by NH_4NO_3. This is in contrast to the results for polysaccharides in HR of *Althaea officinalis*, (Ionkova and Alfermann

1994) or for diterpene formation in HR of *Salvia miltiorrhiza* (Hu and Alfermann 1993).

2.3 Saponins and Polysaccharide Content of the Hairy Root Clones

The hairy root clones transformed with different bacteria were cultured individually. The *Agrobacterium* strain had some effect on growth and content of biologically active compounds. The clones infected by ATCC 15834 and LBA 9402 grew the fastest and the biomass increased two- to three fold more as compared to other clones after 28 days of cultures. Moreover, different *Agrobacterium* strains have an effect on total saponin content in older hairy roots. Maximum saponin content was found in HR of *A. mongholicus* transformed with LBA 9402 bacteria. Our results are similar to those obtained for some solanaceous species (Shimomura et al. 1991; Vanhala et al. 1995).

The results of Tappe et al. (1994) and those of Hirotani et al. (1994a) demonstrate that the saponin production by the hairy root cultures is typical of that found in many species of *Astragalus* (Ionkova 1995). The transformed roots produce cicloartane saponins, which do not differ significantly from those produced by the parent plants. Further work concerning cycloartane triterpeneglycosides from the hairy root cultures of *A. membranaceus* has been reported by Hirotani et al. (1994b). Agroastragalosid II, a new astragaloside, was isolated from these hairy roots Its structure was established as 3-O-β-(2'-O-acetyl)-D-xylopyranosyl-6-O-β-D-glucopyranosyl-(24S)-3β,6α,16β,24,25-pentahydroxy-9,19-cyclolanostane on the basis of various spectroscopic data.

After column chromatography, the sterol fraction in HR cultures of *A. membranaceus* was separated into five compounds by GC-MS. The compounds were identified as: (1) palmitin acid (molecular ion at m/z 256, base peak m/z 41), (2) stearinic acid (molecular ion at m/z 284, base peak m/z 43), (3) campesterol (molecular ion at m/z 400, base peak m/z 43), (4) stigmasterol (molecular ion at m/z 412, base peak m/z 55), and (5) β-sitosterol (molecular ion at m/z 414, base peak m/z 43 (Ionkova 1995). Heterogeneous acid hydrolysis of the total mixture of saponins isolated from selected HR of *A. membranaceus* yielded three aglycons. On the basis of spectral NMR, the structure of aglicons was assigned as 9,19-cyclolanostane (cycloastragenol), one artifact, lanost-9(11)-ene (astragenol), which was secondarily formed during acid hydrolysis from cycloastragenol (molecular ion at m/z 490, base peak at m/z 143), and soya sapogenol B with molecular ion at m/z 458 base peak at m/z 234. The data are in full agreement with the literature (Ionkova 1983; Kitagawa et al. 1983; Nikolov et al. 1985).

Analysis of saponin after 1 month of culture of *A. mongholicus*-HR was performed after lyophilization, extraction, and column chromatography, as described earlier (Ionkova 1983, 1995). In order to confirm the structure of saponins, NI-FAB, MS/MS, EI-MS, ^{1}H- and ^{13}C-NMR (1D and 2D) were carried out (Tappe et al. 1994; Ionkova 1995). *A. mongholicus* HR produces cycloastragenol-saponins: astragalosid I, astragalosid II, and astragalosid III (Fig. 7).

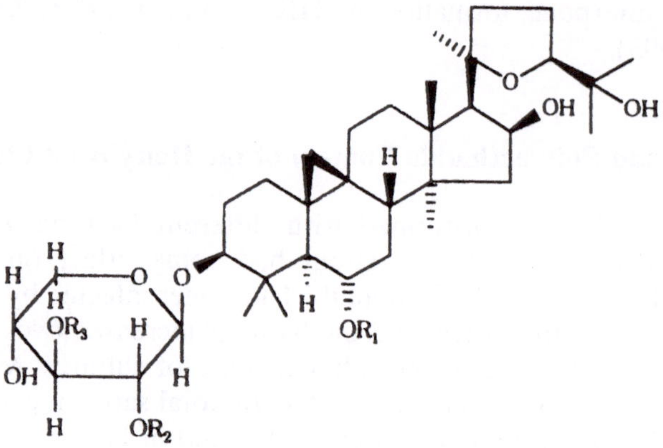

	R_1	R_2	R_3
F1 = astragaloside I	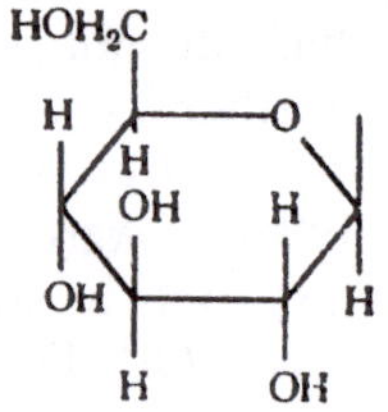 HOH₂C structure	CH_3CO	CH_3CO
F2 = astragaloside II	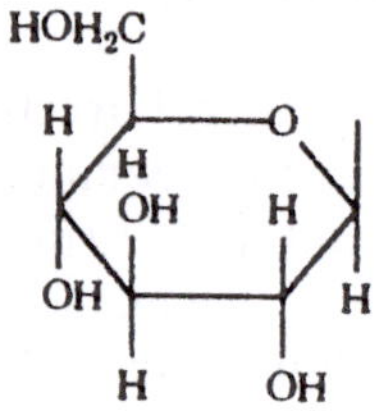 HOH₂C structure	CH_3CO	H
F3 = astragaloside III	H	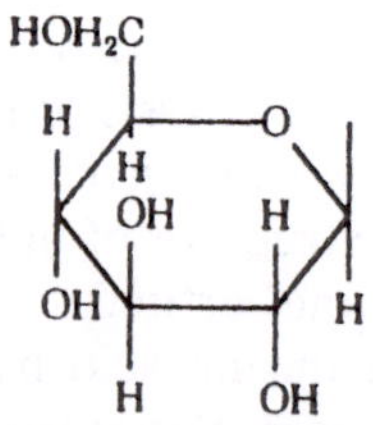 HOH₂C structure	H

Fig. 7. Structures of saponins isolated from hairy root cultures of *Astragalus mongholicus* Bge. (Tappe et al. 1994)

HR cultures of *Astragalus* spp. released part of the saponin product (about 16–20% of the total saponin) into the medium (Fig. 3). This is essential in order to establish continuous production of saponins. The most apparent effect in saponin production was seen when sucrose level of the medium was modified. A MS containing 2% sucrose increased overall saponin yield, but the growth was very poor. The optimal medium for both yield and growth was supplemented with 4% sucrose (Fig. 5). The saponin content was not impaired significantly at high levels of nitrogen (media A and B in Fig. 4). These results suggest that biosynthetic regulation is affected by altered cell organization in roots, as might be expected. HR cultures of *Astragalus* spp. can therefore be an interesting system by which fast growth of the biomass as well as relatively high saponin production can be achieved, facilitating further studies on tritrpene saponin biosynthesis.

Dedifferentiated callus and suspension yield very few polysaccharides (Ionkova 1992; Ionkova and Alfermann 1994). However, HR of *A. gummifer, A. mongholicus*, and *A. membranaceus* produce significant amounts of polysaccharides, which are secreted into the liquid growth media (Fig. 2), and account for the viscosity of the media. The presence of macromolecules in the culture media has been described by a number of investigators (Isa 1991; Ionkova 1992); experiments with different *Astragalus* species showed similar results. *A. mongholicus* secreted water-soluble polysaccharides 8.5 wt%, *A. gummifer* 19.5 wt%, and *A. membranaceus* 7.3 wt%. These polysaccharides can be isolated from the medium by simple ethanol precipitation (Ionkova 1995).

The monosaccharide composition of the macromolecules in various cultures and parent plants has been investigated. Differences have been found in the sugar composition of polysaccharides between HR and intact plants. Furthermore, the amounts of each neutral monosaccharide produced in the various HR cultures were not uniform, and composition varied widely. The polysaccharides of the mother plant contained more compounds different to the mucilage of HR. This difference can be attributed to the different stages of development in the transformed roots.

It is known that *Agrobacterium rhizogenes* produces intercellular and extracellular β-D(1–2)-glucan with considerable viscosity, composed entirely of glucose (Zorrquietta and Ugalde 1986). From our investigation it is clear that *Astragalus* HR mucilages are composed of several types of neutral monosaccharides. Thus it is reasonable to suggest that the polysaccharides secreted into the medium are different in structures and pharmacological effects from those obtained in HR and mother plants. For *Astragalus* plants it has been shown that the use of differentiated cultures results in a considerable production of polysaccharides.

2.4 Production of Swainsonine and Related Hydroxyindolizidine Alkaloids by Hairy Roots

Some *Astragalus* plants contain moderate levels of swainsonine and its N-oxides (Molyneux et al. 1991). Because of their medicinal potential, detailed in the past few years, there is much current interest in optimizing the yield of the alkaloids from natural sources. Recently, in our laboratory, it was observed that transformed root cultures of *A. sulcatus, A. missouriensis, A. oxyglotis, A. australis, A. canadensis, A. englerianus,* and *A. falcatus* grew faster, and at the optimal stage of growth (28 days) some of them produced swainsonine almost eight times that in roots of intact plants (*A. sulcatus*, TLC densitometry). Ermayanthi et al. (1994) investigated *Swainsona galegifolia* and found similar differences. The levels of swainsonine, 62 mg/g of transformed root cultures, are eight times those of intact plants. The generally high water solubility of such alkaloids precludes their purification by normal partioning techniques using organic solvents. Ninhydrin spray reagent has most commonly been used to detect these alkaloids on thin layer chromatographic (TLC) plates, but the reagent is nonspecific and insensitive, reacting to give a yellow-brown spot only when high concentrations of the compounds are present (Hohenschutz et al. 1981). In our experiments, TLC was performed by a method described by Molyneux et al. (1991), which has been used routinely to screen *Astragalus* and *Oxytropis* species for the presence of swainsonine and its N-oxides. The method utilizes the pyrrol-specific Ehrlich's reagent, which reacts with a pyrrolic moiety generated from the pyrrolisidine ring portion of the bicyclic indolizidine ring system by dehydration with acetic anhydride. This technique is congeneric and compatible with that previously developed for the detection of hepatotoxic pyrrolidine alkaloids. The swainsonine became visible as a purple spot. The N-oxide also gave rise to a purple spot, but this appeared at an earlier stage of the heating process.

In order to stimulate the biosynthesis of these alkaloids in hairy roots, two approaches have been used (1) change of the medium constituents, and (2) optimization of the culture condition. Transformed roots treated with copper or cadmium salts showed an accumulation of compounds of interest (Christen and Roberts, 1993). The low NH_4 NO_3 concentration and addition of copper sulphate in the medium of *Astragalus* roots stimulate the production of alkaloids. The latter is in agreement with the results of Ermayanti et al. (1994) for *Swainsona* hairy roots. Reducing the pH of the medium from 5.7 to 2.7 shortly before harvest, and supplements of malonic acid or pipecolic acid, significantly stimulated production of swainsonine in transformed roots. These changes also promoted the release of swainsonine into the culture medium.

Because the highest amounts of alkaloids appeared to occur in the aerial parts (leaves, pods, seeds) of *Astragalus* spp. (Molyneux et al. 1991), hairy roots were investigated in the light, where they turned green. Light is a factor that may increase production for certain cell lines (Ionkova 1995). The green color in these roots may indicate a partial photosynthetic capacity of the roots. Transformed green roots have also been obtained in a few other species of the Solanaceae, Asteraceae, and Cucurbitaceae families (Vanhala et al. 1995). Green hairy roots are known to produce some metabolites that are normally synthesized in the green part of plants (Saito et al. 1992, cited in Vanhala et al. 1995). Chloroplast-dependent reactions are a vital part of certain metabolic pathways and could result in a novel pattern of compounds produced by roots (Signs and Flores 1990). In our investigations, a significant effect on the production has been achieved. It is clear that some interesting possibilities to increase production are still to be explored.

3 Summary and Conclusions

Hairy roots of *Astragalus membranaceus and A. englerianus* on transfer to solidified MS medium (1/2 concentration) containing 1.2 mg/l BAP, 0.1 mg/l IAA, 2 g/l casein, and 2% sucrose and incubated at 25 °C in light, developed calli, turned green, and started to differentiate. There are at present no examples of such procedures being applied to the improvement of valuable secondary metabolite formation in a plant returned to the field. Important factors to be considered in this connection will be biosynthetic rate, accumulation site, and catabolism. This knowledge is also of interest in connection with studies on the role of secondary metabolism for plants, and may contribute to a better understanding of resistance of plants to diseases and various herbivores.

A major problem associated with organized cultures remains their growth on a large scale. The morphological structure of roots presents a unique

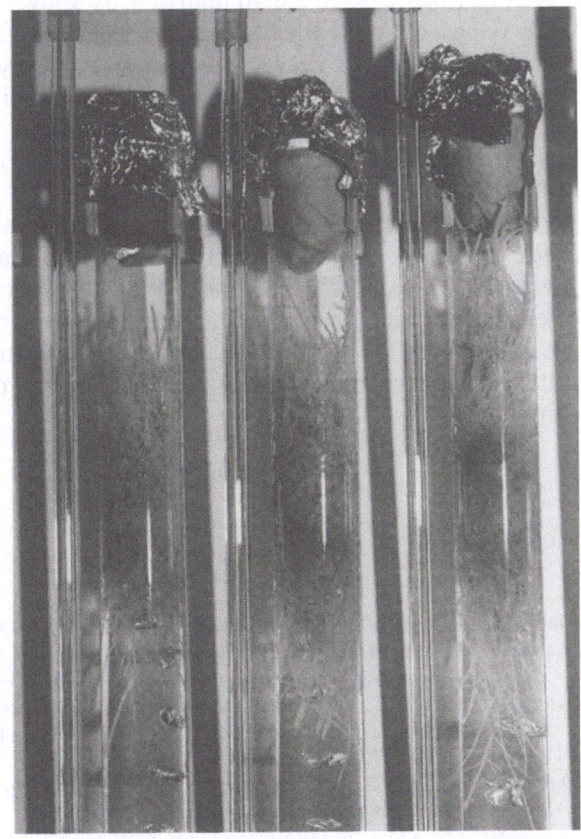

Fig. 8. Hairy roots of *Astragalus membranaceus* in air-lift-type fermenter, 10 days after inoculation

engineering challenge for industrial-scale application of root culture. It is well recognized that plant systems have a slow growth rate and that their cell and tissue chemical transport system is unique and different from that of typical microorganisms. Their air or oxygen requirements are lower than those of microorganisms and may be amply supplied by air-lift fermenters (Staba 1985). A rapidly growing root line of *A. membranaceus* was selected for cultivation of roots in an air-lift fermenter. It was observed that the inner glass tube air-lift type restricted the free movement of growing roots. The meristem-dependent growth of hairy root cultures of *A. membranaceus* established a distinct spatial and temporal order. They are relatively insensitive to oxygen levels. Growth in liquid medium results in a root "nucleus" with young, rapidly growing roots on the periphery and a core of older tissue inside. Restriction of nutrient and oxygen delivery to the center of the mass gives rise to a growing pocket of senescent tissue, as described by Sign and Flores (1990). In our experiments, partial success has been achieved, leading to the formation of dense beds of roots (Fig. 8) with root hairs which provide a large surface for nutrient absorption and release of products into the medium. The biomass concentrations showed a high final density above 17 g dry roots/l. The saponins were released by *A. membranaceus*-transformed roots in growth medium, reaching a maximum after 35 days, while the hairy roots retained high internal saponin levels and remained viable. The possibility of controlling this release of root saponins into the medium is important for *Astragalus* HR scaleup. Corry et al. (1993) have proposed different solutions for the large-scale culture of hairy roots.

A variety of experimental approaches have been used to increase the efficiency of the production of saponins, polysaccharides, and alkaloids in *Astragalus* roots. The excretion of these compounds into the liquid medium offers possibilities for continuous cultures. In general, *Astragalus* hairy roots are preferable to conventional cultures, because of their higher genetic stability, higher growth rates, and production of pharmacologically active compounds.

References

Alania MD (1985) Cycloartanic triterpenoids of plants. Rast it Resur 4:510–523
Asamizu T, Akiyama K, Yasuda I (1990) Aminobutyric acid production by hairy root culture in *Astragalus membranaceus*. Shoyakugaku Zasshi 44(4):311–315
Cardarelli M, Spanol L, De Paolis A, Mauro ML, Nitali G, Constantino P (1985) Identification of the genetic locus responsible for non-polar root induction by *Agrobacterium rhizogenes* 1855. Plant Mol Biol 5:385–391
Chu DT, Lepe-Zuniga J, Wong WL, La Pushin R, Mavligit GM (1988) Fractionated extract of *Astragalus menbranaceus,* a Chinese medicinal herb, potentiates LAK cell cytotoxicity generated by a low dose of recombinant interleukin-2. J Clin Lab Immunol 26(4):183–787
Corry JP, Reed LW, Curtis WR (1993) Enhanced recovery of solavetivone from *Agrobacterium*-transformed root cultures of *Hyoscyamus muticus* using integrated product extraction, Biotechnol Bioeng 42:503–508

Cristen P, Roberts MF (1993) Transformation in *Datura* species. In: Bajaj YPS (ed) Biotechnology in agriculture and forestry, vol 22, Plant Protoplasts and genetic engineering III, Springer, Berlin Heidelberg New York, pp 157–171

Dorling PR, Colegate SM, Huxtable CR (1989) Toxic species of the plant genus *Swainsona*. In: James LF, ELbein AD, Molyneux RJ, Warren CA (eds) Swainsonine and related glycosidase inhibitors. Iowa State University Press, Ames, Iowa, pp 14–22

Elbein AD, Molyneux RJ (1987) The chemistry and biochemistry of simple indolisidine and related polyhydroxy alkaloids. In: Pelletier SW (ed) Alkaloids: chemical and biological perspectives, vol 5. John Wiley, New York, pp 1–54

Ermayanti TM, McComb JA, O Brien PA (1994) Production of swainsonine in root cultures of *Swainsona galedifolia*. J Exp Bot 45:633–637

Gubanev I, Krilova I, Tichonova V (1976) *Astragalus*. In: Agrar plants in the USSR. Misl, Moscow, pp 193–202

Hirotani M, Zhou Y, Rui H, Furuya T (1994a) Cycloartane triterpene glycosides from the hairy root cultures of *Astragalus membranaceus*. Phytochemistry 36:665–669

Hirotani M, Zhou Y, Rui H, Furuya T (1994b) Cycloartane triterpene glycosides from the hairy root cultures of *Astragalus membranaceus*. Phytochemistry 37(5):1403–1407

Hohenschutz LD, Bell EA, Jewess PJ, Leworthy DP, Pryce RJ, Arnold E, Clardy J (1981) Castanospermine, a 1,6,7,8-tetrahydroxyoctahydroindolizidine alkaloid from seeds of *Castanospermum australe*. Phytochemistry 20:811–814

Hu Z, Alfermann AW (1993) Diterpenoid production in hairy root cultures of *Salvia miltiorrhiza*. Phytochemistry 32(3):699–703

Ionkova I (1983) Phytochemical investigation of saponins, sapogenins and flavonoids species of the Fabaceae family. Diss, Medical Academy, Fac Pharm, Sofia, Bulgaria

Ionkova (1991) Production of saponins by conventional and transformed root cultures of *Astragalus hamosus*(Fabaceae). Probl Pharm Pharmacol Sofia 5:31–37

Ionkova I (1992) A method for rapid accumulation of saponin producing biomass in callus and suspension cultures of *Astragalus hamosus*. Bulg Authorship Cert N 94 913/29.07.1991

Ionkova I (1993) Cycloartane-derived saponins from transformed root cultures of *Astragalus membranaceus* Bge. ASP, 34th Annu Meet, 18–22 July, San Diego, California

Ionkova I (1995) *Astragalus* species (milk vetch): in vitro culture and production of saponins, astragaline, and other biologically active compounds. In: Bajaj YPS (ed) Biotechnology in agriculture and forestry, vol 33 Medicinal and aromatic plants VIII. Springer, Berlin Heidelberg New York, PP 97–138

Ionkova I, Alfermann AW (1990) Transformation of *Astragalus* species by *Agrobacterium rhizogenes* and their saponin production. Planta Med 56:634

Ionkova I, Alfermann AW (1994) *Althaea officinalis* L. (marshmallow): in vitro cultures and production of biologically active compounds. In: Bajaj YPS (ed) Biotechnology in agriculture and forestry, vol 28. Medicinal and aromatic plants VII. Springer, Berlin Heidelberg New York, pp 13–42

Ionkova I, Hu Z, Alfermann A (1993) Polysaccharide production by hairy root cultures of some higher medicinal plants. In: Narstead A (ed) 41st Annu Congr Med Plant Res, Dusseldorf. Thieme, Stuttgart, p 85

Isa T (1991) Mucilage production by hairy root cultures of *Astragalus gummifer*. In: Komamine A, Misawa M, Dicosmo F. (eds) Plant cell culture in Japan CMC Co, Copyright, Japan, pp 99–103

Isaev MI, Gorovitz AK, Abubakirov HK (1989) Progress in chemistry of cycloartanes. Chim Prirod Soedin 2:156–157

Kitagawa I, Wang HK, Tagaki A, Fuchida M, Miura I, Yishikawa (1983) Saponin and sapogenol. XXXV. Chemical constituents of Radix Astragali (*A. membranaceus* Bge). Astragalosides I,II,IV, Acetylastragalosid and Isoastragalosides I and II. Chem Pharm Bull 31(2):698–708

Michael JP (1995) Indolizidine and quinolizidine alkaloids. Nat Prod Rep: 535–552

Molyneux RJ, James LF, Panter KE, Ralphs MH (1991) Analysis and distribution of swainsonine and related polyhydroxyindolizidine alkaloids by thin layer chromatography. Phytochem Anal 2:125–129

Nikolov S, Ionkova I, Panova D, Budziliewicz H, Sroeder E (1985) Soyasapogenol B from *Astragalus angustifolius*. CR Acad Bulg Sci 38(7):875–979

Rhodes MJC, Robins JR, Hamil JD, Paar AJ, Walton NJ (1987) Secondery product formation using *Agrobacterium rhizogenes* transformed "hairy roots" cultures. Newsletter N53:2–15

Rizk A (1986) *Astragalus*. In: Rizk A (ed) The phytochemistry of the flora of Qatar, A Copyright, Qatar, pp 246–254

Shimomura K, Sauerwein M, Ishimaru K (1991) Tropane alkaloids in the adventitious and hairy root cultures of solanaceous plants. Phytochemistry 30:2275–2278

Sign MW, Flores HE (1990) The biosynthetic potential of plant roots. Bioessay 12 (N1):7–13

Staba JE (1985) Milestones in plant tissue culture systems for the production of secondary products. J Nat Prod 48(2):203–209

Stepashkina KI (1959) *Astragalus* and their application in clinic practice. Gandjia IM (ed) Med, Kiev, pp 5–104

Stojanov N (1972) *Astragalus glycyphyllos*. In: Ularova K (ed) Our medicinal plants. Nauka and Iskustvo, part I, Sofia, pp 165–166

Tappe R, Budzikiewicz H, Ionkova I, Alfermann AW (1994) Triterpene glycosides from transformed root cultures of *Astragalus mongholicus* Bge. Spectroscopy (IOS Press), 12(1):1–8

Trease GE (1983) Pharmacognosy. Bailliere Tindall, London, PP 358–359

Turova AD (1974) *Astragalus dsyanthus* Pall. In: Medvedev B (ed) Medicinal plants in the USSR and their application. Medicina, Moscow, pp 181–183

Vanhala L, Hiltunen R, Oksman-Caldentey MK (1995) Virulence of different *Agrobacterium* strains on hairy root formation of *Hyoscyamus muticus*. Plant Cell Rep 14:236–240

Verpoorte R, Van der Heijden, Schripsema J (1993) Plant cell biotechnology for the production of alkaloids: present status and prospects, J Nat Prod 56(2):186–207

Zhon QJ (1985) Chinese medicinal herbs in the treatment of viral hepatitis. In: Chang et al. (eds) Advances in Chinese medicinal materials research. World Sci, Singapore, pp 215–219

Zhou Y, Hirotani M, Rui H, Furuya T (1995) Two triglycosidic triterapene astragalosides from hairy root cultures of *Astragalus membranaceus*. Phytochemistry 38:1407–1410

Zorrequieta A, Ugalde R (1986) Formation in *Rhizobium and Agrobacterium* spp. of 235-kilodalton protein intermediate in β-D-(1–2)-glucan synthesis. J Bacteriol 167(3):947–952

V Genetic Transformation of *Atropa belladonna*

M. Jaziri[1], K. Yoshimatsu[2], and K. Shimomura[2]

1 Introduction

Leaves of *Atropa belladonna* L. (Solanaceae) are mainly used as a source for the tropane alkaloids, hyoscyamine and scopolamine, which are not easily produced by chemical synthesis. Thus because of its high morphogenetic potential, the in vitro culture of *A. belladonna* has been actively investigated as an alternative to field cultivation and as a possible source for tropane alkaloid production (for review see Bajaj and Simola 1991). Several studies have confirmed that the product levels in unorganized *A. belladonna* callus and suspension cultures are generally very low and that root morphology is required for synthesis of tropane alkaloid in vitro. However, molecular genetic technology is now available to achieve transformation and regeneration of transgenic *A. belladonna* plants in which morphological traits and/or particular metabolic step(s) are artificially modified. The hairy root system is demonstrated to be much more genetically stable than dedifferentiated normal (callus) or transformed (crown gall) plant cells. In this chapter some important recent studies related to the genetic transformation of *A. belladonna* are discussed.

2 Genetic Transformation

Genetic transformation techniques using *Agrobacterium* sp. have been achieved either to obtain specific organ cultures (hairy root or shooty teratomas) or to affect the metabolism of the plant. These studies are summarized in Table 1.

[1] Université Libre de Bruxelles, Laboratory of Plant Biotechnology, Chaussée de Wavre 1850, B-1160 Brussels, Belgium
[2] Tsukuba Medicinal Plant Research Station, National Institute of Health Sciences, 1 Hachimandai, Tsukuba, Ibaraki, 305 Japan

Biotechnology in Agriculture and Forestry, Vol. 45
Transgenic Medicinal Plants (ed. by Y.P.S. Bajaj)
© Springer-Verlag Berlin Heidelberg 1999

Table 1. Summary of recent studies on genetic transformation of *Atropa belladonna*

Agrobacterium (vector gene)	Objectives and observations	Reference
A. rhizogenes 15834	Tropane alkaloid production	Kamada et al. (1986)
A. tumefaciens B6S3, Bo542, C58	Differentiation, tropane alkaloid production	Ondrej and Protiva (1987)
A. tumefaciens C58C1 (pRiA4b)	Differentiation, tropane alkaloid production	Ondrej and Protiva (1987)
A. rhizogenes 8196	Differentiation, tropane alkaloid production	Ondrej and Protiva (1987)
A. rhizogenes 8196	Kanamycin resistance	Ondrej and Valsak (1987)
A. tumefaciens Bo542, Bo542 (pGA472)	Kanamycin resistance	Ondrej and Valsak (1987)
A. rubi ATCC 13335	Differentiation	Ondrej et al. (1987)
A. rhizogenes A4, 8196	Biomass, tropane alkaloid production	Jung and Tepfer (1987)
A. rhizogenes A4, TR105	Biomass, tropane alkaloid production	Sharp and Doran (1990)
A. tumefaciens C58C1	Promotion of transformation using acetosyringone	Mathews et al. (1990)
A. rhizogenes (pRi 15834 + pARK5)	*bar* gene integration, herbicide-resistant plantlet	Saito et al. (1992)
A. tumefaciens (pAL4404 + pSK223)	*rolC* gene integration, flowering and morphological alteration	Kurioka et al. (1992)
A . tumefaciens LBA4404 (pHY8)	Transfer and expression of the hyoscyamine 6β-hydroxylase Gene in *A. belladonna*	Yun et al. (1992)
A. rhizogenes 15834	Enhancement of scopolamin production	Hashimoto et al. (1993)
A. rhizogenes 15834, MAFF03-01724	Double transformation, morphology and tropane alkaloids	Jaziri et al. (1994)
A. tumefaciens T37, *A. rhizogenes* A4	Synthesis, translocation, biotransformation of metabolites	Subroto et al. (1996a,b)
A. rhizogenes A4 and *A. tumefaciens* R 1000 (pRiA4b)	Genetic transformation of haploid plant	Yoshimatsu et al. (1997)

2.1 Transgenic Organ Cultures

2.1.1 Hairy Root Culture, Growth, and Alkaloid Production

Experiments on transformation of *A. belladonna* using *A. rhizogenes* started in the 1980s with the aim of biomass stimulation and tropane alkaloid production. Kamada et al. (1986) first reported production of hyoscyamine and scopolamine in hairy root cultures of *A. belladonna* at levels the same or higher than in the normal plants grown in the field. The axenic transgenic cultures proliferated 60-fold, based on the initial weight in 1-month culture. Jung and Tepfer (1987) also observed considerable increase in biomass and tropane alkaloid accumulation in transformed root cultures. While there are numerous reports describing growth and tropane alkaloid production in small-scale cultures of transformed roots, the suitability of bioreactor systems (2.5-1 air-lift bioreactor) for cultivation of the *A. belladonna* hairy roots was first demonstrated by Sharp and Doran (1990).

2.1.2 Induction and Culture of Shooty Teratomas

Genetically transformed shoots (shooty teratomas) have been developed for several solanaceneous species, including *A. belladonna* (Saito et al. 1989, 1991; Alvarez et al. 1994; Subroto et al. 1996a). Shooty teratomas are a potential source of chemicals normally synthesized in the aerial parts of the plants. On the other hand, exogenous growth regulators are not required for rapid growth in these organ cultures. Shooty teratomas can be easily obtained in *A. bella-donna* by infecting plants with nopaline strains of *A. tumefaciens* such as T37 or C58 wild-type strains. Genetically modified strains of *A. tumefaciens*, such as disarmed C58 containing a binary vector with the coding sequence of the *ipt* gene fused to the constitutive CaMV 35S promoter with duplicate upstream enhancer sequence, were also successfully used to induce shooty teratomas in *A. belladonna*.

As in hairy root formation, it is likely that an imbalance in the cytokinin/auxin ratio occurs in *A. belladonna* shooty teratomas due to the increase in the expression of the gene coding for indole-3-lactate synthase. The resulting indole-3-lactate may therefore compete with auxin for auxin-binding proteins, which results in the apparent increase in the transformed tissue of the ratio of cytokinin/auxin (Subroto et al. 1996a).

Strains of *A. tumefaciens* bearing mutations in the auxin genes may also induce shooty teratoma formation due to an imbalance in the cytokinin/auxin ratio. Saito et al. (1991) have in fact developed shooty teratomas of *A. belladonna* by using an *aux*⁻ mutant of *A. tumefaciens*. The cultures thus obtained failed to produce tropane alkaloids at a high level, but had the ability to store and metabolize alkaloids. Similarly, Subroto et al. (1996b) demonstrated that shooty teratomas have the ability to convert hyoscyamine to scopolamine.

Interestingly, it is also found that the development of roots significantly improved hyoscyamine synthesis by *A. belladonna* shooty teratomas. Coculture of shooty teratomas and hairy roots in the same hormone-free medium was therefore investigated by Subroto et al. (1996b) as a means of providing a continuous source of hyoscyamine for conversion to scopolamine. The maximum scopolamine concentration in cocultures of shooty teratomas was 0.84 mg/g dry weight, 3–11 times the average levels reported for leaves of the whole plant (Hartmann et al. 1986; Simola et al. 1988). The coculture of genetically transformed organs of *A. belladonna* was described as an effective technique for improving tissue-specific secondary metabolite biosynthesis and biotransformation.

2.2 Transgenic *A. belladonna* Plants

Transgenic *A. belladonna* plantlets have been regenerated from the hairy roots in which wild-type plasmids or artificially engineered genes were induced using *Agrobacterium* strains (Ri-plasmid). Although the regeneration of transgenic plantlets from crown gall tissue is reported to be quite difficult to achieve, *A. belladonna* hairy root cultures have a tendency to generate shoots spontaneously. The phenotypic features of the transformants were found to be stable under successive subcultivation. By using *A. tumefaciens* strain R1000 (pRiA4b), the regenerated plants had wrinkled leaves, shortened internodal length, and decreased apical dominance; the plants regenerated after infection with *A. tumefaciens* LBA4404 (pSK223) showed extensive flowering and extremely decreased apical dominance (Kamada et al. 1986).

2.2.1 Transgenic Herbicide-Resistant A. belladonna *Plants*

The *bar* gene, from *Streptomyces hygroscopicus*, encoding phosphinothricin acetyltransferase, was successfully transferred and expressed in transgenic plants from several plant species including *A. belladonna* (Saito et al. 1992). Phosphinothricin acetyltransferase inactivates the synthetic herbicide phosphinothricin (Basta, Hoechst), which acts as an analogue of L-glutamic acid in plant cells and consequently inhibits glutamine synthase resulting in cell death. The antibiotic herbicide Bialaphos (Herbiace, Meiji Seika Kaisha Ltd.) is a tripeptide containing phosphinothricin and two L-alanine residues produced *by S. hygroscopicus* possessing the *bar* gene. Bialaphos releases phosphinothricin by endogenous peptidase in plant cells and inhibits glutamine synthase (Saito et al. 1992). For the construction of the chimeric *bar* gene for expression in *A. belladonna* plants, the original and unusual start codon (GTG) of the *bar* gene in *by S. hygroscopicus* was replaced with ATG. Then the entire coding sequence of the *bar* gene in the BamHI-SacI fragment of pARK2 was inserted into the corresponding site of the binary vector pBI121 to yield an expression binary vector pARK5. In the T-DNA region of pARK5, the authors have placed the *bar* gene between the CaMV 35S promoter and the *nos* terminator. As a reporter gene for transformation, the chimeric NPT-II gene was included in the T-DNA. Finally, this plasmid was introduced in *A. rhizogenes* harboring a wild Ri plasmid, pRi15834, for plant transformation. *A. rhizogenes* was selected because, unlike crown galls, which are chimeric, hairy roots consist exclusively of transformed cells. In most species, fertile plants can be regenerated from hairy roots.

After the plant transformation procedure, hairy roots were selected in a Bialaphos (5 mg/l)-containing medium. From the selected Bialaphos-resistant hairy root, shoots were regenerated, and then rooted to form plantlets, were transferred on culture soil. The morphological examination of the regenerated plants showed characteristic features caused by T-DNA genes, such as wrinkled leaves and short internodes. The confirmation of the transformation was proved by DNA-blot hybridization and the expression of the *bar* gene in transformed R0 tissues and in backcrossed F_1 progeny with nontransformant and self-fertilized progeny was indicated by enzymatic activity of the acetyltransferase. The transgenic *A. belladonna* plants showed resistance towards Bialaphos and Phosphinotricin. The study of Saito et al. (1992) represents the first successful application of conferring an agronomically useful trait to medicinal plants (in this case, *A. belladonna*) by an Ri plasmid vector. In addition, it was suggested that the expression of the *bar* gene can be an excellent selectable marker in plant cells in which kanamycin selection does not work well.

2.2.2 Transgenic A. belladonna *with an Improved Alkaloid Composition*

With regard to the tropane alkaloid pathway, *A. belladonna* has been used as the transgenic host. The metabolic engineering method developed by Yamada et al. (1994) is a good illustration.

As illustrated in Fig. 1, scopolamine is formed from hyoscyamine via hyoscyamine 6β-hydroxylase (H6H) (EC 1.14.11.11). This enzyme catalyzes the hydroxylation of hyoscyamine to 6β-hydroxyhyoscyamine, as well as the epoxydation of 6β-hydroxyhyoscyamine to scopolamine. Although the epoxydation activity of H6H is much lower than its hydroxylation activity, it was suggested that the epoxydation reaction may not be a limiting factor in planta. Therefore H6H is a promising target enzyme which, if expressed strongly in hyoscyamine-rich plants, would result in increased scopolamine content in the transgenic plants. This strategy was developed by Yun et al. (1992).

The H6H gene of *Hyoscyamus niger* was induced to hyoscyamine-rich *A. belladonna* by a binary vector system using *A. rhizogenes* (Hashimoto et al. 1993). *A. belladonna* leaf explants were infected with *A. rhizogenes* pHY8 (Fig. 2; Yun et al. 1992), and the resulting kanamycin-resistant hairy roots were analyzed by PCR and immunoblot assay. The transformed root clones were found to have the 35S-H6H transgene and elevated levels of H6H polypeptide (six times more H6H polypeptide than wild-type *A. belladonna* hairy root). Alkaloid analysis demonstrated that the levels of scopolamine also increased two- to fivefold and the level of 6β-hydroxyhyoscyamine also increased to some extent. From the established hairy root cultures, the authors have regenerated a transgenic *A. belladonna* plant that expressed the 35S-H6H gene constitutively. The transgenic *A. belladonna* plants expressed relatively high levels of H6H in almost all parts of the plant.

Although the authors used *A. belladonna* as the transgenic host, the metabolic engineering method developed by Yun et al. (1992) is not specifically limited to this plant species. Indeed, several medicinal plants, long known

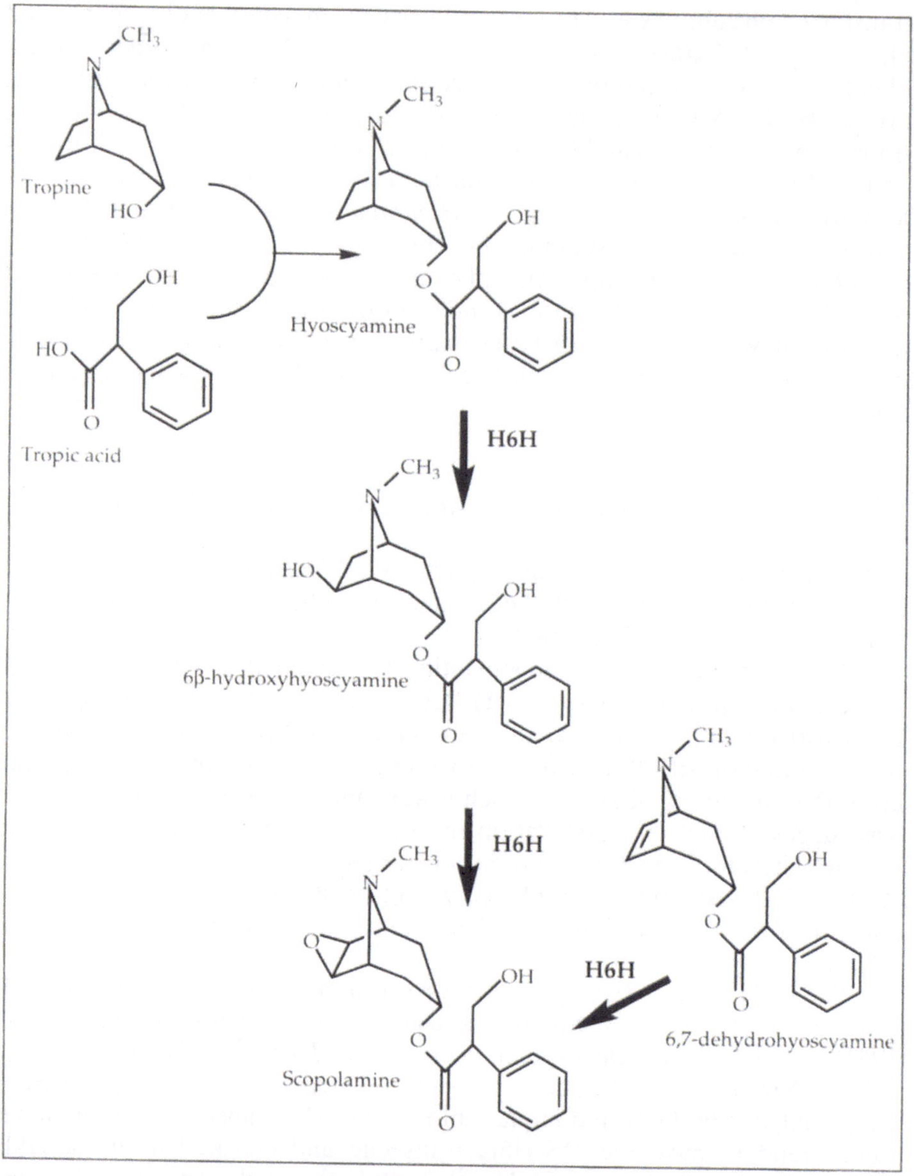

Fig. 1. Hyoscyamine 6β-hydroxylase (H6H), a key enzyme for the conversion of hyoscya mine and derivatives to scopolamine. H6H catalyzes the hydroxylation of hyoscyamine to 6β-hydroxyhyoscyamine, as well as the epoxidation of 6β-hydroxyhyoscyamine to scopolamine. The enzyme was also found to be able to convert 6,7-dehydrohyoscyamine to scopolamine

as a rich source of hyoscyamine but considered unattractive for commercial exploitation because of their low scopolamine contents, may now become promising candidates as sources for scopolamine. In addition, feeding of hyoscyamine to genetically transformed cell or root cultures that overexpress

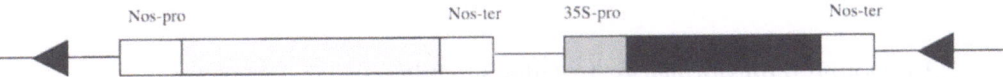

Fig. 2. Binary vector pHY8. This vector was constructed first by placing the H6H cDNA between the CaMV 35S promoter and the nopaline synthase terminator (*nos*), and then by transferring the 35S-H6H chimeric gene to the T-DNA region of the plant-transforming vector pGA482 (Pharmacia). The pHY8 vector contains also in its T-DNA region the neomycin phosphotransferase (NPT II) gene driven by the nopaline synthase promoter, which could produce the kanamycin-resistant phenotype after the transformation procedure. The pHY8 was mobilized to *A. rhizogenes* ATCC 15834

the H6H gene might be an alternative biotechnological approach for scopolamine production.

2.2.3 Rol C Gene Expression Affect Development and Morphogenesis in Transgenic A. belladonna

The four T$_L$ oncogenic root loci or *rol* genes (A–D) of *A. rhizogenes* sensitize the plant cells to phytohormones. They influence signal reception and transduction, and thus affect plant development and morphogenesis (Schmülling et al. 1988). The function of the individual *rol* genes has been assessed by making transgenic plants express one of them. This showed that each gene induces specific developmental abnormalities encountered in the hairy root syndrome (Schmülling et al. 1988).

Kurioka et al. (1992) showed that transgenic *A. belladonna* plants overexpressing *rol* C gene have reduced apical dominance, lanceolated leaves with lowered pigmentation and early formation of flowering as an especially stimulated phenotypic alteration. The *rol* C gene was proposed to encode a cytokinin-β-glucosidase, which in vitro can release free cytokinin from the stable conjugated N-glucosides which show no phytohormone activity (Spena et al. 1987; Estruch et al. 1991ab). It was thus hypothesized that *rol* C would increase cytokinin levels in plants. In this respect, the 35S-*rol* C *A. belladonna* plants described by Kurioka et al. (1992) exhibit promotion of flowering in terms of both number of flower buds and time of flowering. The phenotypic alterations apparently induced by the *rol* C gene were observed in the transgenic *A. belladonna* under the control of the CaMV 35S promoter, but not of the native *rol* C promoter. Northern hybridization analyses showed that the *rol* C gene driven by the CaMV 35S promoter was expressed strongly in leaves, flowers, stems, and roots, while the *rol* C gene with its own promoter was expressed only at low levels in all organs examined. The authors conclude that not only the altered cytokinin balance is at the origin of the flowering promotion but additional or other effects of *rol* C gene are implicated. This observation is quite interesting: indeed, further detailed studies related to the expression pattern of the *rol* C gene in tissues, especially in meristems for

flowering, could be important for the occurrence of the observed phenotype in
A. belladonna.

2.2.4 *Doubly Transformed* A. belladonna *Cultures*

The morphogenetic events associated with the genetic transformation using
Agrobacterium are due to the transfer, integration, and expression of the T-
DNA, which cause metabolic changes involved in phytohormone synthesis
and metabolism (Zambryski et al. 1989). Two regions, TL-and TR-DNA, were
found to be integrated and stably conserved in the plant genome (Jouanin
1984). The TL-and/or the TR-DNA may be present in a different copy
number(s) (Merlo et al. 1980; Thomashow et al. 1980). Until now the exact
number of the T-DNA fragments as well as their position in the plant genome
have not been clarified. Theoretically, a large number of T-DNA copies can be
integrated into the plant genome, but the possible (maximum) number of
integrations is still obscure. Recently, Jaziri et al. (1994) have contributed
toward the solution of the questions of whether the number of the T-DNA
copies in the plant genome can be increased and additional traits can be
induced by retransformation with *A. rhizogenes* and how the double transformation may affect the primary and secondary metabolism of plant cells. *A.
belladonna* has been selected and used as model plant because of its
morphogenetic potential.

Hairy root cultures of *A. belladonna* were established by infection with
either *A. rhizogenes* ATCC 15834 or MAFF 03-01724, and transgenic plants
were obtained from both hairy root cultures (Fig. 3). The transgenic plants
showed some morphological differences as compared to the nontransformed
plants: thick and wrinkled leaves with apparent high chlorophyll content and
short internodes for the transgenic plants from the ATCC 15834 strain;
lanceolated and smaller pale green-colored leaves for those obtained after
infection with the MAFF 03-01724 strain (Fig. 4).

The transformed cultures were then used for the establishment of
the doubly transformed cultures. In order to confirm the integration of the
plasmid DNA after the double transformation, the bacteria selected for
the second infection are different from the first bacteria used to obtain the
transgenic plants.

The genetic transformation of the doubly transformed cultures was
proved by the opine assay and PCR analysis (Table 2). The opine assay
showed that only one hairy root clone out of six clones tested was doubly
transformed, and it contained agropine, mannopine (for *A. rhizogenes* ATCC
15834), and mikimopine (for *A. rhizogenes* MAFF 03-01724). The adventitious shoots and viviparous leaves were also found to be doubly transformed.
The PCR analysis of doubly transformed clone show that T-DNA from both
bacteria used was integrated.

Interestingly, the analysis of phytohormones in the cultures obtained
showed that no trace of the free and conjugated form of IAA was detected
in the doubly transformed cultures while simply transformed culture contain

Fig. 3. Transgenic *A. belladonna* shoots regenerated from hairy root cultures

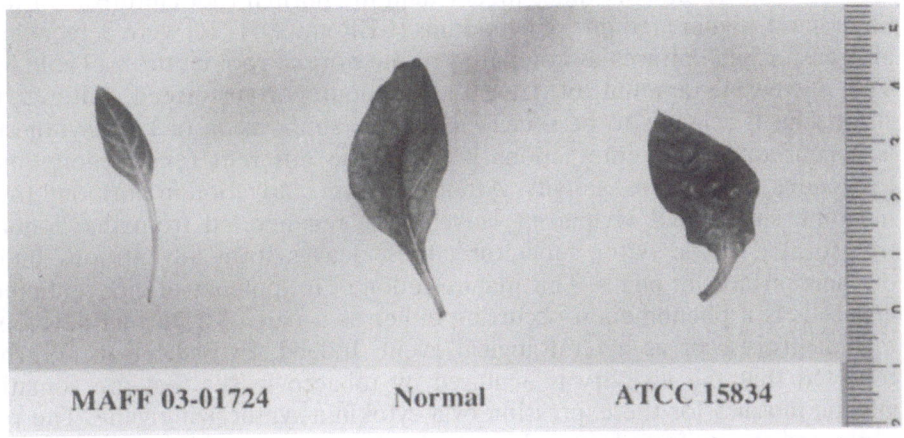

Fig. 4. Alteration in the leaf morphology of transgenic *A. bellabonna* plants transformed by two different *A. rhizogenes* strains

Table 2. Summary of data concerning the genetic transformation of normal, transformed, and doubly transformed root clones of *A. belladonna*

Root clone	Opine assay		PCR analysis		
	agr[a]/man[b]	mik[c]	T-DNA MAFF	TL-DNA ATCC	TR-DNA ATCC
Normal root	−	−	−	−	−
Root transformed with *A. rhizogenes* ATCC 15834	+	−	−	+	−
Root transformed with *A. rhizogenes* MAFF 03-01724	−	+	+	−	−
Doubly transformed cultures	+	+	+	+	−

+/−: presence/absence.

[a] agr: agropine; [b] man: mannopine; [c] mik: mikimopine.

Table 3. IAA and t-ZR content in normal and transformed root cultures of *A. belladonna* (pmol/g fr wt.)

Root strain	IAA	t-ZR
Normal root[a]	23.2	2.79
Root transformed with *A. rhizogenes* ATCC 15834	57.3	25.86
Root transformed with *A. rhizogenes* MAFF 03-01724	77.7	14.86
Doubly transformed root cultures	Id[b]	23.00

[a] Normal root cultured on 1/2 MS solid medium containing NAA (0.5 mg/l).

[b] 1d: value lower than the limit detection of the assay.

level of IAA 2.5 and 3.4 times higher than the normal root cultures. On the other hand, higher amounts of cytokinins (t-ZR and DHZR) were detected in all transformed cultures as compared to the normal root cultures (Table 3). The decreasing amount of IAA in the doubly transformed cultures is surprising. It remains to be seen whether the suppression of IAA synthesis is a consequence of interactions between the different (endogenous and exogenous) IAA genes' activity. After long-term cultivation in hormone-free medium, shoots and viviparous leaves were regenerated from the doubly transformed roots. After subculture, these leaves form adventitious buds on their surfaces or edges. This manifestation of totipotency of differentiated leaf cells is a phenomenon occurring either as a part of a normal developmental process or as a teratological event. Indeed, Estruch et al. (1991b) reported that the vivipary is acquired by tobacco leaves that are somatic genetic mosaics for the expression of a cytokinin-synthesizing gene. The regenerated shoots from leaves thus obtained after the double transformation of *A. belladonna* exhibited loss of apical dominance and were unable to root

Fig. 5. Doubly transformed regenerated *A. belladonna* plant; the plants exhibited loss of apical dominance

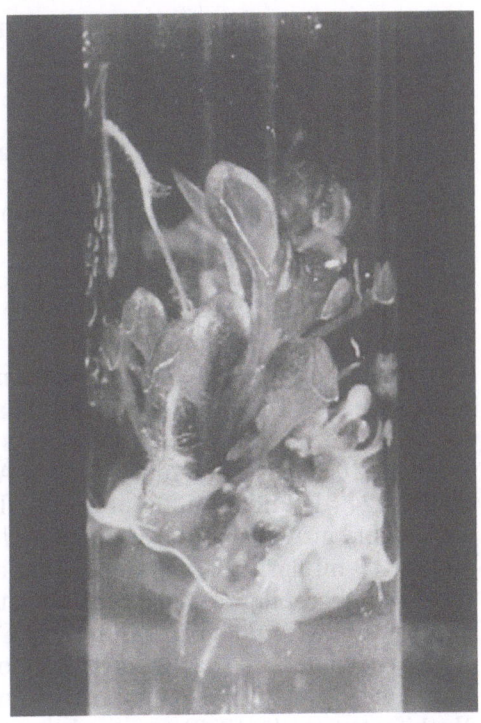

(Fig. 5). This observation is also reported in *Nicotiana* spp. in which the expression of the *ipt* gene under the control of the 35S RNA promoter increases the cytokinin content up to 137 times that in transgenic shoots (Estruch et al. 1991a). In the case of the doubly transformed *A. belladonna* cultures, the formation of the viviparous leaves is a direct consequence of a phytohormone imbalance which might be caused by transcriptional inactivation of the different IAA genes expression and/or by a change in T-DNA by methylation and expression which could be associated with the phenotypic variation observed (vivipary) and phytohormone content (especially the decrease in the IAA level).

The growth and the alkaloid content in hairy roots of *A. belladonna* transformed with *A. rhizogenes* ATCC 15834 and/or MAFF 03-01724 were determined after 3 weeks' culture in hormone-free 1/2 MS liquid medium. The highest content of alkaloids, particularly hyoscyamine, was found in hairy root strain transformed with *A. rhizogenes* ATCC 15834. On the other hand, the growth of this strain was very slow (0.63 g dry wt./flask) as compared to the hairy root transformed with *A. rhizogenes* MAFF 03-01724, and to the doubly transformed cultures strain (2.35 g dry wt./flask for both strains). The integration of the T-DNAs from both bacteria affect the growth and the alkaloid production of the hairy root cultures. The doubly transformed roots demonstrated alkaloid contents as high as the hairy roots (*A. rhizogenes*

ATCC 15834) and growth as good as the hairy roots (*A. rhizogenes* MAFF 03-01724).

2.2.5 Transgenic Haploid A. belladonna *Plants*

Although experiments of haploid cultures of *A. belladonna* have been done (Bajaj et al. 1978), until now most transformed *A. belladonna* cultures reported in the literature were initiated from a diploid plant material. Recently, Yoshimatsu et al. (1997) reported on morphological traits and tropane alkaloid production in haploid transgenic *A. belladonna* organ cultures and plants. The Ri plasmid of *A. rhizogenes* (strain A4) and the insertion and deletion mutants have been used for the establishment of haploid and diploid transformed cultures. The haploid hairy roots were induced by R1000 (wild-type T-DNA), R1022 (insertion mutant of *rol* A), R1023 (insertion mutant of *rol* C), and R1224 (deletion mutant of *rol* D). The haploid hairy root clones showed satisfactory growth and the accumulated tropane alkaloids ranged from 200 to 400 µg/100-ml flask. After examination and description of the morphological traits of the regenerated plantlets, the authors demonstrated that the coexistence of the two *rol* loci (*rol* B and *rol* C) determined expression of the hairy root syndrome. Especially dwarf shoots are reported to be much more pronounced in haploid than in diploid transgenic plants. Furthermore, Yoshimatsu et al. (1997) reported that the diploid *rol* B and -*rol* C plantlets showed nontransformant-like shoots, while the haploid -*rol* B and -*rol* C plantlets demonstrated a strong hairy root syndrome. These results indicate that the hairy root syndrome caused by T-DAN was expressed strongly and was not suppressed in the haploid transgenic plants, as previously observed in the diploid plants. The authors concluded that the T-DNA of the Ri plasmid was introduced in only one chromosome. Therefore *Agrobacterium*-mediated transformation of haploid *A. belladonna* plants may be more promising for molecular breeding.

3 Summary and Conclusions

Because of its high morphogenetic potential, *A. belladonna* has been used as a model system in several laboratories. Genetic transformation techniques using *Agrobacterium* sp. have been investigated either to establish particular organ cultures (such as hairy root or shooty teratomas) or to affect the metabolism of the plant (conferring agronomically useful traits or improving the secondary metabolism biosynthetic pathway). Experiments on transformation of *A. belladonna* started in the 1980s; wild-type Ti or Ri plasmids were used with the aim of biomass and tropane alkaloid production. Especially genetically transformed hairy roots have been researched extensively for the synthesis of hyoscyamine and closely related compounds. More recently, genetically transformed shoots of *A. belladonna* (shooty teratomas) have been developed. The alkaloid production in the shooty teratomas cultures was limited; development of roots significantly improved hyoscyamine synthesis.

Transgenic plantlets were successfully regenerated from hairy roots. They showed altered phenotypes such as high root growth rate, wrinkled leaves and other specific morphological features according to the bacteria used for the transformation. In this respect, the morphological and biochemical traits of doubly transformed *A. belladonna* organ cultures and regenerated plantlets were strongly affected; viviparous leaves were observed associated with an increase in cytokinin levels and a significant decrease in endogenous auxin levels.

Artificially engineered genes were also induced using *Agrobacterium* strains. Transgenic *A. belladonna* conferred with a herbicide-resistant trait (chimeric construct, pARK5 containing the *bar* gene encoding phosphinotricin acetyltransferase flanked with the CaMV 35S promoter) was obtained by transformation with an Ri plasmid binary vector and plant regeneration from hairy roots. The transgenic plants showed resistance towards Pialaphos and Phosphinotricin.

Transgenic *A. belladonna* with an improved tropane alkaloid composition was also obtained following the introduction of an hydroxylase gene from *Hyoscyamus niger* under the control of the CaMV 35S promoter into a hyoscyamine-rich *A. belladonna* by the use of the *Agrobacterium*-mediated transformation system. In the transformant and its selfed progeny that inherited the transgene, the alkaloid contents of the leaf and stem were almost exclusively scopolamine. Finally, the results obtained with the transgenic haploid *A. belladonna* plants indicate that the T-DNA of the Ri plasmid was introduced in one chromosome.

The results discussed above enable to understand the developmental and physiological processes involved in the pathogenesis of crown gall and hairy root disease and to a large extent to the application of the recombinant DNA technology in improving agronomic traits and yields of useful secondary products in *A. belladonna* plants.

Acknowledgments. Mondher Jaziri is an associate reasearcher of the Fonds National de la. Recherche Scientifique, Belgium.

References

Alvarez MA, Talou JR, Paniego NB, Giulietti AM (1994) Solasodine production in transformed organ cultures (roots and shoots) of *Solanum eleagnifolium* Cav. Biotechnol Lett 16:393–396

Bajaj YPS, Simola LK (1991) *Atropa belladonna* L.: in vitro culture, regeneration of plants, cryopreservation, and the production of tropane alkaloids. In: Bajaj YPS (ed) Biotechnology in agriculture and forestry, vol 15. Medicinal and aromatic plants III. Springer Berlin Heidelberg New York, pp 1–23

Bajaj YPS, Gosch G, Ottma M, Weber A, Gröbler A (1978) Production of polyploid and aneuploid plants from anthers and mesopyll protoplasts of *Atropa belladonna* and *Nicotiana tabacum*. Indian J Exp Biol 16:947–953

Estruch JJ, Chriqui D, Grossmann K, Schell J, Spena A (1991a) The plant oncogene *rol C* is responsible for the release of cytokinins from glycoside conjugates. EMBO J 10:2889–2895

Estruch JJ, Prinsen E, Van Onckelen H, Schell J, Spena A (1991b) Viviparous leaves produced by somatic activation of an active *Agrobacterium rhizogenes* influence plant development. EMBO J 7:2621–2629

Hartman T, Witte L, Oprach F, Toppel G (1986) Reinvestigation of the alkaloid composition of *Atropa belladonna* plants, root cultures, and cell suspension cultures. Planta Med 52:390–395

Hashimoto T, Yun D-Jin, Yamada Y (1993) Production of tropane alkaloids in genetically engineered root cultures. Phytochemistry 32:713–718

Jaziri M, Yoshimatsu K, Homès J, Shimomura K (1994) Traits of transgenic *Atropa belladonna* doubly transformed with different *Agrobacterium rhizogenes* strains. Plant Cell Tissue Organ Cult 38:257–262

Jouanin L (1984) Restriction map of agropine-type Ri plasmid and its homologies with Ti plasmids. Plasmid 12:91–102

Jung G, Tepfer D (1987) Use of genetic transformation by the Ri T-DNA of *Agrobacterium rhizogenes* to stimulate biomass and tropane alkaloid production in *Atropa belladonna* and *Calystegia sepium* roots grown in vitro. Plant Sci 50:145–151

Kamada H, Okamura N, Satake M, Harada H, Shimomura K (1986) Alkaloid production by hairy root cultures in *Atropa belladonna*. Plant Cell Rep 5:239–242

Kurioka Y, Suzuki Y, Kamada H, Harada H (1992) Promotion of flowering and morphological alterations in *Atropa belladonna* transformed with a CaMV 35S-rolC chimeric gene of the Ri plasmid. Plant Cell Rep 12:1–6

Mathews H, Bharathan N, Litz RE, Narayanan KR, Rao PS, Bhatia CR (1990) The promotion of *Agrobacterium*-mediated transformation in *Atropa belladonna* L. by acetosyringone. J Plant Physiol 136:404–409

Merlo DJ, Nutter RC, Montaya AL, Garfinkel DJ, Drummond MH, Chilton MD, Gordon MP, Nester EW (1980) The boundaries and copy numbers of Ti plasmid T-DNA vary in crown gall tumors. Mol Gen Genet 177:637–643

Ondrej M, Protiva J (1987) In vitro culture of crown gall and hairy root tumors of *Atropa belladonna*: differentiation and alkaloid production. Biol Plant 29:241–246

Ondrej M, Valsak J (1987) Expression of kanamycin resistance introduced by *Agrobacterium* binary vector into *Nicotiana tabacum* and *Atropa belladonna*. Biol Plant 29:161–166

Ondrej M, Matousek J, Valsak J (1987) Differentiation of transformed plants from tumors induced by *Agrobacterium rubi* ATCC 13335. J Plant Physiol 126:397–407

Saito K, Murakoshi I, Inzé D, Van Montagu M (1989) Biotransformation of nicotine alkaloids by tobacco shooty teratomas induced by a Ti plasmid mutant. Plant Cell Rep 7:607–610

Saito K, Yamazaki M, Kawaguchi A, Murakoshi I (1991) Metabolism of solanaceous alkaloids in transgenic plant teratomas integrated with genetically engineered genes. Tetrahedron 47:5955–5968

Saito K, Yamazaki M, Anzai H, Yoneyama K, Murakoshi I (1992) Transgenic herbicide-resistant *Atropa belladonna* using an Ri binary vector and inheritance of the transgenic trait. Plant Cell Rep 11:219–224

Schmülling T, Schell J, Spena A (1988) Single genes from *Agrobacterium rhizogenes* influence plant development. EMBO J 7:2621–2629

Sharp JM, Doran PM (1990) Characteristics of growth and tropane alkaloid synthesis in *Atropa belladonna* roots transformed by *Agrobacterium rhizogenes*. J Biotechnol 16:171–186

Simola LK, Nieminen S, Huhtikangas A, Ylinen M, Naaranlahti T, Lounasmaa M (1988) Tropane alkaloids from *Atropa belladonna*, Part II. Interaction of origin, age, and environment in alkaloid production of callus cultures. J Nat Prod 51:234–242

Spena A, Schmülling T, Konez C, Schell JS (1987) Independent and synergic activity of rol A, B and C loci in stimulating abnormal growth in plant. EMBO J 6:3897–3899

Subroto MA, Hamill JD, Doran PM (1996a) Development of shooty teratomas from several solanaceous plants: growth kinetics, stoichiometry and alkaloid production. J Biotechnol 45:45–57

Subroto MA, Kwok KH, Hamill JD, Doran PM (1996b) Coculture of genetically transformed roots and shoots for synthesis, translocation, and biotransformation of secondary metabolites. Biotechnol Bioeng 49:481–494

Thomashow MF, Nutter R, Postle K, Chilton MD, Chilton FR, Blattner A, Powell MP, Gordon MP, Nester EW (1980) Recombination between higher plant DNA and Ti plasmid of *Agrobacterium tumefaciens*. Proc Natl Acad Sci USA 77:6448–6452

Yamada Y, Yun DJ, Hashimoto T (1994) Genetic engineering of medicinal plants for tropane alkaloid production. Adv Plant Biotechnol 4:83–93

Yoshimatsu K, Jaziri M, Kamada H, Shimomura K (1997) Production of diploid and haploid transgenic *Atropa belladonna* plants: morphological traits and tropane alkaloid production. Bel J Bot 130:38–46

Yun DJ, Hashimoto T, Yamada Y (1992) Metabolic engineering of medicinal plants: transgenic *Atropa belladonna* with an improved alkaloid composition. Proc Natl Acad Sci USA 89:11799–11803

Zambryski P, Tempé J, Schell J (1989) Transfer and function of T-DNA genes from *Agrobacterium* Ti and Ri plasmids in plants. Cell 56:193–201

VI Genetic Transformation of *Catharanthus roseus* (Periwinkle)

F. Garnier[1], S. Hamdi[1], P. Label[2], and M. Rideau[1]

1 Introduction

1.1 Distribution and Importance of the Plant

Catharanthus roseus (L.) G. Don (= *Vinca rosea* L.), the Madagascar periwinkle, is a pantropical erect subshrub, belonging to the family Apocynaceae. The plant accumulates more than 100 indole alkaloids, mainly the monomeric alkaloids ajmalicine, serpentine, vindoline, and catharanthine. Ajmalicine is an antihypertensive drug used in the treatment of circulatory diseases. Vindoline and catharanthine are the precursors of dimeric alkaloids, such as vinblastine and vincristine, the importance of which in the treatment of acute leukemia was established in the 1960s (Van Tellingen et al. 1992). Indeed, these two alkaloids were the first plant products to be approved by the FDA for cancer treatment (Noble 1990). They accumulate in aerial parts of the plant but only in very low amounts; thus they are very costly.

A tremendous amount of information has been generated concerning the growth of cell suspensions, the regulation of their indole alkaloid production, the signals that trigger the alkaloid biosynthesis, and the capacities of the cells to biotransform exogenous substrates; comprehensive reviews have been published on these subjects in the past decade (Van der Heijden et al. 1989; Ganapathi and Kargi 1990; Moreno et al. 1995). At present, the production of dimeric alkaloids has not been achieved (or only in traces) in vitro, due to the inability of the nonorganogenic cultures to produce vindoline. However, monomeric alkaloids such as ajmalicine, serpentine, or catharanthine can be produced by suspension-cultured cells, although in a noneconomically viable process. On the other hand, cell cultures of periwinkle are very easy to grow in vitro and therefore are highly suitable for plant physiological studies because of the availability of a number of biochemical and molecular data on these cells.

[1] Laboratoire de Biologie Moléculaire et Biochimie Végétale, EA 2106, Faculté de Pharmacie, 31 avenue Monge, 37200 Tours, France
[2] INRA, Centre de Recherches d'Orléans, Station d'Amélioration des Arbres Forestiers, 45160 Orléans, France

Biotechnology in Agriculture and Forestry, Vol. 45
Transgenic Medicinal Plants (ed. by Y.P.S. Bajaj)
© Springer-Verlag Berlin Heidelberg 1999

1.2 Need for Genetic Transformation

Until now, transformed tissues of periwinkle have been investigated with at least three objectives.

1. To compare the physiological and biochemical traits of the tumorous tissues (obtained by transformation with virulent strains of *Agrobacterium tumefaciens*) with those of normal tissues. Differences in cell permeability were early documented (Wood and Braun 1965; Lenz et al. 1979). One crown-gall line (A6 line) was also proved to be a very suitable system for investigating the biosynthesis and metabolism of cytokinins (CKs) (Miller 1974; Stuchbury et al. 1979; McGaw and Horgan 1983; Palni and Horgan 1983; Palni et al. 1983). Two phenolic molecules (dehydrodiconiferyl alcohol glucosides A and B) have been characterized as CK-like factors in periwinkle crown-gall tumors (Lynn et al. 1987).

2. To investigate whether tumorous periwinkle tissues can accumulate the pharmacologically important indole alkaloids usually found in the whole plant. Boder et al. (1964) were the first to detect small amounts of indole alkaloids in a suspension culture derived from a crown-gall callus. Then, Eilert et al. (1987) also found traces of alkaloids in a crown-gall line and proved that the autonomous cells became indifferent to treatments that trigger alkaloid biosynthesis in normal cells. In contrast, Kodja et al. (1989) showed that exogenously applied CK induced alkaloid accumulation in a tumorous callus line of periwinkle. Recently, O'Keefe et al. (1997) has provided evidence of a stable production of vindoline in several undifferentiated lines of periwinkle after transformation with *A. tumefaciens* (strains A 281 and B 0542) or *A. rhizogenes* (strain K599). On the other hand, periwinkle hairy roots produce a wide variety of indole alkaloids (Toivonen et al. 1991; Ciau-Uitz et al. 1994), including vindoline (Bhadra et al. 1993) and catharanthine (Islas et al. 1994; Vasquez-Flota et al. 1994), and the production can be enhanced by growth factors (Bhadra et al. 1993; Vasquez-Flota et al. 1994) as well as by elicitor treatments combined with the adsorption of the alkaloids released in the medium on XAD resins (Sim et al. 1994). The cell suspension / hairy root interchange system described by Jung et al. (1995) is of interest because it combines high biomass (in the form of cells) and high alkaloid production (in the form of hairy roots).

3. To obtain cell suspensions, organs, or whole plants transgenic for a gene of interest. Overexpression of specific genes in periwinkle may represent a good procedure to improve the alkaloid production, or create new pathways for the synthesis of new pharmaceutical products. In a recent work, Goddijn et al. (1995) showed that the tryptophane decarboxylase transcripts were rather undetectable in crown-gall cultures of periwinkle. They attempted to bypass the putative limitation in tryptamine by generating crown-gall tissues with an *A. tumefaciens* strain carrying a *tdc* gene under control of a strong constitutive promoter. However, overexpression of the *tdc* cDNA in these tumorous tissues resulted in the accumulation of tryptamine, but not of indole alkaloids.

2 Genetic Transformation

2.1 General Account and Brief Review

The first transformation of *Catharanthus roseus* was achieved by Braun in 1943 and White in 1945. Young plants were inoculated by multiple needle punctures with a suspension of a highly virulent strain of *Agrobacterium tumefaciens*. In this report, the growing crown galls were rendered bacteria-free by thermal treatment of the inoculated plants at 46–47 °C for 5 days. After growing the plants for a further 10 weeks at 25 °C, the galls were excised and the tumorous tissues were transferred onto White's medium and maintained in vitro as callus cultures. The exposure to high temperature and humidity was reported to kill most of the *Agrobacteria* and to decrease the virulence of those organisms that remain viable (Theis et al. 1950). A similar method was used by Manasse and Lipetz (1971) to obtain crown-gall tissue cultures of *C. roseus* but, later, addition of antibiotics to the growth medium was preferred for killing the bacteria (Okada et al. 1985; Eilert et al. 1987; Kodja et al. 1989). As shown in Table 1, hairy roots have been obtained by several investigators, after transformation of leaves, stems, or normal roots by the soil microorganism *Agrobacterium rhizogenes*.

Periwinkle has the reputation of being recalcitrant to transformation by disarmed *A. tumefaciens* strains, therefore, other procedures have been proposed for genetic engineering of this species. Hasezawa et al. (1981) and Okada et al. (1985) gave evidence for the transformation of protoplasts by spheroplasts of *A. tumefaciens* after treatment with polyethyleneglycol or polyvinyl alcohol and subsequent washings with a high pH–high Ca buffer. Transient and stable transformation of suspension-cultured cells by particle bombardment have been reported by Van der Fits and Memelink (1997), who showed that several CaMV 35S promoter derivatives can confer high levels of expression in periwinkle cells.

The objective of the present work was the following. We previously reported that adding CK to the culture medium of the periwinkle cell line C20 enhanced alkaloid production (Décendit et al. 1992). Our aim was to investigate whether introduction of a CK-biosynthetic gene into the periwinkle genome could also increase the alkaloid content. We also attempted to transform the suspension cells by coculture with an *Agrobacterium tumefaciens* carrying the *ipt* gene, but this was prevented by a severe hypersensibility response that completely prohibited the recovery of viable cells (Garnier et al. 1996a). Therefore, as new material for the establishment of normal (untransformed) cultures as well as *ipt*-transgenic cultures, periwinkle cotyledons were used.

Table 1. Summary of various transformation studies conducted on *Catharanthus roseus* (L.) G. Don (= *Vinca rosea* L.)

Vector used	Explant/tissue culture used	Results/remarks	Reference
Brown-Peach strain of *A. tumefaciens*	"Young healthy plants"	– Bacteria-free tissues obtained by thermal treatment (46–47 °C; 5 days) – Cult. in vitro: PGR-free White's medium	Braun (1943); White (1945)
A. tumefaciens strains B6 and B6[3e]	Greenhouse-grown plants 10–15 cm tall var. "Purity white"	– Thermal treatment (41 °C, 7 days) – Cult. in vitro: 10× PGR-free White's medium	Manasse and Lipetz (1971)
A. tumefaciens strain A208 (nopaline-type Ti plasmids)	a) Stems (development of crown galls)	– Elimination of bacteria: vancomycin 100 mg/l – Cult. in vitro: PGR-free MS medium	Okada et al. (1985)
	b) Mesophyll protoplasts	– Introduction of bacterial spheroplasts into protoplasts via PEG or PVA treatments and high pH–high Ca buffer washings – Transf evid: nopaline synthase activity; Southern hybridization	
A. tumefaciens strain Ti C58 (nopaline-type Ti plasmids)	Disks (5 mm diam.) from young fully expanded leaves	– Cult. in vitro: PGR-free B5 medium with carbenicillin (500 mg/l) – Transf evid: nopaline synthase transcript level	Eilert et al. (1987)
A. tumefaciens strain C58-587/A (nopaline-type Ti plasmids)	8-week-old axenic plantlets var. Little Pinkie (development of crown galls)	– Cult. in vitro: PGR-free B5 medium with cefotaxime (500 mg/l) – Transf evid: nopaline synthase assay	Kodja et al. (1989)
A. rhizogenes strains	Leaves and stems from 2-month-old seedlings Development of hairy roots	– Elimination of bacteria: carbenicillin or cephatoxime – Cult. in vitro of hairy roots: 0.5 × B5 medium – Trans evid: Southern blot, the cultures produced indole alkaloids	Ciau-Uitz et al. (1994)
A. rhizogenes strain 1855 pBI 121.1	Roots (development of hairy roots)	– Cult. in vitro of hairy roots: 0.5 × B5 medium	Islas et al. (1994)
A. rhizogenes strain 15834	Seedling segments	– Initiation of thizogenic callus from hairy root (Schenk and Hildebrandt medium with 1 mg/l NAA + 0.1 mg/l kinetin) – Transf evid: production of opines	Jung et al. (1995)

Table 1. *Continued*

Vector used	Explant/tissue culture used	Results/remarks	Reference
		The rhizogenic callus can give hairy roots or cell suspension. Hairy roots produced catharanthine at high level	
Oncogenic *A. tumefaciens* strain LBA 1010 (pTi B6 plasmid in a C58-C9 background) electroporated with *tac* cDNA linked to Ca MV 355 promoter and terminator	Hypocotyls from 3-week-old seedlings (development of crown galls)	– Cult. in vitro: LS medium with kanamycin (100 mg/l) as selection agent – Transf evid: Western blot analysis of *tdc* level transcripts	Goddijn et al. (1995)
A. tumefaciens strain containing PGV 2492 or PVG2488 plasmids (Beinsberger et al. 1991)	Cotyledons from 15-day-old seedlings (var. Little Pinkie)	– Cult. in vitro: B5 medium with 2,4-D (1 mg/l), cefotaxime, (500 mg/l) for elimination of bacteria and geneticin (15 mg/l: selection agent) – Transf evid: *ipt* transcript level	Garnier et al. (1996a)
A. tumefaciens (strains A281 and BO542) *A. rhizogenes* (strain K599)	Leaves from field-grown plants	– Cult. in vitro: PGR-free B5 medium with 1000 mg/l carbenicillin – transf evid: Southern blot analysis	O'Keefe et al. (1997)
A mixture of *gus* A-reporter gene and a hygromycin-selection gene (pGL2 containing 35S-*hph*)	Cell suspensions	– Transformation by particle bombardment (stably transformed lines) – Selection agent: hygromycin B (50 mg/l) – Transf evid: measurement of GUS and CAT activity	Van der Fits and Memelink (1997)

Abbreviations: Cult. in vitro: culture in vitro; *gus*: bacterial β-glucuronidase gene; GUS: β-glucuronidase; *ipt*: isopentenyl transferase gene; PEG: polyethylene glycol; PGR: plant growth regulator; PVA: polyvinyl alcohol; Transf evid: evidence for transformation; *tdc*: tryptophane decarboxylase gene.

2.2 Methodology

2.2.1 Establishment and Maintenance of Untransformed Callus Cultures

Cotyledons excised from 15-day-old aseptically grown seedlings of periwinkle (cv. Little Pinkie) were cultured on a callus-induction B5 medium (Gamborg et al. 1968) containing 58 mM sucrose, 4.5 µM 2,4-dichlorophenoxyacetic acid (2,4-D), 5 µM 6-benzylaminopurine and 0.8% (w/v) agar, at 24 °C. After 1 month, calli were transferred onto the above medium without CK, and then subcultured every month for about 2 years before analysis.

2.2.2 Establishment and Maintenance of ipt-Transgenic Callus Cultures

Two *Agrobacterium tumefaciens* strains with either the recombinant Ti plasmid pGV2488 or the recombinant Ti plasmid pGV2492 (Beinsberger et al. 1991) were used. The plasmid pGV2492 carries the nopaline-*ipt* gene under the control of its genuine promoter and the neomycine phosphotransferase II gene under the control of the nopaline synthase promoter. The recombinant Ti plasmid pGV2488 contains the same neomycine phosphotransferase II gene and the octopine-*ipt* gene linked to the light-inducible promoter of the gene encoding a small subunit protein (ssu) of ribulose 1,5-bisphosphate carboxylase/oxygenase from *Pisum sativum*. The bacteria were grown overnight in LB liquid medium, the suspension was then diluted with B5 medium to a concentration of 10^8 bacteria/ml.

The transformation procedure was as follows: cotyledons excised from 15-day-old periwinkle seedlings were floated on diluted suspension of bacteria for 2 days, then washed four times with B5 medium. Excess medium was shaken off before the cotyledons were transferred onto B5 agar medium supplemented with 4.5 µM 2,4-D, 500 mg/l cefotaxime and 15 mg/l geneticin. Geneticin was used as selection agent instead of kanamycin because using the latter at high concentration (300 mg/l) inhibited by only 50% the growth of untransformed tissues. Geneticin-resistant calli that formed within the next 3 weeks were individually transferred to the same medium as above, then subcultured every month, in the dark. Antibiotics were removed after 6 months. At that time, 2,4-D was also removed from the medium because it was found to slow down tissue growth. The transgenic callus cultures containing the Pssu-*ipt* construction were referred to as 92/2, 92/3, 92/4, and 92/5; those containing the Pipt-*ipt* construction were referred to as 88/1 and 88/2. All the transgenic lines were subcultured on auxin-free medium, at 24 °C, for a further 18-month period before analysis.

2.2.3 DNA Isolation and Southern Hydridization

Total genomic DNA from putatively transformed and untransformed tissues was isolated according to Graham (1978). Twenty µg of genomic DNA was digested with *BamH*I overnight at 37 °C. DNA fragments were separated by electrophoresis in 0.8% (w/v) agarose gels and transferred to Hybond N⁺ filters (Amersham). The probe for the *ipt* gene was the 850-bp *BamH*I/*Hpa*I *ipt* gene fragment released from the recombinant Ti plasmid pGV2492. The filters were prehybridized in a solution composed of 6x SSC, 0.5% (w/v) SDS, 5x Denhardt, and 100 µg/ml single-stranded herring sperm DNA, then hybridized for 18 h at 65 °C with a ^{32}P-labeled probe (specific activity, about 3 × 10^9 cpm/µg). Labeling was done by the multiprime method (Amersham). After hybridization, the filters were washed and radioactivity was visualized by autoradiography with Kodak XR-OMAT films.

2.2.4 Monitoring of Expression of the ipt Gene

Ipt transgenic and untransformed tissues were subcultured on a 2,4-D-free B5 agar medium, and grown either in darkness or in continuous white light (Sylvania Grolux tubes, 18 µmol/m^2/s) at 24 °C. Tissues were harvested on the 11th day for Northern analysis and endogenous CK quantitation. They were frozen in liquid nitrogen immediately after harvesting and stored at −80 °C (Northern analysis), or freeze-dried (CK quantitation).

2.2.5 RNA Isolation and Northern Hybridization

Total RNA was extracted from 3 g (fresh mass) callus tissues as described in Dean et al. (1985). RNA (20 µg) was separated by electrophoresis in 1.6% (w/v) agarose gel containing 2.2 M formaldehyde. Equal loading of total RNA was confirmed by equal staining intensity in all lanes after ethidium bromide staining. Transfer to Hybond N$^+$ membrane, prehybridization, and hybridization were achieved according to the Amersham manual. Prehybridization was performed in 6x SSC, 0.5% (w/v) SDS, 5x Denhardt, 100 µg/ml single-stranded herring sperm DNA and 50% (v/v) formaldehyde, at 42 °C. Hybridization of Northern blots with the ipt probe used in Southern hybridizations was performed in the solution described above without Denhardt. Blots were washed in 1-0.1x SSC with 0.1% (w/v) SDS, and exposed to autoradiography.

2.2.6 Determination of Cytokinin Concentrations

CKs were extracted by incubating and stirring the freeze-dried callus samples in methanol/40 mM acetic acid (80/20, v/v) with 200 µM butylhydroxytoluene (60 h, dark, 4 °C). Quantitative analysis was performed as described in Schwartzenberg et al. (1994). Briefly, extracts were purified through a Sep-Pak C18 cartridge (Waters) and concentrated by rotary film evaporation. CKs were then separated by HPLC (Beckman system) using a Merck Lichrospher 100 RP-18 encapped 5 µm column (250 × 4 mm) and a linear gradient between 40 mM acetic acid and acetonitrile for 60 min. The CK concentrations were determined by an indirect competitive ELISA test with polyclonal antibodies raised against isopentenyl adenosine and zeatin riboside as described in Maldiney et al. (1986) and Sotta et al. (1987). Zeatin was quantified with the antibodies raised against zeatin riboside because of the cross-reactivity of polyclonal antibodies.

2.2.7 Treatment of Untransformed Callus Culture with Exogenous Cytokinin

Untransformed tissues were subcultured on a 2,4-D-free B5 agar medium (containing or not containing CK), and then maintained in darkness or light. Zeatin and zeatin riboside solutions were filter-sterilized (0.22 μm, Analypore OSI) before addition to the autoclaved B5 medium.

2.2.8 Determination of Alkaloid Contents

Callus tissues were harvested after 3 weeks of culture and then freeze-dried. Ajmalicine and serpentine were extracted with 5 ml methanol from 100 mg (dry mass) tissues and quantified by spectrofluorometry (Mérillon et al. 1989).

3 Results

3.1 Transformation by the *ipt* Gene

Cotyledons from seedlings were transformed with an *A. tumefaciens* strain carrying the *ipt* gene and a geneticin-selection gene. The geneticin-resistant tissues did not require any exogenous PGR. Southern blotting confirmed the presence of the *ipt* gene in six independent callus lines that had been chosen

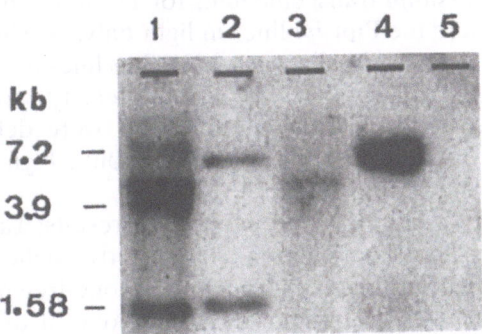

Fig. 1. Southern hybridization of callus cultures obtained from *A. tumefaciens*-infected cotyledons (*lanes 1–4*) and from normal cotyledons (*lane 5*) of *C. roseus* (cv. Little Pinkie). Callus cultures 92/2, 92/3, 92/4, and 92/5 (*lanes 1–4*) were transformed by an *Agrobacterium* strain pGV2492 yielding the nopaline *ipt* gene under control of its genuine promoter. Growth conditions: growth regulator-free medium for transformed callus, or medium supplemented with 4.5 μM 2,4-D for untrans-formed callus, 24 °C, continuous white light. Total DNA was digested by *BamH* I and the fragments were separated in 0.8% (w/v) agarose (20 μg per lane). The 850 bp *ipt* fragment was labeled with (α^{32}P) dCTP to high specific activity (3 × 10^9 cpm/μg) by the multiprime method (Amersham), and used as a probe. *Pst* I fragments of lambda phage DNA were used as molecular weight standards. (Garnier et al. 1996a)

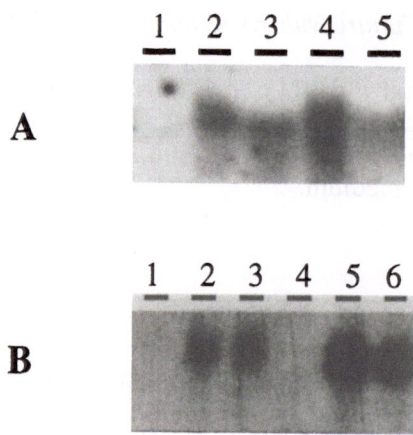

Fig. 2A,B. Northern analysis of transgenic callus cultures yielding the *ipt* gene under the control of either a light-inducible promoter (*Pssu*-ipt constructs: **A**) or its genuine promoter (P*ipt*-ipt constructs: **B**). The lines are the following: in **A**, *lanes 1* and *4*: untransformed tissues; *lanes 2* and *5*: 88/1; *lanes 3* and *6*: 88/3. In **B**, *1* 92/2; *2* 92/3; *3* 92/4; *4* 92/5; *5* untransformed cultures. Culture medium: B5 agar medium without growth substances. Culture conditions: the lines 88/1 and 88/2 as well as untransformed cultures were grown either under continuous white light (**A** *lanes 1–3*) or in darkness (**A** *lanes 4–6*); the lines 92/2, 92/3, 92/4, and 92/5 were grown under continuous fluorescence white light (**B** *lanes 1–5*). Total RNA (20 µg) was loaded in each lane and blots were hybridized with 850 bp *ipt*-gene fragment. (Garnier et al. 1996a,b)

for experiments. For example, the four P*ipt*-*ipt* lines contained two to four copies of the introduced *ipt* gene (Fig. 1).

Northern analysis and CK quantitation were achieved after growing the P*ssu*-*ipt* transgenic lines for 11 days either in continuous light or in darkness, and the P*ipt*-*ipt* lines in light only. As shown in Fig. 2, the *ipt* transcripts were detected in all transgenic callus lines but not in the untransformed callus lines used as control. The signal intensity varied from one P*ipt*-*ipt* line to another (Fig. 2A). The *ipt* transcripts were detected in P*ssu*-*ipt* lines, even under noninductive light conditions, but a higher level of *ipt* transcripts was found in the light-grown calli (Fig. 2B).

In accordance with these results, Table 1 shows that all the transformed callus lines had a significantly higher CK content compared with the untransformed lines. On hormone-free medium, total CK level of the control tissues was about 100–200 pmol/g of dry mass irrespective of light or dark condition, whereas the content of CK in the P*ipt*-*ipt* tissue lines varied from 6000 to 32000 pmol/g of dry mass. The zeatin riboside concentration in the P*ssu*-*ipt* 88/1 line grown either in darkness or in light was up to 700 and 1700 pmol/g of dry mass, respectively.

3.2 Alkaloid Accumulation in Nontransformed Callus Cultures

Untransformed callus cultures were obtained from cotyledons and then maintained in darkness on B5 medium with 2,4-D as the sole plant growth regula-

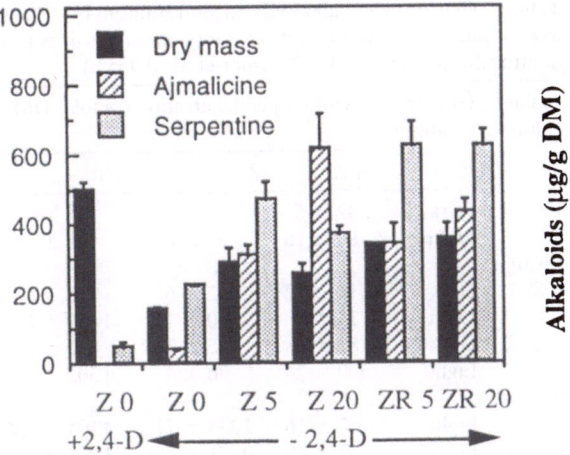

Fig. 3. Effect of applying exogenous CK on growth and alkaloid accumulation in an untransformed callus line initiated from periwinkle cotyledons. *DM* Dry mass; *Z* zeatin; *ZR* zeatin riboside. (Garnier et al. 1996b)

tor. Tissues accumulated very low amounts of serpentine (about 50 µg/g of dry mass) and no ajmalicine. Removing 2,4-D from the medium for one passage severely decreased the dry mass, but induced ajmalicine synthesis and increased the serpentine content fivefold. Adding 5 µM zeatin (Z) or zeatin riboside (ZR) to the 2,4-D-free medium slightly antagonized the inhibition of growth caused by 2,4-D removal, and increased the ajmalicine content eight- to tenfold (Fig. 3). A higher concentration of CK (20 µM) did not further enhance the alkaloid production. Light treatment resulted in a sevenfold increase in the serpentine amount, while the ajmalicine content was only half decreased.

3.3 Alkaloid Accumulation in Transformed Callus Cultures

Indole alkaloids were quantified in transgenic lines after 3 weeks of growth on 2,4-D-free medium, either in darkness for the Pssu-*ipt* lines, or in light for both Pssu and Pipt lines (Table 2). Ajmalicine and serpentine amounts were much lower in all the transformed tissues than in the untransformed tissues used as control. Moreover, light treatment did not change the alkaloid concentration in the Pssu-*ipt* lines.

4 Summary and Conclusions

The untransformed callus line responded to CK application by an elevated level of alkaloids. In comparison, the transformed callus lines in which the *ipt*

Table 2. Cytokinin and alkaloid concentrations in two Pssu-*ipt* transgenic callus lines (referred to as 88/1 and 88/2), in three Pipt-*ipt* transgenic callus lines (numbered 92/2, 92/3, and 92/4) and in a nontransformed one (NT). (Garnier et al. 1996a,b)

Callus culture	Growth condition	Cytokinin concentration (pmol/g DM)			Alkaloid concentration (μg/g DM)		
		iPA	Z	ZR	Total	Ajmalicine	Serpentine
NT	Dark	42 ± 5	25	44 ± 1	111	267 ± 34	152 ± 56
	Light	80 ± 16	30 ± 6	55 ± 16	165	157 ± 31	1064 ± 110
Pssu-*ipt*							
88/1	Dark	114 ± 22	42 ± 6	720 ± 88	876	12 ± 2	17 ± 2
	Light	73 ± 9	60 ± 11	1680 ± 282	1813	12 ± 3	36 ± 1
88/2	Dark	183 ± 9	185 ± 7	492 ± 21	860	26 ± 4	53 ± 9
	Light	300 ± 24	76 ± 4	1302 ± 71	1678	16 ± 12	60 ± 5
Pipt-*ipt*							
92/2	Light	7 ± 16	2377 ± 71	4104 ± 156	6488	<5	87 ± 14
92/3	Light	557 ± 89	10871 ± 554	21963 ± 4507	33391	<5	77 ± 11
92/4	Light	213 ± 28	4710 ± 619	8578 ± 231	13501	<5	64 ± 8

Calli were maintained on a 2,4-D-free B5 agar medium, in light or in darkness. The cultures were harvested at day 11 or day 21 of culture for determining cytokinin and alkaloid concentrations, respectively. Endogenous CK concentrations are given as the four to ten replicates at the error level of 20%; alkaloid concentrations as the mean ± standard deviation of three replicates; DM: dry mass; iPA: isopentenyladenosine; Z: zeatin; ZR: zeatin riboside.

gene was monitored either by its own constitutive Pipt promoter or by a light-inducible Pssu promoter accumulated a higher level of both *ipt* transcripts and endogenous CKs, but a lower level of ajmalicine and serpentine (Table 2). It can, therefore, be concluded that endogenously produced CK does not mimic the effect of exogenously applied CK on alkaloid production in periwinkle. The two following hypotheses may explain this discrepancy.

First, the endogenous CK level may be too high, reaching "toxic" values in the transformed tissues. In the Pssu-*ipt* construction, the *ipt* gene was monitored by a light-inducible weak promoter whose expression is obviously "leaky" under our experimental conditions. In Pssu-*ipt* transgenic tobacco plants, Thomas et al. (1995) showed that the levels of expression were fairly low, even at high intensity light. Differences between these results and those obtained in the present chapter might be explained by species-specific expression, as it was reported for two promoters introduced into *C. roseus* and *N. tabacum* (Van der Fits and Memelink 1997). There is some evidence that CK response in the cells may be saturated with relatively modest increases in CK level (Smigocki 1991; Ainley et al. 1993); furthermore, even an undetectable increase in *ipt* transcription has been found to trigger some changes in the cells (Smart et al. 1991). Previous observations with the suspension cell line C20 showed that 5 μM zeatin was optimal for alkaloid accumulation, whereas higher CK concentrations led to lower alkaloid accumulation (unpubl.). Therefore, the endogenous ZR concentrations in the Pssu-*ipt* callus cultures (Table 2) might be supraoptimal for alkaloid synthesis. However, despite at least a fourfold increase in endogenous CK, the Pipt-*ipt* tissues accumulated higher amounts of serpentine than did the Pssu-*ipt* tissues (Table 2), and thus this hypothesis may be ruled out.

Second, and as reported for several *ipt*-transgenic tobacco lines (Binns et al. 1987; Smigocki and Owens 1989; Beinsberger et al. 1991; Hamdi et al. 1995), the *ipt*-transgenic periwinkle lines were found to be fully habituated (Garnier et al. 1996b). More than 20 years ago, Kurz and Constabel (1979) noted that "cell lines known for auxin autotrophy appeared to lack secondary products", thus habituation may be another cause for the low alkaloid production in the transformed lines. In this connection, we found that a habituated line of periwinkle (line C20A), selected from the 2,4-D-dependent, alkaloid-accumulating line C20), had lost its ability to produce alkaloids (Mérillon et al. 1989). An argument against this hypothesis is that not all the habituated Apocynaceae tissues failed to accumulate alkaloids (Eilert et al. 1987; Dagnino et al. 1993). However, the effect of habituation on alkaloid biosynthesis is not fully understood, and may differ from one cell line to another. Obviously, the *ipt*-transgenic periwinkle tissues exhibited alterations of cell sensitivity to various factors. Neither light condition (Table 2) nor exogenous CK condition (unpubl.) enhanced the alkaloid content in the two Pssu-*ipt* transgenic lines. This is consistent with the increased tolerance of *ipt*-transgenic tobacco cultures to exogenous auxin (Li et al. 1994).

To conclude, periwinkle cotyledons were transformed with *A. tumefaciens* strains carrying the *ipt* gene, and several transgenic callus lines were obtained that can grow on PGR-free medium. In spite of enhanced levels of endogenous CKs, the lines did not accumulate higher contents of indole alkaloids. Since other investigators have found that transformation of periwinkle with disarmed *A. tumefaciens* strains is difficult, we do not know whether the transformation of cotyledons was facilitated by the presence of the *ipt* gene. If not, the procedure used should be adopted in the future to obtain successful alkaloid-producing lines of periwinkle when genes encoding key enzymes of the indole alkaloid pathway are available.

Acknowledgments. Financial support of this work by grants from the Ministère de la Recherche et de la Technologie via Biotechnocentre (France) is gratefully acknowledged. The authors wish to thank Prof. Dr. D. Inzé and Dr. R. Deblaere (Laboratory of Genetics, Gent, Belgium), who kindly provided the *Agrobacterium tumefaciens* strains, Dr. D. Cornu, who provided facilities in the INRA Laboratory of Ardon, the Roussel Uclaf Compagny for gift of macerozyme, as well as Dr. D. Hallard for his collaboration.

References

Ainley WM, McNeil KJ, Hill JW, Lingle WL, Simpson RB, Brenner ML, Nagao RT, Key JL (1993) Regulatable endogenous production of cytokinins up to "toxic" levels in transgenic plants and plant tissues. Plant Mol Biol 22:113–123

Beinsberger SE, Valcke RL, Deblaere RY, Clijsters HM, De Greef JA, Van Onckelen HA (1991) Effects of the introduction of *Agrobacterium tumefaciens* T-DNA *ipt*-gene in *Nicotiana tabacum* L. cv. Petit Havana SR1 plant cells. Plant Cell Physiol 32:489–496

Bhadra R, Vani S, Shanks JV (1993) Production of indole alkaloids by selected hairy root lines of *Catharanthus roseus*. Biotechnol Bioeng 41:581–592

Binns AN, Labriola J, Black RC (1987) Initiation of auxin autonomy in *Nicotiana glutinosa* cells by the cytokinin-biosynthesis gene from *Agrobacterium tumefaciens*. Planta 171:539–548

Boder GB, Gorman N, Johnson JS, Simpson PJ (1964) Tissue culture studies of *Catharanthus roseus* crown gall. Lloydia 27:328–333

Braun AC (1943) Studies on tumor inception in the crown-gall disease. An J Bot 30:674–677

Ciau-Uitz R, Miranda-Ham ML, Coello-Coello J, Chi B, Pacheco M, Loyola-Vargas VM (1994) Indole alkaloid production by transformed and non-transformed root. In Vitro Cell Dev Biol 30P:84–88

Dagnino D, Schripsema J, Peltenburg A, Verpoorte R (1993) Capillary gas chromatographic analysis of indole alkaloids: investigation of the indole alkaloids present in *Tabernaemontana divaricata* cell suspension cultures. Phytochemistry 32:325–330

Dean C, Elzen P, Tamaki S, Duinsmuir P, Bedbrook J (1985) Differential expression of the eight genes of the *Petunia* ribulose bisphosphate carboxylase small subunit multigene family. EMBO J 5:3055–3061

Décendit A, Liu D, Ouelhazi L, Doireau P, Mérillon JM, Rideau M (1992) Cytokinin-enhanced accumulation of indole alkaloids in *Catharanthus roseus* cell cultures: the factors affecting the cytokinin response. Plant Cell Rep 11:400–403

Eilert U, De Luca V, Kurtz WGW, Constabel F (1987) Alkaloid formation by habituated and tumorous cell suspension cultures of *Catharanthus roseus*. Plant Cell Rep 6:271–274

Gamborg OL, Miller RA, Ojima K (1968) Nutrient requirements of suspension cultures of soybean root cells. Exp Cell Res 50:151–158

Ganapathi G, Kargi F (1990) Recent advances in indole alkaloid production by *Catharanthus roseus* (periwinkle). J Exp Bot 41:259–267

Garnier F, Label P, Hallard D, Chénieux JC, Rideau M, Hamdi S (1996a) Transgenic periwinkle tissues overproducing cytokinins do not accumulate enhanced levels of indole alkaloids. Plant Cell Tissue Organ Cult 45:223–230

Garnier F, Carpin S, Label P, Crèche J, Rideau M, Hamdi S (1996b) Effect of cytokinin on alkaloid accumulation in periwinkle callus cultures transformed with a light-inducible *ipt* gene. Plant Sci 120:47–55

Goddijn OJM, Pennings EJM, Van der Helm P, Schilperoort RA, Verpoorte R, Hoge JHC (1995) Overexpression of a tryptophan decarboxylase cDNA in *Catharanthus roseus* crown-gall calluses results in increased tryptamine levels but not in increased terpenoid indole alkaloids. Transgen Res 4:315–323

Graham DE (1978) The isolation of high molecular weight DNA from whole organism or large tissue masses. Anal Biochem 85:609–613

Hamdi S, Crèche J, Garnier F, Mars M, Décendit A, Gaspar T, Rideau M (1995) Cytokinin involvement in the control of coumarin accumulation in *Nicotiana tabacum*. Investigations with normal and transformed tissues carrying the isopentenyltransferase gene. Plant Physiol Biochem 33:283–288

Hasezawa S, Nagata T, Syono (1981) Transformation of *Vinca* protoplasts mediated by *Agrobacterium* spheroplasts. Mol Gen Genet 182:206–210

Islas I, Loyola-Vargas VM, Miranda-Ham ML (1994) Tryptophan decarboxylase activity in transformed roots from *Catharanthus roseus* and its relationship to tryptamine, ajmalicine, and catharanthine accumulation during the culture cycle. In Vitro Cell Dev Biol 30P:81–83

Jung KH, Kwak SS, Choi CY, Liu JR (1995) An interchangeable system of hairy root and cell suspension cultures of *Catharanthus* for indole alkaloid production. Plant Cell Rep 15:51–54

Kodja H, Liu D, Mérillon JM, Andreu F, Rideau M, Chénieux JC (1989) Stimulation of indole alkaloid accumulation in suspension cell cultures of *Catharanthus roseus* (G Don) by cytokinins. CR Acad Sci Paris 309 (III):453–458

Kurz WGW, Constabel F (1979) Plant cell cultures, a potential source of pharmaceuticals. Adv Appl Microbiol 25:209–240

Lenz CB, Hodges TK, Matthysse AG (1979) Uptake of potassium ions by normal and crown gall-tumor cells of *Vinca rosea* grown in tissue culture. Planta 146:113–117

Li Y, Shi X, Strabala TJ, Hagen G, Guilfoyle TJ (1994) Transgenic tobacco plants that overproduce cytokinins show increased tolerance to exogenous auxin and auxin transport inhibitors. Plant Sci 100:9–14

Lynn DG, Chen RH, Manning KS, Wood HN (1987) The structural characterization of endogenous factors from *Vinca rosea* crown-gall tumors that promote cell division of tobacco cells. Proc Natl Acad Sci USA 84:615–619

Maldiney R, Leroux B, Sabbagh I, Sotta B, Sossountzov L, Miginiac E (1986) A biotin-avidin based enzyme immunoassay to quantify three phytohormones: auxin, abscissic acid and zeatin-riboside. J Immunol Methods 90:151–158

Manasse RJ, Lipetz J (1971) A simplified method for isolating bacteria-free crown-gall tissue from *Vinca rosea*. Can J Bot 49:1255–1257

McGaw BA, Horgan R (1983) Cytokinin oxidase from *Zea mays* kernels and *Vinca rosea* crown gall tissue. Planta 159:30–37

Mérillon JM, Ouelhazi L, Doireau P, Chénieux JC, Rideau M (1989) Metabolic changes and alkaloid production in habituated and non-habituated cells of *Catharanthus roseus* grown in hormone-free medium. Comparing hormone-deprived non-habituated cells with habituated cells. J Plant Physiol 134:54–60

Miller CO (1974) Ribosyl-*trans*-zeatin, a major cytokinin produced by crown-gall tumor tissue. Proc Natl Acad Sci USA 71:334–338

Moreno PRH, Van der Heijden R, Verpoorte R (1995) Cell and tissue cultures of *Catharanthus roseus*: a literature survey II. Updating from 1988 to 1993. Plant Cell Tissue Organ Cult 42:1–25

Noble RL (1990) The discovery of the *Vinca* alkaloids-chemotherapeutic agents against cancer. Biochem Cell Biol 68:1344–1351

Okada K, Hasezawa S, Syono K, Nagata T (1985) Further evidence for the transformation of *Vinca rosea* protoplasts by *Agrobacterium tumefaciens* spheroplasts. Plant Cell Rep 4:133–136

O'Keefe BR, Mahady GB, Gills JJ, Beecher CWW (1997) Stable vindoline production in transformed cell cultures of *Catharanthus roseus*. J Nat Prod 60:261–264

Palni LMS, Horgan R (1983) Cytokinins in transfer RNA of normal and crown-gall tissue of *Vinca rosea*. Planta 159:178–181

Palni LMS, Horgan R, Darrall NM, Stuchbury T, Wareing PF (1983) Cytokinin biosynthesis in crown-gall tissues of *Vinca rosea*. The significance of nucleotides. Planta 159:50–59

Schwartzenberg von K, Doumas P, Jouanin L, Pilate G (1994) Enhancement of the endogenous cytokinin concentration in poplar by transformation with *Agrobacterium* T-DNA gene *ipt*. Tree Physiol 14:27–35

Sim SS, Chang HN, Liu JR, Jung KH (1994) Production and secretion of indole alkaloids in hairy-root cultures of *Catharanthus roseus*: effects of in situ adsorption, fungal elicitation and permeabilization. J Ferm Bioeng 78:229–234

Smart CM, Scofield SR, Bevan MW, Dyer TA (1991) Delayed leaf senescence in tobacco plants transformed with *tmr*, a gene for cytokinin production in *Agrobacterium*. Plant Cell 3:647–656

Smigocki AC (1991) Cytokinin content and tissue distribution in plants transformed by a reconstructed isopentenyl transferase gene. Plant Mol Biol 16:105–115

Smigocki AC, Owens LD (1989) Cytokinin-to-auxin ratios and morphology of shoots and tissues transformed by a chimeric isopentenyl transferase gene. Plant Physiol 91:808–811

Sotta B, Pilate G, Pelese F, Sabbagh I, Bonnet M, Maldiney R (1987) An avidin biotin solid phase ELISA for fentomole isopentenyladenine and isopentenyladenosine measurements in HPLC purified plant extracts. Plant Physiol 84:571–573

Stuchbury T, Palni LMS, Horgan R, Wareing PF (1979) The biosynthesis of cytokinins in crown-gall tissue of *Vinca rosea*. Planta 147:97–102

Theis TN, Riker AJ, Allen ON (1950) The destruction of crown-gall bacteria in periwinkle by high temperature with high humidity. Am J Bot 37:792–801

Thomas JC, Smigocki AC, Bohnert HJ (1995) Light-induced expression of *ipt* from *Agrobacterium tumefaciens* results in cytokinin accumulation and osmotic stress symptoms in transgenic tobacco. Plant Mol Biol 27:225–235

Toivonen L, Ojala M, Kauppinen V (1991) Studies on the optimization of growth and indole alkaloid production by hairy-root cultures of *Catharanthus roseus*. Biotechnol Bioeng 37:673–680

Van der Fits L, Memelink J (1997) Comparison of the activities of CAMV35S and FMV34S promoter derivatives in *Catharanthus roseus* cells transiently and stably transformed by particle bombardment. Plant Mol Biol 33:943–946

Van der Heijden R, Verpoorte R, Ten Hoopen HJG (1989) Cell and tissue cultures of *Catharanthus roseus*: a literature survey. Plant Cell Tissue Organ Cult 18:231–280

Van Tellingen O, Sips JHM, Beijnen JH, Bult A, Nooijen WJ (1992) Pharmacology, bio-analysis and pharmacokinetics in the *Vinca* alkaloids and semi-synthetics derivatives (review). Anticancer Res 12:1699–1716

Vasquez-Flota F, Moreno-Valenzuela O, Miranda-Ham ML, Coello-Coello J, Loyola-Vargas VM (1994) Catharanthine and ajmalicine synthesis in *Catharanthus roseus* hairy root cultures. Medium optimization and elicitation. Plant Cell Tissue Organ Cult 38:273–279

White PR (1945) Metastatic (graft) tumors of bacteria-free crown-galls on *Vinca rosea*. Am J Bot 32:237–241

Wood HN, Braun AC (1965) Studies on the net uptake of solutes by normal and crown-gall tumors cells. Proc Natl Acad Sci USA 54:1532–1538

VII Genetic Transformation of *Datura* Species

V.M. Loyola-Vargas

1 Introduction

The commercial value of *Datura* (Solanaceae), together with other solanaceous species already revised in this series (Strauss 1989; Bajaj and Simola 1991, Scholten et al. 1992), comes almost entirely from the alkaloids hyoscyamine and scopolamine (Fig. 1), extracted from plant roots. Members of the genus *Datura* are particularly rich in scopolamine, which is considered the most hallucinogenic of the solanaceous alkaloids (Simpson and Conner-Ogorzaly 1986).

The genus *Datura* consists of native herbs from dry, temperate, and subtropical regions, though many have become worldwide weeds (Haegi 1976, Fuentes 1980; Hammer et al. 1983). The characteristics of the plants support the hypothesis of placing Mexico and the southwestern United States as the center of origin and evolution of the genus *Datura*. However, there are also species native to Asia and Europe.

This genus is also rich in tradition, folklore, and names (Table 1), especially in India and Mexico. In India, according to its ancient literary traditions, as mentioned in Vamana Purana, *Datura metel* is believed to have arisen from the breast of the God Shiva (Mehra 1989). In Mexico it was and is still used in religious ceremonies.

Datura comprises a large number of species, the most important ones from the biological or economical point of view being: *D. alba, D. arborea, D. candida, D. discolor, D. fastuosa, D. fatuosa, D. ferox, D. inermis, D. innoxia, D. kymatocarpa, D. leichhardtii, D. metel, D. meteloides, D. pruinosa, D. reburra, D. sanguinea, D. stramonium, D. suaveolens,* and *D. wriightii.*

In general, *Datura* may be large, annual herbs, shrubs, or small trees, 1 to 1.5 m tall, glabrous or pubescent. Leaves are alternate and oval-shaped, petiolate, dark green in color, growing up to 20 cm tall with short leaf stem. Fruit is about 5 cm long, small seeds, brownish black in color. The inflorescence consists of a solitary flower, funnel-shaped, white, yellow, pink, or red, 2.5 to 12.5 cm long, blooming usually in late spring (Nee 1986; Petri and Bajaj 1989). Karyotype analysis of different *Datura* species show a somatic chromosome number 2n = 24 (Palomino et al. 1988).

Unidad de Biología Experimental, Centro de Investigación Científica de Yucatán, Apdo. Postal 87, CP 97310, Cordemex, Yucatán, México

Biotechnology in Agriculture and Forestry, Vol. 45
Transgenic Medicinal Plants (ed. by Y.P.S. Bajaj)
© Springer-Verlag Berlin Heidelberg 1999

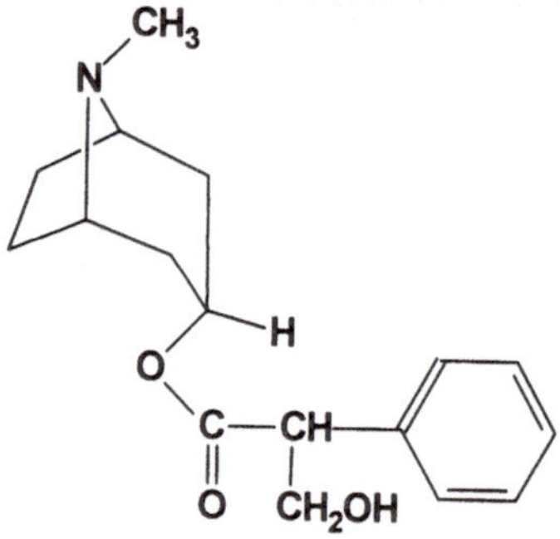

HYOSCYAMINE 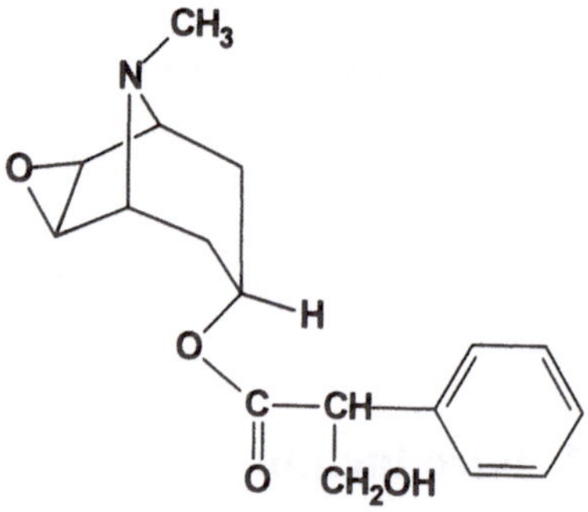   **SCOPOLAMINE**

Fig. 1. Structures of the major alkaloids in *Datura* spp.

Table 1. Common names for the more important *Datura* spp. (Mendieta and Amo 1981)

Datura sp.	Common name	*Datura* sp.	Common name
D. arborea	Angel's trumpet	D. suaveolens	Campana de Paris
	Floripondio		Floripon
D. alba	Man T'O Lo	D. stramonium	Buenas tardes
	Tyosen-Asagao		Chamico
D. candida	Campana de Paris		Chamisco
	Campana		Cocombre zombie
	Lipa-Ca-Tu-Ue		Cojon del diablo
	Palpanichim		Concombre zombie
D. discolor	Desert thornapple		Cornicopio
D. fastuosa	Dhatoora		Daturah
	Ketjoeboeng		Doornappelkruid
	Safeer AI Soltan		Estramoni
	Thornapple		Estramonio
D. ferox	Chinese thornapple		Feng Ch'leh Erh
D. inermis	Toge-Nasi-Yo-Shutyo		Figuiero do inferno
D. innoxia	Angel's trumpet		Floribunda
	Dhatoora		Galurt
	Chamico		Gemeiner stechapfel
	Nohol-X-Tuhk'U		Jimson
	Toloachi		Jimson eed
D. meteloides	Amerika-Tyosen-Asagao		Jinsonweed
	Nongue		Man T'O Lo
	Toloache		Mehenxtohk U
D. metel	Dhatoora		Nafeer
	Gawz Mathil		Noce del Diavolo
	Horn of plenty		Nongue
	Ke-Tyosen–Asagao		Opium tropical
	Kidi Ganian		Pomme epineuse
	Metel		Pomme poison
	Tatula		Stinkblaren
	Kachubang		Stramoine
	Kechubong hitam		Tapate
	Kechubong puteh		Xtohk U
	Kechubong	D. wightii	Sacred thornapple
	Kuchubong		

The tropane alkaloids are widely used as anesthetics and spasmolytics and are obtained directly from the roots. However, ecophysiological factors, among others, can influence the alkaloid content of *Datura* species. This phenomenon has been extensively studied by Cosson et al. (1966, 1978) and Cosson (1966, 1969, 1976, 1981). In *Datura metel* and *D. tatula,* yields of scopolamine were dependent only on light. Altitude can also influence the alkaloid content of *D. metel* (Kamick and Saxena 1970), the total alkaloids being comparatively lower at sea level than at high altitudes. However, opposite results were found in *D. innoxia* (Pelt et al. 1967a,b). Another climatic factor that can generally increase the alkaloid content is water stress (Lebeau and Janot 1956; Ikenaga et al. 1977; Waller and Kowacki 1978; Yaniv and Palevitch 1982).

In general, secondary metabolites are present in the plants in very small amounts, and vary widely with climatic and edaphic conditions (Waller and Kowacki 1978). Plant tissue cultures such as callus or suspension cultures (Petri and Bajaj 1989) or hairy roots (Flores and Medina-Bolívar 1995; Loyola-Vargas and Miranda-Ham 1995) have been used for the production of tropane alkaloids, especially hyoscyamine and scopolamine.

It is desirable for both industry and popular medicine that *Datura's* tropane alkaloid content be more stable, independent of the environmental conditions. An alternative to this problem is the use of genetic engineering. For example, the introduction of the hyoscyamine 6β-hydroxylase (EC 1.14.11.11) gene from *Hyoscyamus niger,* under the control of the constitutive caulifower mosaic virus 35S promoter into *Atropa belladonna* by the use of an *Agrobacterium tumefaciens* transformation system, increased the alkaloid content in the leaves and stem; most importantly, the alkaloid identified was exclusively scopolamine (Yun et al. 1992). Similar results were obtained in engineering *A. belladona* hairy roots (Hashimoto et al. 1993). A similar methodology could prove useful also in the case of *Datura.*

2 Genetic Transformation

2.1 Earlier Studies

The Nomycin phosphotransferase II (NPT-II) gene was introduced in protoplasts from a pantothenate-requiring auxotrophic *D. innoxia* cell line (Saxena et al. 1990); however, no regenerate plants were obtained. Later, one of the lines obtained was used as a "universal hybridizer" for somatic cell fusion (Du et al. 1991).

Schmidt-Rogge et al. (1993) used protoplasts as a system for *D. innoxia* transformation. Using a PEG-mediated direct transfer of plasmid DNA, harboring a neomycin phosphotransferase II gene or a hygromycin B phosphotraspherase into haploid protoplast, they were able to recuperate regenerated haploid transgenic plants. Maize transposable Ac element has also been introduced into haploid protoplasts of *D. innoxia* (Schmidt Rogge et al. 1994).

Coculture of leaf disks of *D. innoxia* with *A. tumefaciens*, carrying binary vectors pGS Glucl, pGSTRN943, pGV2260, and pBI121, yielded putative transformed callus and vegetative buds, and plants regenerated from them. The frequency of transformation was higher than with *Nicotiana*. The transformants were phenotypically normal and the progeny showed Mendelian segregation on kanamycin resistance (Sangwan et al. 1991).

Another approach to obtain transgenic *Datura* plants has been the *Agrobacterium*-mediated transformation of pollen embryos of *D. innoxia* (Sangwan et al. 1993). Among all the parameters examined for transformation, stage of embryo development prior to infection was of critical importance and the transgenes kanamycin and GUS were coinherited as dominant Mendelian traits.

2.2 Hairy Root Cultures and factors affecting alkaloid Production

In order to overcome the slow growth of normal root cultures, a system that involves the generation of fast-growing adventitious roots or hairy roots, which are the product of the infection of different tissues with *A. rhizogenes* has been developed. Very promising results have been reported for the use of hairy root cultures as a source of chemicals (Flores and Medina-Bolívar 1995; Loyola-Vargas and Miranda-Ham 1995). Due to their fast growth rate, genetic/biochemical/molecular stability, and, compared with cell cultures, their high productivity, hairy root cultures could serve as an interesting system to study production and biosynthesis of important natural products typical from plant roots, such as tropane alkaloids.

The genus *Datura* has been thoroughly studied in relation to the production of tropane alkaloids by hairy root cultures (Table 2). In general, hyoscyamine was a major component of the alkaloid fraction and accounted for at least 0.3% of the dry matter, comparable to the pot-grown plants from which the cultures were initiated, but no scopolamine was found (Payne et al. 1987). However, the alkaloids and the amount measured depended on the strain of *A. rhizogenes* used to produce the hairy root lines; i.e., Jaziri et al. (1988) generated several *D. stramonium* hairy root lines with no hyoscyamine at all, but they produced scopolamine between a trace and 0.56% dry weight. Other laboratories have generated *Datura* hairy root lines that produce both alkaloids, hyoscyamine and scopolamine (Knopp et al. 1988; Christen et al. 1989a,b; Rhodes et al. 1990; Shimomura et al. 1991; Maldonado-Mendoza et al. 1993; Hilton and Wilson 1995).

Hairy roots are less susceptible to manipulation by changes in medium composition than callus and cell suspension cultures. However, when the individual components of the medium are modified one by one, several of them modified the yield of *D. stramonium* hairy root cultures (Sáenz-Carbonell and Loyola-Vargas 1996).

The carbon source is a major component of the culture medium and many studies have investigated its contribution to the tropane alkaloid biosynthesis. The use of glucose, instead of sucrose, was deleterious not only for root

Table 2. Hairy root cultures of *Datura* generated for production of secondary metabolites

Datura sp.	Major metabolite	Reference
D. candida × *D. aurea*	Hyoscyamine, scopolamine	Robins et al. (1990); Hilton and Wilson (1995)
D. candida × *D. candida*	Hyoscyamine, scopolamine	Christen et al. (1989a,b); Kephalas et al. (1989); Homeyer et al. (1991)
D. chlorantha, *D. fastuosa*, *D. ferox*	Hyoscyamine, scopolamine	Knopp et al. (1988)
D. innoxia	Hyoscyamine, scopolamine	Flores (1987); Knopp et al. (1988); Doerk et al. (1989); Ionkova et al. (1989, 1994); Kamada et al. (1989); Ohkawa et al. (1989); Shimomura et al. (1991); Boitel Conti et al. (1995)
D. metel, *D. meteloides*	Hyoscyamine, scopolamine	Knopp et al. (1988)
D. quercifolia	Hyoscyamine, scopolamine	Knopp et al. (1988); Dupraz et al. (1993)
D. rosei, *D. sanguinea*	Hyoscyamine, scopolamine	Knopp et al. (1988)
D. stramonium	Hyoscyamine, scopolamine	Flores and Filner (1985); Payne et al. (1987); Aird et al. (1988); Jaziri et al. (1988); Knopp et al. (1988); Mugnier (1988); Wilson et al. (1989); Burbidge et al. (1990); Hilton and Rhodes (1990, 1993, 1994); Furze et al. (1991); Sáenz-Carbonell et al. (1991, 1993); Flores et al. (1993); Maldonado-Mendoza et al. (1993); Maldonado-Mendoza and Loyola-Vargas (1995)
D. wrightii	Hyoscyamine, scopolamine	Hilton and Wilson (1995)

growth, but also for hyoscyamine accumulation in *D. stramonium*. In contrast, an increase in sucrose concentration (from 3 to 5%) produced an increase of 32% in the hyoscyamine content, but reduced growth by 26% (Payne et al. 1987; Hilton and Rhodes 1993). Maximum yields of hyoscyamine (6.5 mg/flask) have been obtained in full-strength B5 medium with 5% sucrose at either 20 or 25 °C (Hilton and Rhodes 1993). In another *D. stramonium* line, the response to the change in the sucrose concentration was even more pronounced, i.e., the hyoscyamine content increased eightfold (Hilton and Rhodes 1993). The reduction from 3 to 2% produced a tenfold decrease in the scopolamine content in a hairy root line of *D. stramonium*, which did not produce hyoscyamine (Jaziri et al. 1988). However, in our laboratory, using the same approach an increase of up to 200% was found. In general, the yield of the alkaloids incremented as a consequence of a major growth (Fig. 2A; Sáenz-Carbonell and Loyola-Vargas 1996).

Since alkaloids are nitrogen-containing molecules, much work has been done in the field of modification of the nitrogen content of the culture medium; however, there is no general consensus about the effects of the nitrogen source in the changes that are produced in the cells, for which reason each culture must be tested individually. The use of ammonium, alone or in combination with di- or tricarboxylic acids, in *D. stramonium* hairy roots, produced a

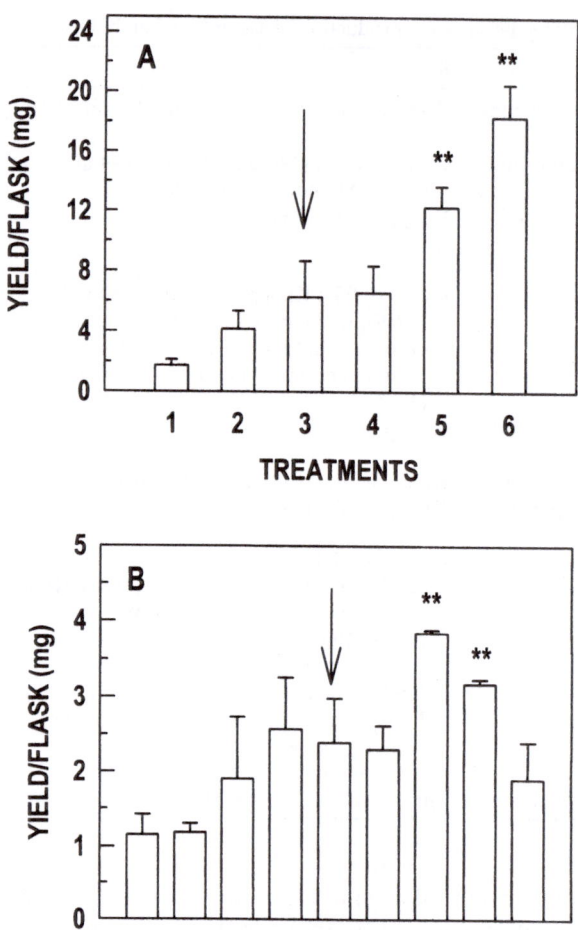

Fig. 2. A Effect of different sucrose concentration on the alkaloid content yield/flask of *D. stramonium* hairy roots. Treatments: *1* sucrose 1%; *2* sucrose 2%; *3* sucrose 3% (control); *4* sucrose 4%; *5* sucrose 5%; *6* sucrose 6%. **B** Effect of the nitrate/ammonium relationship on the alkaloid yield/flask of *D. stramonium* hairy roots. Treatments: *1* 20/5; *2* 21.4/4; *3* 22.5/3; *4* 23.6/2; *5* 24.7/1 (control); *6* 30/1.22; *7* 40/1.63; *8* 50/2; *9* 60/2.4. The *arrows* indicate the control sucrose or nitrogen concentration. (******) highly significant. (Data after Sáenz-Carbonell and Loyola-Vargas 1996)

significant increase in the root hyoscyamine content. By contrast, the amount of scopolamine decreased in all the treatments tested (Sáenz-Carbonell and Loyola-Vargas 1996).

The amount and kind of the nitrogen source are very important; however, it appears that the ratio between the amount of these two nitrogen sources is more important. Increase in the ratio between oxidized and reduced nitrogen caused growth decrease. Use of 41.63 mM of total nitrogen instead of 25.8 mM

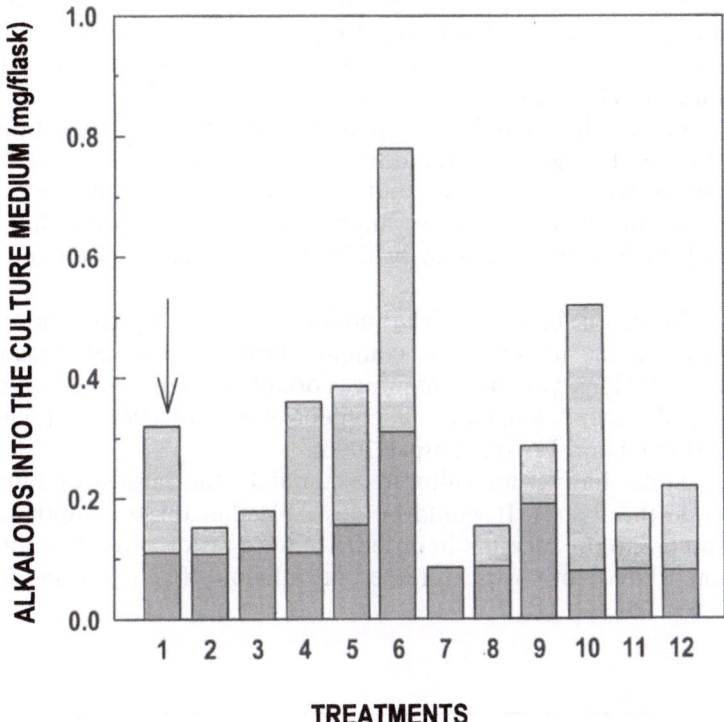

Fig. 3. Effect of different macronutrients on the alkaloids excreted into the culture medium of *D. stramonium* hairy roots. Treatments: *1* control; *2* without Mg; *3* 0.5-fold Mg; *4* 2-fold Mg; *5* 4-fold Mg; *6* Without P; *7* 0.5-fold P; *8* 2-fold P; *9* 4-fold P; *10* without Ca; *11* 0.5-fold Ca; *12* 2-fold Ca. (□) hyoscyamine and (□) scopolamine. The *arrow* indicates the control macronutrient concentration. (Data after Sáenz-Carbonell and Loyola-Vargas 1996)

(control) significantly increased the amount of hyoscyamine in the roots. As a result of better growth and hyoscyamine production, the yield of hyoscyamine per flask was around 55% greater at higher total nitrogen (Fig. 2B; Sáenz-Carbonell and Loyola-Vargas 1996).

In the presence of 0.5 mM phosphate, scopolamine concentration can reach 0.56% on a dry weight basis (Jaziri et al. 1988), whereas hyoscyamine content can reach levels as high as 0.384 mg/g fresh weight (Payne et al. 1987). By contrast, in other hairy root lines of *D. stramonium*, absence of phosphorus increased the alkaloid excretion into the culture medium by more than 130% (Fig. 3); a similar response, but to a lesser extent, was provoked by the absence of calcium. The decrease in magnesium provoked a decrease in the alkaloid content (Fig. 3).

On the other hand, although increase in the mineral salts had little effect on the hyoscyamine content of transformed roots, total yields increased due to growth stimulation (Hilton and Rhodes 1994). The treatment of root cultures of *D. stramonium* with copper and cadmium salts at external concentrations of approximately 1 mM induced rapid accumulation of high levels of sesquiterpenoid defensive compounds, notably lubimin and 3-hydroxylubilim (Furze et al. 1991).

Gibberellic acid (GA$_3$) applied exogenously at concentrations between 10 ng/l and 1 mg/l accelerated the increase in fresh weight of *D. innoxia* hairy roots as well as elongation and lateral branching (Ohkawa et al. 1989). GA$_3$ treatment lowered the hyoscyamine content by 35% and enhanced the content of 6β-hydroxyhyoscyamine by 527% and scopolamine by 400% (Kamada et al. 1989).

Methyl jasmonate (MeJa) addition to the hairy root cultures had a great effect on the hyoscyamine content, showing a typical dosis-response curve (Fig. 4). Hyoscyamine content was doubled when compared to the control at 0.1 μM MeJa. Scopolamine content was, however, not changed (Sáenz-Carbonell and Loyola-Vargas 1996).

Some hairy root cultures accumulate the largest amounts of alkaloids inside the tissues. It would be desirable that these compounds could be released into the medium in order to allow a faster and more efficient recuperation. The use of low pH released the alkaloids from *D. stramonium* hairy root

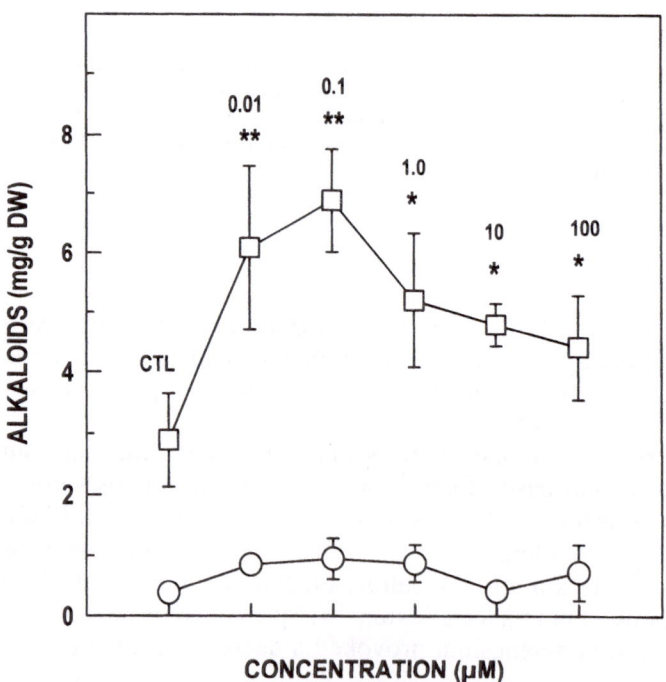

Fig. 4. Effect of MeJa on the alkaloid content of *D. stramonium* hairy roots. Hyoscyamine (□) and scopolamine (○). (*) significant; (**) highly significant. (Data after Sáenz-Carbonell and Loyola-Vargas 1996)

cultures (Sáenz-Carbonell et al. 1993); 100% of scopolamine (14.87 mg/l) and 70% of hyoscyamine (31.58 mg/l) were found in the medium (Fig. 5A).

Tween 20, a detergent, makes the membranes of the cell roots permeable and allows the release of hyoscyamine and scopolamine into the culture medium of *D. innoxia* hairy root cultures (Boitel Conti et al. 1995). This treatment, as well as the pH treatment, also lead to an increase in the total yield of alkaloids (5B), suggesting that probably their excretion into the medium liberates a feedback regulation metabolite allowing the synthesis of more compounds. It can be hypothesized that if an adsorbent or a continuous elimination of the released alkaloids is used, more alkaloids must be produced. The addition of 1 g/l Amberlite XAD-4 resin 10 days after subculture increased hyoscyamine release fourfold; 80% of hyoscyamine was bound to

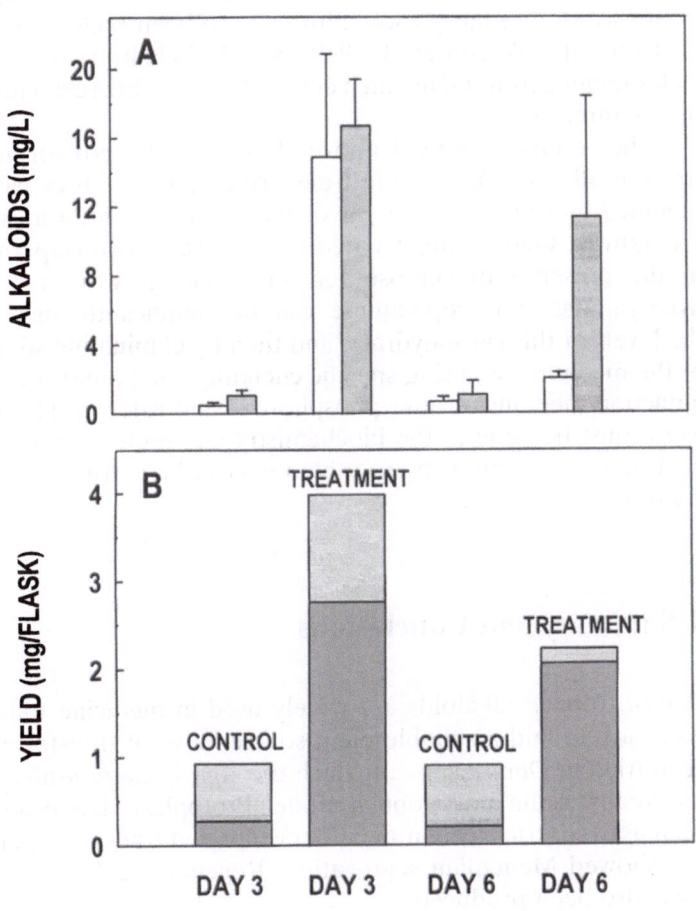

Fig. 5A,B Effect of pH on the release of alkaloids from *D.stramonium* hairy roots. **A** Hyoscyamine (□) and scopolamine (□) in the medium. **B** Total yield in the medium (□) and in the tissue (□). (V.M. Loyola-Vargas and L. Sáenz-Carbonell, unpubl)

the resin without the use of any other addition to the cultures (Dupraz et al. 1993).

Temperature can be another factor that affects the liberation of alkaloids into the medium. At 30 °C significant amounts of hyoscyamine (up to 15% w/w of the total) are released into the medium between 7 and 28 days of culture (Hilton and Rhodes 1993). In a 14-l stirred tank bioreactor operating at 30 or 35 °C, there was a sevenfold increase in hyoscyamine liberation compared with the alkaloid released at 25 °C (Hilton and Rhodes 1990).

D. stramonium hairy roots have also been adapted to photoautotrophy by growing them in the presence of light and CO_2 added (2%) in order to study the alkaloid biosynthesis. The ratio scopolamine/hyoscyamine increased threefold in the photoautotrophic cultures, revealing a correlation between the development of the photosynthetic apparatus and the increase in scopolamine (Flores et al. 1993; Maldonado-Mendoza and Loyola-Vargas 1995).

If this technology is to be used commercially, a scaleup system must be available. *Datura* hairy root cultures have been scaled-up in a bioreactor of 1.5-l capacity (Wilson et al. 1989), and 14-l (Hilton and Rhodes 1990), with yields ranging from 1.4 in batch culture to 8.2 mg hyoscyamine/l day in continuous culture.

The results presented above show that the growth and production of tropane alkaloids by *Datura*-transformed root cultures can be significantly modified by some components of the culture medium in which the cultures are grown. One strategy could be to use a two-step methodology, first in the presence of sucrose for increasing growth, since the amount of hyoscyamine and scopolamine can be significantly increased by altering the levels of this carbohydrate, and then by eliminating some of the minerals of the medium, i.e., using specific cuelating compounds or by changing to an induction medium without phosphorous and calcium. However, much more work must be done at the biochemistry and molecular level before the production of compounds by root cultures can be carried out without empirical control.

3 Summary and Conclusions

Datura tropane alkaloids are widely used in medicine today. This makes it necessary to find a suitable plant source. Several transforming systems have been tried in *Datura* spp., of which the *Agrobacterium*-mediated transformation system is the most popular model. Protoplasts, leaf disks, and pollen have been used for transformation and regenerated transgenic plants and the progeny showed Mendelian segregation. Regenerated haploid transgenic plants have also been produced.

Future work on *Datura* genetic engineering must focus on plants with more useful characteristics, i.e., tropane alkaloid synthesis independent of the environmental conditions, or plants harboring fungi and virus resistance.

Acknowledgments. The author thanks Drs. Ingrid Olmsted, Teresa Hernández, and lvon Ramírez for revision of the English version of the manuscript. The work of the author was supported by CONACYT. México, grants 0429N9108 and P122CCOT894672.

References

Aird ELH, Hamill JD, Rhodes MJC (1988) Cytogenetic analysis of hairy root cultures from a number of plant species transformed by *Agrobacterium rhizogenes*. Plant Cell Tissue Organ Cult 15:47–57

Bajaj YPS, Simola LK (1991) *Atropa belladonna* L.: in vitro culture, regeneration of plants, cryopreservation, and the production of tropane alkaloids. In: Bajaj YPS (ed) Biotechnology in agriculture and forestry vol 15. Medicinal and aromatic plants III. Springer Berlin, Heidelberg New York, pp 1–23

Boitel Conti M, Gontier E, Laberche JC, Ducrocq C, Sangwan Norreel BS (1995) Permeabilization of *Datura innoxia* hairy roots for release of stored tropane alkaloids. Planta Med 61:287–290

Burbidge A, Gartland KMA, Jenkins RO, Woolley JG, Elliot MC (1990) Accumulation of tropane alkaloids by hairy root cultures of *Datura stramonium*. J Exp Bot 41:P5–13 (Abstr)

Christen P, Roberts MF, Phillipson JD, Evans WC (1989a) High-yield production of tropane alkaloids by hairy-root cultures of a *Datura candida* hybrid. Plant Cell Rep 8:75–77

Christen P, Roberts MF, Phillipson JD, Evans WC (1989b) GC/MS examination of hairy roots of a *Datura candida* hybrid for alkaloids. Planta Med 55:595 (Abstr)

Cosson L (1966) Effect of light on the alkaloid content in *Datura*. Effect on the biogenesis of alkaloids. Herba Bung 5:157

Cosson L (1969) Effect of light on the ontogenic variations in scopolamine and hyoscyamine levels of *Datura mentel* leaves. Phytochemistry 8:2227

Cosson L (1976) Importance of climatic factors and development stages on the production of tropanic alkaloids. In: Chouard P, Jacques R (eds) Études de biologié végétale, Paris, 483 pp

Cosson L (1981) Some aspects of the metabolism of alkaloids in *Datura tatula* L. Plant Med Phytother 2:269

Cosson L, Chouard P, Paris P (1966) Influence of light on ontogenic variations of alkaloids in *Datura tatula*. J Nat Prod 29:19

Cosson L, Escudero-Morales A, Cougoul N (1978) Ecophysiological regulation of the metabolism of tropane alkaloids (hyoscyamine and scopolamine). Plant Med Phytother 12:319

Doerk K, Jonkova l, Witte L, Alfermann AW (1989) Synthesis of tropane alkaloids in hairy root cultures of *Datura* and *Hyoscyamus* species. Planta Med 55:688–689 (Abstr)

Du XD, Mourad GS, Williams D, King J (1991) Somatic hybrids derived from fusion of a universal hybridizer and a wild-type line of *Datura*. Plant Sci 79:111–118

Dupraz JM, Christen P, Kapetanidis I (1993) Tropane alkaloid production in *Datura quercifolia* hairy roots. Planta Med 59:A659(Abstr)

Flores HE (1987) Use of plant cells and organ culture in the production of biological chemicals. In: LeBaron HM, Mumma RO, Honeycutt RC, Duesing JH, Phillips JF, Haas MJ (eds) Biotechnology in agricultural chemistry. Symp Series 334, Amer Chemical Soci, Washington, DC pp 66–86

Flores HE, Filner P (1985) 'Hairy roots' of Solanaceae as a source of alkaloids. In Vitro 21:Pt2 53A (Abstr)

Flores HE, Medina-Bolívar F (1995) Root culture and plant natural products: unearthing the hidden half of plant metabolism. Plant Tissue Cult Biotechnol 1:59–74

Flores HE, Yao-rem D, Cuello JL, Maldonado-Mendoza IE, Loyola-Vargas VM (1993) Green roots: photosynthesis and photoautotrophy in an underground plant organ. Plant Physiol 101:363–371

Fuentes V (1980) Solanaceas de Cuba. I. *Datura* L. Rev Jard Bot Nac Cuba 1:61–81

Furze JM, Rhodes MJC, Parr AJ, Robins RJ, Withehead IM, Threlfall DR (1991) Abiotic factors elicit sesquiterpenoid phytoalexin production but not alkaloid production in transformed root cultures of *Datura stramonium*. Plant Cell Rep 10:111–114

Haegi L (1976) Taxonomic account of *Datura* L. (Solanaceae) in Australia with a note on *Brugmansia* Pers. Aust J Bot 24:415–435

Hammer K, Romeike A, Tittel C (1983) Vorarbeiten zur monographischen Darstellung von Wildpflanzensortimenten: *Datura* L., Sectiones *Dutra* Bernh., Ceratocaulis Bernh. et *Datura*. Kulturpflanze 31:13–75

Hashimoto T, Yun D-J, Yamada Y (1993) Production of tropane alkaloids in genetically engineered root cultures. Phytochemistry 32:713–718

Hilton MG, Rhodes MJC (1990) Growth and hyoscyamine production of hairy root cultures of *Datura stramonium* in a modified stirred tank reactor. Appl Microbiol Biotechnol 33:132–138

Hilton MG, Rhodes MJC (1993) Factors affecting the growth and hyoscyamine production during batch culture of transformed roots of *Datura stramonium*. Planta Med 59:340–344

Hilton MG, Rhodes MJC (1994) The effect of varying levels of Gamborg's B5 salts and temperature on the accumulation of starch and hyoscyamine in batch cultures of transformed roots of *Datura stramonium*. Plant Cell Tissue Organ Cult 38:45–51

Hilton MG, Wilson DG (1995) Growth and the uptake of sucrose and mineral ions by transformed root cultures of *Datura staramonium, Datura candida* × *D. aurea, Datura wrightii, Hyoscyamus muticus* and *Atropa belladonna*. Planta Med 61:345–350

Homeyer BC, Fry CZ, Roberts MF (1991) Alkaloid variations in hairy root clones of a *Datura candida* hybrid. J Pharm Pharmacol 43:21 (Abstr)

Ikenaga T, Saisho Y, Ohashi H (1977) Studies on the production of *Duboisia* in Japan. V. The effect of soil water on the growth and the alkaloid content of *Duboisia myoporoides* R. Br. and the amount of water absorption from its roots. Nettai Nogyo 21:11

Ionkova I, Witte L, Alfermann AW (1989) Production of alkaloids by transformed root cultures of *Datura innoxia*. Planta Med 55:229 (Abstr)

Ionkova I, Witte L, Alfermann AW (1994) Spectrum of tropane alkaloids in transformed roots of *Datura innoxia* and *Hyoscyamus* × *györffyi* cultivated in vitro. Planta Med 60:382–384

Jaziri M, Legros M, Homes J, VanHaelen M (1988) Tropine alkaloids production by hairy root cultures of *Datura stramonium* and *Hyoscyamus niger*. Phytochemistry 27:419–420

Kamada H, Ogasawara T, Harada H (1989) Effects of gibberellin A_3 on growth and tropane alkaloid synthesis in Ri transformed plants of *Datura innoxia*. In: Takahashi N, Phinney BO, MacMillan J (eds) Gibberellins. Springer, Berlin Heideberg New York, pp 241–248

Kamick CR, Saxena MD (1970) On the variability of alkaloids production in *Datura* species. Planta Med 18:266 (Abstr)

Kephalas TA, McLennan CJ, Evans WC, Phillipson JD, Roberts MF (1989) Production of hyoscyamine and scopolamine in transformed roots of *Datura candida* hybrid. J Pharm Pharmacol 41:68 (Abstr)

Knopp E, Strauss A, Wehrli W (1988) Root induction on several Solanaceae species by *Agrobacterium rhizogenes* and the determination of root tropane alkaloid content. Plant Cell Rep 7:590–593

Lebeau p, Janot MH (1956) Traité de pharmacie chimique. Masson, Paris, pp 2765

Loyola-Vargas VM, Miranda-Ham ML (1995) Root culture as a source of secondary metabolites of economic importance. In: Arnason JT, Mata R, Romeo JT (eds) Recent advances in phytochemistry, vol 29. Phytochemistry of medicinal plants. Plenum Press, New York, pp 217–248

Maldonado-Mendoza IE, Ayora-Talavera T, Loyola-Vargas VM (1993) Establishment of hairy root cultures of *Datura stramonium*. Characterization and stability of tropane alkaloid production during long periods of subculturing. Plant Cell Tissue Organ Cult 33:321–329

Maldonado-Mendoza IE, Loyola-Vargas VM (1995) Establishment and characterization of photosynthetic hairy root cultures of *Datura stramonium*. Plant Cell Tissue Organ Cult 40:197–208

Mehra KL (1989) Ethnobotany of old world *Solanaceae*. In: Hawkes JG, Lester RN, Skelding AD (eds) The biology and taxonomy of the Solanaceae. Academic Press, London, pp 161–170

Mendieta RM, Amo SRd (1981) Plantas Medicinales del Estado de Yucatán. INIREB/CECSA, Mexico D.F., pp 1–428

Mugnier J (1988) Establishment of new axenic hairy root lines by inoculation with *Agrobacterium rhizogenes*. Plant Cell Rep 7:9–12

Nee M (1986) Flora de Veracruz. Solanaceae I. INIREB, Xalapa, pp 1–191

Ohkawa H, Kamada H, Sudo H, Harada H (1989) Effects of gibberellic acid on hairy root growth in *Datura innoxia*. J Plant Physiol 134:633–636

Palomino G, Viveros R, Bye RAJ (1988) Cytology of five Mexican species of *Datura* L. (Solanaceae). Southwest Nat 33:85-90

Payne J, Hamill JD, Robins RJ, Rhodes MJC (1987) Production of hyoscyamine by "hairy root" cultures of *Datura stramonium*. Planta Med 53:474–478

Pelt JM, Younos C, Hayon JC (1967a) On the alkaloid constitution of some Solanaceae from Afghanistan. I. Ann Pharm Fr 25:59 (Abstr)

Pelt JM, Younos C, Hayon JC (1967b) On the alkaloid constitution of some Solanaceae from Afghanistan. II. Ann Pharm Fr 25:101 (Abstr)

Petri G, Bajaj YPS (1989) *Datura* spp. in vitro regeneration and the production of tropanes. In: Bajaj YPS (ed) Biotechnology in agriculture and forestry vol 7. Medicinal and aromatic plants II. Springer, Berlin Heidelberg New York, pp 135–161

Rhodes MJC, Robins RJ, Hamill JD, Parr AJ, Hilton MG, Walton NJ (1990) Properties of transformed root cultures. In: Charlwood BV, Rhodes MJC (eds) Secondary products from plant tissue culture. Oxford University Press, Oxford, pp 201–225

Robins RJ, Parr AJ, Payne J, Walton NJ, Rhodes MJC (1990) Factors regulating tropane-alkaloid production in a transformed root culture of a *Datura candida* × *D. aurea* hybrid. Planta 181:414–422

Sáenz-Carbonell L, Loyola-Vargas VM (1996) *Datura stramonium* hairy roots tropane alkaloid content as a response to changes in Gamborg's B5 medium. Appl Biochem Biotechnol 61:321–337

Sáenz-Carbonell L, Maldonado-Mendoza IE, Loyola-Vargas VM (1991) Effect of the nitrogen-source in tropane alkaloid production in *Datura stramonium* hairy root cultures. Plant Physiol 96:96 (Abstr)

Sáenz-Carbonell L, Maldonado-Mendoza IE, Moreno V, Ciau-Uitz R, López-Meyer M, Oropeza C, Loyola-Vargas VM (1993) Effect of the medium pH on the release of secondary metabolites from roots of *Datura stramonium, Catharanthus roseus* and *Tagetes patula* cultured in vitro. Appl. Biochem Biotechnol 38:257–267

Sangwan RS, Ducrocq C, Sangwan Norreel BS (1991) Effect of culture conditions on *Agrobacterium*-mediated transformation in *Datura*. Plant Cell Rep 10:90–93

Sangwan RS, Ducrocq C, Sangwan Norreel B (1993) *Agrobacterium*-mediated transformation of pollen embryos in *Datura innoxia* and *Nicotiana tabacum*: production of transgenic haploid and fertile homozygous dihaploid plants. Plant Sci 95:99–115

Saxena PK, Hammerlindl J, Crosby WL, King J (1990) Introduction of resistance to kanamycin into the protoplasts from a pantothenate-requiring auxotrophic cell line of *Datura innoxia* P. Mill. via direct gene transfer. Plant Sci 70:105–114

Schmidt-Rogge T, Meixner M, Srivastava V, Guha-Mukherjee S, Schieder O (1993) Transformation of haploid *Datura innoxia* protoplasts and analysis of the plasmid integration pattern in regenerated transgenic plants. Plant Cell Rep 12:390–394

Schmidt Rogge T, Weber B, Boerner T, Brandenburg E, Schieder O, Meixner M (1994) Transposition and behavior of the maize transposable element Ac in transgenic haploid *Datura innoxia* Mill.Plant Sci 99:63–74

Scholten HJ, Batterman S, Visser JF (1992) *Scopolia* spp. in vitro culture and the production of scopolamine and hyoscyamine. In: Bajaj YPS (ed) Biotechnology in agriculture and forestry, vol 21. Medicinal and aromatic plants IV. Springer, Berlin Heidelberg New York, pp 314–325

Shimomura K, Satake M, Kamada H (1986) Production of useful secondary metabolites by hairy roots transformed with Ri plasmid. VII Int Congr Plant Cell Tissue Organ Cult Minneapolis, 250 pp

Shimomura K, Sauerwein M, Ishimaru K (1991) Tropane alkaloids in the adventitious and hairy root cultures of solanaceous plants. Phytochemistry 30:2275–2278

Simpson BB, Conner-Ogorzaly M (1986) Economic botany. Plants in our World. McGraw-Hill, New York, pp 1–640

Strauss A (1989) *Hyoscyamus* spp.: in vitro culture and the production of tropane alkaloids. In: Bajaj YPS (ed) Biotechnology in agriculture and forestry, vol 7. Medicinal and aromatic plants II. Springer, Berlin Heidelberg New York, pp 286–314

Waller GR, Kowacki EK (1978) Alkaloid biology and metabolism in plants. Plenum Press, New York, pp 1–294

Wilson PDG, Hilton MG, Robins RJ, Rhodes MJC (1989) Fermentation studies of transformed root cultures. In: Moody GW, Baker PB (eds) Bioreactors and biotransformations. Elsevier, New York, pp 38–51

Yaniv Z, Palevitch D (1982) Effect of drought on the secondary metabolites of medicinal and aromatic plants. In: Atal CK, Kapur BM (eds) A review, in cultivation and utilization of medicinal plants. United Printing Press, New Delhi, pp 2

Yun D-J, Hashimoto T, Yamada Y (1992) Metabolic engineering of medicinal plants: transgenic *Atropa belladonna* with an improved alkaloid composition. Proc Natl Acad Sci USA 89:11799–11803

VIII Genetic Transformation of *Duboisia* Species

T. Muranaka[1], Y. Kitamura[2], and T. Ikenaga[3]

1 Introduction

1.1 Distribution and Importance of *Duboisia*

The genus *Duboisia* (family Solanaceae) is indigenous to Australia and comprises three species. Each species is located in a distinct area: *Duboisia myoporoides* along the eastern seaboard in Australia, *D. leichhardtii* in a very restricted area of Southeast Queensland known locally as the Southwest Burnett, and *D. hopwoodii* in Central and Western Burnett (Griffin 1967).

D. *myoporoides* contains hyoscine or hyoscyamine as the dominant alkaloid, and minor alkaloids which have been isolated include nicotine, anabasine, valeroidine, valtropine, tigloidine, tropine, butropnine, and acetyltropine. *D. myoporoides* appears to exist in distinct chemical varieties or chemical types, depending on the growing location (Coulson and Griffin 1967).

D. *leichharditii* contains scopolamine and hyoscyamine together with minor alkaloids similar to those found in *D. myoporoides*. It shows less alkaloid variability than *D. myoporoides* (Barnard 1952; Griffin 1967).

Leaves of *D. hopwoodii*, one of the few alkaloid-yielding plants, are used by Australian aborigines, who call them pituri. Unlike the other species, the alkaloids of *D. hopwoodii* are derived from pyridine rather than tropane. Nornicotine and nicotine were shown to be the major alkaloids (Luanratana and Griffin 1983).

D. *myoporoides* and *D. leichharditii* have been the major source of scopolamine and atropine for over 40 years. Some 1200 tons/year of the harvested, dried, and powdered leaves have been exported to Germany, Switzerland, and Japan.

[1] Biotechnology Laboratory, Sumitomo Chemical Co., Ltd., Takarazuka, Hyogo 665-8555, Japan
[2] School of Pharmaceutical Sciences, Nagasaki University, Nagasaki, 852-8521 Japan
[3] Faculty of Environmental Studies, Nagasaki University, Nagasaki, 852-8521 Japan

Biotechnology in Agriculture and Forestry, Vol. 45
Transgenic Medicinal Plants (ed. by Y.P.S. Bajaj)
© Springer-Verlag Berlin Heidelberg 1999

1.2 Significance of Transgenic *Duboisia*

Many approaches that include conventional breeding and tissue cultures to increase the productivity of alkaolids of *Duboisia* have been investigated. Ikenaga et al. (1979) succeeded in establishing artificial interspecific F_1 hybrids between *D. leichhardtii* with *D. myoporoides* which yielded more tropane alkaloids than the parent plants. Additionally, tissue culture systems for *Duboisia* sp., which include cell suspension culture (Yamada and Endo 1984; Kitamura et al. 1985a,b), root culture (Endo and Yamada 1985; Yoshioka et al. 1989; Yoshimatsu et al. 1990; Kitamura et al. 1991, 1992; Yukimune et al. 1994a,b,c,d), multiple-shoot culture (Luanratana et al. 1990; Lin and Griffin 1992a,b), protoplast culture (Kitamura 1993), and somatic hybrid culture (Endo et al. 1987, 1988, 1991) were established. The greatest effort was aimed at investigating the possibility of producing tropane alkaloids by tissue cultures of *Duboisia* sp. Tropane alkaloids were found in callus or suspension cells. However, the contents were very low compared with those found in intact plants (Yamada and Endo 1984). Endo and Yamada (1985) found that cultured roots of three species, *D. leichhardtii*, *D. myoporoides*, and *D. hopwoodii*, produced both tropane- and pyridine-type alkaloids. Cultured roots of *D. leichhardtii* showed the highest level of tropane alkaloid production, and repeated selection led to obtaining a high alkaloid-producing line.

Although there are a number of reports on transformed hairy roots of various species of *Duboisia* (see Table 1), the regeneration of transgenic plants from hairy roots is reported here. This is a breakthrough in improving the traits of *Duboisia* plants; to increase the production of tropane alkaloids, to produce only scopolamine (without hyoscyamine), and to promote disease resistance.

2 Genetic Transformation

2.1 Brief Review of Transformation Work

2.1.1 Transformation by Agrobacterium rhizogenes

Root cultures of three species of *Duboisia* were established as described in Section 1.2. An alternative method to maintain the roots in vitro is touoe the causative agent of hairy root disease in plants. Although Deno et al. (1987) reported that the scopolamine content was lower in haiy roots of *D. myoporoides* than in un-transformed roots, this seems to be the result of the poor selection of elite hairy root clones. Mano et al. (1989) established hairy root clones of *D. leichhardtii*. Forty five hairy root clones were cultured for at least five subcultures on hormone-free modified Heller's (HF) medium. It was found that the clones varied considerably in growth rate, alkaloid content, and productivity. One of the hairy root clones, DL-34, produced the highest scopolamine yield and was stable for more than 2 years. We found that another hairy root clone of *D. leichardtii*, DL47-1, released scopolamine into the medium (Muranaka et al. 1992). Among media examined, HF medium with 37 mM nitrate and no ammonium was suitable for scopolamine release. Scopolamine in the medium was recovered efficiently by the use of an Amberlite XAD-2 column.

2.1.2 Transformation by *Agrobacterium tumefaciens*

In *Duboisia* sp., only shooty teratomas and crown galls of a *Duboisia* hybrid were induced by infection with a wild-type *A. tumefaciens* (Subroto et al. 1996a,b).

Table 1 summarizes transformation work on *Duboisia* sp. As described above, no genetically engineered chimeric genes were induced in *Duboisia* plants. In this study, we first established a Ri-binary system for *Duboisia* transformation.

2.2 Methodology/Protocol

The Ri-binary system depends on the fact that the T-DNA derived from a Ti plasmid of *A. tumefaciens* can be mobilized in *trans* by *vir* gene products of the Ri plasmid of *A. rhizogenes* (Simpson et al. 1986).

2.2.1 Construction of Binary Vectors for the Ri-Binary System

Prepare plasmid vectors that have replication origins both for *E. coli* and *Agrobacterium*, T-DNA derived form Ti or Ri plasmid, and selection markers

Table 1. Transformation of *Duboisia* species

Species	Method	Introduced genes	Selection marker	Culture	Reference
Dm	*A. rhizogenes*	T-DNA	–	HR	Deno et al. (1987)
D hybrid	*A. rhizogenes*	T-DNA	–	HR	Knopp et al. (1988)
Dl	*A. rhizogenes*	T-DNA	–	HR	Mano et al. (1989); Mano (1993) Muranaka et al. (1992, 1993a,b)
D hybrid	*A. rhizogenes*	T-DNA	–	HR	Shimomura et al. (1991)
D hybrid	*A. tumefaciens*	T-DNA	–	Shooty teratoma Crown gall	Subroto et al. (1996a,b)
Dl	*A. rhizogenes*	T-DNA/H6H	Km^r	HR, regeneration	This chapter

Dl, *D. leichhardtii*; Dm, *D. myoporoides*; D hybrid, *Duboisia* hybrid (Dl × Dm); Km^r, kanamycine resistance; H6H, hyoscyamine-6β-hydroxylase; HR, hairy root.

in both *Agrobacterium* and plants as shown in Table 2. Both pBI101 and pBI121 were purchased from CLONTECH Laboratories, Inc. (Internet; http://www.clontech.com/).

2.2.2 Introduction of Binary Vectors into A. rhizogenes by Electroporation

1. Grow *A. rhizogenes* ATCC15834 [This bacterium is purchased from American Type Culture Collection (ATCC), Internet; http://www.atcc.org/] in 5 ml of liquid LB medium (Table 3) overnight at 30 °C.
2. Transfer 0.5 ml of cultured *A. rhizogenes* obtained from step 1 to 50 ml of liquid YEB medium (Table 3) for 6–9 h at 30 °C. Allow the culture to attain a density of 1–2 × 10^8 cells/ml.
3. Transfer the culture broth to the centrifuge tube and keep on ice. Sediment the bacteria by spinning in a Hitachi high speed refrigerated centrifuge (Model SCR20B) at 6000 rpm for 5 min at 4 °C. Decant the supernant, and suspend the pellet in 20 ml of ice-cold 10% glycerol. Repeat this step at least three times to remove the culture broth completely.
4. After centrifugation, suspend the pellet in 125 µl of 10% glycerol. Divide each 40 µl of the suspension into 1.5 ml of centrifugation tube. Keep the

Table 2. Binary vectors for Ri binary system

Plasmid	Selection marker	Cloning site	Reference
pBI101 pBI121	Km^r	*Eco*RI, *Sst*I, *Sma*I, *Bam*HI, *Xba*I, *Sal*I, *Hind*III	Jefferson et al. (1987)
pGA492	Km^r	*Hind*III, *Xba*I, *Sst*I, *Hpa*I, *Kpn*I, *Cla*I	An (1987)
pGAH	Km^r, Hg^r	*Hind*III, *Xba*I, *Sst*I, *Hpa*I, *Kpn*I, *Cla*I	Onouchi et al. (1991)

Km^r, kanamycin resistance; Hg^r, hygromycine resistance.

Table 3. Medium for experiments

LB		HF	
Bacto tripton	10 g	$CaCl_2$ $2H_2O$	75 mg
Yeast extract	5 g	$CuSO_4$ $5H_2O$	0.03 mg
NaCl	5 g	FeNaEDTA	30 mg
1 N NaOH	2 ml	H_3BO_3	1 mg
YEB		NaH_2PO_4 $2H_2O$	140 mg
Bacto peptone	5 g	KI	1 mg
Beef extract	5 g	KCl	750 mg
Bacto yeast extract	1 g	$NaNO_3$	600 mg
Sucrose	5 g	$MgSO_4$ $7H_2O$	250 mg
$MgSO_4$	2 mM	$MnSO_4$ $4H_2O$	0.1 mg
Adjust to pH 7.2 by		$ZnSO_4$ $7H_2O$	1 mg
1 N NaOH		Sucrose	30 mg
		Thiamine HCl	1 mg

samples in the freezer at $-80\,°C$ until use. They can be kept for 1 to 2 years.

5. Thaw the suspension cells on ice, and add 1 µl of 50 ng/µl binary plasmid to the tube. Mix gently with an ice-cold sterile glass pipet and transfer to an ice-cold 0.2-cm cuvette.
6. Set the Gene Pulser apparatus (Bio-Rad Laboratories, Internet; http://www.biorad.com/) at 25 µF. Set the Pulse Controller to 200 Ω, the Gene Pulser apparatus to 2.5 kV.
7. Pulse once at the above setting and quickly transfer the suspension to a culture tube containing 1 ml of YEB. Incubate for 1 h at $30\,°C$.
8. Dilute the culture broth with YEB. Streak 1/100 diluted cells (100 µl) on LB plate (1.5% agar) containing suitable antibiotics (pBI vector; 200 µl/ml kanamycine, pGAH; 10 µl/ml tetracycline). Incubate at $30\,°C$ for 2–3 days. Usually more than 100 colonies result.
9. Pick up a single colony and streak on the LB plate with suitable antibiotics (as step 8) for single-cell purification.
10. To confirm the introduction of the binary vector, prepare plasmid DNA from *A. rhizogenes* cells by alkaline lysis.

2.2.3 Transformation of Duboisia Plants

Steps 1–7 described below are to be performed under aseptic conditions using sterile equipment and reagents. Plant tissue should always be handled under a laminar flow hood.

From our results, young stem is very active in producing hairy roots, but only few hairy roots could be obtained from other tissues. Plants grown in a greenhouse can be used for the experiments, but we strongly recommend establishing a tissue culture line first to avoid bacteria or fungi contamination and damage during sterilization with sodium hypochlorite. To establish tissue culture lines, refer to Kitamura et al. (1985a) and Luanratana et al. (1990).

1. Prepare suspensions of *Agrobacterium* with binary vector in 5 ml of liquid YEB containing suitable antibiotics (see Sect. 2.2.2, step 7) at $25\,°C$ overnight.
2. Isolate young stems of tissue-cultured *Duboisia* sp. and cut into 1-2-cm-long pieces. Place the stem segments on HF medium solidified with Gellan gum.
3. Inoculate *Agrobacterium* cells to the cutting point of the stem segments by a needle. Incubate for 1–2 months at $25\,°C$ under dim light to allow hairy root induction.
4. Transfer the induced hairy roots to MS medium (Murashige and Skoog 1962) containing antibiotics for selection for transgenic hairy roots and to eliminate *Agrobacterium*; 50 mg/l kanamycin and 150 mg/l carbenisilline are most suitable. When *Agrobacterium* is growing again, transfer the roots promptly to fresh medium with antibiotics to avoid contamination of the bacteria. It is very difficult to eliminate *Agrobacterium* once it has taken hold.

5. In step 4, some of the isolated roots grow on the medium. Isolate root tips (0.5 cm long) of the growing roots and place them on fresh MS medium, and keep at 25 °C in the dark.
6. After 3–5 months, adventitious shoots or callus appear from some of the root clones. Transfer the root clones containing adventitious shoots and/or callus to fresh MS medium with 0 to 5 mg/l of BA and culture under a 16-h light and 8-h dark cycle at 25 °C for another 2–4 months to allow shoot differentiation.
7. Transfer the developed shoots onto MS medium containing 2 mg/l IBA to allow root to form under a light (16 h) / dark (8 h) cycle at 25 °C for about 1 week, then transfer again to hormone-free MS medium and keep for 1–2 months.
8. Gently break up the solidified MS medium and wash it from the rooted transgenic shoots. Plant the shoots in the soil. Cover the plantlets with transparent plastic for 1–2 weeks. Confirm the transformants by genomic PCR as described below.

Option to isolate transgenic hairy root clones for production of secondary metabolites.

1–3. As the protocol for transgenic plants.
4. Transfer induced hairy roots to MS medium containing antibiotics for selection for transgenic hairy roots and to eliminate *Agrobacterium*; 50 mg/l kanamycin and 150 mg/l carbenisilline are most suitable.

   When *Agrobacterium* is growing again, transfer the roots promptly to a fresh medium with antibiotics to avoid contamination of the bacteria. It is very difficult to eliminate *Agrobacterium* once it takes a hold.
5. In step 4, some of the isolated roots are growing on the medium. Isolate root tips (0.5 cm long) of the growing roots and place on fresh HF medium and keep at 25 °C in the dark.
6. Repeat Step 5 three times to select fast-growing hairy roots.
7. Transfer the selected fast-growing hairy root clones to fresh HF medium. Confirm the transformants by genomic PCR as described below.
8. Transfer the roots to liquid HF medium and measure the alkaloid content.

2.2.4 Confirmation of Transformants by Genomic PCR

A rapid method to confirm the introduction of T-DNA from Ri plasmid, and foreign genes from the binary vector to the genome of transformants based on genomic PCR is described here. This protocol is used for both roots and regenerated plants.

   A nonpolar strain of *A. rhizogenes*, such as ATCC 15834, contains TL- and TR-DNA on its Ri plasmid. *rolB* was first found in TL-DNA; the homologue of *rolB* was also cloned from TR-DNA. We attempted to use both sequences for the marker to distinguish TL- and TR-DNA and found that it was possible to distinguish *rolB* in TL- and TR-DNA by using specific PCR primers (Ikenaga et al. 1995).

Table 4. Oligonucleotides for PCR amplification

Primer		Expected size (bp)
ROLB-TL1	5'-TAGCCGTGACTATAGCAAACCCCTCC-3'	670
ROLB-TL3R	5'-GGCTTCTTTCTTCAGGTTTACTGCAG-3'	
ROLB-TR1	5'-CAACCAGTCCTTTCGTGGTGAGATAA-3'	673
ROLB-TR3R	5'-AGCGCTCGTCCGTGTCTCCCCTGGCC-3'	
H6H-F	5'-ATGGCTACTTTTGTGTCGAACTGGTCTACT-3'	850
H6H-R2	5'-CCCTGTCTCTTGTTGGATCTGTCACTACCC-3'	

1. Prepare oligonucleotides as shown in Table 4 to identify the insertion of the TL-DNA or TR-DNA of Ri plasmid into plant genome by PCR amplification. ROLB-TL1 and ROLB-TR1 each corresponds to the N-terminal coding region of *rolB* in TL-DNA (Slightom et al. 1986) or *rolB* in TR-DNA (Bouchez and Camilleri 1990), ROLB-TL3R and ROLB-TR3R each corresponds to complementary sequence of the C-terminal coding region of *rolB* in TL-DNA (Slightom et al. 1986) or *rolB* in TR-DNA (Bouchez and Camilleri 1990).
2. Prepare adequate oligonucleotides to identify the insertion of the foreign genes on the binary vector into the plant genome by PCR amplification.
3. Extract genomic DNA from independent hairy root clones or shoots according to the protocols in several publications (Murray and Thompson 1980; Liu et al. 1995) or by using a DNA isolation kit ISOPLANT (Nippon Gene, Toyama, Japan).
4. Amplify in a 50-μl reaction volume with 1-μM oligonucleotide primers, 200 μM dNTPs, and 0.5 μ TaKaRa Taq polymerase (Takara Shuzo, Kyoto, Japan) in a buffer containing 10 mM Tris HCl (pH 8.3), 50 mM KCl, 1.5 mM $MgCl_2$, and 0.001% gelatin. Carry out PCR amplification in a programmed temperature-control system (model PC-800, Astec, Tokyo, Japan) using the following parameters: 5 min at 95 °C, 2 min at 55 °C, and 2 min at 72 °C for one cycle, 1 min at 95 °C, 1 min at 55 °C, and 1 min at 72 °C for 29 cycles, followed by an extra cycle with a 10-min extension step at 72 °C.
5. Apply the amplified DNA to 4% NuSieve 3:1 agarose (FMC) gel electrophoresis 100 V for 30 min, then stain with EtBr to identify the amplified band.

2.3 Results and Discussion

2.3.1 Construction of Binary Vector pAH6 for H6H Production

The *parAt* gene was first isolated from tobacco mesophyll protoplasts that were treated with auxin (Takahashi et al. 1990). The gene was highly expressed in both intact roots and hairy roots (Niwa et al. 1994). From the results of *parAt*-GUS hairy root experiments, the *parAt* promoter has a potential for a high level of expression of foreign genes in hairy roots (Muranaka et al. 1994). On the other hand, cDNA that encodes hyoscyamine 6β-hydroxylase

(H6H), that catalyzes the conversion from hyoscyamine to scopolamine, was cloned (Matsuda et al. 1991). For the high level of expression of the H6H cDNA in the hairy roots, the translational fusion gene of *parAt*-H6H was constructed, as shown in Fig. 1. The *Bam*HI site was generated immediately upstream of the initiator codon of the H6H reading frame by the PCR technique; pBH43 (Matsuda et al. 1991) was amplified by Ampli Taq DNA polymerase (Perkin-Elmer Cetus) using two primers that anneal the transcription site of H6H and the pBLUESCRIPT sequence, primer 1; 5'-GG*GGATCC*ATGGCTACTTTTGTGTCC-3', and primer 2; 5'-CAGGAAACAGCTATGAC-3'. The PCR-amplified fragment was digested with *Bam*HI and *Xho*I and ligated to the *Bam*HI and *Xho*I site of pBLUESCRIPT KS+ to generate pMBH. The promoter region of *parAt* (−849 to 20bp) was isolated from P1M (Niwa et al. 1994) as a *Bam*HI fragment and then ligated to the *Bam*HI site of pMBH to generate pMPH. The chimeric *parAt*-H6H gene was then excited as a *Spe*I and *Kpn*I (blunt-ended by T4 DNA polymerase) fragment and subcloned in a plant vector pBI101 (Clontech), the resulting binary vector being referred to as pAH6 (Fig. 1).

2.3.2 Transformation of Duboisia by A. rhizogenes Harboring pAH6 and Establishment of Hairy Root Clones

Binary vector pAH6 was transferred to *A. rhizogenes* ATCC 15834 by electroporation. *A. rhizogens* harboring pAH6 was inoculated on the top of the cutting site of the stem segment on solid HF medium containing 300mg/l carbenisilline and 100mg/l kanamycine. After 1 to 2 months, adventitious roots were induced, as shown in Fig. 2a. We also tried to isolate adventitious roots from leaf segments by both coculture with bacteria and also by direct inoculation of the bacteria. In these experiments, the leaf segment turned brown rapidly and no roots were obtained. Mano et al. (1989) also reported that they could not isolate hairy roots from leaf disks. Stems are suitable for induction of hairy roots in *D. leichhardtii*.

The adventitious roots were transferred to solid HF medium containing 50mg/l of kanamycine and 150mg/l of carbenisilline. The root tips (0.5cm) of the hairy root clones grown under these conditions are transferred to solid MS medium. Two hairy root clones designated A-1 and A-2 grew rapidly on the medium and were used for further experiments.

2.3.3 Confirmation of Introducing Genes by PCR

To confirm the introduction of both pAH6 and pRi, PCR amplification of the genomic DNA from the hairy root clones, A-1 and A-2, was performed by using specific primers as shown in Table 4 that recognize H6H cDNA, *rolB* in TR-DNA, and *rolB* in TL-DNA. As shown in Fig. 3, a 850-bp fragment that was amplified by the primer combination H6H-F/H6H-R2 which recognizes H6H, and a 670-bp band for *rolB* in TL-DNA (primer combination; TL1/

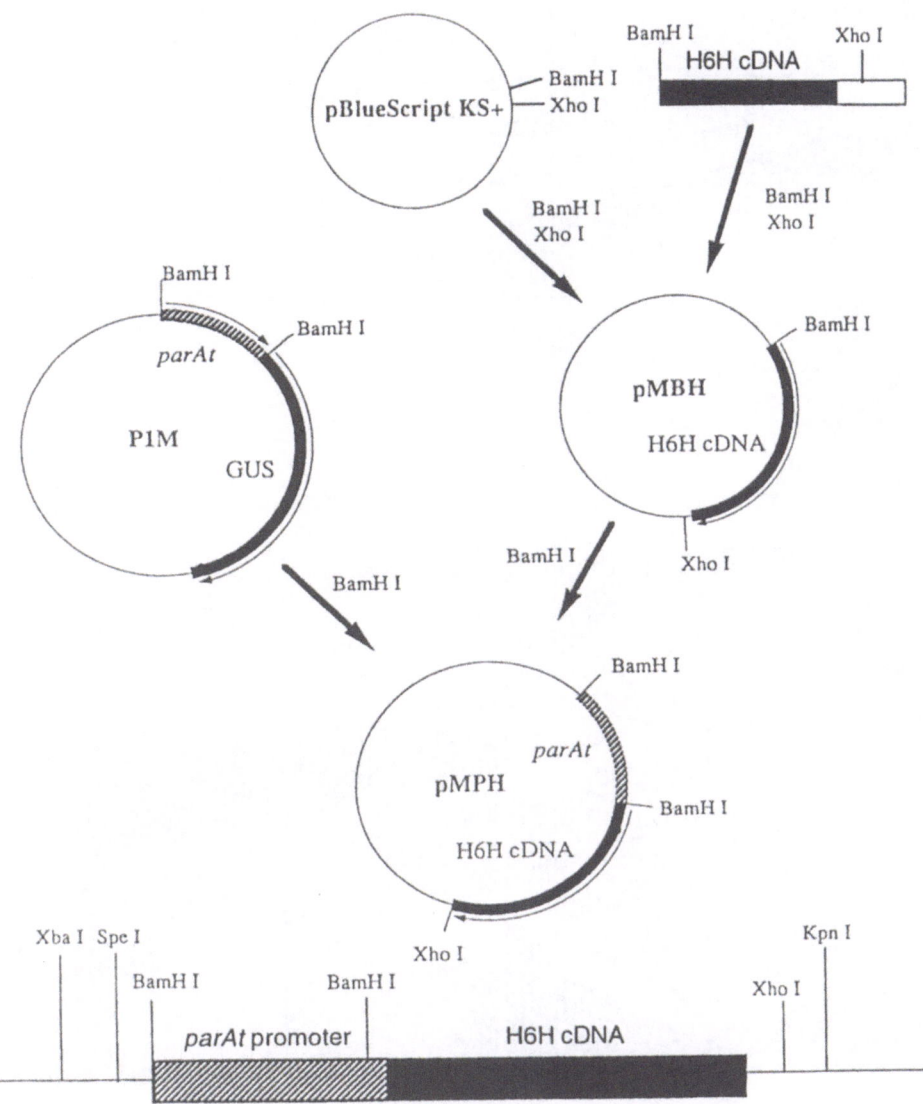

Fig. 1. Construction of the chimeric *parAt*-H6H gene. PMH was digested by *Spe*I and *Kpn*I, then the fragment was subcloned in a binary vector pBI101, the resulting binary vector being referred to as pAH6

TL3R) were found in both A-1 and A-2. In contrast, a 673-bp band for *rolB* in TR-DNA (primer combination; TR1/TR3R) was not detected. These results revealed that both H6H cDNA from binary vector pAH6 and at least TL-DNA from the Ri plasmid of ATCC15834 were integrated into the plant genome of hairy root clones A-1 and A-2. In other experiments, some of the hairy root clones of *D. leichhardtii* possess both TL- and TR-DNA from the PCR-based identification (data not shown). There was also a case where

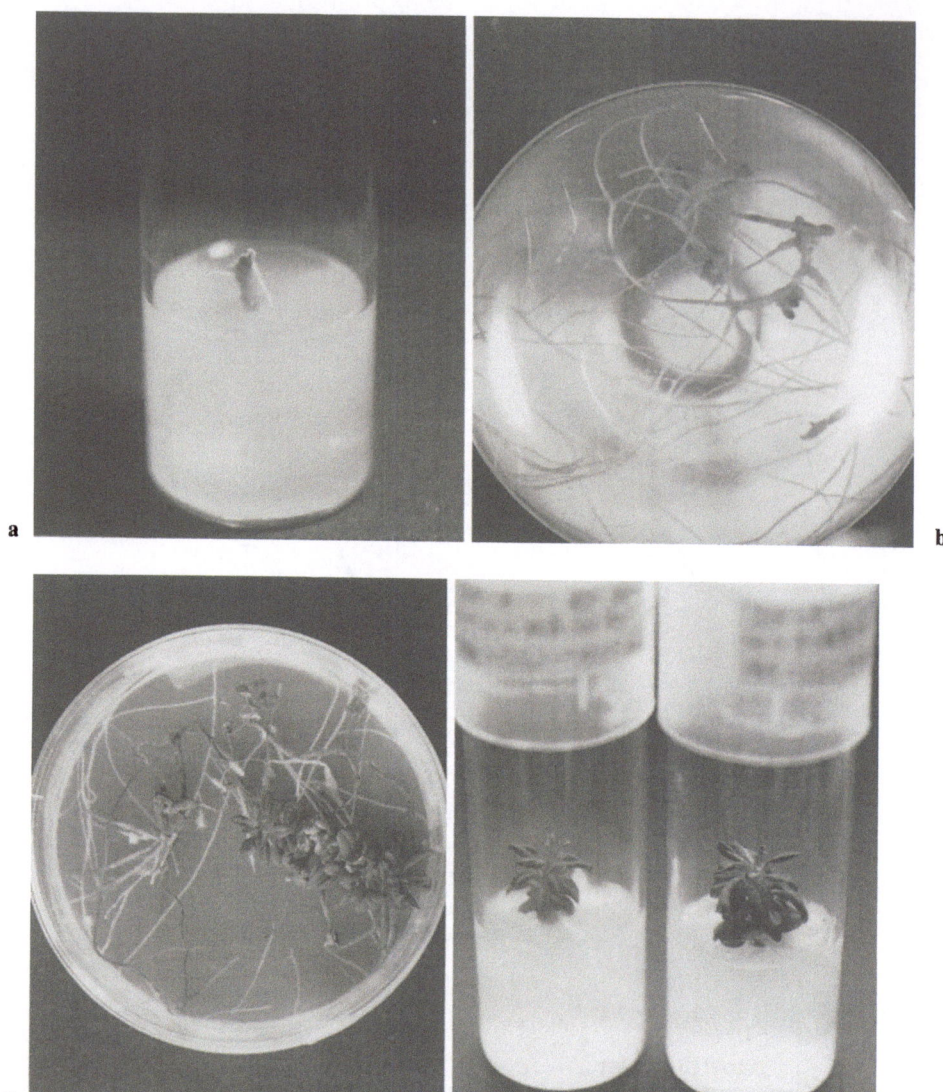

Fig. 2a–d. Transformation of *D. leichhardtii* by Ri binary system. **a** Adventitious root formation. **b** Adventitious bud formation (clone A-1). **c** Shoot development (clone A-1). **d** Regenerated plants (clone A-1)

both TL- and TR-DNA were integrated in the hairy root clones of *Solanum aculeatissimum* (Ikenaga et al. 1995).

2.3.4 Regeneration of Plants from Transgenic Hairy Root Clones

After 4 months' culture of both A-1 and A-2 hairy root clones on solid MS medium, adventitious buds appeared from part of the browned roots, as shown

Fig. 3. PCR amplification of H6H and T-DNA regions from genomic DNA of transgenic hairy root clone of A-1 (*lanes 1, 3, 5*) and A-2 (*lanes 2, 4, 6*). Primer combinations are H6H-F/H6H-R2 to detect H6H cDNA (*lanes 1, 2*), ROLB-TL1/ROLB-TL3R to detect *rolB* in TL-DNA (*lanes 3, 4*), and ROLB-TR1/ROLB-TR3R to detect *rolB* in TR-DNA (*lanes 5, 6*). M; φX174 *Hae*III DNA size marker

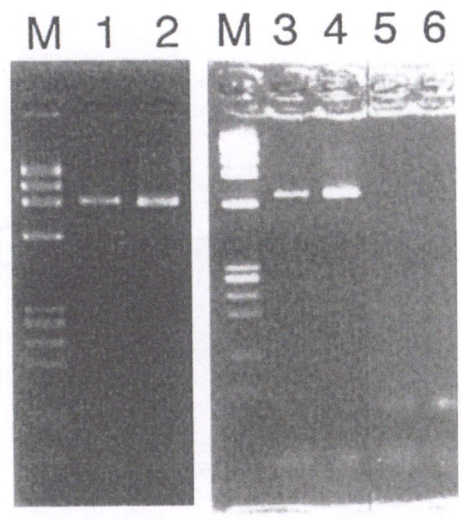

in Fig. 2b. Then the roots with adventitious buds were cultured under the light/dark condition to enhance bud formation and greening (Fig. 2c). Buds were transferred to fresh MS medium containing 1 mg/l BA. After 2 months' culture on the medium, shoots developed in both A-1 and A-2 clones, and were then transferred to MS medium containing 2 mg/l IBA to allow root formation. After 10 to 14 days, roots were formed, then transferred to hormone-free MS medium to enhance root elongation (Fig. 2d). The phenotype of regenerated plants from hairy roots showed the short internode that is typical of the presence of T-DNA of the Ri plasmid (Tepfer 1984), and round-shaped leaves.

The regenerated plants on agar medium were transferred to the soil and maintained in a greenhouse. After 2 months, the average height of the plants was about 20 cm. The tropane alkaloid contents of these regenerated plants is being investigated now.

2.3.5 Bioreactor Studies

A prototype of a reactor system consisting of a 2-l air-lift reactor and a 25-ml column packed with Amberlite XAD-2 was constructed for production of scopolamine by culture of the hairy root clone of *D. leichhardtii*. The culture medium was passed through the column and the eluent from the column was backed into the reactor continuously using a low-pressure pump (Fig. 4a). When the hairy root clone DL47-1 was cultured in the reactor, 245 mg/l of scopolamine was released into the medium during 6 weeks, and 97% scopolamine in the medium was recovered by the column. Scopolamine production was about five times higher in the column-combined reactor than in the reactor without the column. Scopolamine was recovered as the hydrobromide salt with more than 90% purity (Muranaka et al. 1993a).

A modified bioreactor system for continuous production of scopolamine was established to increase productivity. The medium used was continuously

a

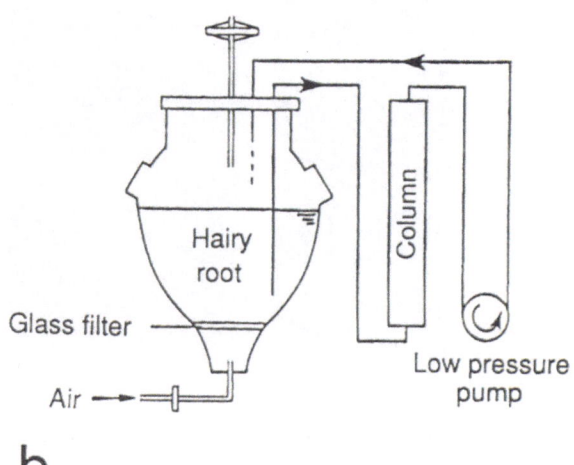

b

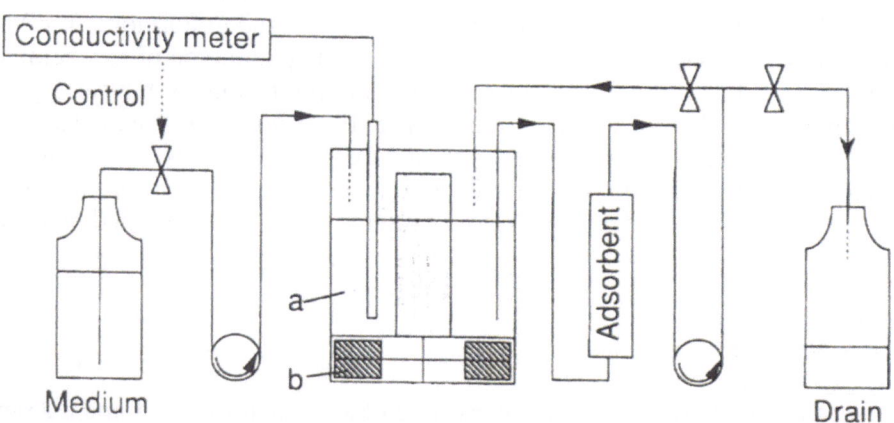

Fig. 4a,b. Bioreactor system for continuous production of scopolamine by culture of *D. leichhardtii* hairy root clones. **a** Column-combined air-lift reactor (Muranaka et al. 1992). **b** Column-combined turbine-blade reactor (Muranaka et al. 1993a). *a* Part for growing hairy roots; *b* part for turbine blades

exchanged during culture to hold the electric conductivity of the medium constant. On culture of the hairy roots in the system for 11 weeks, 0.5 g/l of scopolamine was obtained on the column. When the roots were cultured in the reactor system containing a support such as polyurethane foam and stainless steel mesh to hold the growth of the hairy roots, scopolamine recovery was increased. Thereafter, a two-stage culture, the first stage in the medium for scopolamine release, was carried out in this system by using a turbine-blade reactor with stainless steel mesh as a support (Fig. 4b). Under these conditions, 1.3 g/l of scopolamine was recovered during 11 weeks' culture in the medium for scopolamine release. This value is the highest so far published. This bioreactor system appears suitable for the production of various plant metabolites by culture of transgenic hairy roots.

3 Summary

An Ri-binary vector system for *Duboisia* transformation was first established. The 5'-upstream region of *parAt* that is active in root tissues at a high level was fused to the coding sequence of the hyoscyamine 6β-hydroxylase (H6H) gene to generate the *parAt*-H6H fusion gene, which was introduced into the binary vector for *Agrobacterium*. Hairy roots that carried the fusion gene were obtained by infecting the stem of *D. leichhardtii* with *A. rhizogenes* carrying the fusion gene in the binary vector. Integration of the H6H gene and TR-DNA were confirmed by genomic PCR, using the specific primers for H6H and *rolB* in TR-DNA. After 4 months' culture of hairy root clones on solid MS medium, adventitious buds appeared from part of the browned roots. The roots with adventitious buds were cultured under the light/dark condition to enhance bud formation and greening. After 2 months' culture of buds in MS medium containing 1 mg/l BA, shoots developed, then were transferred to MS medium containing 2 mg/l of IBA to allow the roots to form. Finally, regenerated plants were transferred to the soil and maintained in a greenhouse.

We also established a system by which products released from the cultured hairy root of *Duboisia* into medium can be continuously recovered. This system can be applied to recover a product of the culture of transgenic hairy roots carrying a transgene under the control of the promoters that have a high level of activity in root tissues. By culturing transgenic hairy roots carrying an engineered gene of interest under the control of the *parAt* promoter, (1) a bioconversion system for conversion of a substrate added to the culture medium to a product of interest can be developed, and (2) enhancement of a de novo biosynthesis of a useful substrate from an endogenous precursor substrate to a product can be established.

Acknowledgments. The authors wish to thank Drs. T. Hashimoto and Y. Yamada (Nara Institute of Science and Technology, Japan.) for providing pBH43, and Mr. K. Amano (Nagasaki University, Japan) for performing transformation experiments.

References

An G (1987) Binary Ti vectors for plant transformation and promoter analysis. Methods Enzymol 153:292–305

Barnard C (1952) The *Duboisia* of Australia. Econ Bot 6:3–17

Bouchez D, Camilleri C (1990) Identification of a putative *rolB* gene on the TR-DNA of the *Agrobacterium rhizogenes* A4 Ri plasmid. Plant Mol Biol 14:617–619

Chilton M, Tepfer D, Petit A, Chantal D, Casse-Delbart F, Tempé J (1982) *Agrobacterium rhizogenes* inserts T-DNA into the genomes of the host plant root cells. Nature 295:432–434

Coulson JF, Griffin WJ (1967) The alkaloids of *Duboisia myoporoides*. Planta Med 15:459–466

Deno H, Yamagata H, Emoto T, Yoshioka T, Yamada Y, Fujita Y (1987) Scopolamine production by root cultures of *Duboisia myoporoides*. II. Establishment of a hairy root culture by infection with *Agrobacterium rhizogenes*. J Plant Physiol 131:315–323

Endo T, Yamada Y (1985) Alkaloid production in cultured roots of three species of *Duboisia*. Phytochemistry 24:1233–1236

Endo T, Komiya T, Masumitsu Y, Morikawa H, Yamada Y (1987) An intergeneric hybrid cell line of *Duboisia hopwoodii* and *Nicotiana tabacum* by protoplast fusion. J Plant Physiol 129: 453–459

Endo T, Komiya T, Mino M, Nakanishi K, Fujita S, Yamada Y (1988) Genetic diversity among sublines originating from a single somatic hybrid cell of *Duboisia hopwoodii* + *Nicotiana tabacum*. Theor Appl Genet 76:641–646

Endo T, Hamaguchi N, Eriksson T, Yamada Y (1991) Alkaloid biosynthesis in somatic hybrids of *Duboisia leichhardtii* F. Muell. and *Nicotiana tabacum* L. Planta 183:505–510

Griffin WJ (1967) The alkaloids of *Duboisia leichhardtii* tetrametyl putrescine. Aust J Pharm 48:520–521

Ikenaga T, Saisho Y, Ohashi (1979) Alkaloid contents in leaves of an artificial interspecific F_1 hybrid between *Duboisia myoporoides* and *D. leichhardtii*. Planta Med 35:51–55

Ikenaga T, Oyama T, Muranaka T (1995) Growth and steroidal saponin production in hairy root cultures of *Solanum aculeatissimum*. Plant Cell Rep 14:413–417

Jefferson R, Kavanagh T, Bevan M (1987) GUS fusions: β-glucuronidase as a sensitive and versatile gene fusion marker in higher plants. EMBO J 6:3901–3907

Kitamura Y (1993) Regeneration of plants from protoplasts of *Duboisia*. In: Bajaj YPS (ed) Biotechnology in agriculture and forestry, vol 23. Plant protoplasts and genetic engineering IV. Springer, Berlin Heidelberg New York, pp 18–31

Kitamura Y, Miura H, Sughi M (1985a) Alkaloid composition and atropine-esterase activity in callus and differentiated tissues of *Duboisia myoporoides* R. BR Chem Pharm Bull 33: 5445–5448

Kitamura Y, Miura H, Sugii M (1985b). Change of alkaloid distribution in the regenerated plants of *Duboisia myoporoides* during development. Planta Med 6:489–491

Kitamura Y, Sugimoto Y, Samejima T, Hayashida K, Miura H (1991) Growth and alkaloid production in *Duboisia myoporoides* and *D. leichardtii* root cultures. Chem Pharm Bull 39:1263–1266

Kitamura Y, Taura A, Kajiya Y, Miura H (1992) Conversion of phenylalanine and tropic acid into tropane alkaloids by *Duboisia leichardtii* root cultures. J Plant Physiol 140:141–146

Knopp E, Strauss A, Wehrli W (1988) Root induction on several Solanaceae species by *Agrobacterium rhizogenes* and the determination of root tropane alkaloid content. Plant Cell Rep 7:590–593

Lin GD, Griffin WJ (1992a) Organogenesis and a general procedure for plant regeneration from callus culture of a commercial *Duboisia* hybrid (*D. leichardtii* × *D. myoporoides*). Plant Cell Rep 11:207–210

Lin GD, Griffin WJ (1992b) Scopolamine content of a *Duboisia* hybrid in callus cultures. Phytochemistry 31:4151–4153

Liu YG, Mitsukawa N, Oosumi T, Whittier RF (1995) Efficient isolation and mapping of *Arabidopsis thaliana* T-DNA insert junctions by thermal asymmetric interlaced PCR. Plant J 8:457–463

Luanratana O, Griffin WJ (1983) Alkaloids of *Duboisia hopwoodii*. Phytochemistry 21:449–451

Luanratana O, Lertthamapravit Y, Noymuang W, Buasang W (1990) Micropropagation of *Duboisia myoporoides* by tissue culture methods. Planta Med 56:500–501

Mano Y (1993) Transformation in *Duboisia* spp. In: Bajaj YPS (ed) Biotechnology in agriculture and forestry, vol 22. Plant protoplasts and genetic engineering I. Springer, Berlin Heidelberg New York, pp 190–201

Mano Y, Nabeshima S, Matsui C, Ohkawa H (1986) Production of tropane alkaloids by hairy root cultures of *Scopolia japonica*. Agric Biol Chem 50:2715–2722

Mano Y, Ohkawa H, Yamada Y (1989) Production of tropane alkaloids by hairy root cultures of *Duboisia leichhardtii* transformed by *Agrobacterium rhizogenes*. Plant Sci 59:191–201

Matsuda J, Okabe S, Hashimoto T, Yamada Y (1991) Molecular cloning of hyoscyamine 6β-hydfroxylase, a 2-oxoglutarate-dependent dioxygenase, from cultured roots of *Hyoscyamus niger*. J Biol Chem 266:9460–9464

Mizukami H, Ohbayashi K, Kitamura Y, Ikenaga T (1993) Restriction fragment length polymorphisms (RFLPs) of medicinal plants and crude drugs. I. RFLP probes allow clear identification of *Duboisia* interspecific hybrid genotypes in both fresh and dried tissues. Biol Pharm Bull 16:388–390

Muranaka T, Ohkawa H, Yamada Y (1992) Scopolamine release into media by *Duboisia leichardtii* hairy root clones. Appl Microbiol Biotechnol 37:554–559

Muranaka T, Ohkawa H, Yamada Y (1993a) Continuous production of scopolamine by a culture of *Duboisia leichhardtii* hairy root clone in a bioreactor system. Appl Microbiol Biotechnol 40:219–423

Muranaka T, Kazuoka T, Ohkawa H, Yamada Y (1993b) Characteristics of scopolamine-releasing hairy root clones of *Duboisia leichhardtii*. Biosci Biotechnol Biochem 57:1398–1399

Muranaka T, Niwa Y, Machida Y (1994) A model for a bioconversion system with the promoter of the *parAt* gene, which confers a high level of expression of a transgene in hairy roots. Appl Microbiol Biotechnol 40:841–845

Murashige T, Skoog F (1962) A revised medium for rapid growth and bioassays with tobacco tissue cultures. Physiol Plant 15:473–497

Murray M, Thompson WF (1980) Rapid isolation of high molecular weight plant DNA. Nucleic Acids Res 8:4321–4325

Niwa Y, Muranaka T, Baba A, Machida Y (1994) Organ-specific and auxin-inducible expression of two tobacco *parA*-related genes in transgenic plants. DNA Res 1:213–221

Onouchi H, Yokoi K, Machida C, Matsuzaki H, Oshima Y, Matsuoka K, Nakamura K, Machida Y (1991) Operation of an efficient site-specific recombination system of *Zygosaccharomyces rouxii* in tobacco cells. Nucleic Acids Res 19:6373–6378

Shimomura K, Sauerwein M, Ishimaru K (1991) Tropane alkaloids in the adventitious and hairy root cultures of solanaceous plants. Phytochemistry 30:2275–2278

Simpson R, Spielmann A, Margossian L, McKnight T (1986) A disarmed binary vector from *Agrobacterium tumefaciens* functions in *Agrobacterium rhizogenes*: frequent co-transformation of two distinct T-DNAs. Plant Mol Biol 6:403–415

Slightom J, Durand-Tardif M, Jouanin L, Tepfer D (1986) Nucleotide sequence analysis of TL-DNA of *Agrobacterium rhizogenes* agropine type plasmid. Identification of open reading frames. J Biol Chem 261:108–121

Subroto MA, Hamill JD, Doran PM (1996a) Development of shooty teratomas from several solanaceous plants: growth kinetics, stoichiometry and alkaloid production. J Biotechnol 45:45–57

Subroto MA, Kwok KH, Hamill JD, Doran PM (1996b) Coculture of genetically transformed roots and shoots for synthesis, translocation, and biotransformation of secondary metabolites. Biotechnol Bioeng 49:481–494

Takahashi Y, Niwa Y, Machida Y, Nagata T (1990) Location of the *cis*-acting auxin-responsive region in the promoter of the *par* gene from tobacco mesophyll protoplasts. Proc Natl Acad Sci USA 87:8013–8016

Tepfer D (1984) Transformation of several species of higher plants by *Agrobacterium rhizogenes*: sexual transmission of the transformed genotype and phenotype. Cell 37:959–967

Yamada Y, Endo T (1984) Tropane alkaloid production in cultured cells of *Duboisia leichhardtii*. Plant Cell Rep 3:186–188

Yoshimatsu K, Hatano T, Katayama M, Marumo S, Kamada H, Shimomura K (1990) IAA derivative induced tropane alkaloid production in root cultures of a *Duboisia* hybrid. Phytochemistry 29:3525–3528

Yoshioka T, Yamagata H, Itoh A, Deno H, Fujita Y, Yamada Y (1989) Effects of exogenous polyamines on tropane alkaloid production by a root culture of *Duboisia myoporoides*. Planta Med 55:523–524

Yukimune Y, Hara Y, Yamada Y (1994a) Tropane alkaloid production in root cultures of *Duboisia myoporoides* obtained by repeated selection Biosci Biotechnol Biochem 58:1443–1446

Yukimune Y, Tabata H, Hara Y, Yamada Y (1994b) Increase of scopolamine production in high density culture of *Duboisia myoporoides* roots. Biosci Biotechnol Biochem 58:1447–1450

Yukimune Y, Tabata H, Hara Y, Yamada Y (1994c) Scopolamine yield in cultured roots of *Duboisia myoporoides* improved by a novel two-stage culture method. Biosci Biotechnol Biochem 58:1820–1823

Yukimune Y, Yamagata H, Hara Y, Yamada Y (1994d) Effects of oxygen on nicotine and tropane alkaloid production in cultured roots of *Duboisia myoporoides*. Biosci Biotechnol Biochem 58:1824–1827

IX Genetic Transformation of *Fagopyrum* Species (Buckwheat)

F. Trotin[1,2], C. Quettier-Deleu[1], and J. Vasseur[1]

1 Introduction

1.1 Distribution and importance of the plant

Buckwheat (*Fagopyrum esculentum*) is mainly known as a food grain crop, probably originating from Siberia or China. Being tolerant to poor soils, it became in some regions a basal component in rural alimentation. In France it covered large surfaces during the 19th century before being drastically decreased in culture areas, now insufficient for the present needs (Ferault 1984); a similar phenomenon occurred in the USA. (Pomeranz 1983). This is due to different reasons such as the lack of improvement in culture yields compared to other cereals, a declining taste in the 20th century for traditional buckwheat-based foods, digestibility problems (Kreft et al. 1994) bound to phenolics, and a lower demand in animal feed.

Renewed interest is appearing in many countries, accompanied by a growing consumption of noodles, buns, and pancakes in Japan or China, while attention is paid in many countries to the dietetical advantages of buckwheat flour, its low sodium (Ferault 1984) and high lysin contents (Pomeranz and Robbins 1972), and its possible use by gluten-allergic persons (Wieslander 1996). Buckwheat is at present grown in China, Korea, Russia, Canada, the USA, South Africa, Italy, Poland, Slovenia, and France.

In the pharmaceutical field, buckwheat also presents some potentialities. At present, other higher-yielding plants are used (Paris and Moyse 1976) for industrial extraction of the vasculoprotective flavonoid rutin, but buckwheat leaves remain a possible source (Couch et al. 1946) with up to 8% dry weight rutin (Bruneton 1993). The flour and preparations such as soba contain this natural preventive component (Ohsawa and Tsutsumi 1995).

Tannins, found in the seeds and flour (Luthar 1992; Kreft et al. 1994), are responsible with other phenolics for slower digestibility, but should nevertheless be considered for their therapeutic value.

[1] Laboratoire de Physiologie et de Morphogénèse Végétales, Université des Sciences et Technologies de Lille, 59655 Villeneuve d'Ascq Cedex, France
[2] Laboratoire de Pharmacognosie, Faculté des Sciences Pharmaceutiques et Biologiques, BP 83, 59006 Lille Cedex, France

Biotechnology in Agriculture and Forestry, Vol. 45
Transgenic Medicinal Plants (ed. by Y.P.S. Bajaj)
© Springer-Verlag Berlin Heidelberg 1999

More recently, different pharmacological effects of buckwheat extracts have been described. A drug containing buckwheat plant extract and troxerutin (Fagorutin) decreases more efficiently than rutin alone the concentration of blood plasma malondialdehyde in high-cholesterol-fed rabbits (Wójcicki et al. 1995a), and reduces atherosclerotic deposits (Wojcicki et al. 1995b). This extract seems more effective than troxerutin in the prophylaxis of diabetic retinopathy in man (Archimowicz-Cyrylowska et al. 1996). A buckwheat protein-enriched flour extract induces a marked decrease in plasma cholesterol in rats fed with high-cholesterol diets (Kayashita et al. 1995). Antioxidative activities of hulls and seeds from different cultivars without direct correlation to the rutin contents have been demonstrated (Oomah and Mazza 1996).

1.2 Need for Genetic Transformation

As compared to other grain crops, buckwheat productivity has remained at a low level (De Jong 1972), even in selected varieties (Bar and David 1992), due to various causes, including the indeterminate growth habit of the plant and its low seed set (Pomeranz 1983). This last character mainly depends on a strong heteromorphic self-incompatibility (Nešković et al. 1995), so that classical plant breeding methods are inefficient to improve the yields. An alternative approach could be genetic transformation.

As a source of bioactive secondary metabolites, buckwheat transgenic technology, especially through hairy root cultures, could also be relevant.

Rutin production has been investigated in callus cultures of *Fagopyrum esculentum* (Moumou et al. 1987). Whatever the conditions (Moumou et al. 1992a), the contents remained very low but important levels of flavanols (Fig. 1) were found including two catechins: $(-)$epicatechin and $(-)$epicatechin 3-O-gallate, and two dimeric proanthocyanidins: B_2 and B_2-3'-O-gallate (Moumou et al. 1992b,c).

Catechins and oligomeric proanthocyanidins comprising 2 to 5-6 catechin subunits (Haslam et al. 1989) are the active principles of pharmaceuticals used against chronic venous disease, such as those of *Vitis vinifera* extracts (Endotélon), with a better efficacy than rutin or other flavonoids (Masquelier et al. 1979; Tixier et al. 1984). Flavanols exhibit other properties such as elastase, collagenase (Meunier et al. 1994; Maffei Facino et al. 1994), and cyclooxygenase inhibition (Chang and Hsu 1989), together with important antioxidant and radical scavenging effects. In this last case, the dimeric or trimeric oligomers, their galloyl esters, and those of catechins show maximal activity (Ariga et al. 1988; Hatano et al. 1989; Ricardo Da Silva et al. 1991; Maffei-Facino et al. 1994; Bahorun et al. 1994; 1996). Improvement of the synthesis of the most active phenolics in buckwheat tissues by transformation seems a reasonable target.

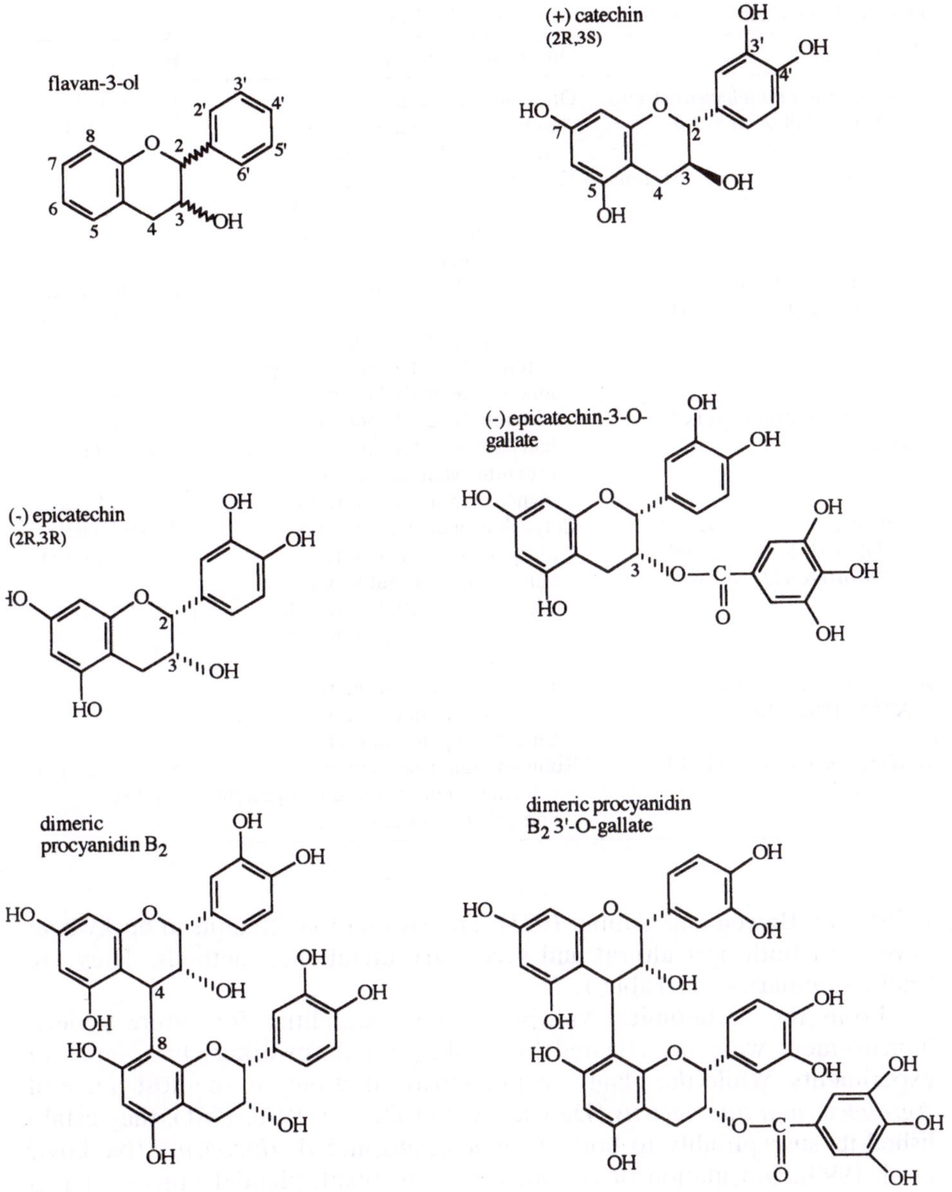

Fig. 1. Structures of catechins and proanthocyanidins in buckwheat hairy roots

2 Genetic Transformation

2.1 General Account

Numerous studies have been conducted on various in-vitro aspects of buckwheat, and the literature reviewed in this Series of books (see Nešković et al.

Table 1. Review table. Genetic transformation in buckwheat

Vector/method	Observations/remarks	Reference
Agrobacterium tumefaciens strains: **A 281**, A 348, Ach 5, A 6	Obtaining viable, transformed, hormone-independent calli. Strain A 281 most virulent	1. Nešković et al. (1990)
Agrobacterium rhizogenes strains: ATCC **15834**, 13332	Hairy root formation (higher efficiency with strain 15834) Susceptibility demonstrated to both *Agrobacterium* species	
A. tumefaciens strains: A 281, **A 281/pGA 472**, A 348, Ach 5	Transformed shoots with A 281/pGA 472, generating transformed fertile plants with heterostyly and Mendelian-type inheritance of NPTII gene	2. Miljuš-Djukić et al. (1992)
A. rhizogenes strains: ATCC **15834**	Noticeable flavanol contents in hairy roots (catechins, dimeric proanthocyanidins, and gallates) higher than in normal roots	3. Trotin et al. (1993)
A. tumefaciens strains: A 281, A 348, Ach 5, A 6, A 208, **A 281/pGA 472**, AC 34-8	Cotyledons inoculated with A 281/pGA 472 regenerate kanamycin-resistant shoots and whole plants with heterostyly and Mendelian-type inherited NPTII gene	4. Nešković et al. (1995)
Agrobacterium rhizogenes strains: ATCC **15834**, 13332	Efficient hairy root formation. *A. rhizogenes* is an alternative vector for foreign gene transfer	
A. rhizogenes strains: MAFF 03-01724	Rutin and important amounts of catechins, dimeric proanthocyanidins and gallates in hairy roots	5. Tanaka et al. (1996)

1986, 1995; Bowen and Cubbin 1993). Transformation attempts in buckwheat have used both agricultural and secondary metabolite methods. They are briefly summarized in Table 1.

From the agronomical viewpoint, the possibilities for future genetic improvement were investigated by Nešković and coworkers in a series of experiments. While the plant was previously cited only in the host range of *Agrobacterium tumefaciens* (De Cleene and De Ley 1976, 1981), they established its susceptibility to both *A. tumefaciens* and *A. rhizogenes* (Nešković et al. 1990). Inoculation of *A. tumefaciens* to plant, plantlet organs, or leaf disks produced long-term cultured, hormone-independent, stable calli with confirmed transformation characters, the most virulent strain being A 281. In further studies, they demonstrated the integration and expression of the neomycin-phosphotransferase (NPT II) gene to buckwheat cells using the binary vector pGA 472 (Miljuš-Djukić et al. 1992; Nešković et al. 1995). Transformed kanamycin-resistant shoot clones gave plantlets displaying NPT II enzyme activity and positive DNA hybridization tests. Regenerated adult flowering plants were obtained which showed heterostyly as in the original plant. After cross-pollination of the transformed clones, the seeds had a Mendelian-type inheritance of kanamycin resistance.

In the method of secondary metabolite, hairy roots were initiated by Trotin et al. (1993) by inoculation of *A. rhizogenes* ATCC 15834 aiming to produce more flavanols than callus cultures. A flavanolic pattern more extensive than in calli was found. The yield in hairy roots was four- to eightfold higher than in normal roots.

Polyphenolics were also investigated in hairy roots induced by *A. rhizogenes* MAFF 03-01724 and grown on different media by Tanaka et al. (1996). A selected clone on MS medium produced rutin (0.24% dry weight, tenfold the content of field-grown roots), together with the same flavanols as in Trotin et al. (1993) at similar levels, plus procyanidin B_1.

2.2 Methodology/Protocol

Obtaining Aseptic Plantlets. Seeds of *Fagopyrum esculentum* Moench from the classical trade (Belgian commercial mark Benelux) were sterilized for 15 min in Ca $(OCl)_2$ (120 g/l), rinsed, impregnated in water for 4–5 h, sown in Petri dishes on solid Heller's medium (Heller 1953) with 10 g/l sucrose. After 5 days in the dark at 22 °C, aseptic plantlets were transferred in tubes for 20 days under a 18/6-h light/dark period.

Bacteria Culture. *Agrobacterium rhizogenes* strain 15834 kept at 4 °C on solid YEB medium was grown for 14–18 h at 28 °C in shaken liquid YEB until absorbance was 0.5–0.8 at 600 nm. A 1/10 v/v dilution of this suspension in the liquid root culture medium (RCM) was used for inoculation.

Hairy Root Culture (HR). Root culture medium (RCM) comprised B_5 minerals (Gamborg et al. 1968), with additional (per liter): myoinositol (100 mg), casein hydrolysate (100 mg), pyridoxine (0.5 mg), nicotinic acid (0.5 mg), thiamine (0.1 mg) (Murashige and Skoog 1962), sucrose 30 g, adjusted to pH 5.6 before sterilization. In some cases, it was solidified with 7 g/l agar.

Aseptic plantlet hypocotyl (ca. 2 cm) and leaf limb fragments (ca. 1 × 2 cm) were scarified, dipped for 2 h in the *Agrobacterium rhizogenes* suspension, and cultured for 2 weeks in the dark at 22 °C on solid RCM. Roots appeared at the wounded zones and were excised. After culture (three transfers, every 4 weeks) on solid RCM containing 300 mg/l carbenicillin, the root tips (2 cm in length) were grown on solid medium without antibiotic for two or three subcultures.

They were finally grown (inoculum ca. 100 mg fr. wt.) in 70 ml liquid RCM in 125-ml flasks on an orbital shaker (100 rpm) in the dark at 22 °C for at least 6 months (transfers every 4 weeks). After positive opine assay, the transformed, fastest-growing and best phenolic-producing line (from hypocotyls) was selected. For analysis, three flasks were harvested every 7 days during a 42-day subculture and either deep frozen or directly studied.

Opine Analysis. Samples of root lines extracted in 1% HCl (Petit et al. 1983) were submitted to paper electrophoresis (Whatman 3 MM; 100 V/cm, 15 min) in formic acid/acetic acid/water buffer (30/60/910, v/v/v) pH 1.9, together with authentic mannopine from Sigma and a sample of agropine kindly supplied by Professor Tempe (Institut des Sciences Végétales, CNRS, Gif-sur-Yvette, France).

Normal Root Culture (NR). Excised roots of 20-day-old aseptic plantlets grown in the dark at 22 °C on solid RCM (four subcultures every 4 weeks) were transferred (inoculum ca. 100 mg fr wt) in flasks on 70 ml liquid RCM, cultivated, and collected as above.

Field-Grown Roots. Seeds from the same origin were sown in open field. Roots from 60-day-old plants were collected, washed, and deep frozen for storage.

Callus Culture. Selected red calli of *Fagopyrum esculentum* previously obtained (Moumou et al. 1987, 1992a) were grown on solid RCM plus 2 mg/l 2,4-D and 0.5 mg/l kinetin at 22 °C under permanent light (transfer every 3 weeks). For phenolic evaluation, 20 calli were harvested after 21 days (Moumou et al. 1992a) and three measures effected.

Polyphenol Analysis. Five g crushed tissue samples were extracted three times with 50 ml methanol during 24 h. After adding 20 ml water, the filtrate was low-pressure evaporated and the aqueous phase extracted by ethyl acetate (4×20 ml). The ethyl acetate phase was dried on Na_2SO_4, concentrated to dryness at 40 °C, and taken in 50 ml methanol (methanolic stock extract).

Thin Layer Chromatography. Extracts were analyzed on silicagel in toluene/acetone/formic acid (3/3/1, v/v/v) according to Lea et al. (1979), where flavanols are separated into main polymerization grade categories. Rf values are in inverse ratio of the subunit number (monomers > dimers > trimers > tetramers > ...). Comparison was made to authentic spots of monomers and dimers (see CLHP) and to the Rf values of Lea et al. (1979). Spots were revealed with anisaldehyde-sulfuric reagent at 105 °C.

TLC chromatograms of hairy and normal roots (day 28) methanolic extracts were scanned and digitalized following the method of Boyer et al. (1993) on a Microtek Color/Gray Scanner connected to a MacIntosh II C Computer, with software freeware program Image (V. 1.41) provided by Dr. Hilbert, Laboratoire de Physiologie Végétale, USTL, Lille. The 100% flavanol value corresponds to the total measured peak areas.

High Performance Liquid Chromatography. The methanolic stock extract filtered on Millipore 0.45 µm was injected (20 µl) on a Merck Lichrosorb RP 18 column (0.5 µm; 4.6 mm id $\times$ 150 mm) and analyzed with an LKB HPLC apparatus (Controller 2152, Pump 2150, Detector 2151) and a Shimadzu CR 5A integrator (Moumou et al. 1992c). Elution (flow rate 0.7 ml/min.) was performed in the order: 0–20 min = 0–13% B in A; 20–40 min = 13–13% B in

A; 40–50 min = 13–25% B in A; 50–80 min = 25–100% B in A (solvent A = acetonitrile/water: 1/9, v/v, pH 2.6; solvent B = acetonitrile/water: 1/1, v/v, pH 2.6). The column temperature was 30 °C, and detection at 280 nm.

Individual flavanols were compared to standard solutions of (+) catechin and (−)epicatechin (Extrasynthese, France) and of (−)epicatechin-3-O-gallate, dimeric procyanidin B_2, and procyanidin B_2-3′-O-gallate previously obtained from buckwheat callus cultures and indentified by Moumou et al. (1992c).

2.3 Results and Discussion

Obtaining Hairy and Normal Roots. A positive response to *Agrobacterium* appeared on plantlet fragments within 2 weeks. The elongating roots from hypocotyl or leaf were excised and cleared of bacteria, then transferred to hormone-free liquid RCM during at least 6 months for this study. Normal roots from the plantlets were treated under the same conditions. During this period, hairy root lines were compared for growth capacity and polyphenolic production. The hairy, fast-growing lines were tested for transformation. Figure 2 shows the opine electrophoregram of four hypocotyl-issued lines (1, 2, 3, 4) compared to authentic agropine and mannopine. Lines 2, 3, and 4 were characterized by similar flavanolic productions and elevated proliferation rates. The fastest-growing line (3), was selected for further studies.

All the analyses on hairy and normal roots were performed after the first 6-month period. At present, the hairy root line issued from hypocotyls has been maintained on hormone-free medium for more than 6 years with the same high-proliferating capacity and a slightly lowered phenolic production.

Hairy and Normal Root Growth. Maximal growth in hairy roots (Fig. 3) was reached at day 21 (720 mg dry wt./flask) representing a ca. 70 times increase, and for normal roots, the increase was ca. 18 times (180 mg dry wt./flask).

Flavanols in Hairy and Normal Roots. Flavanols are localized in the hairy root tips, as shown by the histochemical red coloration with vanillin-hydrochloric reagent (Chalker-Scott and Krahmer 1989) and blue-green coloration with 4-dimethylaminocinnamaldehyde (Feucht and Schmid 1983).

The qualitative HPLC profile is similar in normal and hairy roots with three different catechins: (+)catechin, (−)epicatechin, (−)epicatechin-3-O-gallate, and two dimeric procyanidins: procyanidin B_2 and procyanidin B_2-3′-O-gallate. An analytical graph of hairy root extract is given in Fig. 4. All these compounds were previously identified in the buckwheat calli (Moumou et al. 1992b,c).

Catechins. The production kinetics of individual catechins measured by HPLC during a 42-day subculture are given in Fig. 5A for hairy roots (HR)

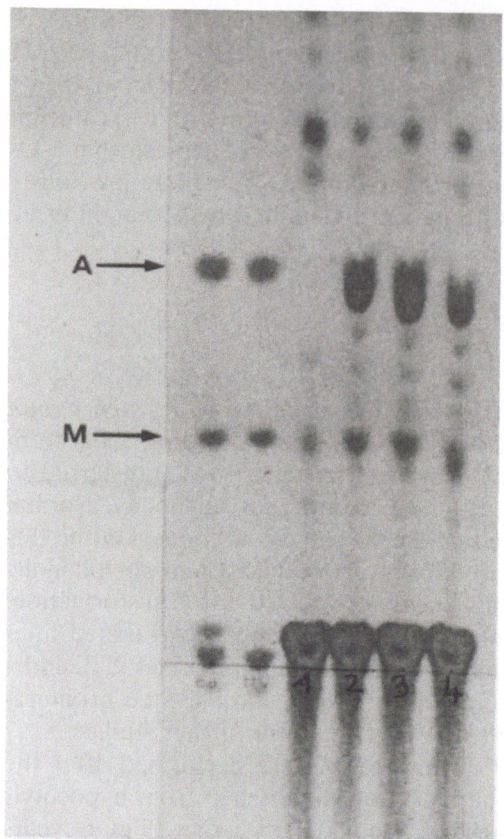

Fig. 2. Opine analysis in extracts of four transformed root lines (*1, 2, 3, 4*) induced from hypocotyls. *A* Agropine; *M* mannopine. (unpubl.)

and Fig. 5B for normal roots (NR). Maximal yields were reached at day 28 for normal roots and sooner (day 21) in hairy roots at the exception of (+)catechin (day 28). Hairy and normal roots (Table 2) present a similar total catechin content and a clear dominance of the esterified derivative (−)epicatechin-3-O-gallate (10.5 mg/g dry wt in normal roots and 10.1 mg/g dry wt in hairy roots, respectively). On the other hand, the nongalloylated catechin synthesis differ; more (−)epicatechin is found in normal roots and more (+)catechin in hairy roots.

Dimers. The highest dimeric procyanidin B_2 values (Fig. 6) were sooner reached (day 21) in normal roots than in hairy ones (day 35). Maximal content in procyanidin B_2-3′-O-gallate was simultaneously found at day 28 in both tissues. In both root types (Table 3, Fig. 6), this galloylated dimer was the major compound, while B_2 dimer production was weak; but hairy roots clearly

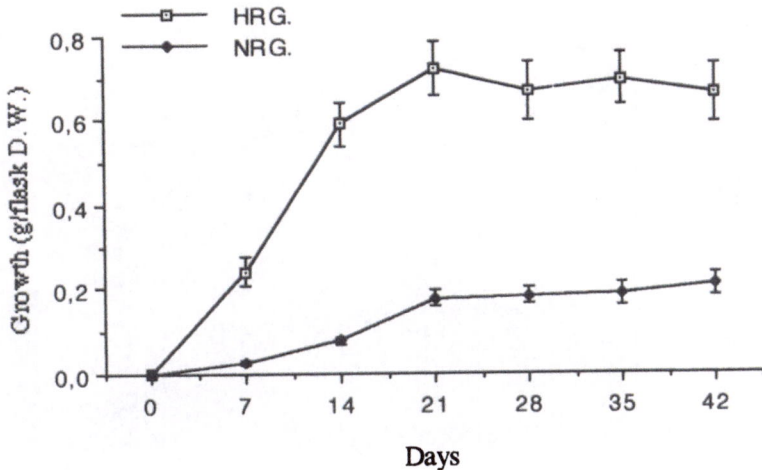

Fig. 3. Growth time-course of buckwheat hairy (*HR G.*) and normal roots (*NR G.*) during a 42-day subculture. (After Trotin et al. 1993)

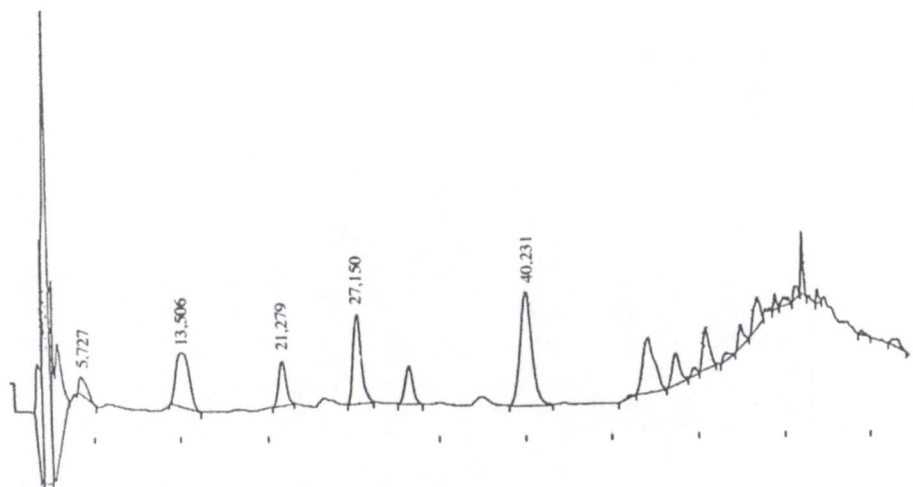

Fig. 4. HPLC analysis of buckwheat hairy root methanolic extract. Peak identification (RT min): *5.727* (+)catechin; *13.506* (−)epicatechin; *21.279* procyanidin B$_2$; *27.150* procyanidin B$_2$-3′-O-gallate; *40.231* (−)epicatechin-3-O-gallate (unpubl.)

Table 2. Maximal catechin production (mg/g dry weight) in hairy and normal roots

	(+)Catechin	(−)Epicatechin	(−)Epicatechin-3-O-gallate	Total
Hairy roots	7.8	2.1	10.1	20
Normal roots	3.6	6.4	10.5	20.5

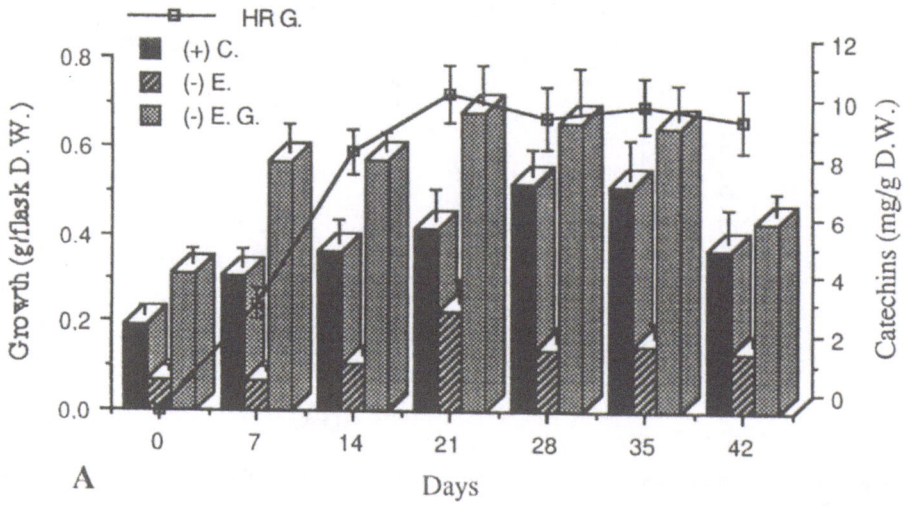

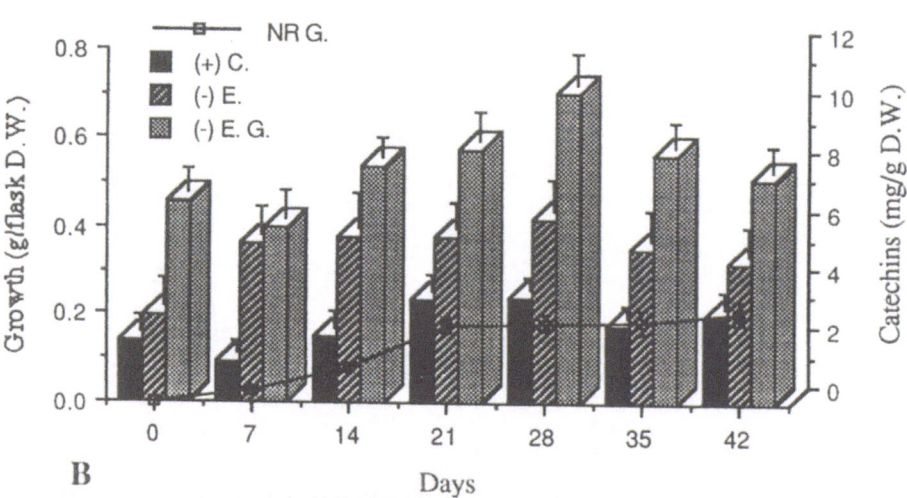

Fig. 5A,B. Catechin production (mg/g dry wt.) and growth (g/flask day wt.) of **A** hairy root (*HR*) and **B**: normal root (*NR*). Legend: (+)C. (+)catechin; (−)E. (−)epicatechin; (−)E.G. (−)epicatechin-3-O-gallate; *HR G*. hairy root growth; *NR G*. Normal root growth. (After Trotin et al. 1993)

Table 3. Maximal dimeric procyanidin production (mg/g dry wt.) in hairy and in normal roots

	B_2 dimer	B_2-3′-O-gallate dimer	Total
Hairy roots (HR)	0,8	8,6	9,4
Normal roots (NR)	0,8	4,2	5

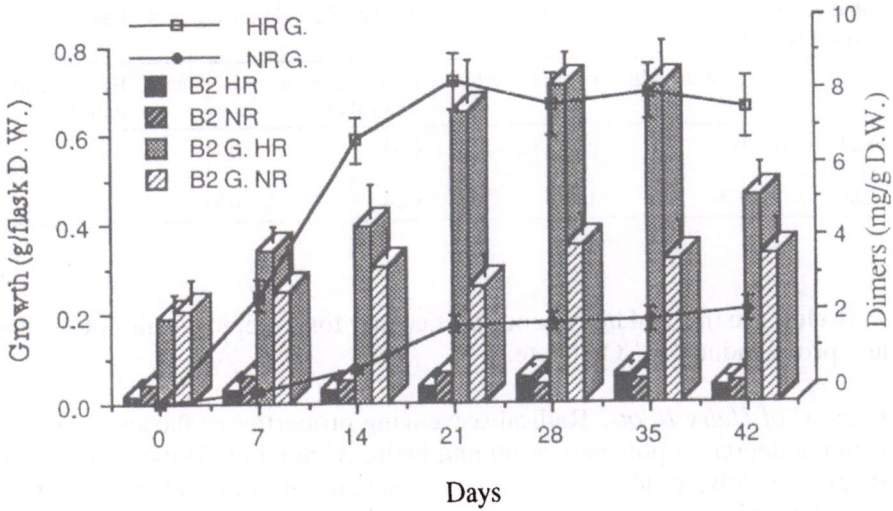

Fig. 6. Production of dimeric procyanidins (mg/g dry wt) and growth (g/flask dry wt) of hairy (*HR*) and normal roots (*NR*). B_2 Procyanidin B_2; B_2 G. procyanidin B_2-3′-O-gallate; *HR G*. hairy root growth; *NR G*. normal root growth. (After Trotin et al. 1993)

Table 4. Respective proportions of the main flavanolic categories in hairy and normal roots methanolic stock extracts determined by TLC densitometry. (unpubl.)

	Hairy roots (HR) (%)	Normal roots (NR) (%)
Catechins	10.65	11.03
Dimers	16.68	9.12
Trimers	16.17	8.25
Tetramers	14.37	12.90
Higher polymers	42.13	58.70
Total	100	100

differed with nearly twice more procyanidin B_2-3′-O-gallate (8.6 mg/g dry wt) and total dimers.

Main Flavanolic Classes. The TLC densitometric evaluation of the main flavanolic classes showed (Table 4) that the more active dimeric and trimeric procyanidins are more abundant in hairy roots than in normal roots (1.82- and 1.96-fold respectively) while monomer (catechin) contents are analogous. The same observation was made in HPLC for total catechins and dimers (Tables 2, 3).

Composition of Calli and Field-Grown Roots. The comparative contents of these tissues are presented in Table 5. Field-grown roots contained far less flavanols than hairy an normal roots (Tables 2, 3) while the levels in red calli

Table 5. Flavanol production (mg/g dry wt) in 60-day-old field-grown buckwheat root, and selected red calli

	(+)Catechin	(−)Epicatechin	(−)Epicatechin-3-O-gallate	B$_2$ dimer	B$_2$-3′-O-gallate dimer
Field-grown roots	0.13	0.20	0.43	0.11	0.47
Calli	1.34	4.01	6.24	0.84	8.35

were closer to those of hairy roots, particularly for (−)epicatechin-3-O-gallate and procyanidin B$_2$-3′-O-gallate.

Interests of Hairy Roots. Radical scavenging properties of flavanols increase with the degree of polymerization and hydroxyl number (Hatano et al. 1989; Ricardo da Silva et al. 1991). Moreover, gallates are more efficient than the nonesterified corresponding compound, so that the activities are in the order: monomers (catechins) < monomer gallates # dimers < dimer gallates < trimers < tetramers.

Field-grown roots were poor in these phenolics. Calli presented higher contents in total flavanols and gallates. Better productions were measured in hairy and normal roots with similar values of (−)epicatechin-3-O-gallate and total catechins. Nevertheless, hairy roots showed together the optimal contents in total dimeric procyanidins, in procyanidin B$_2$-3′-O-gallate and in the both gallates (18.7 mg/g dry wt.).

Taking account of the differential growth rates, the interest of hairy roots becomes more obvious. The growth rate of calli was too low (ca. seven fold the inoculum value in 21 days). The maximal productivity yields in hairy roots expressed per flask were, in particular (Fig. 7) for (+)catechin ca. eightfold higher (5.4 vs. 0.7 mg), for (−)epicatechin-3-O-gallate ca. fourfold higher (7.2 vs. 1.9 mg) and (Fig. 8) for procyanidin B$_2$-3′-O-gallate ca. eightfold higher (5.9 vs. 0.7 mg) than in normal roots. Thus, hairy roots appear to be a better source of active flavanolic structures.

3 Summary and Conclusions

In order to improve the levels of bioactive flavanols already produced by a buckwheat selected callus line, hairy roots were induced by Agrobacterium from aseptic plantlets. Their growth and flavanolic production were compared to those of in vitro normal roots, field-grown roots, and calli. While field-grown roots were poor, the catechin and proanthocyanidin contents per weight unit were higher in calli, normal, and hairy roots, with a slight advantage for these last.

The flavanolic pattern, as in the calli line, comprised (+)catechin, (−)epicatechin, (−)epicatechin-3-O-gallate, procyanidin B$_2$, and procyanidin

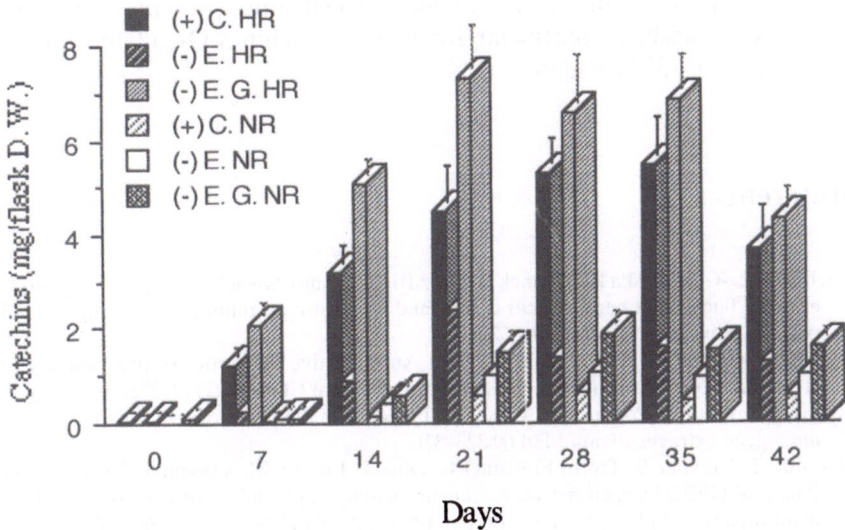

Fig. 7. Catechin productivity (mg/flask dry wt) in hairy (*HR*) and normal roots (*NR*) cultures of buckwheat. Legend, see Fig. 5. (After Trotin et al. 1993)

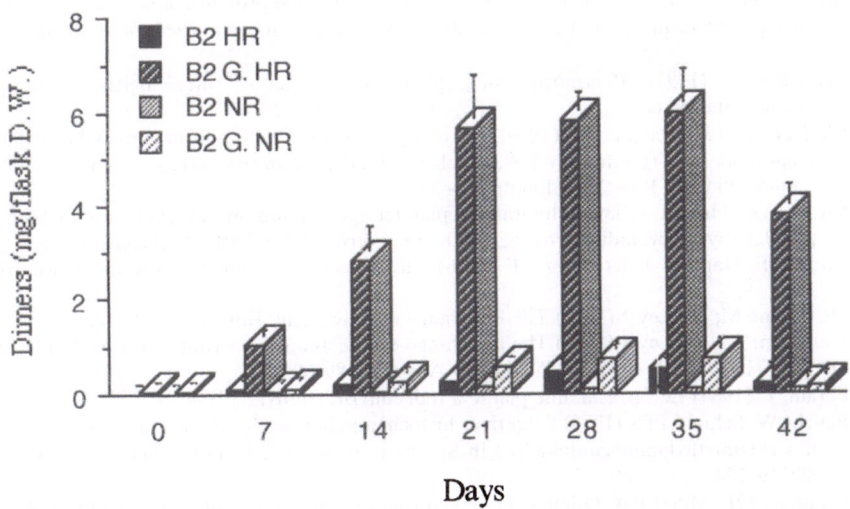

Fig. 8. Dimeric procyanidins productivity (mg/flask dry wt) of hairy (*HR*) and normal roots (*NR*). Legend see Fig. 6. (After Trotin et al. 1993)

B_2-3′-O-gallate. Epicatechin-3-O-gallate was the major monomer produced in both hairy and normal roots while their respective contents in (+) catechin and (−) epicatechin differed. Hairy roots produced nearly twice as much dimeric procyanidin B_2-3′-O-gallate (8.6 mg/g dry wt.) and total dimers (9.4 mg/g dry wt.) than the normal roots.

The hairy root line, with its higher growth rate, presented the best productivity by flask, in particular for (−)epicatechin-3-O-gallate and dimeric procyanidin B_2-3′-O-gallate.

References

Archimowicz-Cyrylowska B, Adamek B, Drozdzik M, Samochowiek L, Wójcicki J (1996) Clinical effect of buckwheat herb, *Ruscus* extract and troxerutin on retinopathy and lipids in diabetic patients. Phytother Res 10:659–662

Ariga T, Koshiyama I, Fukushima D (1988) Antioxidative properties of procyanidins B-1 and B-3 from Azuki beans in aqueous systems. Agric Biol Chem 52:2717–2722

Bahorun T, Trotin F, Pommery J, Vasseur J, Pinkas M (1994) Antioxidant activities of *Crataegus monogyna* extracts. Planta Med 60:323–328

Bahorun T, Gressier B, Trotin F, Brunet C, Dine T, Luyckx M, Vasseur J, Cazin M, Cazin JC, Pinkas M (1996) Oxygen species scavenging activity of phenolic extracts from hawthorn fresh plant organs and pharmaceutical preparations. Arzneim Forsch 46:1086–1089

Bar C, David H (1992) Petites céréales: avoine, seigle, épeautre, sarrasin: variétés, débouchés, qualité. Perspect Agric 171:69–86

Bowen IH, Cubbin IJ (1993) *Fagopyrum esculentum* Moench. (Buckwheat): In vitro culture and the production of rutin. In: Bajaj YPS (ed) Biotechnology in agriculture and forestry, vol 24. Medicinal and aromatic plants V. Springer, Berlin Heidelberg New York, pp 202–228

Boyer C, Hilbert JL, Vasseur J (1993) Embryogenesis-related protein synthesis and accumulation during early acquisition of somatic embryogenesis competence in *Cichorium*. Plant Sci 93:41–53

Bruneton J (1993) Pharmacognosie, phytochimie, plantes medicinales. Technique et Documentation ed, Paris

Chalker-Scott L, Krahmer RL (1989) Microscopic studies of tannin formation and distribution in plant tissues. In: Hemingway RW, Karchesy JJ (eds) Chemistry and significance of condensed tannins. Plenum Press, London pp 345–368

Chang WC, Hsu FL (1989) Inhibition of platelet aggregation and arachidonate metabolism in platelets by procyanidins. Prostaglandins Leukotrienes Ess Fatty Acids 38:181–188

Couch JF, Naghski J, Krewson CF (1946) Buckwheat as a source of rutin. Science 103:197–198

De Cleene M, De Ley J (1976) The host range of crown gall. Bot Rev 42:389–486

De Cleene M, De Ley J (1981) The host range of infectious hairy root. Bot Rev 47:147–194

De Jong H (1972) Buckwheat. Field Crop Abstr 25:389–395

Ferault C (1984) Le sarrasin: une plante à redécouvrir. Cultivar 172:44–45

Feucht W, Schmid PPS (1983) Selektiver histochemischer Nachweis von Flavanen (Catechinen) mit p-Dimethylaminozimtaldehyd in Sprossen einiger Obstgehölze. Gartenbauwissenschaft 48:119–124

Gamborg OL, Miller RA, Ojima K (1968) Nutrient requirement suspensions cultures of soybean root cells. Exp Cell Res 50:151–158

Haslam E, Lilley TH, Cai Y, Martin R, Magnolato D (1989) Traditional herbal medicines: the role of polyphenols. Planta Med 55:1–8

Hatano T, Edamatsu R, Hiramatsu M, Mori A, Fujita Y, Yasuhara T, Yoshida T, Okuda T (1989) Effects of the interaction of tannins with co-existing substances. VI. Effects of tannins and related polyphenols on superoxide anion radical, and on 1, 1-diphenyl-2-picrylhydrazyl radical. Chem Pharm Bull 37:2016–2021

Heller R (1953) Recherche sur la nutrition minérale des tissus végétaux cultivés in vitro. Ann Sci Nat Bot Biol Veg 14:1–233

Kayashita J, Shimaoka I, Nakajoh M (1995) Hypocholesterolemic effect of buckwheat protein extract in rats fed cholesterol-enriched diets. Nutr Res 15:691–698

Kreft I, Bonafaccia G, Zigo A (1994) Secondary metabolites of buckwheat and their importance in human nutrition. Prehrambeno-Technol Biotechnol Rev 32:195–197

Lea AGH, Bridle P, Timberlake CF, Singleton VL (1979) The procyanidins of white grape and wines. Am J Enol 30:289–300

Luthar Z (1992) Polyphenol classification and tannin content of buckwheat seeds (*Fagopyrum esculentum* Moench). Fagopyrum 12:36–42

Maffei-Facino R, Carini M, Aldini G, Bombardelli E, Morazzoni P, Morelli R (1994) Free radical scavenging action and anti-enzyme activities of procyanidines from *Vitis vinifera*. Arzneim Forsch/Drug Res 44:592–601

Masquelier J, Dumon MC, Dumas J (1979) Stabilisation du collagène par les oligomères procyanidoliques. Acta Ther 7:101–104

Meunier MT, Villie F, Bastide P (1994) Etude de l'interaction des oligomères proanthocyanidoliques de *Cupressus sempervirens* L. sur l'élastase et les élastines. J Pharm Belg 49:453–461

Miljuš-Djukić J, Nešković M, Ninković S, Crkvenjakov R (1992) *Agrobacterium*-mediated transformation and plant regeneration of buckwheat (*Fagopyrum esculentum* Moench.). Plant Cell Tissue Organ Cult 29:101–108

Moumou Y, Trotin F, Pinkas M, Dubois J, Vasseur J (1987) Production de polyphénols par des colonies tissulaires de Sarrasin. Ann Pharm Fr 45:255–260

Moumou Y, Trotin F, Dubois J, Vasseur J, El-Boustani E (1992a) Influence of culture conditions on polyphenol production by *Fagopyrum esculentum* tissue cultures. J Nat Prod 55:33–38

Moumou Y, Vasseur J, Trotin F, Dubois J (1992b) Catechin production by callus cultures of *Fagopyrum esculentum*. Phytochemistry 31:1239–1241

Moumou Y, Trotin F, Vasseur J, Vermeersch G, Guyon R, Dubois J, Pinkas M (1992c) Procyanidin production by *Fagopyrum esculentum* callus cultures. Planta Med 58:516–519

Murashige T, Skoog F (1962) Revised medium for rapid growth and bio-assays with tobacco tissue cultures. Physiol Plant 15:473–497

Nešković M, Srejovic V, Vujicic R (1986) Buckwheat: (*Fagopyrum esculentum* Moench.). In: Bajaj YPS (ed) Biotechnology in agriculture and forestry, vol 2. Crops I. Springer, Berlin Heidelberg New York, pp 579–593

Nešković M, Vinterhalter B, Miljuš-Djukić J, Ninković S, Vinterhalter D, Jovanović V, Kneževič J (1990) Susceptibility of buckwheat (*Fagopyrum esculentum* Moench.) to *Agrobacterium tumefaciens* and *A. rhizogenes*. Fagopyrum 10:57–61

Nešković M, Miljuš-Djukić J, Ninković S (1995) Genetic transformation in *Fagopyrum esculentum* (buckwheat). In: Bajaj YPS (ed) Biotechnology in agriculture and forestry, vol 34. Plant protoplasts and genetic engineering VI. Springer, Berlin Heidelberg New York, pp 171–182

Ohsawa R, Tsutsumi T (1995) Intervarietal variations of rutin content in common buckwheat flour (*Fagopyrum esculentum* Moench). Euphytica 86:183–189

Oomah BD, Mazza G (1996) Flavonoids and antioxidative activities in buckwheat. J Agric Food Chem 44:1746–1750

Paris RR, Moyse H (eds) (1976) Precis de matière médicale, vol 1. Masson, Paris

Petit A, David C, Dahl GA, Ellis JG, Guyon P (1983) Further extension of the opine concept: plasmids in *Agrobacterium rhizogenes* cooperate for opine degradation. Mol Gen Genet 190:204–214

Pomeranz Y (1983) Buckwheat: structure, composition and utilization. CRC Crit Rev Food Sci Nutr 19:213–257

Pomeranz Y, Robbins GS (1972) Amino acid composition of buckwheat. J Agric Food Chem 20:270–274

Ricardo Da Silva JM, Darmon N, Fernandez Y, Mitjavila S (1991) Oxygen free radical scavenger capacity in aqueous models of different procyanidins from grape seeds. J Agric Food Chem 39:1549–1552

Tanaka N, Yoshimatsu K, Shimomura K, Ishimaru K (1996) Rutin and other polyphenols in *Fagopyrum esculentum* hairy root. Nat Med 50:269–272

Tixier JM, Godeau G, Robert AM, Hornebeck W (1984) Evidence by in vivo and in vitro studies that binding of pycnogenols to elastin affects its role of degradation by elastases. Biochem Pharmacol 33:3933–3939

Trotin F, Moumou Y, Vasseur J (1993) Flavanol production by *Fagopyrum esculentum* hairy root and normal root cultures. Phytochemistry 32:929–931

Wieslander G (1996) Review on buckwheat allergy. Allergy 51:661–665

Wójcicki J, Samochowiec L, Gonet B, Juźwiak S, Dabrowska-Zamojcin E, Katdońska M, Tustanowski S (1995a) Effect of buckwheat extract on free radical generation in rabbits administered high fat diet. Phytother Res 9:323–326

Wójcicki J, Barcew-Wisniewska B, Samochowiec L, Rożewicka L (1995b) Extractum Fagopyri reduces atherosclerosis in high-fat diet fed rabbits. Pharmazie 50:560–562

X Genetic Transformation of *Glycyrrhiza uralensis* (Licorice) and Related Species

M. YAMAZAKI and K. SAITO

1 Introduction

Licorice (liquorice) consists of the dried roots and stolons of plants in the genus *Glycyrrhiza* (family Leguminosae) and is one of the oldest herbs, known for 3000 years (Stuart 1979; Trease and Evans 1983). Among 30 species in this genus, a few are of economical interest, for example, *Glycyrrhiza glabra* L., *G. glabra* L. var. *glandulifera* Reg. et Herd., *G. uralensis* Fisch. et D.C., and *G. echinata* L. (Henry et al. 1991). These plants are native to temperate regions and are distributed from Spain, central and eastern Europe, Russia, and the near East to eastern China.

In the food and tobacco industries, the dried roots of these plants, Glycyrrhizae Radix, are widely used as a sweetener. The sweet taste is mainly due to glycyrrhizin, a triterpenoid saponin, produced specifically in the plants of the genus *Glycyrrhiza*. The taste of glycyrrhizin is 150-fold sweeter than that of sucrose. Licorice is also used as a traditional crude drug in China and Japan. A number of traditional Chinese/Japanese prescriptions (KANPO) contain Glycyrrhizae Radix as a main component.

The antiinflammatory and antiulcerative activities of licorice are attributed essentially to the actions of glycyrrhizin and its aglycone glycyrrhetinic acid. Antiallergic and antihepatitic properties are also known. Recently, licorice has become a candidate for treatment of AIDS (acquired immune deficiency syndrome); (Hotta et al. 1989).

G. uralensis Fisch. et D.C. (Manchurian licorice) is distributed mainly in Siberia, and the northeastern part of China. This plant contains as much glycyrrhizin as the other varieties.

Since licorice plants can grow only in limited regions in the world in the temperate zone of the Eurasian Continent, some countries need to import large quantities of licorice. The USA is the main importer, with 20000 tons handled by only one company. (Henry et al. 1991). Japan imports 6000–8000 tons/year, the greatest amount of all crude drugs imported into Japan. In future, it will be necessary to enhance the producibility of the specific secondary metabolites, for instance, glycyrrhizin and flavonoids, and to confer agronomically useful traits such as herbicide resistance and disease resistance, by

Faculty of Pharmaceutical Sciences, Chiba University, Yayoi-cho 1-33, Inage-ku, Chiba 263-8522, Japan

Biotechnology in Agriculture and Forestry, Vol. 45
Transgenic Medicinal Plants (ed. by Y.P.S. Bajaj)
© Springer-Verlag Berlin Heidelberg 1999

means of genetic engineering. The transformation method, therefore, is needed for the purpose of introduction of foreign genes.

2 Genetic Transformation

The binary vector system based on the *Agrobacterium*-Ri plasmid can be used efficiently to produce transformed hairy roots containing the T-DNAs of the helper Ri plasmid and of a vector based on a disarmed mini Ti plasmid (Hamill et al. 1987). This technique depends on the fact that the T-DNA derived from the Ti plasmid can be mobilized by *vir* gene products of the Ri plasmid. This method provides a simple means to obtain transgenic tissues containing any desired foreign genes, because double transformants can be expected in 60% of the cases (Stougaard et al. 1987). A number of transgenic medicinal plants have been obtained by an Ri binary vector (Saito et al. 1990a,b, 1991a,b,c, 1992a,b; Saito 1993).

2.1 Protocol

Plasmids and Bacteria. The plasmid pGSGluc1 (from Plant Genetic Systems, Gent, Belgium) is a binary vector containing chimeric *neo* and *gus* genes driven by the T-DNA TR 1′ and 2′ promoters, respectively (Fig. 1). This plasmid was conjugatively transferred by triparental mating using pRK2013 as a helper plasmid (Figurski and Helinski 1979) into *A. rhizogenes* Rif[R]

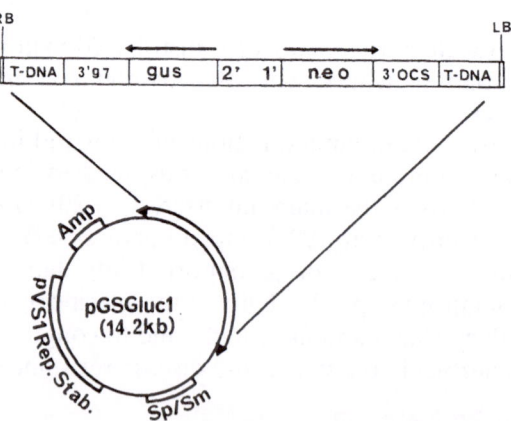

Fig. 1. Schematic presentation of a binary vector pGSGluc1. The chimeric *neo* and *gus* genes are driven by TR 1′ and 2′ promoters, respectively. *3′OCS* 3′ region of octopine synthase gene; *3′g7* 3′ region of gene 7; *RB* right border; *LB* left border

harboring an agropine-type Ri plasmid, pRi15834. The transconjugant was selected on a YEB plate (beef extract 5 g/l, yeast extract 1 g/l, peptone 5 g/l, sucrose 5 g/l, MgSO$_4$ 2 mM, pH 7.2) supplemented with 50 mg/l riflampicin and 100 mg/l spectinomycin. Before infection, *Agrobacterium* was cultured in liquid minimal A medium (Miller 1972) at 28 °C for 2 days.

Transformation and Tissue Culture of G. uralensis. The seeds of *G. uralensis* were surface sterilized and germinated on an agar plate containing 1/2 MS (Murashige and Skoog 1962) salts, 1% sucrose, and 0.8% agar. After 4 days, the hypocotyls of seedlings were infected with *Agrobacterium* by scratching with a needle. Hairy roots appeared from the site of infection after 1 week. These roots were taken off and maintained on B5 medium (Gamborg et al. 1968) supplemented with 200 mg/l Claforan (Hoechst). Claforan could be omitted after three to four transfers.

Opine Assay. Agropine and mannopine were detected by high-voltage paper electrophoresis as described previously (Petit et al. 1983).

Plant DNA Isolation and Southern Analysis. DNA was extracted from the transformed hairy roots as described by Dellaporta et al. (1983) and further purified with a Qiagen tube (Qiagen Inc.). DNA was digested with EcoRV and electrophoresed in a 0.7% agarose gel, transferred to Hybond-N filter (Amersham), and hybridized with random prime labeled ^{32}P probes (Takara) by the procedures recommended by the suppliers. Purified BamHI-EcoRV (1.3 kb) and NcoI-HindIII (1.4-kb) fragments of pGSGluc1 were used as the probes A and B, respectively. The final wash was carried out in 0.1× SSC at 65 °C.

GUS and NPT-II Enzymatic Assay. Protein was extracted from plant tissue with extraction buffer containing 50 mM Tris-HCl (pH 6.8) and 2% 2-mercaptoethanol. The same protein extracts were used for both GUS and NPT-II enzymatic assays. GUS activity was determined with 4-methylumbelliferone β-glucuronide as substrate, as described by Jefferson et al. (1987). NPT-II activity was assayed by *in-situ* reaction on native polyacrylamide gel as reported (Reiss et al. 1984) and quantified with a densitometer after exposure to X-ray film.

Histochemical Localization of GUS Activity. The histochemical localization of GUS activity was determined by the method of Jefferson et al. (1987) with some modifications.

2.2 Results

Transformation. Aseptic seedlings of *G. uralensis* were scratched with *A. rhizogenes* Rif^R harboring pRi15834 and a binary vector pGSGluc1. After several days, small white callus appeared, followed by fine hairy roots at the

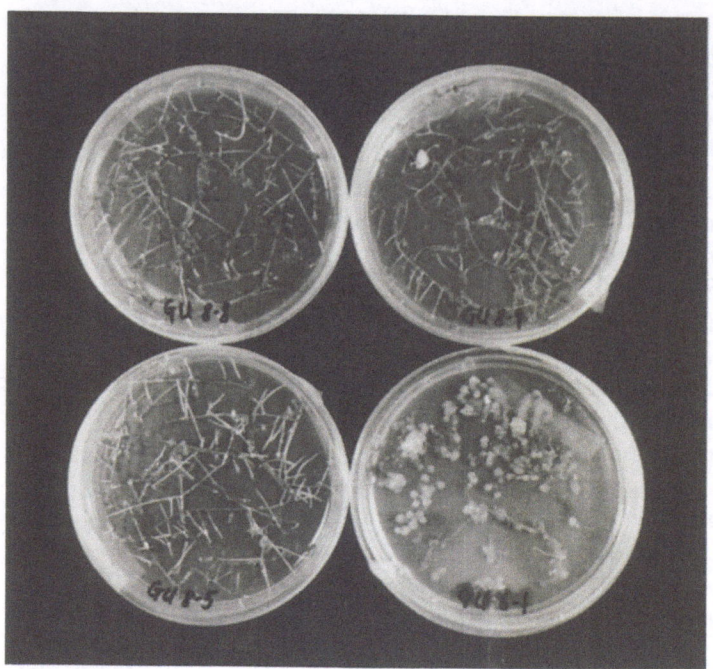

Fig. 2. The established licorice tissues transformed with *A. rhizogenes* (pRi15834; pGSGluc1). *GU8-8* Clone 8-8; *GU8-9* clone 8-9; *GU8-5* clone 8-5; *GU6-1* clone 6-1

site of infection. The hairy roots were cultured on the B5 agar plate supplemented with Claforan (200 mg/l) to remove *Agrobacterium*. Out of 26 clones of hairy roots examined, 20 were positive for agropine and/or mannopine; of these 20 clones, 16 were positive for the GUS enzymatic assay after 2–3 months of the infection. Figure 2 shows the four transformed clones which were chosen for further study and maintained. Morphologically, clones 8-8, 8-9, and 8-5 showed fine root structures. However, clone 6-1 spontaneously gave callus without any addition of phytohormones. All four clones produced agropine and mannopine (Table 1), indicating the integration of T-DNA from pRi15834.

Southern Blot Analysis. The integration of T-DNA of pGSGluc1 was analyzed by Southern hybridization. The copy number of the T-DNA integrated into plant DNA was determined by using the restriction enzyme EcoRV which cut the T-DNA region at a unique site to create a composed fragment with T-DNA and plant genomic DNA (Table 1).

Clones 8-8 and 8-9 emerged independently from the same infected site and were established separately. However, these clones gave exactly the same hybridization bands with *gus* and *neo* probes, respectively. This suggested that these clones were derived from the same infection event and contained one T-DNA copy in plant genomic DNA. Clone 8-5 gave two hybridization bands

Table 1. Summary of transformed hairy roots of *G. uralensis*

Clone	Opines produced	Copy number of T-DNA insertion	Enzymatic activities (nmol/min/mg)	
			NPT-II	GUS
8-8	Agropine, mannopine	1	27.0	134
8-9	Agropine, mannopine	1	34.3	159
8-5	Agropine, mannopine	2	28.7	157
6-1	Agropine, mannopine	3	71.0	3.3
HR	Agropine, mannopine	nd	nd	nd

8-8, 8-9, 8-5, 6-1, transformed hairy roots with *A. rhizogenes* (pRi15834; pGSGluc1); HR, control hairy root transformed with *A. rhizogenes* (pRi15834); nd, not detected.

with each probe, suggesting two copies of the T-DNA. Clone 6-1 contained three copies of the T-DNA of pGSGluc1. The control hairy root transformed with pRi15834 showed no hybridization signals with pGSGluc1 T-DNA.

Expression of Chimeric Genes in Transgenic Licorice Roots. The chimeric *neo* and *gus* are controlled by TR 1′ and 2′ promoters, respectively, of TR-DNA of an octopine Ti plasmid (Velten et al. 1984). All the established clones showed both NPT-II and GUS activities (Table 1), while the control hairy root of *G. uralensis*, transformed only with pRi15834, had no detectable NPT-II and GUS activities. The expression level varied with each clone. Clone 6-1, in particular, showed lower GUS activity (3.3 nmol/min/mg) but much higher NPT-II activity (71.0 nmol/min/mg) than the other clones.

Histochemical Staining of GUS Activity. The histochemical analysis of expression of the GUS gene was carried out as shown in Fig. 3. Staining was observed in the phloem and pericycle cells in transverse sections of the roots (Fig. 3A) and the root caps (data not shown). In the callus tissue, spontaneously formed from hairy roots of clone 6-1, the redifferentiating tissues from unorganized callus were strongly stained (Fig. 3B). These results suggested that the TR 2′ promoter shows tissue specificity.

Application to Other Species of Glycyrrhiza. The same protocol of genetic transformation using a binary vector based on *Agrobacterium*-Ri plasmid was applied to another *Glycyrrhiza* species, *G. glabra* (Asada et al. 1998). A binary vector, pBI121, and an Ri plasmid, pRi15834, were used for transformation. The transformed root tissues produced prenyl flavonoids as the original plants.

3 Discussion

The method employed by us for the transformation of licorice is based on a Ri plasmid binary system. The frequency of double transformation with pRi15834 and a mini Ti was rather high, up to 80% (16/20 clones). This high

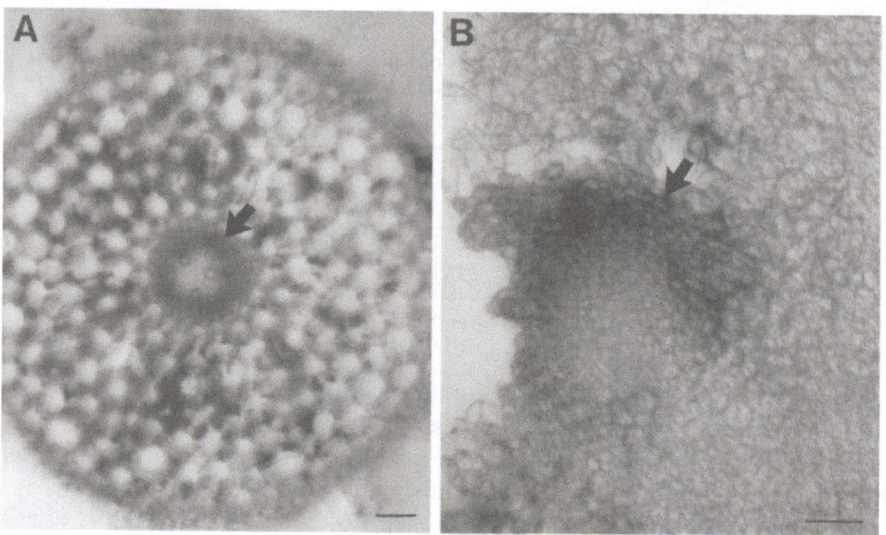

Fig. 3A,B. Histochemical assay of GUS activity in transgenic licorice. Thin sections of plant material were cut by hand with a razor blade and stained with X-glucuronide by the method of Jefferson et al. (1987). **A** Transverse section of transformed root of clone 8-9. Staining is most intense in the phloem and the pericycle tissues (*arrow*). **B** Thin section of callus spontaneously formed from clone 6-1. The redifferentiating tissue is stained (*arrow*), but the unorganized tissue is not. *Bar* 100 μm

ratio has also been reported in other plant species (Shahin et al. 1986; Simpson et al. 1986; Hamill et al. 1987; Stougaard et al. 1987; Saito 1993 and references cited therein). By this method one can easily obtain transgenic roots integrated with any desirable foreign genes without selection with growth inhibitors (antibiotics and herbicides). This technique will be applicable to other pharmaceutically important plants where selection conditions by means of growth inhibitors have not been established. In particular, this technique is most suitable for genetic manipulation of secondary metabolism of hairy roots which produce metabolites in high yields.

The TR 1′ and 2′ genes that provided the dual TR 1′-2′ promoter are part of the TR-DNA of pTiAch5, and encode mannopine synthases (Velten et al. 1984). It has been reported that the expression of this promoter is higher in roots than in leaves and inducible by wounding (Teeri et al. 1989). Our histochemical study indicated that TR 2′ promoter is specifically expressed in phloem and pericycle tissues of licorice hairy roots. Almost the same pattern of expression is observed also in transgenic tobacco and *Digitalis*. The expression of TR 1′-2′ promoter is coordinately enhanced by stress, such as wounding and addition of plant growth regulators, in licorice, tobacco, and *Digitalis* (Saito et al. 1991a).

The dry roots of mature licorice contain large amounts of the sweet saponin, glycyrrhizin. We examined the production of glycyrrhizin in those transgenic roots, and so far no production has been observed by either HPLC or the sweetness test, although the plant from which these lines were derived contains glycyrrhizin.

Recently, some progress has been made on genetic characterization of *Glycyrrhiza* plants by random amplified polymorphic DNA (RAPD) and restriction-enzyme fragment polymorphism (RFLP; Yamazaki et al. 1994). The genetic relationships among several species of *Glycyrrhiza* (*G. uralensis*, *G. glabra*, *G. echinata*, and *G. pallidiflora*) were determined from the phylogenic trees constructed by RAPD and RFLP profiles. These results indicated that *G. uralensis* and *G. glabra*, rich in glycyrrhizin, are more closely related to each other than to *G. echinata* or *G. pallidiflora*. RAPD analysis is also possible by using dried licorice roots as a source of DNA, indicating the possibility of DNA identification of the dried crude drug (Yamazaki et al. 1995).

4 Summary and Conclusions

The pharmaceutically important plant, licorice (*Glycyrrhiza uralensis* Fisch. et D.C.), was transformed with a binary vector system of an Ri plasmid, pRi15834, and a mini Ti vector, pGSGluc1, containing chimeric *neo* and *gus* genes. The transgenic state of transformed roots was confirmed by detection of agropine and mannopine and by Southern blot hybridization with T-DNA of pGSGluc1. One to three copies of T-DNA of pGSGluc1 were integrated into the genomic DNA of *G. uralensis*. The expression of chimeric *neo* and *gus* genes driven by TR 1′ and 2′ promoters, respectively, was demonstrated by enzymatic assays. Histochemical analysis showed that the chimeric TR 2′-*gus* gene was expressed specifically in phloem and pericycle tissues of the transformed licorice roots. The protocol established by this study is applicable for further genetic improvement of licorice by means of DNA transformation.

References

Asada Y, Li W, Yoshikawa T (1998) Isoprenylated flavonoids from hairy root cultures of *Glycyrrhiza glabra*. Phytochemistry 47:389–392

Dellaporta SL, Wood J, Hicks JB (1983) A plant DNA minipreparation: version II. Plant Mol Biol Rep 1:19–21

Figurski DH, Helinski DR (1979) Replication of an origin-containing derivative of pRK2 dependent on a plasmid function provided in *trans*. Proc Natl Acad Sci USA 76:1648–1652

Gamborg OL, Miller RA, Ojima K (1968) Nutrient requirements of suspension culture of soybean root cells. Exp Cell Res 50:151–158

Hamill JD, Prescott A, Martin C (1987) Assessment of the efficiency of cotransformation of the T-DNA of disarmed binary vectors drived from *Agrobacterium tumefaciens* and the T-DNA of *A. rhizogenes*. Plant Mol biol 9:573–584

Henry M, Edy AM, Desmarest P, Du Manoir J (1991) *Glycyrrhiza glabra* L. (Licorice): Cell culture,regeneration, and the production of glycyrrhizin. In: Bajaj YPS (ed) Biotechnology in agriculture and forestry 15, Medicinal and aromatic plants III. Springer, Berlin Heidelberg New York, pp 270–282

Hotta M, Ogata K, Nitta A, Hoshikawa K, Yanagi M, Yamazaki K (eds) (1989) Useful plants of the world. Heibonsha LTD, Tokyo

Jefferson RA, Kavanagh TA, Bevan MW (1987) GUS fusions: β-glucuronidase as a sensitive and versatile gene fusion marker in higher plants. EMBO J 6:3901–3907

Miller JH (ed) (1972) Experiments in molecular genetics. Cold Spring Harbor Laboratory, New York

Murashige T, Skoog F (1962) A revised medium for rapid growth and bio-assays with tobacco tissue cultures. Physiol Plant 15:473–497

Petit A, David C, Dahl GA, Ellis JG, Guyon P, Casse-Delbart F, Tempé J (1983) Further extension of the opine concept: plasmids in *Agrobacterium rhizogenes* cooperate for opine degradation. Mol Gen Genet 190:204–214

Reiss B, Sprengel R, Will H, Schaller H (1984) A new sensitive method for quantitative and qualitative assay of neomycin phosphotransferase in crude cell extracts. Gene 30:211–218

Saito K (1993) Genetic engineering in tissue culture of medicinal plants. Plant Tissue Cult Lett 10:1–8

Saito K, Kaneko H, Yamazaki M, Yoshida M, Murakoshi I (1990a) Stable transfer and expression of chimeric genes in licorice (*Glycyrrhiza uralensis*) using an Ri plasmid binary vector. Plant Cell Rep 8:718–721

Saito K, Yamazaki M, Shimomura K, Yoshimatsu K, Murakoshi I (1990b) Genetic transformation of foxglove (*Digitalis purpurea*) by chimeric foreign genes and production of cardioactive glycosides. Plant Cell Rep 9:121–124

Saito K, Yamazaki M, Kaneko H, Murakoshi I, Fukuda Y, Van Montagu M (1991a) Tissue-specific and stress-enhancing expression of the TR promoter for mannopine synthase in transgenic medicinal plants. Planta 184:40–46

Saito K, Yamazaki M, Kawaguchi A, Murakoshi I (1991b) Metabolism of solanaceous alkaloids in transgenic plant teratomas integrated with genetically engineered genes. Tetrahedron 47:5955–5968

Saito K, Noji M, Ohmori S, Imai Y, Murakoshi I (1991c) Integration and expression of a rabbit liver cytochrome P-450 gene in transgenic *Nicotiana tabacum*. Proc Natl Acad Sci USA 88:7041–7045

Saito K, Yamazaki M, Anzai H, Yoneyama K, Murakoshi I (1992a) Transgenic herbicide-resistant *Atropa belladonna* using an Ri binary vector and inheritance of the transgenic trait. Plant Cell Rep 11:219–224

Saito K, Yamazaki M, Murakoshi I (1992b) Transgenic medicinal plants: *Agrobacterium*-mediated foreign gene transfer and production of secondary metabolites. J Nat Prod 55:149–162

Shahin EA, Sukhapinda K, Simpson RB, Spivey R (1986) Transformation of cultivated tomato by a binary vector in *Agrobacterium rhizogenes*: transgenic plants with normal phenotypes harbor binary vector T-DNA, but no Ri-plasmid T-DNA. Theor Appl Genet 72:770–777

Simpson RB, Spielmann A, Margossian L, McKnight TD (1986) A disarmed binary vector from *Agrobacterium tumefaciens* functions in *Agrobacterium rhizogenes*. Plant Mol Biol 6:403–415

Stougaard J, Abildsten D, Marcker K (1987) The *Agrobacterium rhizogenes* pRi TL-DNA segment as a gene vector system for transformation of plants. Mol Gen Genet 207:251–255

Stuart M (ed) (1979) The Encyclopedia of Herbs and Herbalism. Orbis Publishing, London

Teeri TH, Lehvaslaiho H, Framck M, Uotila J, Heino P, Palva ET, Van Montagu M, Herrera-Esterella L (1989) Gene fusion to lacZ reveal new expression patterns of chimeric genes in transgenic plants. EMBO J 8:343–350

Trease GE, Evans WC (1983) Pharmacognosy, 12th edn. Bailliere Tindall, London

Velten J, Velten L, Hain R, Schell J (1984) Isolation of a dual plant promoter fragment from the Ti plasmid of *Agrobacterium tumefaciens*. EMBO J 3:2720–2730

Yamazaki M, Sato A, Shimomura K, Saito K, Murakoshi I (1994) Genetic relationships among *Glycyrrhiza* plants determined by RAPD and RFLP analyses. Biol Pharm Bull 17:1529–1531

Yamazaki M, Sato A, Shimomura K, Inoue K, Ebizuka Y, Murakoshi I, Saito K (1995) Extraction of DNA and RAPD analysis from dried licorice root. Nat Med 49:488–490

XI Genetic Transformation in *Lobelia* Species

K. Ishimaru[1] and K. Shimomura[2]

1 General Account

1.1 Distribution and Importance of *Lobelia* Plants

The genus *Lobelia* (Campanulaceae), comprises about 200 to 300 species of mostly annuals and herbaceous perennials, ranging from partially submerged aquatic field habitats in North America, Europe, and East Asia to a mountainside in East Africa and China, where some giant tree-type lobelias grow (Everett 1981). The bright colored flowers of North American lobelias such as *L. cardinalis* (reddish pink) and *L. siphilitica* (bright blue) are outstandingly beautiful and best adapted for naturalizing at watersides. Many garden varieties of the South African *L. erinus* are popular as dwarf edging plants and for window and porch boxes, garden vases, and other containers. Some species of *L. cardinalis*, *L. dortmanna*, etc. appeal to collectors of aquatic plants to decorate ornamental tanks for tropical fishes.

Lobelia inflata (popular as Indian-tobacco) and some, perhaps most, lo-belias contain a piperidine alkaloid lobeline (Fig. 1) which stimulates the respiratory center of the brain. Owing to this effect, *L. inflata* is applied in cases of asthma, collapse, and gas and narcotic poisoning, but excessive intake is also poisonous, resembling the effect of nicotine from smoking tobacco. Although the giant lobelia group in Africa, such as *L. giberroa*, is not used medicinally, a tree-like lobelia, *L. seguinii*, in China is used for reducing fever (cold). *L. chinensis*, a weed widely grown in Southeast Asian countries such as Japan, China, Korea, Malaysia, etc., is also used to reduce fever (sometimes from tumors) via detoxication and diuresis in China. *L. sessilifolia*, found in temperate and subtropical regions in East Asia, is also well known as containing lobeline, but no detailed chemical studies on the other secondary metabolites have been reported.

---

[1] Department of Applied Biological Sciences, Faculty of Agriculture, Saga University, 1 Honjo, Saga, 840 Japan
[2] Tsukuba Medicinal Plant Research Station, National Institute of Health Sciences, 1 Hachimandai, Tsukuba, Ibaraki, 305 Japan

Biotechnology in Agriculture and Forestry, Vol. 45
Transgenic Medicinal Plants (ed. by Y.P.S. Bajaj)
© Springer-Verlag Berlin Heidelberg 1999

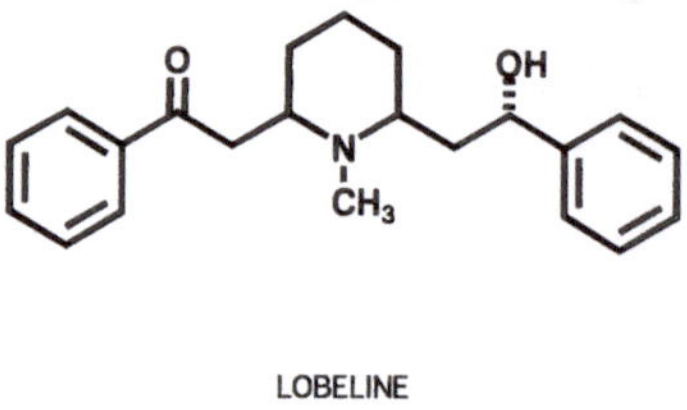

Fig. 1. Structure of lobeline

LOBELINE

1.2 In Vitro Approaches and Secondary Metabolites

1.2.1 Callus and Suspension Cultures

Wysokinska (1977), established callus cultures from shoot tips of *L. inflata*, and examined the influence of some growth regulators on the growth and alkaloid production of the callus. The tissue grew best on culture medium containing 2,4-D (10^{-6}M), and rhizogenesis was induced on medium with NAA (10^{-7}M). The highest alkaloid content was attained under the influence of IAA (10^{-5}M) and NAA (10^{-7}M), however 2,4-D inhibited the formation of lobeline.

Krajewska and Szoke (Szoke 1994) later studied alkaloid production in callus, suspension, and organized cultures of *L. inflata*. The callus grew vigorously on an MS medium (Murashige and Skoog 1962) containing 1 mg/l kinetin and 1 mg/l 2,4-D (AMS-2 medium), and growth and alkaloid production were affected by new synthetic regulators (Sz/11 and Sz/28). When the regulators were applied in combination with alkaloid precursor amino acids (phenylalanine and lysine) to the culture medium, the biomass and alkaloid production of *L. inflata* callus were optimal. The cell suspension cultures in AMS-2 medium containing phenylalanine and lysine as supplements were also found to increase the alkaloid content twofold in response to the alkaloid precursor amino acids. Szoke (1994) also succeeded with organized (shoot) cultures of *L. inflata*, and showed that lobeline content in the shoots increased sevenfold in response to phenylalanine, although it caused fourfold increment by lysine.

1.2.2 Hairy Root Cultures

The hairy root cultures of *L. inflata*, derived from the stem segments by direct infection with *Agrobacterium rhizogenes* ATCC 15834, were established, and the secondary metabolites were analyzed (Yonemitsu et al. 1990). The alkaloid lobeline content of the hairy roots was of the same order of magnitude as that of the roots of the parent plants. The physiological conditions of transformed root cells of *L. inflata* can be changed under the influence of its culture medium. Although hairy roots cultured in NN medium (Nitsch and Nitsch 1967) grew relatively slowly, they produced lobeline in much higher amounts.

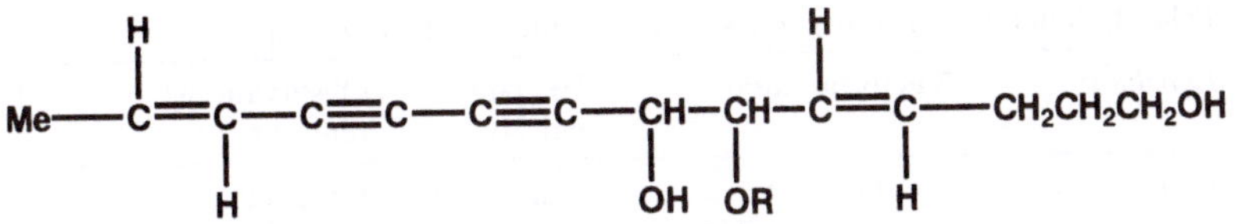

Me—C=C—C≡C—C≡C—CH—CH—C=C—CH₂CH₂CH₂OH

lobetyol : R=H
lobetyolin : R=Glc
lobetyolinin : R=Glc^6-^{1}Glc

Fig. 2. Structures of polyacetylenes, lobetyol, lobetyolin, and lobetyolinin

In the hairy root cultures, two new polyacetylene compounds, lobetyol and lobetyolin (Fig. 2), were isolated, of which lobetyol has not been detected in the parent plants in the field (Ishimaru et al. 1991; Tanaka et al. 1993). The transformed root cultures, therefore, indicated the possibility for production and accumulation of new secondary metabolites which had hitherto escaped detection in the plants due to their low contents. Ishimaru et al. (1992a) also obtained *L. inflata* hairy root cultures, by using the same *Agrobacterium* strain, and isolated a new polyacetylene gentiobioside lobetyolinin (Fig. 2) in high amounts from the tissues. Producing the diglycosylated constituents in the cells, *L. inflata* hairy roots were presumed to have a strong capability for glycosylation in the secondary metabolism which could be available for the biotransformation (glycosylation) of some useful chemicals by using the hairy roots as a bioreactor system.

2 Transformation Studies (Table 1)

2.1 Methods for Transformation and Analysis of Secondary Metabolites

2.1.1 Plant Material, Bacterial Strain, and Transformation Method

Five *Lobelia* species, *L. cardinalis*, *L. chinensis*, *L. inflata*, *L. sessilifolia*, and *L. siphilitica*, were used for the experiments, i.e., establishment of hairy root cultures by infection with two types of *Agrobacterium* strains of *A. rhizogenes* ATCC 15834 and/or *A. rhizogenes* MAFF 03-01724, and determination of the secondary metabolites under various culture conditions.

1. Coculture Method. Agrobacterium strain, subcultured on YEB agar medium (Vervliet et al. 1975), was transferred to YEB liquid medium (20 ml/ flask) and precultured for 1 day in the dark at 25 °C on a rotary shaker (100 rpm). The solution of the bacterium (200 μl) and the segments (leaf and/ or stem) cut from the axenic plantlets were inoculated to 1/2 MS liquid medium (20 ml/flask) and cocultured for 2–4 days in the dark. The infected

Table 1. Summary of transgenic studies conducted on *Lobelia* species

Lobelia sp.	Vector/method	Explant/ culture	Observations/ remarks	Reference
Lobelia inflata	*Agrobacterium rhizogenes* 15834	Hairy root	Lobeline production	1. Yonemitsu et al. (1990)
L. inflata	*A. rhizogenes* 15834	Hairy root	Lobetyolin and lobetyol production	2. Ishimaru et al. (1991)
L. inflata	*A. rhizogenes* 15834	Hairy root	Lobetyolinin production	3. Ishimaru et al. (1992a)
L. inflata	*A. rhizogenes* 15834	Hairy root	Lobeline production	4. Ishimaru et al. (1992b)
L. inflata	*A. rhizogenes* 15834	Hairy root	Effects of media on growth and polyacetylene production	5. Ishimaru et al. (1993)
L. sessilifolia	*A. rhizogenes* 15834, MAFF03-01724	Hairy root	Polyacetylene production	6. Ishimaru et al. (1994)
L. inflata	*A. rhizogenes* 15834	Hairy root	Effects of additives on growth and polyacetylene production	7. Ishimaru et al. (1995)
L. chinensis	*A. rhizogenes* 15834	Hairy root	Polyacetylene production	8. Tada et al. (1995b)
L. sessilifolia	*A. rhizogenes* 15834	Hairy root	Biotransformation of phenolics	9. Yamanaka et al. (1995)
L. chinensis	*A. rhizogenes* 15834	Hairy root	Anthocyanin production	10. Tada et al. (1996)
L. cardinalis	*A. rhizogenes* 15834, MAFF03-01724	Hairy root	Polyacetylene production	11. Yamanaka et al. (1996)
L. sessilifolia, *L. cardinalis*	*A. rhizogenes* 15834, MAFF03-01724	Hairy root	Biotransformation of phenolics	12. Ishimaru et al. (1996)

segments, after being rinsed with sterile water, were transferred to 1/2 MS solid medium (solidified with 0.25% Gelrite containing 0.5 g/l Claforan) and incubated at 25 °C in the dark. After 3 to 4 weeks, several hairy roots appeared on the infected segments. The tips of the hairy roots were cut off and cultured for 1 week on the same medium to eliminate the bacteria. The axenic hairy roots were transferred and maintained in hormone-free MS liquid medium in the dark (100 rpm).

2. Direct Infection. The segments (stem, leaf, and/or petiole) were cut from the axenic plantlets and a piece of *Agrobacterium* was pasted by a needle onto the cut ends. The infected segments were transferred onto 1/2 MS solid medium containing an antibiotic (Claforan) and incubated at 25 °C in the light (ca. 2000 lx, 16-h photoperiod 1 day) until some hairy roots appear on the infected sites. The hairy roots obtained were maintained as above.

2.1.2 Confirmation of the Transformation

The certification of the transformation of the hairy roots is performed by the detection of opines, whose production is originated in the function of the genes in T-DNA regions of the bacteria (agropine and mannopine by *A. rhizogenes* ATCC 15834: White and Sinker 1987, and mikimopine by *A. rhizogenes* MAFF 03-01724: Isogai et al. 1988), using paper electrophoresis (Petit et al. 1983; Tanaka 1990). The electrophoretograms were visualized with alkaline silver nitrate reagent for agropine and mannopine and with Pauly reagent for mikimopine.

2.1.3 Time Course Experiments

Hairy roots (ca. 50–200 mg, fr wt) were transferred into various hormone-free liquid media, MS, 1/2 MS, B5 (Gamborg et al. 1968), woody plant (WP; Lloyd and McCown 1980) and/or root culture (RC) and cultured in the dark or light (ca. 2000–3000 lx) condition at 25 °C (100 rpm). Growth (fr and dry wt) and secondary metabolites (using HPLC analysis) were determined periodically.

2.1.4 Analysis of Secondary Metabolites

1. Quantitative Determination of Lobeline. Samples were lyophilized, mashed (20–200 mg), and sonicated with 4 ml of 0.01 N HCl for 10 min at room temperature. After filtration, the residue was resonicated with equal volume of 0.01 N HCl. The combined filtrates were dried under reduced pressure at 40 °C, dissolved in 2 ml of 0.01 N HCl and injected (60 μl) to HPLC. HPLC conditions were as follows, column: Wakosil 5C$_{18}$-200 (4.6 mm id × 150 mm), mobile phase: MeCN-0.1% TFA (3:7), flow rate: 1 ml/min, column temperature: 40 °C, detection: 254 nm, Rt: lobeline (9.6 min).

2. Quantitative Determination of Polyacetylenes. Lyophilized samples (20–30 mg) were mashed by pestle and extracted with MeOH (2 ml) for 15 h at room temperature. The extract, after filtration with Millipore filter (0.45 μm), was injected (3–10 μl) to HPLC: column, Inertsil ODS (4.6 mm id × 250 mm); mobile phase, MeCN-H$_2$O (1:4→9:1, linear gradient in 30 min); flow rate, 0.68 ml/min; detection, 270 nm (UV); column temperature, 40 °C; R$_t$ (min): lobetyol (15.9), lobetyolin (19.8), and lobetyolinin (23.9).

3. Quantitative Determination of Chlorophylls. Lyophilized samples (10–20 mg) were mashed by pestle and extracted with acetone (1 ml) for 16 h at room temperature in the dark. The extract, after filtration with Millipore filter (0.5 μm), was injected (15 μl) to HPLC: column; Shim-pack CLC-ODS (6 mm id × 150 mm); mobile phase, MeOH; flow rate, 0.97 ml/min; column temperature, 40 °C; detection, 420 nm; Rt (min), chlorophyll b (11.2), and chlorophyll a (17.6).

4. Quantitative Determination of Anthocyanins. Lyophilized samples (ca. 50 mg) were extracted with 1% HCl-MeOH (1 ml) for 1 day at room temperature in the dark. The extract, after filtration through filter paper (ADVANTEC, 5B), was applied (15 μl) to HPLC analysis. HPLC conditions were as follows: column; Inertsil ODS-2 (4.6 mm id × 250 mm), mobile phase; A: 1.5% H_3PO_4, B: 1.5% H_3PO_4–20% CH_3COOH–25% CH_3CN (A:B = 3:1→13:7, in 30 min), flow rate; 1 ml/min, column temperature; 35 °C, detection; 520 nm. Two compounds, Lc-1 (Rt 10.5 min) and Lc-2 (Rt 12 min) observed in the HPLC chromatogram were identified as cyanidin 3-O-glucoside and cyanidin 3-O-rutinoside, respectively.

2.2 Results and Discussion

2.2.1 Lobelia inflata

Hairy roots of *L. inflata* were induced by direct infection with *A. rhizogenes* ATCC 15834. Among three clones which grew extensively on hormone-free MS solid medium, two clones, Li-A and Li-B, were selected by their morphological differences and used for this investigation. On MS (both liquid and solid) medium, Li-A showed extensive lateral branching and profuse root hairs, while Li-B produced poor branching and short root hairs.

The growth (both fresh and dry weight) of these hairy roots began to increase from the beginning of the culture under both light and dark conditions (Ishimaru et al. 1992b). The amount of alkaloid lobeline in Li-A cultured in the light gradually increased from the beginning of the culture (Fig. 3). After it reached a plateau level (12.2 μg/flask) at week 4, it slightly increased until the end of the culture (15.1 μg/flask at week 7). In this culture, light inhibited the formation of lobeline, although the growth of Li-A cultured in the light was comparable to that cultured in the dark. On the other hand, the amount of lobeline in Li-B remarkably increased under both dark and light conditions. Therefore, it is noteworthy that two clones of the hairy roots (Li-A and Li-B) showed a different response to illumination in producing lobeline.

In MS, 1/2 MS, and B5 media, the content (mg/flask) of glycosylated polyacetylenes (lobetyolin and lobetyolinin) roughly paralleled the growth of the hairy roots (Ishimaru et al. 1993). Generally, the amount of monoglucoside lobetyolin exceeded that of diglucoside lobetyolinin, and the content of the free polyacetylene lobetyol was the lowest of the three. The highest levels of both lobetyolin and lobetyolinin among these cultures were observed at week 6 of Li-A in MS culture (lobetyolin, 14 mg; lobetyolinin, 13.2 mg/flask).

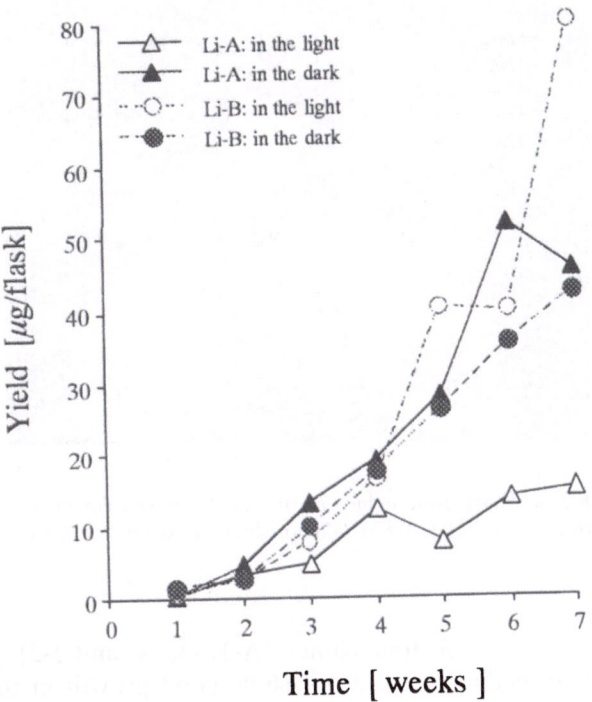

Fig. 3. Lobeline production in hairy roots (*Li-A* and *Li-B*) of *Lobelia inflata* in MS liquid medium. (Ishimaru et al. 1992b)

To develop growth and polyacetylene production of *L. inflata* hairy roots (Li-A and Li-B), the effects of some additives were determined (Ishimaru et al. 1995). When cultured in RC liquid medium with 0.002~0.2 mg/ml myoinositol, the growth and polyacetylene production of Li-B were satisfactorily promoted, the maximum (with 0.1 mg/ml myoinositol) was almost five (growth) or seven (polyacetylene production) times more than those of the control cultured in basal RC medium.

Although the addition of NH_4NO_3, KNO_3, $CdCl_2$, casein hydrolysate, and yeast extract, etc. to the culture medium showed no clear influence, $CuSO_4$ perfectly inhibited only polyacetylene production (showing no effect on the growth) of the hairy roots.

2.2.2 *Lobelia sessilifolia*

For the induction of hairy roots, two types of bacteria, *A. rhizogenes* ATCC 15834 and *A. rhizogenes* MAFF 03-01724, were used (Fig. 4; Ishimaru et al. 1994). Six clones, A-1-4 by *A. rhizogenes* ATCC 15834 and J-1 and 2 by *A. rhizogenes* MAFF 03-01724, were selected and used for the experiment.

Fig. 4. Hairy root induction of *Lobelia sessilifolia* by direct infection with *Agrobacterium rhizogenes* ATCC 15834. (Photo Ishimaru, November, 1991)

Although four clones (A-1, -3, -4, and J-2) grew well in both MS and WP media, A-2 did not show good growth in these media (Fig. 5). When these hairy roots were cultured under light condition, some clones turned green (so-called green hairy root), presumably through the production of some chlorophylls. Coinciding with the appearance (greenish coloration) of the roots, four clones (A-1, 3, -4, and J-1) in MS medium and three clones (A-1, -3, and -4) in WP medium were shown to contain chlorophylls. The highest content of chlorophyll a (0.26% dry weight) was observed in A-3 in MS medium. The clone J-2 produced no detectable level of chlorophylls in both media. It was very interesting that, in MS medium, only the green hairy roots (A-1, -3, -4, and J-1) were promoted in growth and polyacetylene production by illumination (Fig. 5). On the other hand, in WP medium, illumination did not affect the growth of hairy roots, without reference to their chlorophyll production (the growth of A-1, -3, and J-2 was restrained in the light).

The content of polyacetylenes in the intact plant was almost 1/20 (lobetyolin: 0.204%, lobetyolinin: 0.017% as dry weight) of that in the hairy roots. This also showed the usefulness of *L. sessilifolia* hairy roots for the production of these polyacetylenes (especially lobetyolin).

The occurrence of a high content of the glucosides lobetyolin and lobetyolinin in *L. sessilifolia* hairy roots suggested that the roots had a strong capability for glycosylation of secondary metabolites. Therefore, *L. sessilifolia* hairy roots were used for biotransformation of some phenolics, flavan-3-ols [(+)-catechin, (−)-epicatechin, etc.], and C6–C1 phenols (protocatechuic acid, gallic acid, etc.) which were successful in the production of new glucosylated compounds (Yamanaka et al. 1995). The hairy roots also used for the biotransformation of C6–C3 phenolics, such as *trans*-cinnamic

[ ]: **Fresh weight (g / flask)**

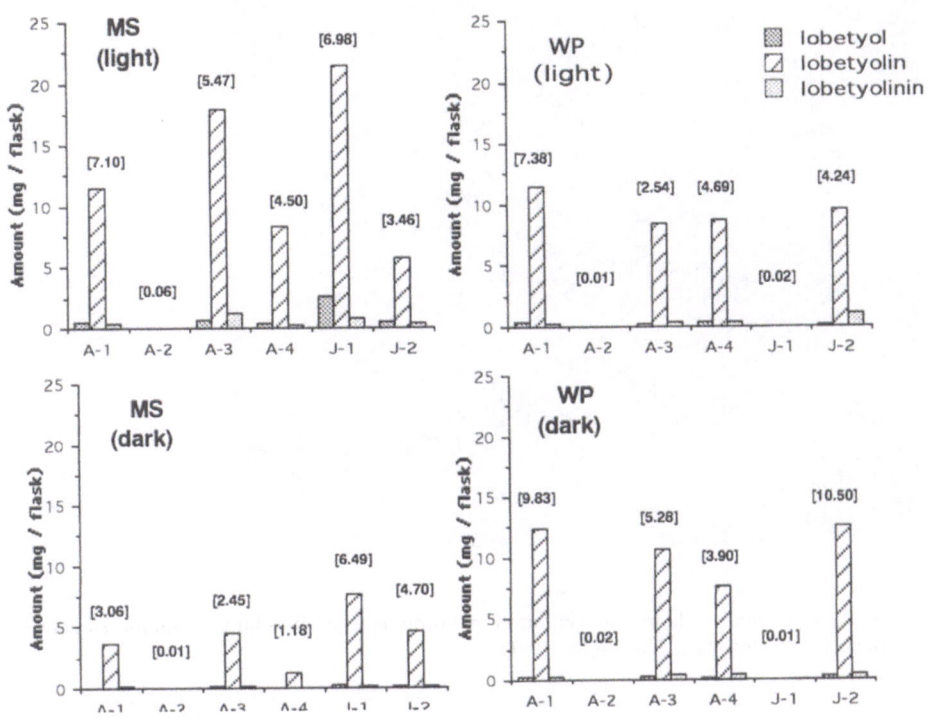

Fig. 5. Growth and polyacetylene production in hairy roots of *Lobelia sessilifolia*. (Ishimaru et al. 1994)

acid, *p*-coumaric acid, and caffeic acid, were successful in selective glycosylation of the substrates (Ishimaru et al. 1996).

2.2.3 *Lobelia chinensis*

L. chinensis hairy roots, induced with *A. rhizogenes* ATCC 15834, were cultured in hormone-free MS (Figs. 6, 7), B5, WP, and RC media in the dark and in light (Tada et al. 1995b). The roots grew particularly well in three media (MS, B5, and WP) both in the dark and in light (Fig. 8). High amounts of three polyacetylenes (lobetyol, lobetyolin, and lobetyolinin) were detected in the hairy roots in these media. Although the growth rates of the hairy roots in light were superior to those in the dark, high amounts of polyacetylenes were observed in the dark.

The hairy roots cultured in the light became greenish, accumulating chlorophylls. The chlorophyll contents (total contents of chlorophylls a and b) observed in B5 and RC media were higher than those in MS and WP media. It was presumed that the hairy roots have a tendency to produce higher

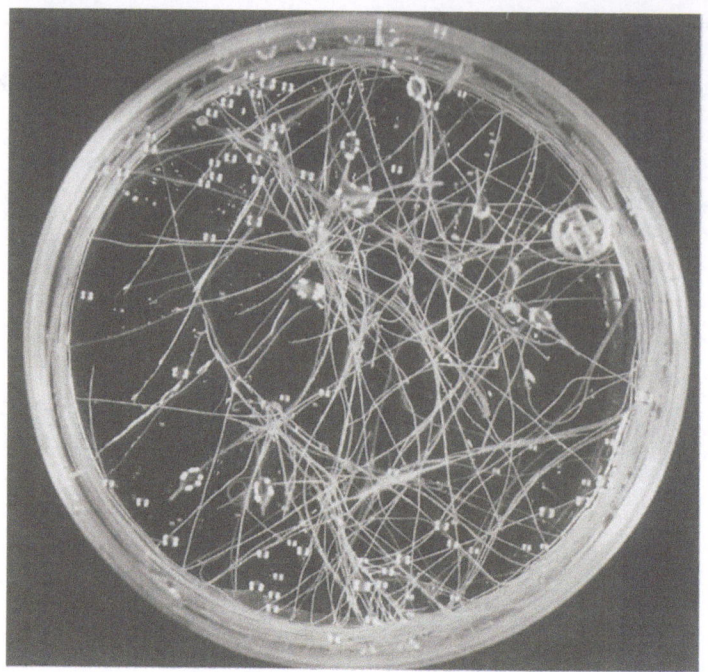

Fig. 6. *Lobelia chinensis* hairy roots cultured on hormone-free MS solid medium for 4 weeks in the dark. (Photo Ishimaru, November, 1991)

Fig. 7. *Lobelia chinensis* hairy roots cultured in hormone-free MS liquid medium for 4 weeks in the dark. (Photo Ishimaru, November, 1991)

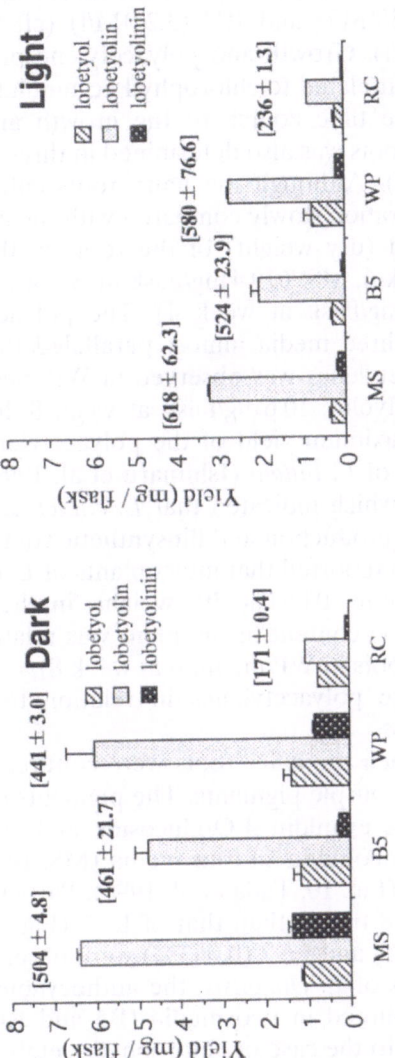

Fig. 8. Dry weight and polyacetylene production of *Lobelia chinensis* hairy roots cultured for 5 weeks in four basal liquid media under dark and light conditions. [ ]: dry weight, mg/flask. means ± standard errors; *bars* standard errors. (Tada et al. 1995b)

contents of chlorophylls in the media with high ratios of NO_3^-/NH_4^+, such as B5 (25/2 M/l) and RC (3.2/0 M/l) (cf. NO_3^-/NH_4^+: MS, 39.4/20.6 M/l; WP, 10/5 M/l). Growth and polyacetylene production in *L. chinensis* hairy roots were unrelated to chlorophyll accumulation.

The time course of the growth and polyacetylene production of the hairy roots was also determined in three media (MS, B5, and WP) in the dark (Fig. 9). Although the hairy roots cultured in MS and WP media started proliferation slowly compared with the cultures in B5 medium, the maximum amount (dry weight) of the roots in these two media (WP: 701.4 mg/flask at week 6, MS: 679.4 mg/flask at week 7) was larger than that in B5 medium (482.2 mg/flask at week 4). The polyacetylene production of the roots in these three media almost paralleled the growth. The highest yield of the polyacetylenes was observed in WP medium (lobetyol, 3.4 mg/flask at week 6; lobetyolin, 10.6 mg/flask at week 8; lobetyolinin, 2.8 mg/flask at week 7). The maximum yield of the polyacetylenes in WP medium was comparable to that of *L. inflata* (Ishimaru et al. 1993) and *L. sessilifolia* (Ishimaru et al. 1994), which indicated that *L. chinensis* hairy roots are also useful material for the production and biosynthetic study of these polyacetylenes. Tada et al. (1995a) reported that intact plants of *L. chinensis* also produce polyacetylene (lobetyolin, 0.064% dry weight, in the leaf portion). Although the polyacetylene content of the plant was relatively low (almost 1/25 of that of the hairy roots in WP medium at week 8), research on the mechanisms of activity of these polyacetylenes in relation to their pharmaceutical effects is in progress.

When the hairy roots were cultured under light, they also yielded some reddish purple pigments. The pigments (Lc-1 and Lc-2) produced were identified as cyanidin 3-O-glucoside and cyanidin 3-O-rutinoside, respectively, and the contents in four media (MS, B5, WP, and RC) were determined by HPLC (Fig. 10; Tada et al. 1996). Particularly the content of Lc-2 was higher (ca. 4~8 times) than that of Lc-1 (Fig. 11). The maximum contents of Lc-1 (0.012%) and Lc-2 (0.063%) were observed in B5 medium. In the hairy root cultures of *L. chinensis*, the anthocyanin production also had a tendency to be promoted in two media (B5 and RC) with a high ratio of NO^{3-}/NH^{4+} similar to the case of chlorophyll metabolism. This is the first example of the production of anthocyanins in transformed tissue cultures of campanulaceous plants, and is also very helpful for future determination of the pigments in the petal portion of *L. chinensis* intact plants whose chemical structures remain obscure.

2.2.4 Lobelia cardinalis

In-vitro plants of *L. cardinalis* were infected with two types of *A. rhizogenes* strains (MAFF 03-01724 and ATCC 15834) (Fig. 12). Among several hairy roots induced from the infected tissues, nine clones (seven clones by MAFF 03-01724; two clones by ATCC 15834) were selected and cultured in hormone-free 1/2 MS liquid medium. Two clones of Lc-A (induced by *A. rhizogenes*

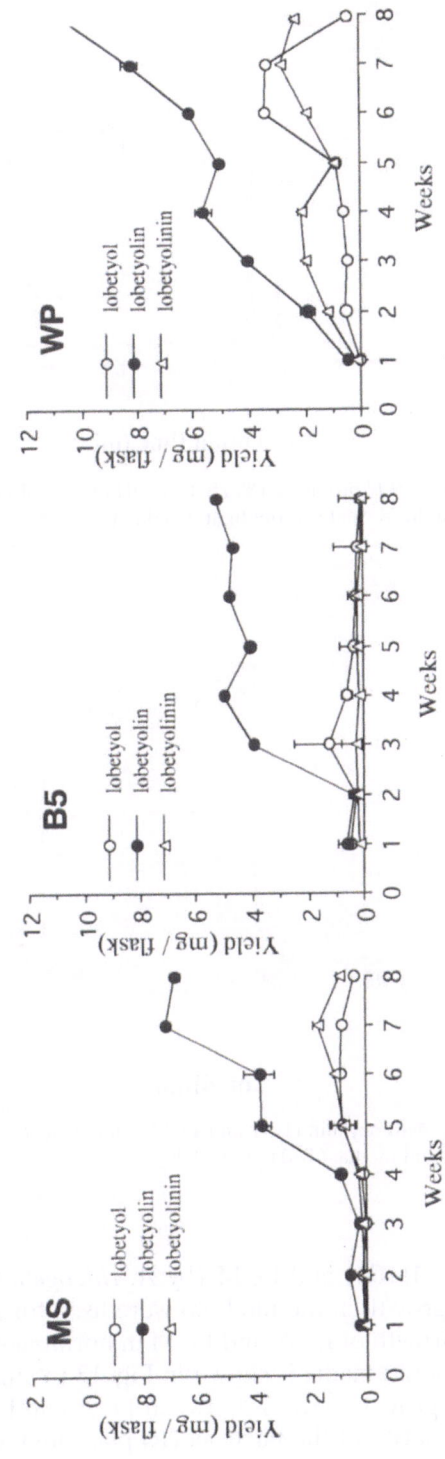

Fig. 9. Polyacetylene production in *Lobelia chinensis* hairy roots cultured in MS, B5, and WP liquid media in the dark. *Bars* standard errors. (Tada et al. 1995b)

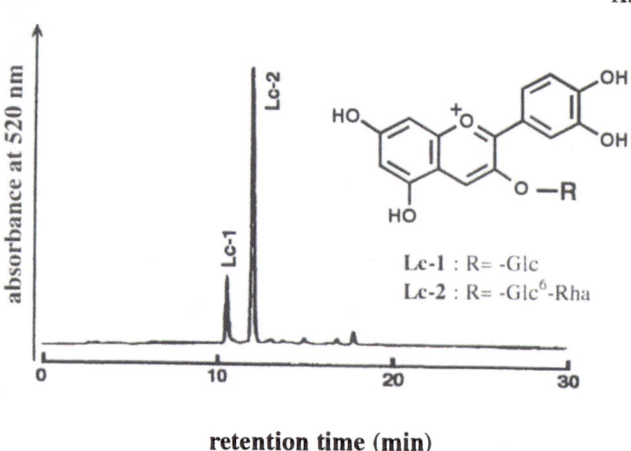

Fig. 10. HPLC profile of 1% HCl-MeOH extract of *Lobelia chinensis* hairy roots cultured in RC medium for 8 weeks in the light. (Tada et al. 1996)

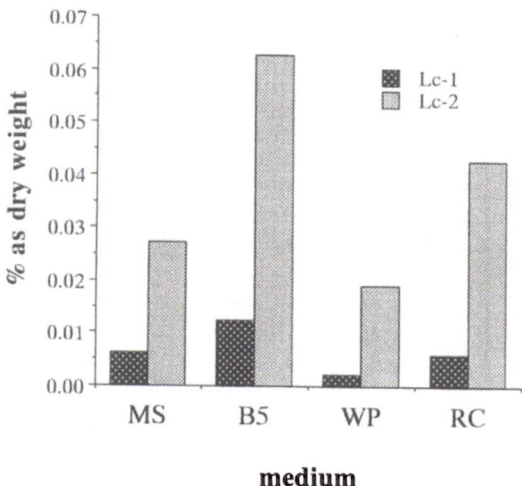

Fig. 11. Anthocyanin (Lc-1 and Lc-2) contents in *Lobelia chinensis* hairy roots cultured in four basal liquid media. (Tada et al. 1996)

ATCC 15834) and Lc-M (by *A. rhizogenes* MAFF 03-01724) which showed good growth in the medium were used for the experiment.

Growth of Lc-A and Lc-M in hormone-free various (MS, 1/2 MS, B5, and WP) liquid media is shown in Fig. 13 (Yamanaka et al. 1996). The difference in the growth patterns between Lc-A and Lc-M seemed to originate from the dissimilarity of the bacteria (Ri plasmids) infected.

In both cultures (Lc-A and Lc-M), the following results were obtained; (1) the maximum amount of polyacetylenes was observed at 6~8 weeks of the cultures (e.g., 24.7 mg/flask of lobetyolinin in Lc-A at week 6 and 18.7 mg/flask

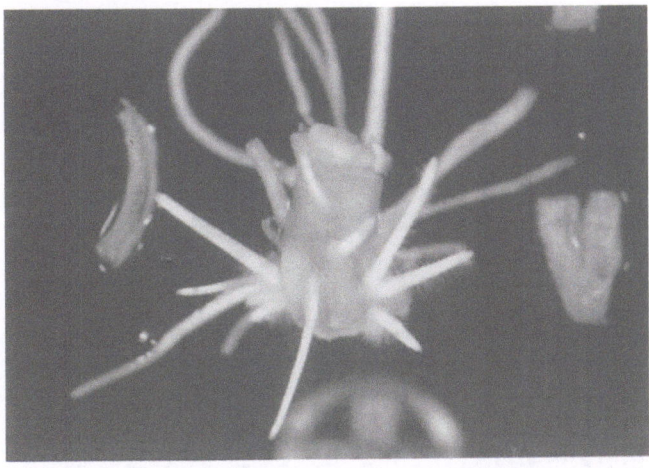

Fig. 12. Hairy root induction of *Lobelia cardinalis* by coculture method infection with *Agrobacterium rhizogenes* ATCC 15834. (Photo Ishimaru, November, 1993)

of lobetyolin in Lc-M at week 8, both in MS medium), (2) the amount of lobetyol was constantly low (below 2 mg/flask), (3) in 1/2 MS, B5, and WP media, the diagrams (amounts of polyacetylenes) of Lc-A and Lc-M were almost identical although Lc-M in B5 medium showed a short lag period (1–4 weeks) (Fig. 14, showing the data of only Lc-A). The high production of lobetyolin and lobetyolinin in these cultures showed the strong capability for glucosylation in these hairy roots. Among several tissue cultures of campanulaceous plants investigated so far, *L. cardinalis* hairy roots seemed so be best for biosynthetic study of glucosyl metabolism, as well as for the production of these polyacetylene glucosides.

2.2.5 *Lobelia erinus*

Callus and cell suspension cultures of *L. erinus* following successful differentiation (shoot regeneration from calli) were established (Fig. 15; Tanaka et al. 1996). In callus cultures, two types of morphologically different shoots (shoot with long internodes, similar to the intact plant, and one with short internodes) were regenerated. The polyacetylene contents in the calli and suspension cells were almost two to four times higher than in the intact plant. The polyacetylene production in the regenerated shoots, morphologically similar to the intact plants, gradually decreased (to the same level as the intact plant) with the subcultures, whereas in the regenerants with short internodes it maintained a high level (ca. ten times that of the intact plant; Fig. 16). This observation indicated the relationship between morphology and secondary metabolism (polyacetylene production) in *L. erinus*. The results show the possibility of inducing

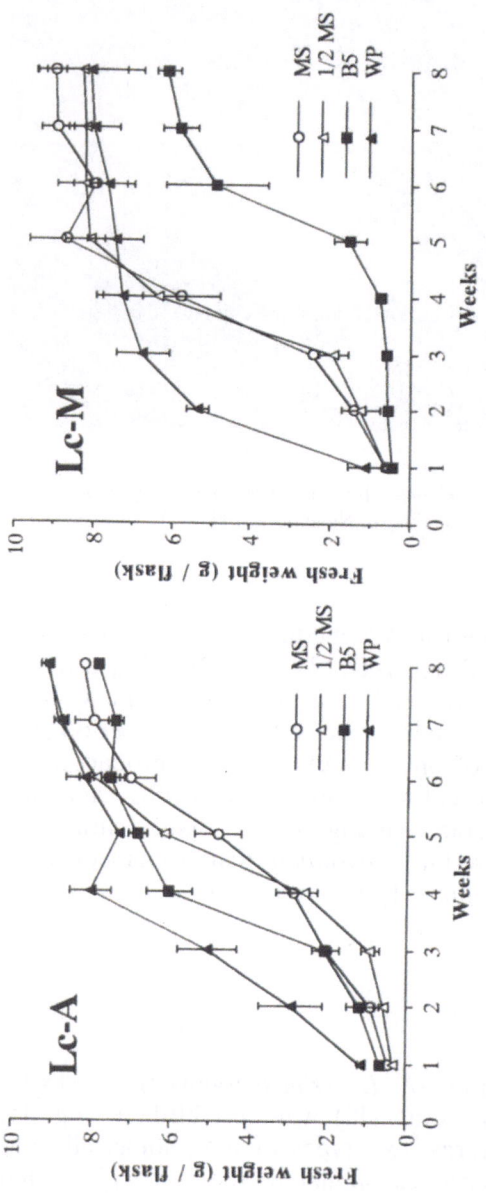

Fig. 13. Growth of *Lobelia cardinalis* hairy roots (*Lc-A* and *Lc-M*) cultured in four basal liquid media (*bar* standard errors). (Yamanaka et al. 1996)

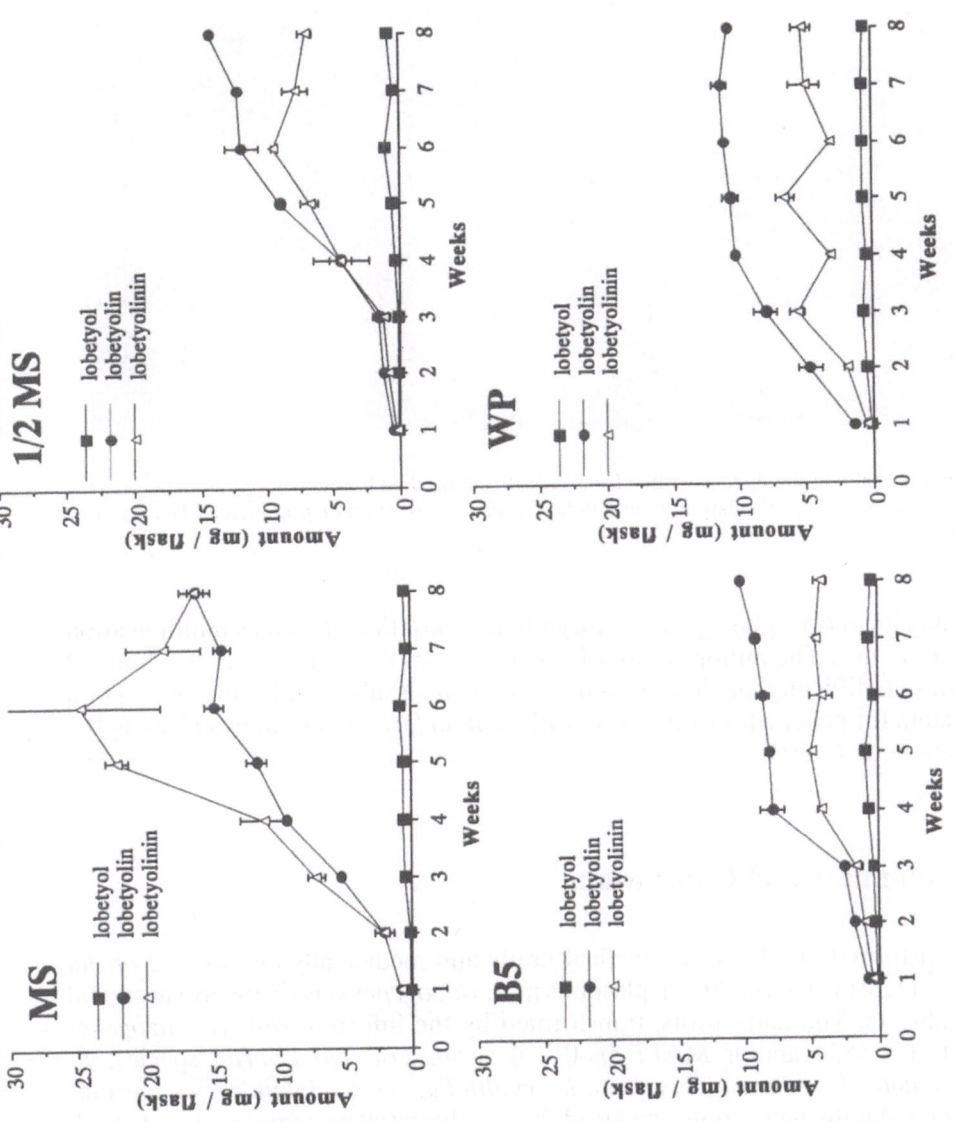

Fig. 14. Polyacetylene production in *Lobelia cardinalis* hairy roots (*Lc-A*) cultured in four basal liquid media (*bar* standard errors). (Yamanaka et al. 1996)

Fig. 15. Multiple shoots regenerated from the calli cultured on hormone-free MS medium in the light, the day after inoculation: *right* (10-day); *center* (15-day) and *left* (20-day). (Tanaka et al. 1996)

somaclonal (morphologic and biosynthetic) variation as well as multiplication of *L. erinus*. The multiple shoot formation from the callus cultures, obtained without difficulty on hormone-free medium, is also applicable as a good system for genetic transformation with *A. tumefaciens* and for establishing the transgenic *L. erinus*.

3 Summary and Conclusions

Transformation of various horticulturally and medicinally important *Lobelia* species, using root-inducing plasmids in *A. rhizogenes* was relatively successful (Table 1). The hairy roots, transformed by the infection with *A. rhizogenes* ATCC 15834 and/or MAFF 03-01724 strains, of five *Lobelia* species, *L. cardinalis, L. chinensis, L. inflata, L. sessilifolia*, and *L. siphilitica*, were established. All the hairy roots produced three polyacetylene compounds, lobetyol, lobetyolin (monoglucoside of lobetyol), and lobetyolinin (gentiobioside of lobetyol), in high amounts. Although some of the hairy roots (not all clones were investigated) also yielded the piperidine alkaloid lobeline, the major constituents in the roots were polyacetylenes (particularly lobetyolin). Concomitant analysis of secondary metabolites in the intact plants of several Campanulaceae (*Lobelia, Adenophora, Campanula, Platycodon, Specularia, Wahlenbergia*, etc.) indicated that these polyacetylenes, produced in all plants investigated, could be one of the most useful and important markers for the chemotaxonomy in campanulaceous plants (Tada et al. 1995a). Some hairy roots (*L. cardinalis, L. sessilifolia*, etc.), producing a large amount of highly glycosylated polyacetylenes (lobetyolinin) by their strong capability for

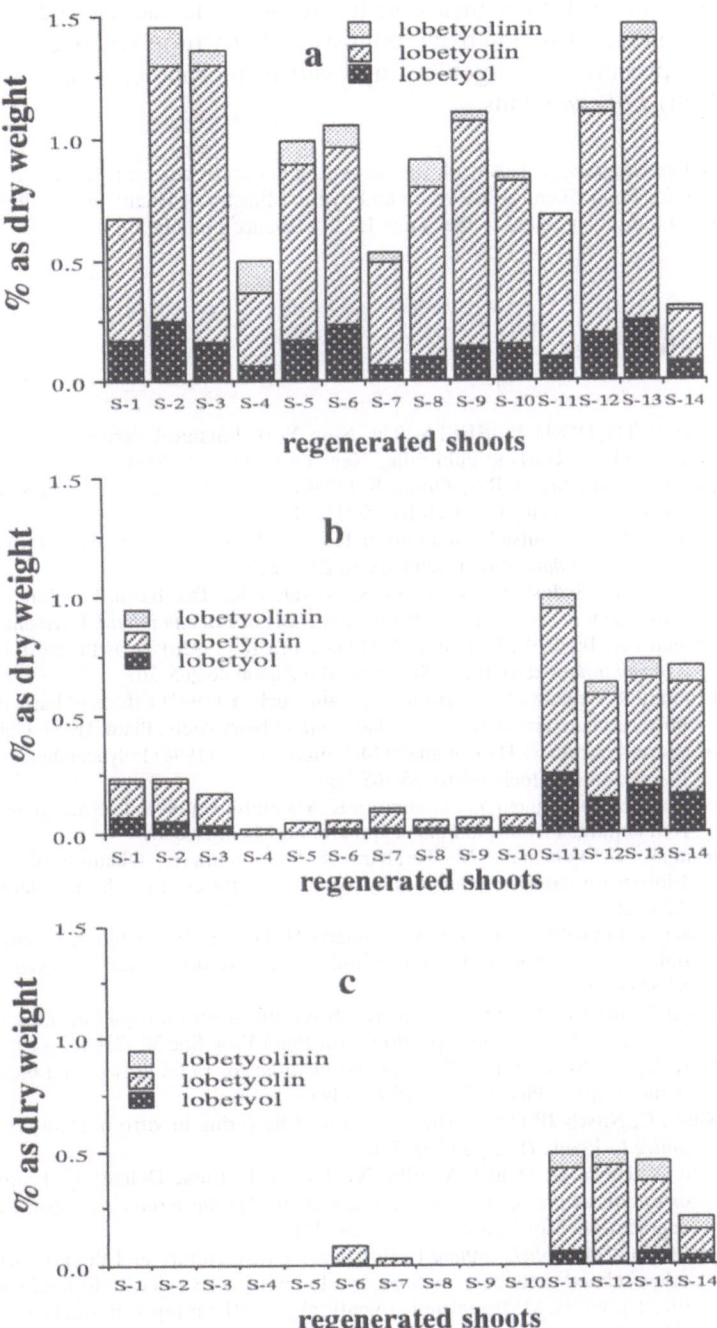

Fig. 16a–c. Polyacetylene content in *Lobelia erinus* regenerants. **a** Shoots regenerated from calli. **b** and **c** Shoots subcultured on hormone-free MS medium; subculture passage = 0 (**a**), 2 (**b**), and 4 (**c**). (Tanaka et al. 1996)

glycosylation, can also be available as a bioreactor system for the glycosylation of some useful chemicals. In future, with the success and establishment of a regeneration system from root tissues, hairy roots will become more important for procuring transgenic (improved in morphology and secondary metabolism) *Lobelia* plants.

Acknowledgments. This work was supported in part by a subsidy from The San-Ei Gen Foundation for Food Chemical Research and by the Ministry of Health and Welfare, Science Research Fund Subsidy, granted to the Japan Health Science Foundation.

References

Everett TH (1981) LOBELIA, The New York botanical garden illustrated encyclopedia of horticulture. Garland Publishing, New York, pp 2048–2050

Gamborg OL, Miller RA, Ojima K (1968) Nutrient requirements of suspension cultures of soybean root cells. Exp Cell Res 50:151–158

Ishimaru K, Yonemitsu H, Shimomura K (1991) Lobetyolin and lobetyol from hairy root cultures of *Lobelia inflata*. Phytochemistry 30:2255–2257

Ishimaru K, Sadoshima S, Neera S, Koyama K, Takahashi K, Shimomura K (1992a) A polyacetylene gentiobioside from hairy roots of *Lobelia inflata*. Phytochemistry 31:1577–1579

Ishimaru K, Ikeda Y, Kuranari Y, Shimomura K (1992b) Growth and lobeline production of *Lobelia inflata* hairy roots. Shoyakugaku Zasshi 46:265–267

Ishimaru K, Arakawa H, Sadoshima S, Yamaguchi Y (1993) Effects of basal media on growth and polyacetylene production on *Lobelia inflata* hairy roots. Plant Tissue Cult Lett 10:191–193

Ishimaru K, Arakawa H, Yamanaka M, Shimomura K (1994) Polyacetylenes in *Lobelia sessilifolia* hairy roots. Phytochemistry 35:365–369

Ishimaru K, Yamaguchi Y, Shimomura K, Yoshihira K (1995) Polyacetylene production in hairy root cultures of *Lobelia inflata*. Jpn J Food Chem 2:80–84

Ishimaru K, Yamanaka M, Terahara N, Shimomura K, Okamoto D, Yoshihira K (1996) Biotransformation of phenolics by hairy root cultures of five herbal plants. Jpn J Food Chem 3:38–42

Isogai A, Fukuchi N, Hayashi M, Kamada H, Harada H, Suzuki A (1988) Structure of a new opine, mikimopine, in hairy root induced by *Agrobacterium rhizogenes*. Agric Biol Chem 52:3235–3237

Lloyd G, McCown B (1980) Commercially feasible micropropagation of mountain laurel, *Kalmia latifolia*, by use of shoot-tip culture. Int Plant Prop Soc 30:421–427

Murashige T, Skoog F (1962) A revised medium for rapid growth and bio-assays with tobacco tissue cultures. Physiol Plant 15:473–497

Nitsch C, Nitsch JP (1967) The induction of flowering in vitro in stem segments of *Plumbago indica* L. Planta (Berl) 72:355–370

Petit A, David C, Dahl GA, Ellis JG, Guyon P, Casse-Delbart F, Tempé J (1983) Further extention of the opine concept: plasmids in *Agrobacterium rhizogenes* cooperate for opine degradation. Mol Gen Genet 190:204–214

Szoke E (1994) *Lobelia inflata* L. (lobelia): in vitro culture and the production of lobeline and other related secondary metabolites. In: Bajaj YPS (ed) Biotechnology in agriculture and forestry, vol 28. Medicinal and aromatic plants VII. Springer, Berlin Heidelberg, New York, pp 253–265

Tada H, Shimomura K, Ishimaru K (1995a) Polyacetylenes in *Platycodon grandiflorum* hairy root and campanulaceous plants. J Plant Physiol 145:7–10

Tada H, Shimomura K, Ishimaru K (1995b) Polyacetylenes in hairy root cultures of *Lobelia chinensis* Lour. J Plant Physiol 146:199–202

Tada H, Terahara N, Motoyama E, Shimomura K, Ishimaru K (1996) Anthocyanins in *Lobelia chinensis* hairy roots. Plant Tissue Cult Lett 13:85–86

Tanaka N (1990) Detection of opines by paper electrophoresis. Plant Tissue Cult Lett 7:45–47

Tanaka M, Yonemitsu H, Shimomura K, Ishimaru K, Mochida S, Endo T, Kaji A (1993) Transformation in *Lobelia inflata*. In: Bajaj YPS (ed) Biotechnology in agriculture and forestry, vol 22. Plant protoplasts and Genetic engineering III. Springer, Berlin Heidelberg, New York, pp 253–265

Tanaka N, Shimomura K, Ishimaru K (1996) In vitro shoot regeneration and polyacetylene production in callus cultures of *Lobelia erinus* L. J Plant Physiol 149:153–156

Vervliet G, Holsters M, Teuchy H, Van Montagu M, Schell J (1975) Characterization of different plaque-forming and defective temperate phages in *Agrobacterium* strains. J Gen Virol 26:33–48

White FF, Sinker VP (1987) Molecular analysis of root induction by *Agrobacterium rhizogenes*. In Hohn T, Schell J (eds) Plant DNA infection agents Springer, Vienna New York, pp 149–177

Wysokinska H (1977) Wystepowanie alkaloidow w hodowli tkankowej *L. inflata* L. Farm Pol 33:725–727

Yamanaka M, Shimomura K, Sasaki K, Yoshihira K, Ishimaru K (1995) Glucosylation of phenolics by hairy root cultures of *Lobelia sessilifolia*. Phytochemistry 40:1149–1150

Yamanaka M, Ishibashi K, Shimomura K, Ishimaru K (1996) Polyacetylene glucosides in hairy root cultures of *Lobelia cardinalis*. Phytochemistry 41:183–185

Yonemitsu H, Shimomura K, Satake M, Mochida S, Tanaka M, Endo T, Kaji A (1990) Lobeline production by hairy root culture of *Lobelia inflata* L. Plant Cell Rep 9:307–310

XII Genetic Transformation of *Papaver somniferum* L. (Opium Poppy) for Production of Isoquinoline Alkaloids

K. Yoshimatsu and K. Shimomura

1 Introduction

Opium poppy (*Papaver somniferum* L., Papaveraceae) is one of the most important medicinal plants and has been cultivated since early times. Opium, the dried cytoplasm of a specialized internal secretory system called the laticifer, is normally collected from the unripe capsule, and is the source for the commercial production of the medicinally important alkaloids, morphine, codeine, thebaine, noscapine, and papaverine (Roberts 1988; Nessler 1990); (Fig. 1). Morphine, which has strong addictive property, is still the most effective analgesic for the treatment of terminal cancer patients in modern medicine, and codeine is commonly used as an antitussive. However, field cultivation of this plant has been limited since 1953 by the United Nations Opium Conference Protocol to prevent narcotic abuse (Evans 1989). Therefore, establishing a tissue culture technique for the production of morphinan alkaloids seems to be desirable not only for medicinal purpose but also to decrease the abuse of the use of opiates.

Many researchers have investigated tissue culture of *P. somniferum* (Nessler 1990; Table 1), and most cultured cells, either as callus or cell suspensions, readily produced isoquinoline alkaloids such as sanguinarine (Furuya et al. 1972; Ikuta et al. 1974; Schuchmann and Wellmann 1983; Tyler et al. 1988; Roberts 1988; Songstad et al. 1989; Williams and Ellis 1993), but have rarely, if ever, produced morphinan alkaloids (Constabel 1985). Kamo et al. (1982, 1988), Schuchmann and Wellmann (1983), and Yoshikawa and Furuya (1985) reported the production of morphinan alkaloids in redifferentiated organs, either shoots or somatic embryos, and their results emphasize the importance of the degree of cell differentiation for the biosynthesis of morphinan alkaloids.

We established transformed cultures of *P. somniferum* by infection with A. *rhizogenes* MAFF 03-01724 (Yoshimatsu and Shimomura 1992, 1996), and found that one of the three transformed clones capable of forming somatic embryos contained considerable quantities of morphinan alkaloids by enzyme-linked immunosorbent assay (ELISA). In this chapter, establishment of transformed cultures of *P. somniferum* and isoquinoline alkaloid

Tsukuba Medicinal Plant Research Station, National Institute of Health Sciences, 1 Hachimandai, Tsukuba, Ibaraki, 305-0843 Japan

Biotechnology in Agriculture and Forestry, Vol. 45
Transgenic Medicinal Plants (ed. by Y.P.S. Bajaj)
© Springer-Verlag Berlin Heidelberg 1999

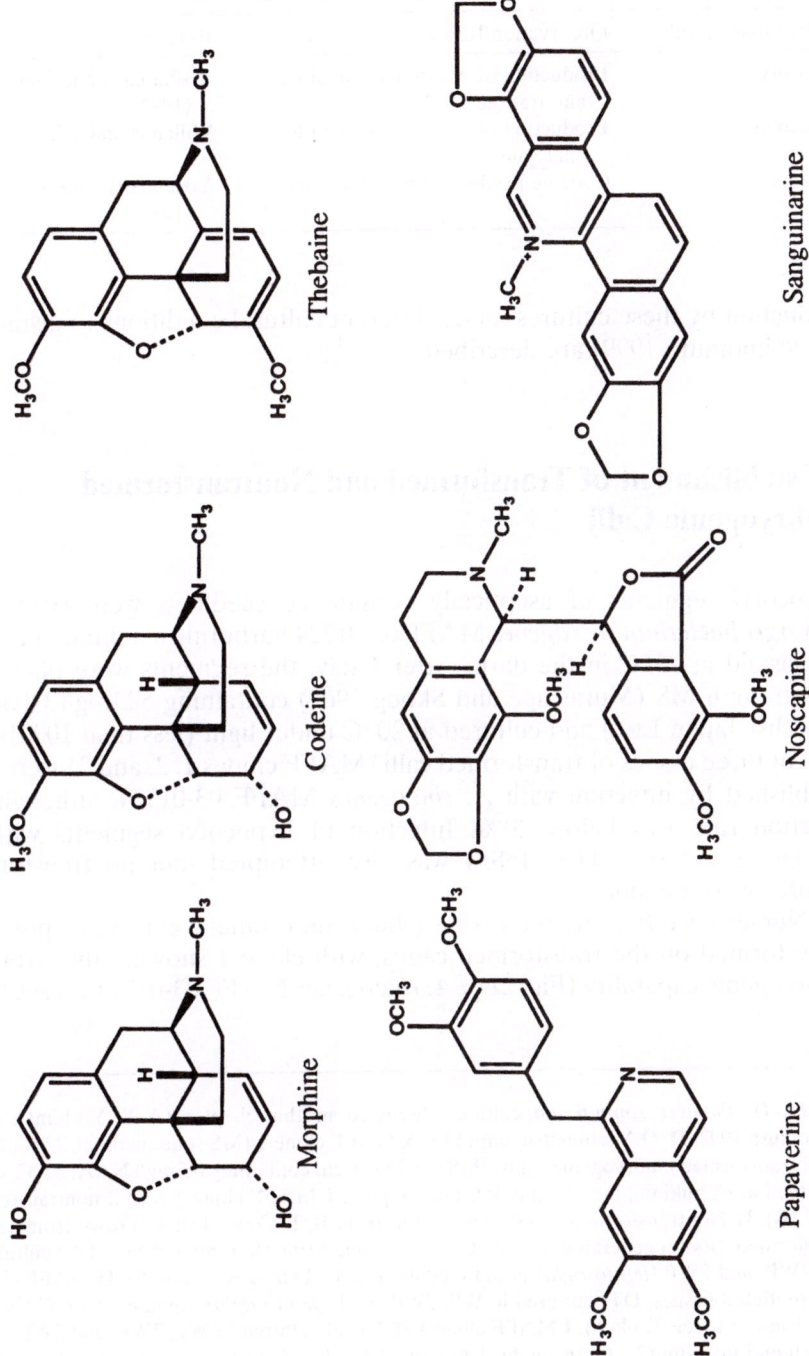

Fig. 1. Isoquinoline alkaloids in *Papaver somniferum* cultures

Table 1. Summary of various studies conducted on transformation of *Papaver*

Explant/tissue used	Observation/Remark	Reference
Hypocotyl	Production of morphinan alkaloids and transgenic plants	Yoshimatsu and Shimomura (1992)
Hypocotyl	Production of sanguinarine and its analogues	Williams and Ellis (1993)
Hypocotyl	Codeine production in a fermenter	Yoshimatsu and Shimomura (1999)

production by these cultures under different cultural conditions (Yoshimatsu and Shimomura 1999) are described.

2 Establishment of Transformed and Nontransformed Embryogenic Calli

Hypocotyl segments of aseptically germinated seedlings were cocultured with *Agrobacterium rhizogenes* MAFF 03-01724 harboring a mikimopine-type Ri plasmid at 25 °C in the dark. After 1 day, the segments were placed on half-strength MS (Murashige and Skoog 1962) containing 500 mg/l Claforan (Hoechst Japan Ltd.) and cultured at 20 °C under light (less than 1000 lx). A total of three clones of transformed calli (MAFF clones 1, 2, and 3) were thus established by infection with *A. rhizogenes* MAFF 03-01724, although the infection rate was below 20%. Infection of hypocotyl segments with *A. rhizogenes* A4 or ATCC 15834 was also attempted, but no transformed cultures were obtained.

Numerous differentiated tissues (shoots and somatic embryos) spontaneously formed on the transformed callus, with clone 1 showing the strongest embryogenic capability (Fig. 2A). *A. rhizogenes* MAFF 03-01724 causes hairy

───▶

Fig. 2A–O. *Papaver somniferum* cultures discussed in this chapter (**A–E** Yoshimatsu and Shimomura 1992; **D–O** Yoshimatsu, unpubl.). **A** MAFF clone 1 (MS solid medium, 22 °C, dark). **B** Nontransformed embryogenic callus (MS solid medium containing 0.5 mg/l NAA, 22 °C, dark). **C** Detection of mikimopine (*M* and *ML* mikimopine; *1* MAFF clone 1 (**A**); *2* nontransformed callus (**B**). **D** Nontransformed shoot regenerated from **B**. **E** Control shoot grown from seed. **F** Transformed shoot regenerated from **A**. **G** MAFF clone 1 (undifferentiated cells, UC) cultured in WP, 2WP, and 3WP (*left to right*) liquid medium (22 °C, 14 h light, 1 month). **H** MAFF clone 1 (differentiated tissues, DT) cultured in WP, 2WP, 3WP (*left to right*) liquid medium (22 °C, 14 h light, 1 month) (see Table 6). **I** MAFF clone 1 (UC) subcultured in WP, 2WP, and 3WP (*left to right*) liquid medium (22 °C, 14 h light, 1 month). **J** MAFF clone 1 (DT) subcultured in 1/4 WP, 1/2 WP, WP, 2WP, and 3WP (*left to right*) liquid medium (22 °C, 14 h light, 1 month); (see Tables 7, 8). **K** MAFF clone 1 culturing in a rotating drum fermenter (RDF). **L** MAFF clone 1 (DT) proliferated in a RDF. **M** MAFF clone 1 (UC) produced in a RDF. **N** MAFF clone 1 shoots grwon in a RDF. **O** MAFF clone 1 embryos produced in a RDF (see Table 9). *Bars* 1 mm (**A, B, O**), 5 cm (**K**), 3 cm (**L, M**), 1 cm (**N**), respectively

roots on the infected site of tobacco plants (Isogai et al. 1988); however, in the case of opium poppy no roots appeared. There is another example of transformation of *P. somniferum* with *A. rhizogenes* by Williams and Ellis (1993), who reported that transformed tissue led to the production of disorganized cell cultures rather than hairy root cultures. Therefore, this callus-forming tendency may be attributed less to the insertion and/or position effects of the T-DNA of Ri plasmid than to intrinsic properties of *P. somniferum*, such as a peculiar response to endo/exogenous phytohormones.

Embryogenic callus was also obtained from nontransformed roots cultured with 0.5 mg/l NAA (Fig. 2B). Mikimopine, however, was detected only in the MAFF clones (Fig. 2C).

3 Morphinan Alkaloid Contents in Transformed and Nontransformed Calli on Solid Medium

The morphinan alkaloid contents of transformed and nontransformed calli was analyzed by ELISA and high-pressure liquid chromatography (HPLC) (Table 2). The morphine-specific antibody used has a high affinity for morphine, codeine, ethylmorphine, dihydromorphine, and dihydrocodeine, less reactivity with dihydromorphinone, dihydrocodeinone, and norcodeine, and almost no reactivity with naloxone and naltrexone (Sawada et al. 1988). More precisely, five alkaloids, including morphine, codeine, papaverine, noscapine, and sanguinarine (Fig. 1) were simultaneously analyzed by HPLC.

MAFF clone 1, which had the strongest embryogenic capability, contained the highest level of morphinan alkaloids by ELISA (1.76 μg morphine equivalents/g fresh wt.) among the MAFF clones cultured in the dark, and codeine and sanguinarine could be detected by HPLC. Kamo et al. (1982) and Yoshikawa and Furuya (1985) had suggested that the biosynthesis of morphinan alkaloids might be closely related to the differentiation of shoots, and this result seems to support their suggestion. However, the

Table 2. Morphinan alkaloid contents in the calli of *Papaver somniferum*. (After Yoshimatsu and Shimomura 1992)

Cultures	Embryogenesis	ELISA (μg/g fr. wt) Morphine equivalents	HPLC (μg/g dry wt)	
			Codeine	Sanguinarine
Nontransformant	+++	0.17 ± 0.06	nd	nd
MAFF – clone 1	+++	1.76 ± 0.87	Trace	29
MAFF – clone 2	–	0.22 ± 0.01	nd	nd
MAFF – clone 3	+	0.08 ± 0.03	nd	nd

+++: vigorous, ++: moderate, +: poor, −: none, nd: not detected.

Calli were cultured on MS solid medium at 22 °C in the dark for 4 weeks.

nontransformed callus contained tenfold less morphinan alkaloids (0.17µg morphine equivalents/g fresh wt.) and no alkaloid was detected by HPLC although it showed almost the same embryogenic capability as MAFF clone 1. In addition, cultures transformed by Williams and Ellis (1993) accumulated sanguinarine and its analogues in high quantities but almost no morphinans. Therefore, the difference in morphinan alkaloid production observed among MAFF clones is related more to clonal line differences (Mano et al. 1986).

4 Culture of Transformed and Nontransformed Shoots

The transformed and nontransformed shoots (MAFF clone 1) formed in the dark were transferred onto phytohormone-free (HF) MS solid medium (30 ml/ 30 i.d. × 150-mm test tube) and their growth and features were compared with intact shoots established from seeds (Fig. 2D–F). The nontransformed shoots showed almost the same growth and appearance as the shoots germinated from seeds, while transformed shoots, which also produced mikimopine, demonstrated aberrant structural features such as wider leaves and longer internodes.

5 Morphinan Alkaloid Contents in Transformed and Nontransformed Shoots

The content of morphinan alkaloids in regenerated shoots was analyzed by ELISA and HPLC (Table 3). The regenerated shoots apparently could accumulate more morphinan alkaloids than their original calli (approximately 1000-fold higher level of morphine equivalents for the nontransformants, 100-fold higher for the transformants). The transformed shoots contained almost the same level of morphinan alkaloids as the nontransformed shoots when analyzed by ELISA, but no morphine could be detected by HPLC. In contrast,

Table 3. Morphinan alkaloid contents in regenerated shoots of *P. somniferum*. (Yoshimatsu and Shimomura 1992)

Cultures	ELISA (µg/g fresh wt) Morphine equivalents	HPLC (µg/g dry wt)	
		Morphine	Codeine
Control (from seeds)	472 ± 155	70 ± 40	2030 ± 330
Nontransformant	182 ± 62	50 ± 10	1310 ± 650
MAFF clone 1	213 ± 4	nd	750 ± 230

nd: not detected.

Shoots were cultured on MS solid medium under 14 h/day light (80 µE m^{-2} s^{-1}) at 22 °C for 6 weeks. Data shown as the mean of three replicates.

both nontransformed shoots regenerated from nontransformed embryos and developed from seeds contained much more codeine (1310–2030 µg/g dry wt.) than morphine (50–70 µg/g dry wt.). Other alkaloids (papaverine, noscapine, and sanguinarine) were detected in both nontransformed and transformed shoots. As a general concept, morphine is biosynthesized from codeine by demethylation (Evans 1989). Yoshikawa and Furuya (1985) observed that codeine was the main morphinan alkaloid at the early shoot stage (0.5–3 cm in height) and the morphine content increased during shoot development. These results imply that the biosynthesis of morphine from codeine is, to some extent, correlated with plant cell maturation.

6 Isoquinoline Alkaloid Production by Liquid Culture of MAFF Clone 1

MAFF clone 1 maintained stable embryogenic capability for years and accumulated morphinan alkaloids, though two of the three clones lost their embryogenic capability after the long culture time. Therefore, MAFF clone 1 was used for the alkaloid production study by liquid culture.

6.1 Effect of Temperature and Basal Medium in the Dark

MAFF clone 1 cells on solid medium were transferred into half-strength MS (1/2 MS), MS, B5 (Gamborg et al. 1968), woody plant (WP); (Lloyd and McCown 1980) and root culture (RC); (Thomas and Davey 1982) liquid media and cultured for 2 months at either 22 or 25 °C in the dark (Table 4). Although the mixture of undifferentiated cells (UC) and differentiated tissues (shoots

Table 4. Influence of temperature and basal media on the growth and alkaloid contents in MAFF clone 1. (Yoshimatsu and Shimomura 1999)

Temperature	Medium	Morphology	Growth index[a]	Sanguinarine (µg/g dry wt.)
22 °C	1/2 MS	UC	14.4	11
	MS	UC	8.5	5
	B5	UC and DT	12.3	2
	WP	UC and DT	13.2	136
	RC	UC	4.4	652
25 °C	1/2 MS	UC	14.3	25
	MS	UC	13.1	nd
	B5	UC	15.1	14
	WP	UC	7.6	3
	RC	UC	4.6	503

[a] Final fresh wt./initial fresh wt. (50–100 mg), UC: undifferentiated cells, DT: differentiated tissues (shoots and embryos), nd: not detected.
Mixture of UC and DT was cultured in liquid medium (20 ml/100-ml flask) in the dark on a rotary shaker (30 rpm) for 2 months.

and embryos) (DT) was used as inoculum, UC predominantly proliferated in the liquid medium. The cells grew satisfactorily in all liquid media except RC medium; however, DT production was restricted to B5 and WP media at 22 °C. In this culture condition, sanguinarine alone was detected, especially in the cells grown in the RC medium.

6.2 Effects of WP Macrosalts Concentration

Since both UC and DT proliferated well and a relatively high content of sanguinarine was detected in them grown in WP liquid medium at 22 °C, the effect of WP macrosalts concentration on growth and alkaloid production was studied by using either UC or DT as inoculum (Tables 5, 6). The growth and alkaloid contents of the cultures under 14 h/day light were superior to those in the dark, though their growth pattern and morphology were similar. When UC were inoculated and cultured in liquid medium, only UC were propagated and no DT formation was observed (Fig. 2G). The UC growth in full, double, and three-times strength macrosalts WP (WP, 2WP, and 3WP) media were better than those in a quarter- and half-strength macrosalts WP (1/4 WP and 1/2 WP) media, but alkaloids were negligible in all the UC cultures. Both UC and DT proliferated in the liquid medium when DT were used as inocula. The UC to DT ratio became higher along with the WP macrosalts concentration (Fig. 2H). DT growth without UC proliferation was observed when the DT was cultured in 1/4 WP (in the light and dark) and the 1/2 WP (in the light) medium and a relatively high concentration of alkaloids was detected. Then the cultures in the light were subcultured under the same conditions (Fig. 2I,J, Tables 6, 7). No alkaloid was detected in all the UC cultures. Further development of DT without proliferation was observed in the 1/4 WP medium (Fig. 2J, Table

Table 5. Effect of WP macrosalts concentration on the growth and alkaloid production in MAFF clone 1 cultured in the dark. (Yoshimatsu and Shimomura 1999)

Material	Macrosalts conc.	Morphology	Growth index[a]	Alkaloid µg/g dry wt	
				Codeine	Sanguinarine
UC	1/4	UC	5.6	nd	nd
	1/2	UC	8.5	3	2
	1	UC	14.4	nd	nd
	2	UC	13.7	nd	nd
	3	UC	13.1	nd	nd
DT	1/4	DT	11.7	11	110
	1/2	DT and UC	6.0	nd	9
	1	DT and UC	12.5	nd	18
	2	DT and UC	39.6	6	7
	3	DT and UC	25.9	nd	8

[a] Final fresh wt./initial fresh wt. (50–100 mg), UC: undifferentiated cells, DT: differentiated tissues (shoots and embryos), nd: not detected.
Either UC or DT was cultured in liquid medium (20 ml/100-ml flask) at 22 °C in the dark on a rotary shaker (50 rpm) for 1 month.

Table 6. Effect of WP macrosalts concentration on the growth and alkaloid production in MAFF clone 1 cultured under 14h/day light (first passage). (Yoshimatsu and Shimomura 1999)

Material	Macrosalts conc.	Morphology	Growth index[a]	Alkaloid µg/g dry wt	
				Codeine	Sanguinarine
UC	1/4	UC	7.8	nd	nd
	1/2	UC	10.8	1	5
	1	UC	17.2	nd	2
	2	UC	23.8	nd	nd
	3	UC	25.5	nd	1
DT	1/4	DT	15.9	67	300
	1/2	DT	30.3	27	60
	1	DT and UC	36.8	29	11
	2	DT and UC	45.7	26	2
	3	DT and UC	53.3	5	1

[a] Final fresh wt./initial fresh wt. (50–100 mg), UC: undifferentiated cells, DT: differentiated tissues (shoots and embryos), nd: not detected.
Either UC or DT was cultured in liquid medium (20 ml/100-ml flask) at 22 °C under 14 h/day light ($80 \, \mu E \, m^{-2} \, s^{-1}$) on a rotary shaker (50 rpm) for 1 month.

Table 7. Effect of WP macro salts concentration on the growth and alkaloid production in MAFF clone 1 cultured under 14h/day light (second passage). (Yoshimatsu and Shimomura 1999)

Material	Macrosalts conc.	Morphology	Growth index[a]	Alkaloid µg/g dry wt.			
				Codeine	Papaverine	Noscapine	Sanguinarine
UC	1/2	UC	6.4	nd	nd	nd	nd
	1	UC	11.7	nd	nd	nd	nd
	2	UC	15.8	nd	nd	nd	nd
DT	1/4	DT	12.0	648	29	48	9
	1/2	DT	6.4	60	nd	59	15
	1	DT	24.5	158	11	3	1
	2	DT	11.4	179	8	10	nd
	3	DT	7.6	6	nd	nd	1

[a] Final fresh wt./initial fresh wt. (50–100 mg), UC: undifferentiated cells, DT: differentiated tissues (shoots and embryos), nd: not detected.
Either UC or DT was subcultured in the same liquid medium (20 ml/100-ml flask) at 22 °C under 14 h/day light ($80 \, \mu E \, m^{-2} \, s^{-1}$) on a rotary shaker (50 rpm) for 1 month.

8) and the highest concentration of codeine (648 µg/g dry wt.) was detected together with other alkaloids (papaverine, noscapine, and sanguinarine) (Table 7). In the 3WP medium, both DT development and proliferation were suppressed (Fig. 2J), resulting in a low concentration of alkaloids (Table 7). Morphine was not detected in any of the cultures.

6.3 Culture of DT in a Rotating Drum Fermenter

Since moderate growth and alkaloid accumulation and proliferation of DT were observed in WP medium, DT were cultured in a rotating drum fermenter (Yoshimatsu and Shimomura 1991) using WP liquid medium (Fig. 2K, Table 9). Both DT and UC proliferated as observed in the liquid culture

Table 8. Effect of WP macro salt concentration on shoot development and rooting of MAFF clone 1 (22 °C, 14 h/day light, second passage). (Yoshimatsu and Shimomura 1999)

Macrosalts conc.	Shoot development	Rooting
1/4	+++	+++
1/2	++	++
1	+++	+
2	+++	+
3	+	−

+++: vigorous, ++: moderate, +: poor, −: none.

Table 9. Growth and alkaloid production by MAFF clone 1 cultured in a RDF. (Yoshimatsu and Shimomura 1999)

Parts	Fresh wt (g)	Dry wt. (g)	Growth index[a]	Alkaloid (µg/g dry wt.)		Alkaloid (µg)	
				Codeine	Sanguinarine	Codeine	Sanguinarine
DT	24.9	2.40	−	81	3	194.4	7.2
UC	27.0	3.65	−	15	4	54.8	14.6
Medium	−	−	−	−	−	79.3	36.8
Total	51.9	6.05	13.6	−	−	328.5	58.6

Final frest wt./initial fresh wt. (initial fresh wt 3.81 g).
DT was cultured in WP liquid medium (2.1 l, rotated at 3 rpm) at 23 °C under 14 h/day light (80 µE m^{-2} s^{-1}) at 200 ml/min vvm for 4 weeks.

in a 100-ml flask (Fig. 2L–O). Codeine and sanguinarine were detected in these cultures and the culture medium, with codeine showing the better productivity. After 4 weeks of culture, ca. 0.3 mg codeine and 0.06 mg sanguinarine were obtained (Table 9).

7 Summary and Conclusions

Transformed cultures of opium poppy (*Papaver somniferum* L.) were established by infecting hypocotyl segments with *Agrobacterium rhizogenes*. Among the bacteria tested, only *A. rhizogenes* MAFF 03-01724, which was recently isolated in Japan and induces mikimopine production in transformed cells (Isogai et al. 1988, 1990), could induce transformed cells. In addition to culture at 20 °C, subsequent coculture with bacteria was crucially important because the segments readily turned brown at 25 °C.

The transformed cells easily formed somatic embryos instead of roots. MAFF clone 1, which was the most embryogenic, accumulated about tenfold higher levels of morphinan alkaloids than nontransformed embryogenic callus (Table 2). Williams and Ellis (1993) reported that their *P. somniferum*-transformed cultures accumulated sanguinarine and its analogues in high quantities, but almost no morphinans. In our transformed cultures, only one clone of the three (MAFF clone 1) maintained stable embryogenic capability for 8 years and accumulated morphinan alkaloids. Therefore, the lack of accumulation of morphinans in their cultures may be due to genetic variation among the clones, inevitably induced by the Ri plasmid.

Nutrient medium composition, temperature, and illumination crucially affected cell morphology and alkaloid accumulation in MAFF clone 1 (Tables 4–8). Lower temperature (22°C), lower inorganic salts concentrations (a quarter- to full-strength macrosalts WP), and illumination were required for the stable production of DT and alkaloids. Higher concentrations of inorganic salts in WP promoted the formation and propagation of unorganized cells that caused diminishing alkaloids, especially codeine, though this phenomenon was reversible. Regeneration of embryos was observed when the unorganized cells obtained in 3WP liquid medium were statically cultured on the solid medium under the light. These results imply that the degree of cell differentiation as shoots or somatic embryos in transformed, as well as in nontransformed cultures, seems to be associated with the formation of morphinans.

Codeine production by transformed cultures was performed using a rotating drum fermenter (Table 9). The yields of alkaloids were not economically feasible; however, the data present possibilities for the continuous production of codeine, as one fourth of codeine produced was detected in the culture medium. In addition, productivity might be improved by further optimization including the selection of appropriate transformed clones and culture apparatus and conditions.

There are two possible biosynthetic pathways for the conversion of thebaine to morphine (Fig. 3); (Wilhelm and Zenk 1997). One is orthodox and known in the literature (Roberts 1988; Evans 1989; Nessler 1990), morphine biosynthesis from thebaine via codeinone and codeine. Another is that demonstrated by Brochmann-Hanssen (1984), biosynthesis via oripavine and morphinone. In the present study, the transformed clone could synthesize codeine but lacked morphine, though the nontransformed clone obtained from the same plant material accumulated morphine at the later developmental stage (Table 3).

8 Protocols

8.1 Induction of Transformed Calli

1. Surface-sterilize the seeds of *Papaver somniferum* L. according to the method of Yoshimatsu and Shimomura (1993), then place onto agar solid medium (0.5% sucrose and 0.5% agar) and keep at 18°C under 14h/day light (80μE/m²/s).
2. After 1 week, excise hypocotyl segments (ca. 0.5cm) from seedlings, inoculate into 1/2 MS liquid medium (20ml/100-ml Erlenmeyer flask) and coculture with *A. rhizogenes* MAFF 03-01724 (ca. 10^5–10^6 bacteria in the liquid medium) at 25°C in the dark on a rotary shaker (100rpm).
3. After 1 day, place the segments onto 1/2 MS solid medium containing 500mg/l Claforan. Approximately 10 days later, white calli will be formed on the segments.

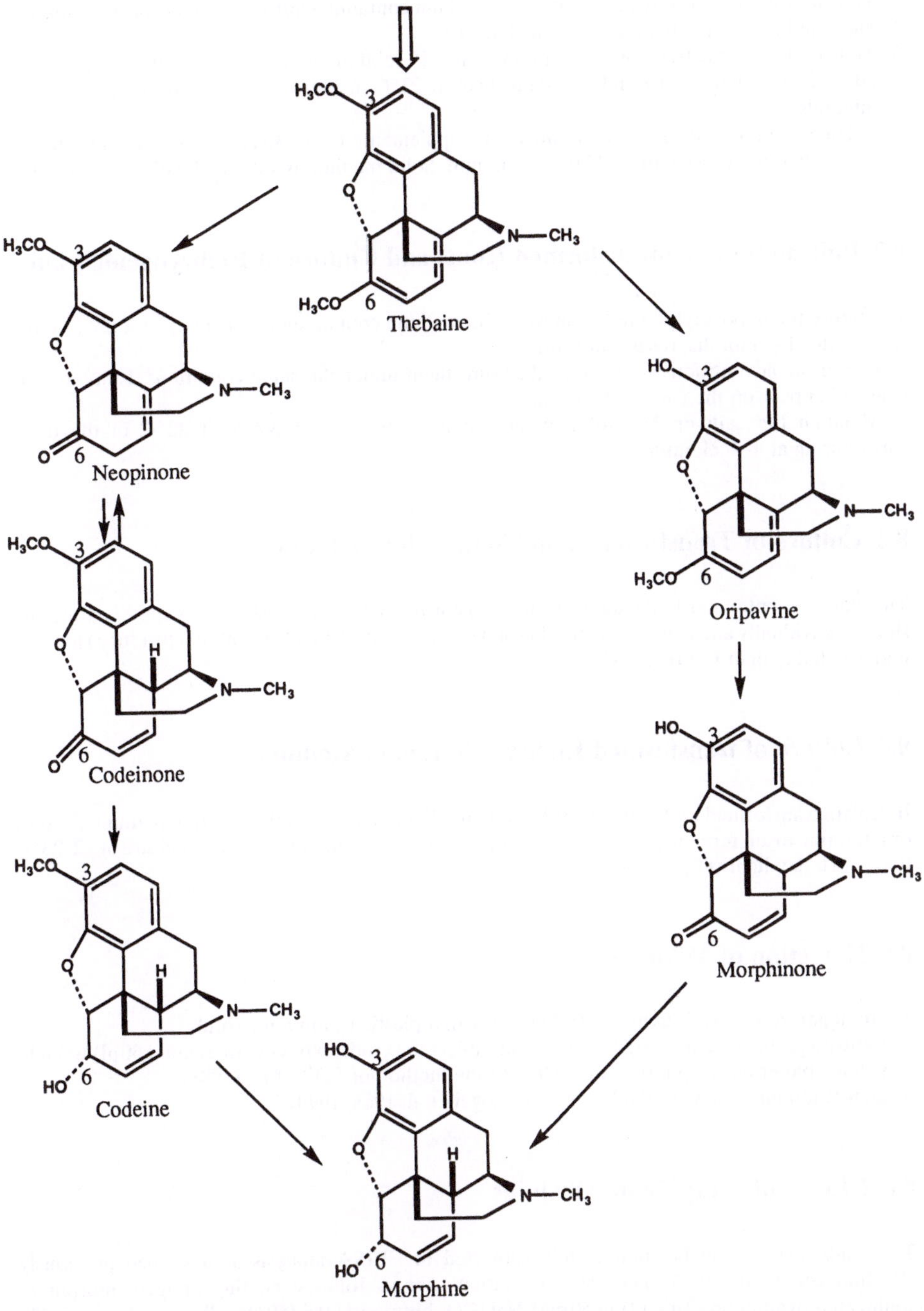

Fig. 3. Two possible biosynthetic pathways for the conversion of thebaine to morphine. Each pathway requires the demethylation of thebaine. *Left route* leads to morphine via codeine after enolether cleavage of thebaine. *Right route* leads to morphine via oripavine resulting from 3-*O*-demethylation of thebaine. (After Wilhelm and Zenk 1997)

4. Transfer the calli onto fresh 1/2 MS solid medium containing antibiotic and repeat this procedure until complete elimination of the bacteria.
5. Transfer the axenic transformed calli onto HF MS solid medium without antibiotics (25 ml/90 i.d. × 20-mm Petri dish) and maintain them at 22 °C in the dark, subculturing at 3–4 week intervals.

The medium used for the experiments is supplemented with 30 g/l sucrose and adjusted to pH 5.7 before autoclaving at 121°C for 15 min. Solid medium is solidified with 0.2% Gelrite.

8.2 Induction of Nontransformed Roots and Culture of Embryogenic Calli

1. Culture the hypocotyl segments on MS solid medium containing 0.5 mg/l NAA at 22 °C in the dark. After 1–2 months, roots will form.
2. Excise the adventitious roots and subculture them under the same conditions. Embryogenic calli will appear on the roots sporadically.
3. Maintain the calli on MS solid medium containing 0.5 mg/l NAA at 22 °C in the dark, subculturing at 4-week intervals.

8.3 Culture of Transformed and Nontransformed Shoots

The shoots spontaneously develop on the transformed and nontransformed calli. Separate the shoots individually and culture on HF MS solid medium (30 ml/30 i.d. × 150-mm test tube) at 22 °C under 14 h/day light (80 µE/m^2/s^1).

8.4 Culture of transformed Embryos in Liquid Medium

Inoculate transformed embryos into WP liquid medium in a 100-ml flask (20 ml medium, 50 rpm) or a rotating drum fermenter (21 working volume, 3 rpm, 200 ml/min vvm) and culture at 22–23 °C under 14 h/day light (80 µE/m^2/s^1).

8.5 Detection of Mikimopine

1. Homogenize the fresh tissue (100–200 mg) with a plastic rod in a microtube.
2. Centrifuge the tube at 6000 g for 1 min and subject the crude extract (supernatant 50 µl) to high-voltage paper electrophoresis according to the method of Petit et al. (1983).
3. Detect mikimopine with Pauly reagent (Isogai et al. 1988, 1990).

8.6 ELISA of Morphinan Alkaloids

The crude extract can be immediately subjected to ELISA analysis as described previously (Yoshimatsu et al. 1990). The assay conditions are as follows: coating antigen, morphine-conjugated ovalbumin (2 µg/ml) in 50 mM NaHCO$_3$ buffer, pH 9.6 (50 µl/well); morphine-specific monoclonal antibody, MOR 131.5.13 (Sawada et al. 1988) diluted × 10000 with 10 mM phosphate-buffered saline containing 1 g/l casein (C-PBS, pH 7.2) (50 µl/well); second antibody-enzyme conjugate, peroxidase-conjugated sheep anti-mouse IgG (BIOSYS, SA, France, BI 3413/8249) diluted × 10000 with C-PBS (50 µl/well); substrate, TMB (Kirkegaard & Perry Laboratories Inc.). Morphine equivalents in samples are calculated from standard curve for morphine.

8.7 Extraction of Alkaloids from Lyophilized Samples

1. Grind lyophilized sample to a fine powder.
2. Weigh the sample accurately (ca. 50 mg) and extract with 1 ml of 5% acetic acid twice by ultrasonication.
3. Wash the aqueous solution with 2 ml of chloroform.
4. Make the aqueous solution alkaline with 28% ammonium hydroxide and then extract with chloroform-isopropanol (3:1) three times (2, 2, 1 ml).
5. Evaporate the combined chloroform extracts in vacuo and dissolve in an appropriate volume of methanol for HPLC analysis. (After Staba et al. 1982).

8.8 Extraction of Alkaloids from Culture Medium

1. Add 50 ml of acetic acid to 1 l culture medium (pH 2–3).
2. Wash the aqueous solution with 100 ml of $CHCl_3$ twice.
3. Make the aqueous solution alkaline with 28% ammonium hydroxide (pH 8–9) and then extract with 100 ml of chloroform-isopropanol (3:1) three times.
4. Evaporate the combined chloroform extracts in vacuo and dissolve in an appropriate volume of methanol for HPLC analysis.

8.9 HPLC Conditions

Alkaloid extracts are injected into a TSK gel ODS-120A column (TOSOH, 4.6 i.d. × 250 mm), mobile phase acetonitrile-10 mM sodium 1-heptanesulphonate (pH 3.5) gradient (25–80% acetonitrile), flow rate 1 ml/min, column temperature 35 °C, UV at 284 nm. In our laboratory, each peak corresponding to morphine, codeine, papaverine, noscapine, and sanguinarine is confirmed by measuring the UV spectrum using a photodiode array detector (Waters 990J).

References

Brochmann-Hanssen E (1984) A second pathway for the terminal steps in the biosynthesis of morphine. Planta Med 50:343–345

Constable F (1985) Morphinan alkaloids from plant cell cultures. In: Phillipson JD, Roberts MF, Zenk MH (eds) The chemistry and biology of isoquinoline alkaloids. Springer, Berlin Heidelberg, New York, pp 257–264

Evans WC (1989) 31. Alkaloids, Opium. In: Pharmacognosy, 13th edn. Baillière Tindall, London, pp 582–591

Furuya T, Ikuta A, Syôno K (1972) Alkaloid from callus tissue of *Papaver somniferum*. Phytochemistry 11:3041–3044

Gamborg OL, Miller RA, Ojima K (1968) Nutrient requirements of suspension culture of soybean root cells. Exp Cell Res 50:151–158

Ikuta A, Syôno K, Furuya T (1974) Alkaloids of callus tissue and redifferentiated plantlets in the Papaveraceae. Phytochemistry. 13:2175–2179

Isogai A, Fukuchi N, Hayashi M, Kamada H, Harada H, Suzuki A (1988) Structure of a new opine, mikimopine, in hairy root induced by *Agrobacterium rhizogenes*. Agric Biol Chem 52:3235–3237

Isogai A, Fukuchi N, Hayashi M, Kamada H, Harada H, Suzuki A (1990) Mikimopine, an opine in hairy roots of tobacco induced by *Agrobacterium rhizogenes*. Phytochemistry 29:3131–3134

Kamo KK, Mahlberg PG (1988) Morphinan alkaloids: biosynthesis in plant (*Papaver* spp.) tissue cultures. In: Bajaj YPS (ed) Biotechnology in agriculture and forestry, vol 4. Medicinal and aromatic plants I. Springer, Berlin Heidelberg New York, pp 251–263

Kamo KK, Kimoto W, Hsu A-F, Mahlberg PG, Bills DD (1982) Morphinan alkaloids in cultured tissues and redifferentiated organs of *Papaver somniferum*. Phytochemistry 21:219–222

Lloyd G, McCown B (1980) Commercially feasible micropropagation of mountain laurel, *Kalmia latifolia*, by use of shoot-tip culture. Int Plant Propag Soc Combd Proc 30:421–427

Mano Y, Nabeshima S, Matsui C, Ohkawa H (1986) Production of tropane alkaloids by hairy root cultures of *Scopolia japonica*. Agric Biol Chem 50:2715–2722

Murashige T, Skoog F (1962) A revised medium for rapid growth and bioassays with tobacco tissue culture. Physiol Plant 15:473–497

Nessler CL (1990) Poppy. In: Ammirato PV, Evans DA, Sharp WR, Bajaj YPS (eds) Handbook of plant cell culture, vol 5. Ornamental species. McGraw-Hill, New York, pp 693–715

Petit A, David C, Dahl GA, Ellis JG, Guyon P, Casse-Delbart F, Tempé J (1983) Further extension of the opine concept: plasmids in *Agrobacterium rhizogenes* cooperate for opine degradation. Mol Gen Gent 190:204–214

Roberts MF (1988) Isoquinolines (*Papaver* alkaloids). In: Constabel F, Vasil IK (eds) Cell culture and somatic cell genetics of plants, vol 5. Academic Press San Diego, pp 315–334

Sawada J, Janejai N, Nagamatsu K, Terao T (1988) Production and characterization of high-affinity monoclonal antibodies against morphine. Mol Immunol 25:937–943

Schuchmann R, Wellmann E (1983) Somatic embryogenesis of tissue cultures of *Papaver somniferum* and *Papaver orientale* and its relationship to alkaloid and lipid metabolism. Plant Cell Rep 2:88–91

Songstad DD, Giles KL, Park J, Novakovski D, Epp D, Friesen L, Roewer I (1989) Effect of ethylene on sanguinarine production from *Papaver somniferum* cell cultures. Plant Cell Rep 8:463–466

Staba EJ, Zito S, Amin M (1982) Alkaloid production from *Papaver* tissue cultures. J Nat Prod 45:256–262

Thomas E, Davey MR (1982) Plant tissue culture media (5) root culture medium. In: EMBO course on Ti plasmids. Riaks Universiteit Gent Belgium, p 109

Tyler RT, Eilert U, Rijnders COM, Roewer IA, Kurz WGW (1988) Semi-continuous production of sanguinarine and dihydrosanguinarine by *Papaver somniferum* L. cell suspension cultures treated with fungal homogenate. Plant Cell Rep 7:410–413

Wilhelm R, Zenk MH (1997) Biotransformation of thebaine by cell cultures of *Papaver somniferum* and *Mahonia nervosa*. Phytochemistry 46:701–708

Williams RD, Ellis BE (1993) Alkaloids from *Agrobacterium rhizogenes*-transformed *Papaver somniferum* cultures. Phytochemistry 32:719–723

Yoshikawa T, Furuya T (1985) Morphinan alkaloid production by tissue cultures differentiated from cultured cells of *Papaver somniferum*. Planta Med 110–113

Yoshimatsu K, Shimomura K (1991) Efficient shoot formation on internodal segments and alkaloid formation in the regenerates of *Cephaelis ipecacuanha* A. Richard. Plant Cell Rep 9:567–570

Yoshimatsu K, Shimomura K (1992) Transformation of opium poppy (*Papaver somniferum* L.) with *Agrobacterium rhizogenes* MAFF 03-01724. Plant Cell Rep 11:132–136

Yoshimatsu K, Shimomura K (1993) *Cephaelis ipecacuanha* A. Richard (Brazilian ipecac): micropropagation and the production of emetine and cephaeline. In: Bajaj YPS (ed) Biotechnology in agriculture and forestry, vol 21. Medicinal and aromatic plants IV. Springer, Berlin Heidelberg, New York pp 87–103

Yoshimatsu K, Shimomura K (1996) Genetic transformation in *Papaver somniferum* L. (opium poppy) for enhanced production of morphinan. In: Bajaj YPS (ed) Biotechnology in agriculture and forestry, vol 38. Plant protoplasts and genetic engineering VII. Springer, Berlin Heidelberg New York, pp 243–252

Yoshimatsu K, Shimomura K (1999) Isoquinoline alkaloid production by transformed cultures of *Papaver somniferum*. Plant Cell Rep.

Yoshimatsu K, Satake M, Shimomura K, Sawada J, Terao T (1990) Determination of cardenolides in hairy root cultures of *Digitalis lanata* by enzyme-linked immunosorbent assay. J Nat Prod 53:1498–1502

XIII Genetic Transformation of *Panax ginseng* (Ginseng)

J.R. Liu, H.S. Lee, and S.W. Kim

1 Introduction

Ginseng (*Panax ginseng* C.A. Meyer) belongs to the Araliaceae family. In China and Korea, ginseng has been used medicinally for over 1000 years. Old literature recounts that ginseng cultivation in Korea began at least around 400 A.D. and in China around 40 B.C.

Ginseng is traditionally considered one of the most potent medicinal plants in the Orient. The shape and the medicinal effects of ginseng are reflected in its botanical name. *Panax* is derived from the Greek word for panacea. Ginseng (or len seng in Mandarin) literally means root of man, so named because the root of this plant resembles the shape of a human body. A Chinese herbalist from 209 B.C. gave the best description when he wrote that ginseng can vitalize the five organs, calm the nerves, stop palpitations due to fright, brighten vision, increase intellect, and, with long-term use, prolong life and make one feel young.

Korean ginseng root has been analyzed as containing a glycoside panaquilon, a saponin panaxin, oils, vitamins B_1 and B_2, alkaloids, polysacharides, etc. These ingredients have various pharmacological effects on the human body as well as antioxidant and anticancer activities, showing that ginseng can be a source of useful and effective drugs.

Various aspects of secondary metabolites (Choi 1988), somatic embryogenesis (Shoyama et al. 1995), and genetic transformation (Inomata and Yokoyama 1996), have already been reviewed in this Series.

Early studies on genetic transformation of ginseng were focused on hairy root formation by *Agrobacterium rhizogenes* (Table 1). Transformed hairy root cultures have been used for the high-level productivity of secondary metabolites because the roots can grow rapidly in defined hormone-free media. However, genetic transformation of ginseng as a possible means for molecular breeding was first reported by Lee et al. (1995), who introduced the *E. coli* β-glucuronidase (GUS) gene into ginseng using *A. tumefaciens*-mediated transformation, and expressed it at the whole plant level. This chapter

Plant Cell and Molecular Biology Research Unit, Korea Research Institute of Bioscience and Biotechnology, KRIBB, P.O. Box 115, Yusong, Taejon, 305, Korea

Biotechnology in Agriculture and Forestry, Vol. 45
Transgenic Medicinal Plants (ed. by Y.P.S. Bajaj)
© Springer-Verlag Berlin Heidelberg 1999

Table 1. Summary of various genetic transformation studies conducted on *Panax ginseng* (see also Inomata and Yokoyama 1996)

Culture system used	Strains/vector used for transformation	Observation/remarks	Reference
Hairy root culture	*A. rhizogenes* A4	Production of ginsenosides	1. Yoshikawa and Furuya (1987)
Hairy root culture	*A. rhizogenes* A4	Production of ginsenosides, adenosine, and guanosine	2. Ko et al. (1989)
Hairy root culture	*A. rhizogenes* A4	Glycosylation of digitoxigenin	3. Kawaguchi et al. (1990)
Hairy root culture	*A. rhizogenes* A4	Glycosylation of (RS)-2-phenyl-propionic acid	4. Yoshikawa et al. (1993)
Hairy root culture	*A. rhizogenes* ATCC15834	Production of ginsenosides	5. Inomata et al. (1993)
Regeneration of transformed cell	*A. tumefaciens* LBA4404/pBI121	*E. coli* β-glucuronidase gene expressed at the whole plant level	6. Lee et al. (1995)

describes a ginseng transformation system using *A. tumefaciens* LBA4404, and the binary vector pBI121 (Jefferson et al. 1987), which carries the CaMV 35S promoter-GUS gene fusion and the neomycin phosphotransferase gene as a selectable marker.

On the other hand, a transgene incorporated into plants may be deleted or impaired by methylation or mutation in meiotic transmission (Schmidt and Willmitzer 1988; Scheid et al. 1991; Topping et al. 1991). The transgene is not necessarily stable during mitosis (Gao et al. 1991), and it may be deleted or inactivated during this process. However, since meiotic stability of the transgene has been determined at the whole plant level, the mitotic stability of the transgene should also be determined at the same level. In this chapter, the stability of the inserted GUS gene in regenerants from isolated protoplasts was also evaluated.

2 Genetic Transformation

2.1 Tissue Culture of Ginseng

2.1.1 Coculture of Ginseng Explants and Agrobacterium

Cotyledons of dehusked mature zygotic embryos of ginseng (*P. ginseng* CA Meyer) were excised, transversely scored with a scalpel, and cocultured with

Agrobacterium for 48 h in Murashige and Skoog's (MS) (1962) liquid medium containing 1 mg/l 2,4-dichlorophenoxyacetic acid (2,4-D) and 0.1 mg/l kinetin. They were rinsed three to four times in the liquid culture medium and placed with the adaxial surfaces down on medium containing 1 mg/l 2,4-D, 0.1 mg/l kinetin, 500 mg/l carbenicillin, and 100 mg/l kanamycin. They were cultured at 25 °C in the dark. In addition, cotyledonary explants which were not exposed to *Agrobacterium* were also cultured in the same manner as the control.

2.1.2 Agrobacterium *Strain, Culture Media, and Solutions*

A binary vector pBI121 carrying the CaMV 35S promoter-GUS gene fusion and the neomycin phosphotransferase gene as a selectable marker was transferred into the disarmed *A. tumefaciens* strain LBA4404 harboring the helper plasmid pTi Bo542, as described by An (1987).

Unless mentioned otherwise, basal MS medium containing 100 mg/l myo-inositol, 0.4 mg/l thiamine·HCl, 3% sucrose, and 0.4% Gelrite was used throughout the experiment. The enzyme solution used for protoplast isolation consisted of 2% (w/v) Cellulase R-10, 0.5% Macerozyme R-10, and 9% mannitol in CPW (Frearson et al. 1976). All media and solutions were adjusted to pH 5.8 before autoclaving.

2.1.3 Protoplast Isolation, Culture, and Plant Regeneration

A few hundred (approximately 1 g) somatic embryos from transformed calli were transversely sliced into 1- to 2-mm-thick segments with a scalpel, and incubated with 10 ml of the enzyme solution in an 87 × 15-mm plastic Petri dish on a gyratory shaker (30 rpm) at 25 °C in the dark overnight. The digested tissues were filtered through 50- and 100-µm stainless steel meshes. The enzyme solution containing isolated protoplasts was transferred to a 15-ml Falcon centrifuge tube and spun at 100 g for 5 min. The protoplasts forming a pellet at the base of the tube were resuspended in 10 ml CPW solution containing 9% mannitol and the centrifugation-resuspension process was repeated three times, but the final resuspension was conducted in protoplast culture medium. The purified protoplasts were plated at a density of 5×10^5 protoplasts/ml in liquid medium containing 60 g/l myo-inositol, 1 mg/l 2,4-D, 0.5 mg/l BA, and 0.5 mg/l kinetin (protoplast culture medium) (Arya et al. 1991) in 55 × 12-mm Petri dishes. The dishes were incubated at 25 °C in the dark and occasionally observed with an inverted microscope. Colonies derived from protoplasts were transferred onto medium containing 1 mg/l 2,4-D and 0.1 mg/l kinetin for continued growth.

To regenerate whole plants, somatic embryos were transferred onto 1/2 MS medium with 1 mg/l BA and 1 mg/l GA_3 (Lee et al. 1990), and cultured in the light (approximately 1000 lx cool-white fluorescent lamps under a 16-h photoperiod).

2.1.4 GUS assay, PCR, and Southern Blot Analysis

GUS assay was performed in somatic embryos, transversely sliced stems and petioles (100 to 200 μm in thickness), squashed root tips, and flowers of regenerants from kanamycin-resistant calli, and in leaves of 50 randomly selected regenerants from transformed somatic embryo-derived protoplasts. As control, somatic embryos from nontransformed calli were subjected to X-glucuronide (X-gluc) treatment according to the CLONTECH (Palo Alto, California, USA) manual.

DNA was extracted from the leaf segments of 50 protoplast-derived regenerants by the cetyltrimethyl ammonium bromide method. PCR was carried out using 21-mer oligonucleotides (for GUS gene detection, 5'-GGTGGGAAAGCGCGTTACAAG-3' and 5'-GTTTACGCGTTGCTTCC GCCA-3') (Hamill et al. 1991). The amplification cycle consisted of denaturation at 93 °C for 1 min, primer annealing at 55 °C for 1 min, and primer extension at 72 °C for 2 min. After 31 repeats of the thermal cycle, amplification products were analyzed on 1.2% agarose gels. Gels were stained with ethidium bromide and visualized with UV light.

Samples of genomic DNA of GUS-positive and GUS-negative regenerants were digested with *Eco*RI/*Bam*HI, *Eco*RI, and *Hin*dIII, respectively, at 37 °C for 5 h and subjected to electrophoresis on 0.8% agarose gel. The DNA was transferred to Nytran membrane and a 2.2-kb probe including the GUS gene as well as the NOS poly (A) region labeled with digoxigenine (DIG; Boehringer Mannheim) was used for Southern hybridization.

2.2 Results and Discussion

After 8 weeks of culture on medium containing 100 mg/l kanamycin, yellowish compact calli formed on the cut surfaces of about 50% of the cotyledonary explants which had been cocultured with *Agrobacterium*. The other explants degenerated due to contamination. All explants which were not exposed to *Agrobacterium* failed to form calli and degenerated due to necrosis (data not shown). Four weeks after transfer to medium containing 1 mg/l 2,4-D and 0.1 mg/l kinetin, the yellowish compact calli gave rise to numerous somatic embryos closely aggregated together (Fig. 1A), as observed by Lee et al. (1990). Clumps of somatic embryos at various developmental stages were transferred onto 1/2 MS medium with 1 mg/l each of BA and GA₃, and most of them developed into plantlets after 8 weeks. After 10 weeks of culture, more than 50% of the plantlets flowered during in-vitro culture (Fig. 1B).

When treated with X-gluc, most of the embryos derived from kanamycin-resistant calli exhibited a GUS-positive response (Fig. 1C), whereas all the embryos from nontransformed calli exhibited a GUS-negative response (Fig. 1D). Also, more than 50% of the regenerants were GUS-positive in petiole, stem, root, and flower (Fig. 1E–H). Epidermal layers of stem and petiole, vascular regions of stem, petiole, and root, and root apical meristem were more densely stained than other tissues. This seems to be a typical expression

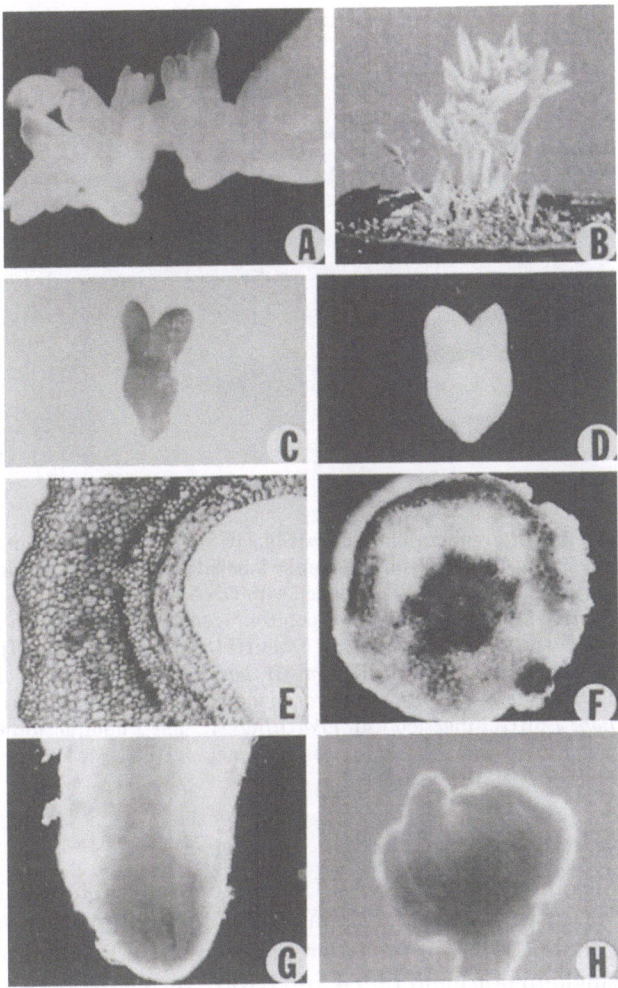

Fig. 1A–H. *Agrobacterium*-mediated transformation and histochemical assay of GUS activity in somatic embryos and regenerants of ginseng. Numerous somatic embryos formed on kanamycin-resistant calli (**A**); flowering plantlets regenerated from somatic embryos (**B**); GUS-positive (**C**) and GUS-negative somatic embryos (**D**); GUS-positive cross-sections of petiole (**E**), stem (**F**), root (**G**), and flower (**H**) of the regenerant. (Lee et al. 1995)

pattern of the CaMV 35S promoter-regulated GUS gene (Jefferson et al. 1987).

The genomic DNA of three randomly selected GUS-positive regenerants was digested with restriction enzymes and subjected to Southern blot analysis using a DIG-labeled GUS NOS poly(A) probe (Fig. 2). Only one band of more than 23 kb in two regenerants (I, II); (Fig. 2A, lane 4; 2B, lane 3), and two bands of 14.4 and 13.3 kb in one regenerant (III) were obtained after *Eco*RI digestion (Fig. 2C, lane 3). This indicates that the GUS gene was incorporated

198   J.R. Liu et al.

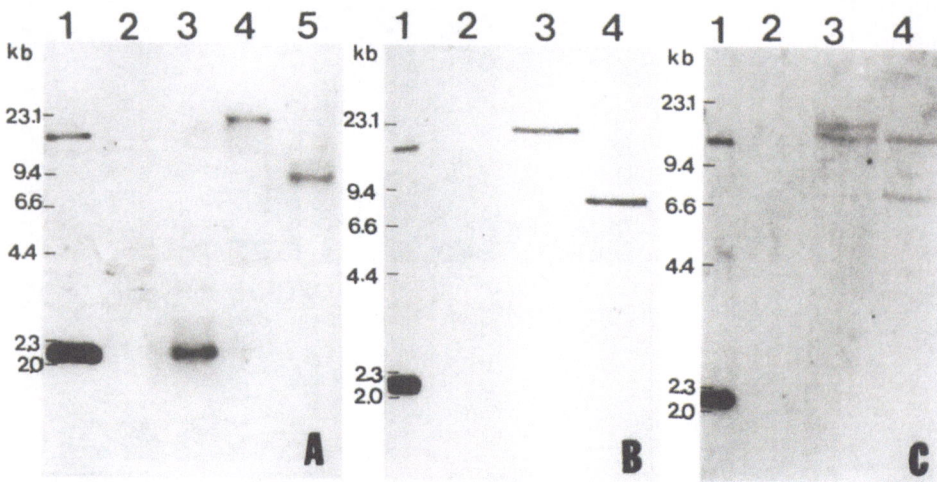

Fig. 2A–C. Southern blot analysis of GUS-positive and GUS-negative regenerants of ginseng. The 2.2-kb GUS NOS poly(A) probe labeled with DIG was hybridized with the genomic DNA of regenerants. Each lane was loaded with DNA as follows: **A** *Lane 1* pBI121 digested with *Eco*RI/*Bam*HI; *lane 2* nontransformed control regenerant digested with *Eco*RI; *lanes 3–5* GUS-positive regenerant I digested with *Eco*RI/*Bam*HI (*lane 3*), *Eco*RI (*lane 4*), and *Hin*dIII (*lane 5*).**B** *Lane 1* pBI121 digested with *Eco*RI/*Bam*HI; *lane 2* nontransformed control regenerant digested with *Eco*RI; *lanes 3–4* GUS-positive regenerant II digested with *Eco*RI (*lane 3*) and *Hin*dIII (*lane 4*).**C** *Lane 1* pBI121 digested with *Eco*RI/*Bam*HI; *lane 2* nontransformed control regenerant digested with *Eco*RI; *lanes 3–4* GUS-positive regenerant III digested with *Eco*RI (*lane 3*) and *Hin*dIII (*lane 4*). (Lee et al. 1995)

into the genomic DNA of the regenerants. These DNA sizes are longer than vector pBI121 (pBI121 has one *Eco*RI site with a total size of 12.7 kb). The genomic DNA of the two regenerants I and II digested with *Hin*dIII yielded a single major band of 9.4 and 8.8 kb, respectively (Fig. 2A, lane 5; 2B, lane 4). Two bands of 13.3 and 7.8 kb were obtained from regenerant III (Fig. 2C, lane 4), indicating that two copies of the GUS gene were incorporated into the genomic DNA of regenerant III. One 2.2-kb band obtained by *Eco*RI/*Bam*HI digestion showed that the incorporated DNA fragment included the intact GUS gene (Fig. 2A, lane 3).

To investigate the mitotic stability of the inserted GUS gene in regenerants from isolated single cells, protoplasts were enzymatically isolated from transformed somatic embryo segments in which the presence of the GUS gene in the genomic DNA was confirmed by Southern blot analysis (data not shown). After 3 months of culture, colonies of about 1 mm in diameter formed at a plating efficiency of 0.01%. Upon transfer onto medium containing 1 mg/l 2,4-D and 0.1 mg/l kinetin, colonies gave rise to numerous somatic embryos. These somatic embryos developed into plantlets when cultured on medium containing 1 mg/l each of BA and GA₃.

The PCR method revealed that 92% of the regenerants retained the GUS gene (Fig. 3), indicating that the GUS gene is retained and is stable through

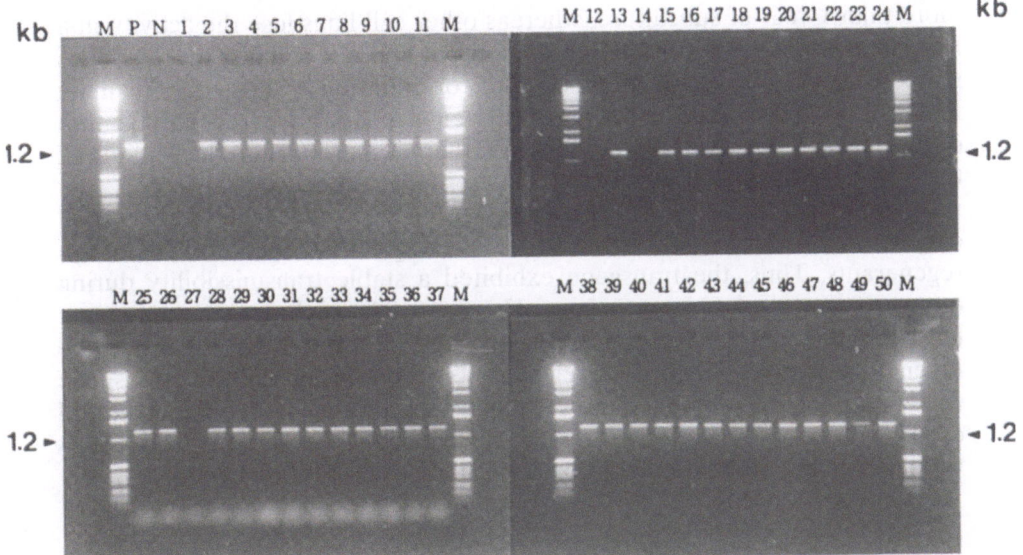

Fig. 3. Detection of the GUS gene by PCR in the genomic DNA of 50 transgenic somatic embryo-derived regenerants. *Lane M* molecular size markers; *P* pBI121 as a positive control; *N* nontransformed regenerant; *1–50* transgenic regenerants. (Lee et al. 1995)

ontogeny. When treated with X-gluc, 78% of the regenerants from protoplasts were GUS-positive, whereas all of the regenerants from nontransformed calli were GUS-negative.

Although somatic (Chang and Hsing 1980; Lee et al. 1990) and zygotic ginseng embryos (Lee et al. 1991) are capable of flowering in vitro after 2 months of culture, this shortened juvenile period has not yet been satisfactorily exploited for breeding purposes. Viable seeds have not been obtained from in-vitro flowering plants. Thus, the *Agrobacterium*-mediated transformation system described in this study may be of practical use for introducing single gene-mediated traits, such as herbicide, insect, and disease resistance into this species. In an attempt to prevent fungal diseases of the ginseng taproot, as demonstrated in a tobacco system (Broglie et al. 1991), introduction of the chitinase gene into ginseng is currently under investigation.

Due to the normal 3-year-long juvenile period of ginseng, the transmissibility of foreign genes in progeny cannot be readily determined. Since in vitro regenerants may be used to assist breeding, such as the recurrent selection, it may be more practical to investigate the mitotic stability of transgenes in regenerants.

The stability of transgenes in the progeny of transgenic plants has been documented: in tobacco only 8 to 36% of the first generation shows transgene-encoded drug resistance (Topping et al. 1991) and in *Arabidopsis* the frequency ranges from 10 to 50% (Schmidt and Willmitzer 1988; Scheid et al. 1991). Some cell lines of tobacco maintain the bacterial GUS activity stably for more than 1 year of subculture, whereas other cell lines lose the activity in a few months (Gao et al. 1991). Both meiotic and mitotic stability of transgenes may be impaired by methylation, deletion, or mutation of the gene. The main factor in progressive inactivation of transgenes in *Arabidopsis* is methylation of the genes (Kilby et al. 1992). However, there is no report on the stability of the transgene in regenerants from isolated single cells.

Although 78% of the regenerants were GUS-positive, the PCR method detected the presence of the GUS gene in the genomic DNA of 92% of the regenerants. Thus, the transgene exhibited a stable transmissibility during ontogeny from isolated protoplasts. However, 8% of the regenerants must have lost the transgene during ontogeny, and in about 15% (GUS-positive regenerants of the PCR-positive regenerants) of the regenerants the transgene was silent. The overall results indicate that the transgene is stably transmitted during ontogeny and stably expressed in most of the regenerants from transformed somatic embryo-derived protoplasts.

3 Summary and Conclusions

Cotyledonary explants of ginseng zygotic embryos were cocultured with *A. tumefaciens* strain LBA4404 harboring the binary vector pBI121 for 48h and transferred onto MS medium supplemented with 1 mg/l 2,4-D, 0.1 mg/l kinetin, and 100 mg/l kanamycin. After 8 weeks of culture, kanamycin-resistant calli formed on the cut surfaces of cotyledonary explants and subsequently gave rise to numerous somatic embryos. Eight weeks after transfer onto medium containing 1 mg/l each of BA and gibberellic acid, most of them developed into plantlets. Southern analysis confirmed that the GUS gene was incorporated into the genomic DNA of regenerants. Protoplasts were enzymatically isolated from transformed somatic embryo segments and cultured in liquid medium containing 60 g/l myo-inositol, 1 mg/l 2,4-D, 0.5 mg/l BA, and 0.5 mg/l kinetin. Plants were regenerated from protoplasts via somatic embryogenesis. The polymerase chain reaction revealed that 92% of the regenerants retained the GUS gene. When treated with X-gluc, 78% of the regenerants showed a GUS-positive response. The overall results indicate that the transgene is stably transmitted during somatic ontogeny and stably expressed in most of the regenerants, whereas it may be deleted or impaired in some portion of them.

References

An G (1987) Binary Ti vectors for plant transformation and promoter analysis. Methods Enzymol 153:292–305
Arya S, Liu JR, Eriksson T (1991) Plant regeneration from protoplasts of *Panax ginsneg* (CA Meyer) through somatic embryogenesis. Plant Cell Rep 10:277–281

Broglie K, Chet I, Hollyday M, Cressman R, Biddle P, Knowltion S, Mauvais CJ, Broglie R (1991) Transgenic plants with enhanced resistance to the fungal pathogen *Rhizoctonia solani*. Science 254:1194–1197

Chang WC, Hsing YH (1980) In vitro flowering of embryoids derived from mature root callus of ginseng (*Panax ginseng*). Nature 284:341–342

Choi KT (1988) *Panax ginseng* CA Meyer: micropropagation and the in vitro production of saponins. In: Bajaj YPS (ed) Biotechnology in agriculture and forestry, vol 4. Medicinal and aromatic plants I. Springer, Berlin Heidelberg New York, pp 484–500

Frearson EM, Power JB, Cocking EC (1976) The isolation, culture and regeneration of *Petunia* leaf protoplasts. Dev Biol 33:130–137

Gao J, Lee JM, An G (1991) The stability of foreign protein production in genetically modified plant cells. Plant Cell Rep 10:533–536

Hamill JD, Rounsley S, Spencer A, Todd G, Rhodes JC (1991) The use of the polymerase chain reaction in plant transformation studies. Plant Cell Rep 10:221–224

Inomata S, Yokoyama Y (1996) Genetic transformation of *Panax ginseng* (CA Meyer) for increased production of ginsenosides. In: Bajaj YPS (ed) Biotechnology in agriculture and forestry, vol 38. Plant protoplasts and genetic engineering VII. Springer, Berlin Heidelberg New York, pp 253–269

Inomata S, Yokoyama M, Gozu Y, Shimizu T, Yanagi M (1993) Growth pattern and ginsenoside production of *Agrobacterium*-transformed *Panax ginseng* roots. Plant Cell Rep 12:681–686

Jefferson RA, Kavanagh TA, Bevan MW (1987) Gus fusion: β-glucuronidase as a sensitive and versatile gene fusion marker in higher plants. EMBO J 6:3901–3907

Kawaguchi K, Hirotani M, Yoshikawa T, Furuya T (1990) Biotransformation of digitoxigenin by ginseng hairy root cultures. PhytoChemistry 29:837–843

Kilby NJ, Ottoline Leyser HM, Furner IJ (1992) Promoter methylation and progressive transgene inactivation in *Arabidopsis*. Plant Mol Biol 20:103–112

Ko KS, Noguchi H, Ebizuka Y, Sankawa U (1989) Oligoside production by hairy root cultures transformed by Ri plasmids. Chem Pharm Bull 37:245–248

Lee HS, Liu JR, Yang SG, Lee YH, Lee K-W (1990) In vitro flowering of ginseng plants regenerated from zygotic embryos-derived somatic embryos of ginseng (*Panax ginseng* CA Meyer) by growth regulators. HortScience 25:1652–1654

Lee HS, Lee K-W, Yang SG, Liu JR (1991) In vitro flowering of ginseng (*Panax ginseng* CA Meyer) zygotic embryos induced by growth regulators. Plant Cell Physiol 32:1111–1113

Lee HS, Kim SW, Lee K-W, Eriksson T, Liu JR (1995) *Arobacterium*-mediated transformation of ginseng (*Panax ginseng*) and mitotic stability of the inserted β-glucuronidase gene in regenerants from isolated protoplasts. Plant Cell Rep 14:545–549

Murashige T, Skoog S (1962) A revised medium for rapid growth and bio-assays with tobacco tissue cultures. Physiol Plant 15:473–497

Scheid OM, Paszkowski J, Potrykus I (1991) Reversible inactivation of a transgene in *Arabidopsis thaliana*. Mol Gen Genet 228:104–112

Schmidt R, Willmitzer L (1988) High efficiency *Agrobacterium tumefaciens*-mediated transformation of *Arabidopsis thaliana* leaf and cotyledon explants. Plant Cell Rep 7:583–586

Shoyama Y, Matsushita H, Zhu XX, Kishira H (1995) Somatic embryogenesis in ginseng (*Panax* species). In: Bajaj YPS (ed) Biotechnology in agriculture and forestry, vol 31. Somatic embryogenesis and synthetic seed II. Springer, Berlin Heidelberg New York, pp 343–356

Topping JF, Wei W, Lindsey K (1991) Functional tagging of regulatory elements in the plant genome. Development 112:1009–1019

Yoshikawa T, Furuya T (1987) Saponin production by cultures of *Panax ginseng* transformed with *Agrobacterium rhizogenes*. Plant Cell Rep 6:449–453

Yoshikawa T, Asada Y, Furuya T (1993) Continuous production of glycosides by a bioreactor using ginseng hairy root culture. Appl Microbiol Biotechnol 39:460–464

XIV Genetic Transformation of *Peganum harmala*

J. BERLIN

1 Introduction

Peganum harmala L. (Zygophyllaceae) has been a well-known plant in folk medicine and is still used in areas where it is endemic. An incredibly high number of quite different medicinal applications of the seeds and the herb are mentioned in herbal books describing native plants of North Africa and India (Boulos 1983; Ambasta 1986). For example, dried seeds or their extracts are sold as antispasmodicum, anthelminticum, antiasthmaticum, or antir-heumaticum at the drug market of Rabat and Algier (Boulos 1983). Though the mode of action of the constituents of *P. harmala* against many of these quite different diseases is not yet scientifically proven, it is clear that the β-carboline alkaloids of the seeds (up to 6% of dry mass) and serotonin are highly effective psychomimetica (Allen and Holmstedt 1980). Despite the presence of pharmacologically active constituents in *P. harmala*, the plant has no pharmaceutical importance in Europe and North America. Consequently, the biotechnological interest in in-vitro cultures of this plant as producer of biologically active compounds was low in the past. Indeed, its structures seem to be chemically too simple to be produced via expensive tissue culture technology. Thus the question arises why it has nevertheless been worthwhile to work on this culture system and why it is not only justified, but also useful to continue these studies with genetically transformed cultures. A short summary of the work on untransformed cultures of *P. harmala* may explain this.

The first report on *P. harmala* callus cultures described the presence of low levels of harmine (Reinhard et al. 1968). Nettleship and Slaytor (1974a) detected the whole spectrum of β-carboline alkaloids (Fig. 1), known as con-stituents of *P. harmala*, in these callus cultures. The levels of the alkaloids, however, remained, quite low in undifferentiated cultures despite all efforts to improve alkaloid production (Nettleship and Slaytor 1974a,b; Sasse et al. 1982a; Berlin and Sasse 1988). When growth rates of the cultures improved, productivity went down or was lost. Freshly initiated cell cultures tended to form aggregates with short, star-shaped root-like structures. Such cultures contained up to 2% β-carboline alkaloids (Berlin 1989). By careful adaptation

GBF – Gesellschaft f. Biotechnologische Forschung mbH, Mascheroder Weg 1, 38124 Braunschweig, Germany

Biotechnology in Agriculture and Forestry, Vol. 45
Transgenic Medicinal Plants (ed. by Y.P.S. Bajaj)
© Springer-Verlag Berlin Heidelberg 1999

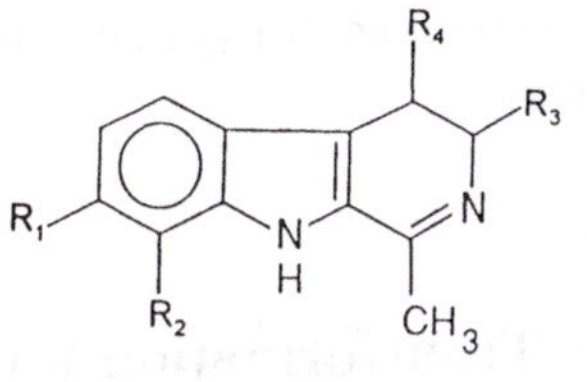

β-Carboline Alkaloids

	R₁	R₂	R₃	R₄
Harmalol	OH	H	H₂	H₂
Harmaline	OCH₃	H	H₂	H₂
Harmol	OH	H	H	H
Harmine	OCH₃	H	H	H
Ruine	OCH₃	Ogluc	H	H

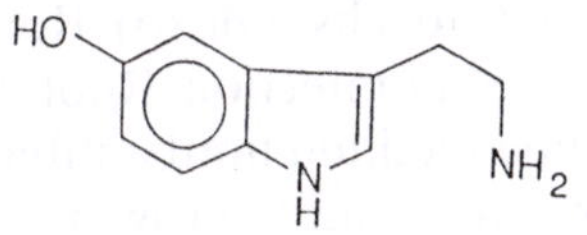

Serotonin

Fig. 1. The chemical structures of the β-carboline alkaloids present in *P. harmala* plants and cell cultures

of the phytohormone levels, it was sometimes possible to maintain such root-forming cultures over a period of several months (ca. ten growth cycles). However, this culture state was unstable and, if the culture had once changed into the undifferentiated state, root formation could not be recovered. Stable root cultures might be obtained only by genetic transformation of *P. harmala* with *Agrobacterium rhizogenes*.

Another reason for establishing transformed cultures of *P. harmala* arises from the observation that all their cell cultures – even those which are unable to biosynthesize serotonin de novo – are able to transform tryptamine into serotonin (Sasse et al. 1982b, 1987; Courtois et al. 1988). It has been demonstrated that the lack or loss of tryptophan decarboxylase activity explains the cells' inability to synthesize serotonin (Sasse et al. 1982b; Berlin et al. 1987). Serotonin is formed in higher plants in two steps from tryptophan – tryptophan decarboxylation is followed by 5-hydroxylation of tryptamine. This means that the biosynthesis of serotonin might easily be restored by engineering a tryptophan decarboxylase activity into *P. harmala* cells because the second and final enzymatic step of serotonin biosynthesis is always highly active in the cells. Restoration of serotonin biosynthesis by genetic engineering seemed to be a useful system for entering the field of metabolic engineering of pathways.

2 Genetic Transformation for Enlarging the Expression of the Biosynthetic Potential of *Peganum harmala*

2.1 Establishment of Hairy Root Cultures

To establish a stable root culture system forming high levels of β-carboline alkaloids, *P. harmala* was infected and transformed with *A. rhizogenes* strains (Table 1). All hairy root cultures initiated up to now (Kuzovkina et al. 1990; Berlin et al. 1992) were created by the same technique. Stems and leaves of 2-month-old seedlings grown under sterile conditions were infected with widely used *A. rhizogenes* strains (e.g., A4 and 15834). Two days (depending upon the extent of bacterial growth) after infection, the plant pieces were washed with and plated on phytohormone-free MS medium containing 500 mg Claforan/l (Hoechst/Glaxo). Root formation at the site of wounding was noted 2–3 weeks after infection. Root tips were cultivated on both antibiotic-free and -containing medium until antibiotic-free controls showed no regrowth of bacteria for two to three growth periods. The appearance of transformed roots (longer, stronger, and more branched) was so different from roots (star-shaped) found during normal callus culture initiation on phytohormone-containing medium, that one could immediately distinguish between normal and transformed roots. Growth of the latter stopped soon due to the lack of phytohormones. The transformation of some of our lines was proved by

Table 1. Summary of publications describing the establishment of transformed cultures of *P. harmala* by infection with *Agrobacterium* strains. All other publications on transformed cultures of *P. harmala* cited in this chapter are taken from the three publications listed below

Explant	Transformed by strain/vector	Culture characteristics	Reference
Sterile-grown seedlings	*A. rhizogenes* A4	Wild-type transformed root culture	Kuzovkina et al. (1990)
Pieces of sterile-grown seedlings	*A. rhizogenes* A4, 15834 and pLTCgus1	Root cultures transformed by wild-type strains and a strain carrying β-glucuronidase gene as marker	Berlin et al. (1992)
Stem and leaves of seedlings	C58CI pRiA4(pTDCs) and LBA4404 (pTDCs)	Root/suspension cultures expressing a tryptophan decarboxylase gene of *Catharanthus roseus*	Berlin et al. (1993b)

Southern analyses (Berlin et al. 1992) using an internal T_L-DNA *Eco*RI fragment from the plasmid pRiHRI (Jouanin 1984). A more convenient proof of the transformation is obtained when, instead of wild-type strains, constructed strains are used for transformation. For example, roots transformed with strain *A. rhizogenes* Gus1, whose construction was described by Berlin et al. (1992), are easily recognized by the β-glucuronidase activity. In addition, this strain carries the *npt-II* gene as selectable marker for kanamycin-tolerant roots (Berlin et al. 1992).

2.2 Results and Discussion of Work with Hairy Root Cultures

Over the years, more than 60 individually established hairy root cultures derived from seed material of different origin were analyzed. A first observation was that the root structure (Fig. 2) was much better in phytohormone-free B5 liquid medium than in MS medium, due to the lower ammonium ion content of the B5 medium. As the biomass formation was increased by higher phosphate and sucrose levels (Berlin et al. 1992), all cultures were maintained on a WM medium, which is, in fact, a B5 medium with the amount of phosphate doubled and 4% sucrose. All root cultures accumulated β-carboline

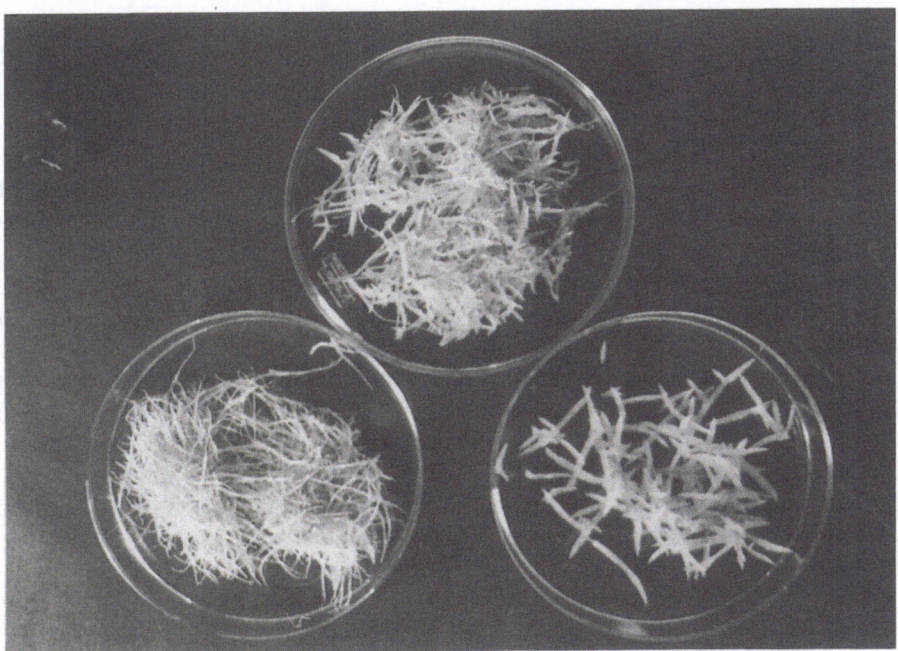

Fig. 2. The appearance of a hairy root culture of *P. harmala* when grown in liquid phytohormone-free B5 or MS medium. *Left* Roots from B5 medium (long, thin, interlaced); right roots permanently maintained on MS (short, thick, loose texture); center root culture was moved from B5 into MS for one cycle (root tips became thicker, seemed to become covered with callus-like material)

Table 2. Effect of ammonium ions on growth and product levels of some hairy root cultures. MS medium has a concentration of 20 mM ammonium, while B5 medium has only 0.5 mM. (Some data from Berlin et al. 1992)

A The roots (1 g fresh mass/35 ml) were grown in the indicated media for 3 weeks before analysis

Medium during cycle			Last biomass increase (-fold)	β-Carbolines	Serotonin
I	II	III		(mg/g dry mass)	
MS	MS	MS	8.4	4.0	5.4
MS	B5	B5	7.3	10.0	4.1
MS	B5	MS	7.2	3.9	5.7
B5	B5	B5	8.6	14.2	5.5
B5	MS	MS	8.3	4.9	4.3
B5	MS	B5	7.0	8.2	4.7

B The roots were grown for 18 days on media supplemented with different concentrations of ammonium

Ammonium (mM)	Last biomass increase (-fold)	β-Carbolines	Serotonin
		(mg/g dry mass)	
0.5 (B5)	6.3	17.2	6.6
0 (MS)	5.6	14.3	4.1
10 (MS)	5.8	4.0	4.3
20 (MS)	7.2	3.7	4.6

alkaloid levels of 0.5 to 2% of dry mass on this medium. The main alkaloid was always harmine (~60%), while harmalol, harmol, and ruine varied between 10 and 20%. Harmaline was hardly quantified in HPLC chromatograms (Berlin et al. 1992), but always detectable by its pale fluorescence. TLC chromatograms showed also the presence of a number of other fluorescent compounds. Two of these were identified as 5- and 6-hydroxytetra-hydronorharmanes, which, however, should not be regarded as intermediates of the main β-carboline alkaloids (Berlin et al. 1993a). In cultures containing lower alkaloid levels, root formation was often disturbed by the presence of callus material. There was some indication that the hairy root cultures derived from one batch of *P. harmala* seeds had very similar product levels, and that apparent differences were due to the morphological and physiological state of the culture at the time of analysis. Although we have not studied it in detail, there was also some evidence that plant material with higher β-carboline content might give better-producing hairy root cultures. Thus, if plant material with substantially different β-carboline contents are available, one should use the best-producing source for the initiation of hairy root cultures. The medium not only affected the appearance of the root cultures, but also influenced alkaloid formation (Table 2A,B). When switching from B5 to MS medium and vice versa, some morphological changes were observed after 10 days (Fig. 2) and concomitant with this, the levels of the β-carboline alkaloids changed, while serotonin contents were not affected (Table 2A). The main reason for the low alkaloid production on MS medium resulted from its high concentra-tion of ammonium ions (Table 2B). While alkaloid formation is repressed in rapidly growing suspension cultures, it is a well-expressed pathway in roots.

Consequently, the higher the growth of the roots, the better is the alkaloid production. Growth inhibition of the root culture of *P. harmala* by phosphate limitation (an often powerful tool for increasing production in suspension cultures) thus had a negative effect on alkaloid formation in B5 medium (Berlin et al. 1992). In contrast, in MS medium the specific alkaloid (not the total) content in phosphate-limited cultures could be enhanced, because the negative effect of ammonium ions became evident only in growing cultures. By growing the root culture on B5 medium containing $2\,\mu M$ 2,4-D, an aggregated callus culture was obtained with alkaloid levels reduced by 90%. When the phytohormone was removed from the medium, the callus culture changed into a root culture with its previous appearance and productivity after three to five growth periods (Berlin et al. 1992). Thus, in contrast to untransformed root cultures, one can grow a transformed root culture as a suspension culture and return to the root state at a desired time. However, callus/suspension cultures derived from the hairy root cultures showed a slower growth rate than the roots. For scalingup purposes, it might therefore be better to grow the roots first on MS medium (Fig. 2) because the roots are shorter and disintegrate more easily. Then the roots should be transferred to the low-ammonium B5 medium with its higher productivity (Table 2A,B).

The main aim of establishing transformed root cultures of *P. harmala* was to have a convenient and stable system for studying the enzymology and regulation of β-carboline alkaloid biosynthesis. First hints were to be obtained from the analyses of feeding experiments. As the root cultures usually contained β-carboline alkaloids and serotonin in a ratio of ~2–3:1 when permanently grown on phytohormone-free B5 medium, it was assumed that the incorporation of potential precursors would be at least as good as found with roots of intact plants. Feeding of labeled tryptophan to roots of intact plants had shown that tryptophan is most likely the primary precursor of the β-carboline alkaloids in *P. harmala* (Stolle and Gröger 1968). The expectation of achieving a significant increase in the alkaloid content by addition of tryptophan to growth medium of the highly productive root cultures was, however, not fulfilled. Tryptophan feeding had no effect on alkaloid levels (Berlin et al. 1993a). As serotonin levels were also not enhanced during these feedings, the result could not be explained by a rapid decarboxylation of tryptophan which might have prevented the incorporation into the β-carboline alkaloids. The low effect on serotonin levels could be explained (Table 2) by the low tryptophan decarboxylase activity in these hairy root cultures which was indeed high (~30 pkat/mg protein) for only 1 or 2 days during the first week of the culture period (Berlin et al. 1993a). When tryptophan-[side chain-$3\text{-}^{14}C$] was used as tracer, 10–15 times more label was found in serotonin than in the alkaloids (Berlin et al. 1993a). Despite de-novo synthesis of the alkaloids during the feeding period, the incorporation rate was so low that one had to conclude that either free tryptophan is not the natural precursor of β-carboline alkaloid biosynthesis or that fed tryptophan cannot reach the site of alkaloid biosynthesis. We also tested whether anthranilic acid (2 mM), the precursor of tryptophan, would have a positive effect on alkaloid levels. However, this compound caused a strong inhibition of root culture growth, so that changes in specific alkaloid levels and serotonin could not be interpreted unambiguously. Although the root cultures have not yet provided more clues about the initial steps of β-carboline alkaloid biosynthesis than the intact plants, it is evident that the hairy roots provide a more convenient system for enzymatic trials.

In contrast to the results with tryptophan, feeding of various β-carboline alkaloids to the root cultures gave some new clues for enzymatic experiments. Harmalol and harmaline were converted quite well into harmol and harmine, respectively. Thus, the aromatization of the dihydro-β-carboline alkaloids is most likely the final step of the biosynthetic sequence (Berlin et al. 1993a). The feeding experiments gave no hints about the methylation step. However, crude enzyme preparations contained enzymes methylating harmalol to harmaline and harmol to harmine. Methylated harmaline was then converted to harmine by the same enzyme extract (Berlin et al. 1994). As 6-hydroxytryptamine, serotonin, or 5-hydroxytryptophan were not methylated, it can be assumed that methylation of the alkaloids occurs only when the β-carboline structure has been built up. Thus, the root cultures seem to be a good source for purifying the enzymes which biosynthesize harmaline, harmol, and harmine from harmalol or a yet unknown basic β-carboline ring structure.

3 Genetic Transformation for Metabolic Engineering of the Serotonin Biosynthetic Pathway

3.1 Establishment of Tryptophan Decarboxylase (*tdc*) Transgenic Suspension and Root Cultures

The construction of the transformation vector pTDCs has been described by Goddijn et al. (1995). The *tdc* cDNA sequence was under the control of the CaMV 35S promoter and terminator in the pBin19-derived binary plant vector pBDH5. The vector contained also the gene encoding neomycin phosphotransferase II conferring resistance to kanamycin. The plasmid was mobilized into *A. tumefaciens* LBA4404 (Hoekema et al. 1983) for initiating suspension cultures of *P. harmala* and into *A. tumefaciens* C58CI pRiA4 (Petit et al. 1983) for establishing root cultures. Tissue pieces of stems, leaves, and hypocotyls of sterile-grown seedlings of *P. harmala* were infected by wounding, placed on MS medium with 2 μM 2,4-D and phytohormone-free B5 medium, respectively, for initiating transgenic callus or root cultures. Removal of the bacteria was achieved by growing the resulting cultures in the presence of up to 500 μg Claforan/ml. Only calli and roots able to grow in the presence of 100 μg kanamycin/ml were further followed. While in the case of root cultures a selective marker may not be so important (see Sect. 2.1), it is absolutely necessary for distinguishing between normal and transgenic callus formation. It is also important to analyze a large number of individually established transformants, because we found no correlation between formation of the tolerance conferring enzyme neomycin phosphotransferase II and expression of tryptophan decarboxylase activity. To find a suspension or a root culture expressing the *tdc* to the highest extent, quite a number of transformants must be screened for the engineered enzyme activity. The integration of the *tdc* gene of *C. roseus* into *P. harmala* cultures was proven by Southern, Northern, and Western blot hybridizations using a pTDC probe or a TDC-specific antibody (Berlin et al. 1993b).

3.2 Results and Discussion of Work with the *tdc* Transgenic Cultures

Roughly one third of all calli obtained after infection of the plant tissue with the *tdc*-transgenic *A. tumefaciens* LBA 4404 strain seemed to be transformed, since they grew continuously on the selective medium. All kanamycin-tolerant callus/suspension cultures were indeed also positive for the *tdc* gene in Southern analyses. These lines were screened for TDC activity. Enzyme activities between 10 and 60 pkat/mg protein were found, while normal lines contained 0–5 pkat/mg. The specific TDC activity remained quite stable over the whole growth period in *tdc* transgenic lines. In normal lines we found often a peak activity 24 h after inoculation, which disappeared or declined rapidly. A good correlation between TDC activity and serotonin accumulation was not always visible because specific serotonin levels were sometimes not stable over the growth cycle but decreased in some lines. Nevertheless, the difference between normal and the *tdc* transgenic cultures was evident. Normal suspension cultures contained between 0.1 and 1 mg serotonin/g dry mass; the highly productive transgenic lines accumulated levels of 10–25 mg/g dry mass.

Although the transformation goal, the establishment of rapidly growing cell suspension cultures producing permanently good levels of serotonin due to the engineered TDC activity, was achieved, it was soon realized that the cells could produce much more of the target compound if more tryptophan were available for decarboxylation. Indeed, even simple pathways like serotonin biosynthesis have never only one site controlling the carbon flux. The increase of any product level by a "one-site" genetic engineering step depends upon the difference between the main limiting step and the second limiting step. In case of serotonin biosynthesis in cell suspension cultures, it appears that overexpression of TDC activity can make the cells produce roughly 2–3% because more tryptophan is usually not available. This is roughly the same level found in the unstable normal star-shaped root cultures. When tryptophan was fed to the *tdc* transgenic cultures, serotonin levels of up to 7–8% of dry mass were noted, while in normal cultures tryptophan did not enhance the product level. Thus, the suspension cultures behaved in the same manner to tryptophan as shown for the root cultures (Table 3). If the internal tryptophan supply in *tdc*-transgenic lines can be increased much higher serotonin content might be achieved. The substantial difference between serotonin levels in tryptophan-fed and -unfed *tdc*-transgenic lines suggests making these lines overproduce tryptophan. Overproduction of tryptophan might be achieved by overexpression of anthranilate synthase activity in the cells or by selection of 5-methyltryptophan-tolerant cell lines. Such tolerant

Table 3. Effect of feeding of tryptophan and tryptamine to a wild-type-transformed root culture PH A4 with low TDC activity and the *tdc*-transgenic root culture R 2/16 with high activity. Initial inoculum 100mg/35ml B5 medium. The roots were analyzed after 10 days. (Some data from Berlin et al. 1993b)

Substrate added (mg/flask)	PH A4		R 2/16	
	Alkaloids	Serotonin	Alkaloids	Serotonin
	(mg/g dry mass)			
None	13.2	5.7	3.9	15.0
10mg tryptophan	13.4	6.0	3.8	55.3
10mg tryptamine	11.5	36.5	3.7	52.8

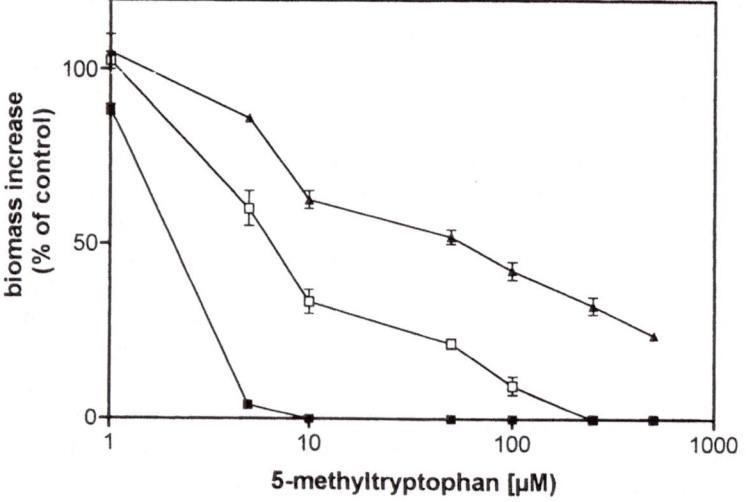

Fig. 3. Comparison of tolerance to 5-methyltryptophan of two lines selected from the mother line S5.2/3. Mother line (■); line F_2 (□); line F_1/y (▲). Initial inoculum 0.5g/35ml medium. Harvest after 14 days. Final biomass: S5.2/3: 7.4g; F2: 8.2; F1/y: 6.6

cell lines have an altered anthranilate synthase which is less feedback-sensitive to tryptophan (Wiholm 1983). We tried first to select amino acid analogue-tolerant cell lines from the *tdc*-transgenic line S5.2/3. This mother line exhibited TDC activities of 80–100pkat/mg protein at the time of selection and average serotonin levels of 16mg/g dry mass. A total of nine 5-methyltryptophan-tolerant lines were isolated and analyzed. Selection had been performed in the presence of 200µg kanamycin/ml. While the original line was extremely sensitive to 5-methyltryptophan, the variant lines grew on concentrations from 5×10^{-5} to 5×10^{-4}M of the amino acid analogue (Fig. 3). However, none of the 5-methyltryptophan-tolerant lines had increased levels of tryptophan or serotonin. On the contrary, the most tolerant line, F_1/y, accumulated only one third to one half of the serotonin levels found in the mother line, while some others contained similar levels (Fig. 4). Interest-

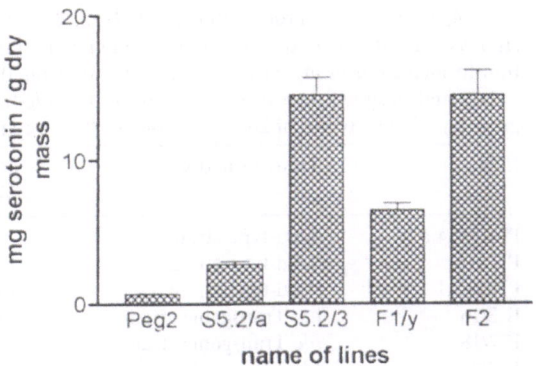

Fig. 4. Comparison of the fluctuations of the serotonin content of various *P. harmala* suspension cultures. *Peg 2* (untransformed control); *S5.2/a* (*tdc*-transgenic line which has lost its initial high productivity); *S5.2/3* (with unchanged serotonin content since 1991); *F1/y* (cell line with the highest tolerance to 5-methyltryptophan selected from S5.2/3); *F2* (medium tolerance against 5-methyltryptophan). The data were calculated from five measurements of 14-day-old cultures during a period of 2 years

ingly, in most tolerant cell lines, TDC activity was much lower than in the unselected culture. In seven tolerant lines, the TDC activity varied from 20–45 pkat. In two lines, TDC activity was nearly lost, with 2 pkat/mg protein remaining. The reason for the 5-methyltryptophan tolerance of the new lines was not further analyzed because none of the selected cultures expressed the desired traits. Therefore, we have only some evidence that the absorption/uptake of 5-methyltryptophan and tryptophan from the medium was substantially retarded in the most tolerant line F_1/y. In general, it seems to be quite difficult to obtain tryptophan-overproducing cell lines in *P. harmala* by selection for tryptophan-analogue tolerance (Berlin et al. 1987). However, it might also be important to realize that the selection approach not only failed with regard to overproduction of tryptophan but also resulted in lines with lower TDC activity. During any selection for a second step to be improved, one may lose, partially or completely, the first, already optimized step. If the genes for both rate-limiting enzymes are available, it seems to be more appropriate to create such lines by engineering both limiting enzymes into the cells during a single transformation. In the case of serotonin biosynthesis, the genes of both rate-limiting steps, anthranilate synthase (Bohlmann et al. 1996; Song et al. 1996) and tryptophan decarboxylase, are now available.

How stable are the productivities of the transgenic lines? The product levels of three of the original lines which were initiated in 1991, were followed for 6 years, and the stabilities of some 5-methyltryptophan-tolerant cell lines over 2 years (Fig. 4). The lines S5.2b and S5.2/3 (Berlin et al. 1993b) have maintained their productivity and levels fluctuated between 12 and 22 mg serotonin/g dry mass. Under stress conditions (growth inhibition) levels of up to 35 mg/g were found. In contrast, line S5.2a lost the initial productivity and fell from 13–15 mg to 3 mg serotonin. TDC activity was reduced in this line to 2–3 pkat/mg protein, while the stable lines showed on the average activities of

Table 4. Comparison of root cultures of *P. harmala* which were transformed either with wild-type strains of *A. rhizogenes* or with a corresponding *tdc*-transgenic strain. TDC activity in wild-type-transformed cultures increased after inoculation from 0–3 pkat/mg protein to 15–30 pkat after 3–5 days and declined then to the low levels. The *tdc*-transgenic lines contained over the whole growth period activities of 20–30 pkat/mg. Cells were analyzed after 10 days of cultivation

Line	Transformed with	β-Carboline alkaloids	Serotonin
		(mg/g dry mass)	
PH15834 a	Wild-type strain	8.5	6.1
PH A4	Wild-type strain	14.2	6.6
PH Gus1	Wild-type strain	13.5	4.5
R 2/16	*tdc*-Transgenic strain	7.9	16.3
R 2/18	*tdc*-Transgenic strain	7.6	20.8
R 3/2	*tdc*-Transgenic strain	6.7	18.3

30–60 pkat/mg protein. The tolerance to the selective antibiotic kanamycin remained stable in all three original lines and in all 5-methyltryptophan-resistant lines. The only visible sign that something had changed in line S5.2a was the improved growth rate. TDC activities and serotonin production of the once-selected 5-methyltryptophan-tolerant cell lines remained stable during the 2-year period of observation. Thus, one can conclude that *tdc*-transgenic lines, as well as the lines derived therefrom, are in general quite stable, as high enzyme activity and product formation were maintained without special measures. So far we have not experienced that a *tdc*-transgenic which has lost its initial high TDC activity returned spontaneously to that state.

Although it was evident from previous experiences that TDC activity is not a major limiting factor of β-carboline alkaloid biosynthesis, *tdc*-transgenic root cultures were established. As expected, overexpression of TDC had no positive effect on β-carboline alkaloid levels. The alkaloid production capacity most likely depended upon the genetic potential of the seeds rather than on the transformation. On the other hand, one cannot exclude that the overproduction of serotonin caused a "negative" effect on alkaloid biosynthesis. The ratio of β-carboline alkaloids: serotonin was quite different in *tdc*-transgenic lines and cultures transformed with wild type strains (Table 4). In contrast to wild-type-transformed root cultures, we found two to three times more serotonin than alkaloids in the *tdc*-transgenic root cell. Feeding of tryptophan and tryptamine to the *tdc*-transgenic lines showed the expected effects (Table 3). Addition of tryptophan caused no effect on alkaloid levels, while serotonin levels were dramatically increased in the transgenic line R2/16.

4 Summary and Conclusions

A convenient system for studying β-carboline alkaloid biosynthesis has become accessible by creating transformed root cultures of *P. harmala*. While in the case of β-carboline alkaloids our knowledge is not yet sufficient to manipulate this pathway by genetic engineering, we were able to demonstrate that the production of serotonin can be greatly enhanced by overcoming the main rate-limiting step in root cultures as well as in suspension cultures through this technique. Future research with transformed cultures of *P. harmala* could go in two directions. Firstly, transformed cell cultures of *P. harmala* could become a useful model system for how to manipulate a secondary pathway. Not very often are the two main limiting steps of any secondary biosynthetic pathway of higher plants so unambiguously clear. The genes for both limiting steps of serotonin biosynthesis (tryptophan supply, tryptophan decarboxylation) are available. Indeed, if serotonin content can dramatically be increased in such double transformants, overexpressing a normal or deregulated anthranilate synthase and a tryptophan decarboxylase, this would be a great stimulus for molecular approaches to other, commercially more important pathways. Secondly, although the secondary metabolites of *P. harmala* seem to be rather simple, many biochemical, regulatory, and molecular questions are still open. While the 5-hydroxylation of tryptamine to serotonin may be studied in normal suspension cultures, transformed root cultures are needed for finding the enzyme responsible for the 6-hydroxyl group of the alkaloids. Whether these hydroxylases are cytochrome-P_{450} enzymes or require other cofactors (tetrahydropteridene?) (Ichiyama and Nakamura 1970) is unknown. If β-carboline alkaloids and serotonin derive from the same primary precursor (tryptophan?), it would be especially interesting to elucidate how the cells manage their channeling/distribution into the two biosynthetic pathways. Thus, a number of questions of general importance concerning regulatory aspects of secondary pathways might be answered by the use of transformed cultures of *P. harmala*.

Acknowledgments. The author would like to thank Drs. JHC Hoge and OJM Goddijn (Plant Molecular Sciences, University of Leiden) for providing the constructs for establishing the *tdc*-transgenic lines and for their cooperation in this project.

References

Allen JRF, Holmstedt BR (1980) The simple β-carboline alkaloids. Phytochemistry 19:1573–1582

Ambasta SSP (ed) (1986) The useful plants of India. Publications & Information Directorate, CSIR, New Delhi

Berlin J (1989) On the formation of secondary metabolites in plant cell cultures – some general observations and some regulatory studies. In: Production of metabolites by plant cell cultures. Proc Eur Symp. APRIA, Paris, pp 89–98

Berlin J, Sasse F (1988) β-Carbolines and indole alkylamines. In: Constabel F, Vasil IK (eds) Cell culture and somatic cell genetics of plants, vol 5. Academic Press, Orlando, pp 357–369

Berlin J, Mollenschott C, Sasse F, Witte L, Piehl HG, Büntemeyer H (1987) Restoration of serotonin biosynthesis in cell suspension cultures of *Peganum harmala* by selection for 4-methyltryptophan-tolerant cell lines. J Plant Physiol 131:225–236

Berlin J, Kuzovkina IN, Rügenhagen C, Fecker L, Commandeur U, Wray V (1992) Hairy root cultures of *Peganum harmala*. II. Characterization of cell lines and effect of culture conditions on the accumulation of β-carboline alkaloids and serotonin. Z Naturforsch 47c:222–230

Berlin J, Rügenhagen C, Greidziak N, Kuzovkina IN, Witte L, Wray V (1993a) Biosynthesis of serotonin and β-carboline alkaloids in hairy root cultures of *Peganum harmala*. Phytochemistry 33:593–597

Berlin J, Rügenhagen C, Dietze P, Fecker LF, Goddijn OJM, Hoge JHC (1993b) Increased production of serotonin by suspension and root cultures of *Peganum harmala* transformed with a tryptophan decarboxylase cDNA clone of *Catharanthus roseus*. Transgen Res 2:336–44

Berlin J, Rügenhagen C, Kuzovkina IN, Fecker LF, Sasse F (1994) Are tissue cultures of *Peganum harmala* a useful model system for studying how to manipulate the formation of secondary metabolites? Plant Cell Tissue Organ Cult 38:289–297

Bohlmann J, Lins T, Martin W, Eilert U (1996) Anthranilate synthase from *Ruta graveolens* – duplicated as alpha genes encode tryptophan-sensitive and tryptophan-insensitive isoenzymes specific to amino acid and alkaloid biosynthesis. Plant Physiol 111:507–514

Boulos L (1983) Medicinal plants of North Africa. Reference Publications, Algonac, Michigan

Courtois D, Yvernel D, Florin B, Petiard V (1988) Conversion of tryptamine to serotonin by cell suspension cultures of *Peganum harmala*. Phytochemistry 27:3137–3141

Goddijn OJM, Pennings EJM, Van der Helm P, Schilperoort RA, Verpoorte R, Hoge JHC (1995) Overexpression of a tryptophan decarboxylase cDNA in *Catharanthus roseus* crown gall calluses results in increased tryptamine levels but not in increased terpenoid indole alkaloid production. Transgen Res 4:315–323

Hoekema A, Hirsch PR, Hooykaas PJJ, Schilperoort RA (1983) A binary vector strategy based on separation of vir- and T-region of the *Agrobacterium tumefaciens* Ti-plasmid. Nature 303:179–180

Ichiyama A, Nakamura S (1970) Tryptophan-5-monooxygenase. Methods Enzymol XVII:449–459

Jouanin L (1984) Restriction map of an agropine-type Ri plasmid and its homologies with Ti plasmids. Plasmid 12:91–102

Kuzovkina IN, Gohar A, Alterman IE (1990) Production of β-carboline alkaloids in transformed cultures of *Peganum harmala* L. Z Naturforsch 45c:727–728

Nettleship L, Slaytor M (1974a) Adaptation of *Peganum harmala* callus to alkaloid production. J Exp Bot 25:1114–1123

Nettleship L, Slaytor M (1974b) Limitations of feeding experiments in studying alkaloid biosynthesis in *Peganum harmala* callus cultures. Phytochemistry 13:735–742

Petit A, David C, Dahl GA, Ellis JG, Guyon P, Casse-Delbart F, Tempé J (1983) Further extension of the opine concept: plasmids in *Agrobacterium rhizogenes* cooperate for opine degradation. Mol Gen Genet 190:204–214

Reinhard E, Corduan G, Volk OH (1968) Nachweis von Harmin in Gewebekulturen von *Peganum harmala*. Phytochemistry 7:503–504

Sasse F, Heckenberg U, Berlin J (1982a) Accumulation of β-carboline alkaloids and serotonin in cell cultures of *Peganum harmala* L.I. Correlation between plants and cell cultures and influence of medium. Plant Physiol 69:400–404

Sasse F, Heckenberg U, Berlin J (1982b) Accumulation of β-carboline alkaloids and serotonin by cell cultures of *Peganum harmala*. II. Interrelationship between accumulation of serotonin and related enzymes. Z Pflanzenphysiol 105:315–322

Sasse F, Witte L, Berlin J (1987) Biotransformation of tryptamine to serotonin by cell suspension cultures of *Peganum harmala*. Planta Med 53:354–359

Song HS, Brotherton JE, Widholm JM (1996) Cloning and characterization of *Nicotiana tabacum* anthranilate synthase. Plant Physiol 11:534

Stolle K, Gröger D (1986) Untersuchungen zur Biosynthese des Harmins. Arch Pharm 301:561–571

Widholm JM (1983) Isolation and characterization of mutant plant cell cultures. Basic Life Sci 22:71–87

XV Genetic Transformation of *Perezia* Species

J. Arellano and G. Hernández

1 Introduction

1.1 Distribution and Importance of the Genus *Perezia*

The genus *Perezia* (Asteraceae family) seems to be confined to the New World, and is divided both taxonomically and geographically into two sections: *Acourtia* and *Euperezia*. The North American section *Acourtia* extends from the coastal region of southern California to El Salvador and eastward to central Texas. Most of this section is restricted to the middle and higher slopes of the Mexican Sierra Madre and to the high plateau region of the mountains in central and northern Mexico.

Perezia species of the *Acourtia* section are erect, mostly glabrate or resinous-punctuate, caulescent perennials, always with the base of the stem more or less rusty-woolly, for the most part 1–1.5 m tall and robust, sometimes several leafy stems below the terminal leafy or leafy-bracteate inflorescence, sometimes unbranched and terminated by a single large head, leaves alternate, mostly sessile, usually ample and thin-chartaceous to rigid and thick-coraceous; heads homogamous, often glomerate, 4–60 flowered; corollas homomorphous, mostly lavender-pink and glabrous, roots fibrous and sometimes tuberous (Bacigalupi 1931).

The woody roots of several *Perezia* species have been used since pre-Hispanic times by the Mexican ethnic groups to prepare a laxative beverage. The active compound was isolated from *Perezia* roots more than a century ago (1852) and was named pipitzahoic acid. Further studies showed that this compound is a sesquiterpene quinone which today is known as perezone. Perezone has been isolated from several *Perezia* species and it is fairly abundant in the roots of *Perezia cuernavacana* Robinson & Greenman. The structural formula of this compound is shown in Fig. 1. It possesses the properties of a quinone: it is a crystalline substance with a deep orange color; it reduces readily, reoxidyzes spontaneously, and yields amino derivatives (Joseph-Natan and Santillan 1986). Perezone can also be used as a pigment and it may form coordinated compounds with metals to produce other pigments of different colors. In addition, it can be used as an antimicrobial agent.

Centro de Investigación sobre Fijación de Nitrógeno-UNAM, Ap. Postal 565-A, Cuernavaca, Morelas México

Biotechnology in Agriculture and Forestry, Vol. 45
Transgenic Medicinal Plants (ed. by Y.P.S. Bajaj)
© Springer-Verlag Berlin Heidelberg 1999

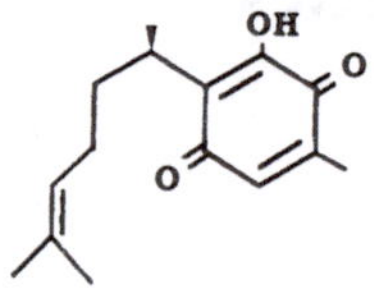

Fig. 1. Structural formula of the sesquiterpene quinone: perezone, produced by *Perezia* roots

PEREZONE

1.2 Need for Genetic Transformation

Perezia species produce and accumulate large amounts of perezone and/or related compounds in their roots. Despite its potential applications, it has not been possible to isolate perezone from *Perezia* in adequate amounts for industrial production because: (1) natural populations of *Perezia* species are very small and in many places they are in the process of extinction; (2) several attempts to cultivate *Perezia* by traditional agricultural methods have been unsuccessful. Therefore, the establishment of genetically transformed cell or organ culture from *Perezia* species with a high production of perezone is an alternative for the adequate production of this secondary metabolite for industrial purposes.

2 Genetic Transformation

2.1 Methodology

Axenic plant material of *P. cuernavacana* R & G and *P. thyrsoidea* Gray was obtained by surface sterilization of seeds collected from field-grown plants. Seeds were germinated on solid hormone-free MS medium (Murashige and Skoog 1962), supplemented with 30 g/l sucrose. Plants and hairy root cultures were incubated in growth chambers with the environmental conditions reported (Arellano et al. 1996). Nodal/internodal segments were prepared from 3-week-old sterile plants by leaf excision and cutting just below each node. The basal extreme of each explant was inoculated with *Agrobacterium rhizogenes* AR12 strain that carried the *gus* gene fused to the 35SCaMV constitutive promoter in the Ri plasmid T-DNA region (Hansen et al. 1989). Inoculated explants were placed distally in solid hormone-free MS medium, supplemented with 300 mg/l Cefotaxime and hairy roots appeared after 4 or 5 weeks. The hairy roots obtained were excised and cultured separately, as reported (Arellano et al. 1996).

The transformed nature of the hairy root cultures was confirmed by genomic integration (PCR and slot blot hybridization) and expression of enzyme activity (Jefferson et al. 1987) of the marker *gus* gene (Arellano et al. 1996).

The production of perezone by hairy roots was evidenced from the organic extract of transformed tissue by silica gel thin layer chromatography (TLC) and by IR spectroscopy (Arellano et al. 1996).

2.2 Results and Discussion

Several types of the explants tested, such as leaves, roots, or shoot segments, did not form hairy roots after cocultivation with the AR12 strain. Under these conditions, only nodal/internodal segments of both *Perezia* species showed hairy root formation in 16% of the explants (Fig. 2A,B). Hairy roots were excised from the explants and cultures were established, which showed active growth in the absence of growth regulators. The hairy root cultures derived from *P. cuernavacana* (Fig. 2C) grew profusely and therefore were further characterized (see below). *P. thyrsoidea*-derived hairy root cultures were also established, but their growth rate was evidently slower (data not shown). It remains to be explored if the exogenous supply of hormones may permit an improvement in the growth rate of the *P. thyrsoidea* hairy root cultures obtained.

The genomic integration and the expression of the marker *gus* gene was confirmed in the putatively transformed *P. cuernavacana* root cultures. Figure 2D shows that the GUS expression in the hairy root culture was increased in the vascular tissue, as expected for a 35SCaMV promoter-driven *gus* gene construct (Benfey et al. 1989). The GUS histochemical assay in control roots from nontransformed *P. cuernavacana* plants showed no GUS expression (data not shown). In addition, the GUS fluorometric assay was performed in hairy roots and in roots from nontransformed plants of *P. cuernavacana* (control). While the control tissue showed a minimum background of GUS-specific activity (217 nmol 4-MU/h/mg prot), the hairy roots showed a very high GUS-specific activity (74638 nmol 4-MU/h/mg prot).

The presence of the *gus* gene in the hairy roots was confirmed by PCR (Fig. 3A). Using *gus* gene-specific primers, a 1.2-kb fragment was amplified from DNA extracted from *P. cuernavacana* hairy roots as template (Fig. 3A). DNA from roots of nontransformed plants did not act as a template for these *gus* gene-specific primers (Fig. 3A). The size of the amplified fragment observed in Fig. 3A, lanes 3 and 4, corresponded with that expected according to the *gus* gene sequence and the primers used (Hamill et al. 1991); it comigrates with the fragment observed when DNA from the pBI121 plasmid carrying the *gus* gene (Jefferson et al. 1987) was used as a template for PCR amplification (data not shown).

The integration of the *gus* gene in hairy roots was also confirmed by slot blot hybridization (Fig. 3B). DNA from hairy roots, was hybridized with a *gus* gene-specific probe; for comparison, DNA from nontransformed roots was used (Fig. 3B). Hybridization showed that the *gus* gene is present only in the genome of hairy root tissue.

Inocula (0.7 g fresh weight) of actively growing hairy roots culture of *P. cuernavacana* were transferred to fresh medium and the fresh weight data was recorded during 28 days of culture. Figure 4A shows the growth curve of such culture. A remarkable characteristic of the transformed root cultures is their potential for biomass production (Flores and Curtis 1992). The *P. cuernavacana* transformed root culture showed a high growth rate with a $m_{max} = 0.111$/day and a doubling time of 6.2 days, which may be considered

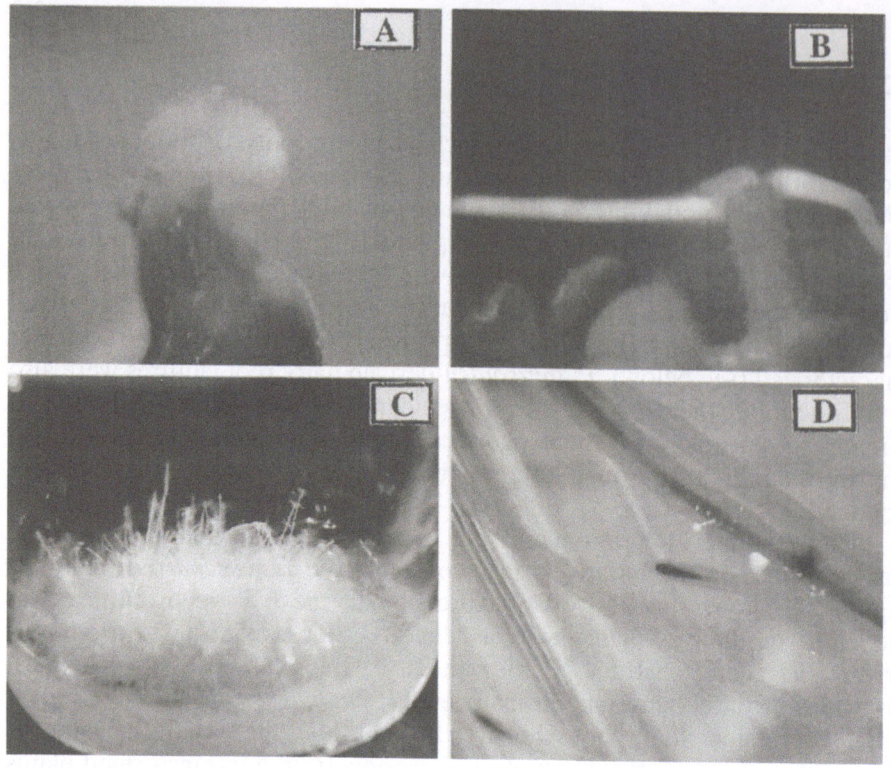

Fig. 2. *P. thyrsoidea* (**A**) and *P. cuernavacana* (**B**) hairy roots formed on nodal-internodal shoot segments inoculated with *A. rhizogenes* AR12. *P. cuernavacana* hairy root culture in MS liquid medium (**C**) and histochemical GUS assay of the culture, (**D**). (Arellano et al. 1996)

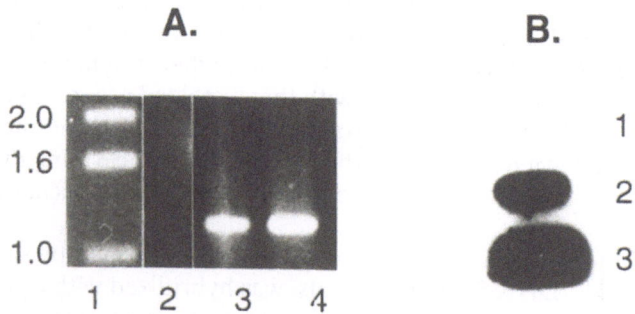

Fig. 3A,B. Integration of the *gus* gene into the *P. cuernavacana* hairy roots genome. **A** Agarose gel electrophoresis of the DNA sequences obtained by PCR amplification of the DNA extracted from: nontransformed roots of *P. cuernavacana* (*lane 2*) or hairy root culture, undigested (*lane 3*) or digested with *Hind*III (*lane 4*). The specific (*gus*-gene) primers used have been reported (Hamill et al. 1991). *Lane 1* BRL 1-kb DNA ladder, bands with indicated kb sizes. **B** Autoradiograph of the slot blot hybridization of DNA extracted from: nontransformed digested with *Hind*III (*1*) or hairy root culture, undigested (*2*) or digested with *Hind*III (*3*). The 2.2-kb *Bam*HI-*Eco*RI (*gus*-gene) fragment from pBI121 (Jefferson et al. 1987) was used as a probl. (Arellano et al. 1996)

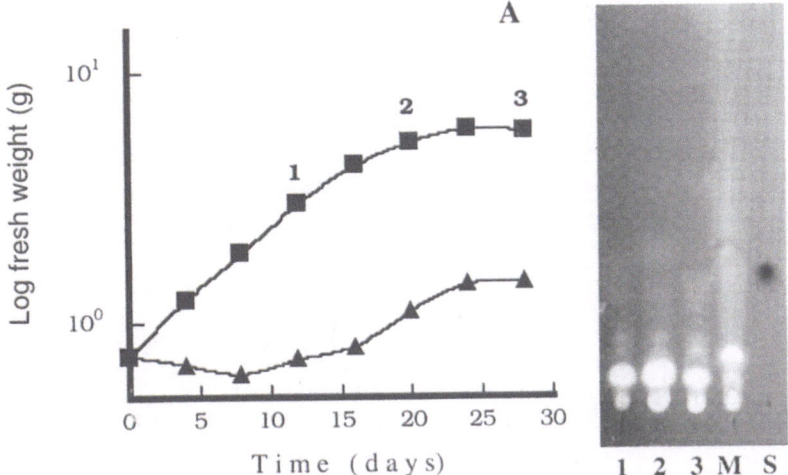

Fig. 4A,B. Growth curves of *P. cuernavacana*-transformed hairy root (■) and untransformed root (▲) cultures. The time points when the hexane extract from the hairy roots was analyzed for perezone production are indicated (*1, 2,* and *3*). **B** TLC of hexane extract from hairy root cultures sampled at the indicated time points (*1, 2,* and *3*) and from the culture medium (*M*); perezone standard (*S*)

important features for perezone production. The specific growth rate of roots derived from nontransformed *P. cuernavacana* plants (Fig. 4A) was evidently smaller (m_{max}= 0.021/days) with a doubling time of 32.3 days, despite the addition of indole acetic acid (0.5 mg/l) to the culture medium.

Since perezone is produced in roots of *Perezia* plants, we tested the transformed hairy root cultures for the production of this compound at different growth phases (Fig. 4B). The extraction procedure from transformed roots was similar to the one previously used from roots of field-grown plants (Walls et al. 1965). The highest perezone production determined by TLC was observed at the stationary growth phase (Fig. 4B). Additionally, we also tested the culture medium for the presence of perezone and observed that perezone is not excreted into the culture medium (Fig. 4B).

The nature of the crystals obtained from the hexane extract from transformed roots was confirmed by IR spectroscopy (Fig. 5). The IR spectrum (KBr) of the sample extracted from transformed roots was identical to that previously obtained for purified perezone (Sanchez et al. 1985). The spectrum showed the characteristic relative positions of peaks at 3285, 1655, 1645, 1620 and 1610/cm (Fig. 5).

The yield of perezone obtained from transformed stationary growth-phase hairy root cultures was 1.3% of the dry weight. The yield of perezone from field-grown *P. cuernavacana* plants ranges from 2 to 8% of the dry weight, depending upon the developmental stage of the plants, the season of the year when the plants are collected, and natural stress conditions (Walls

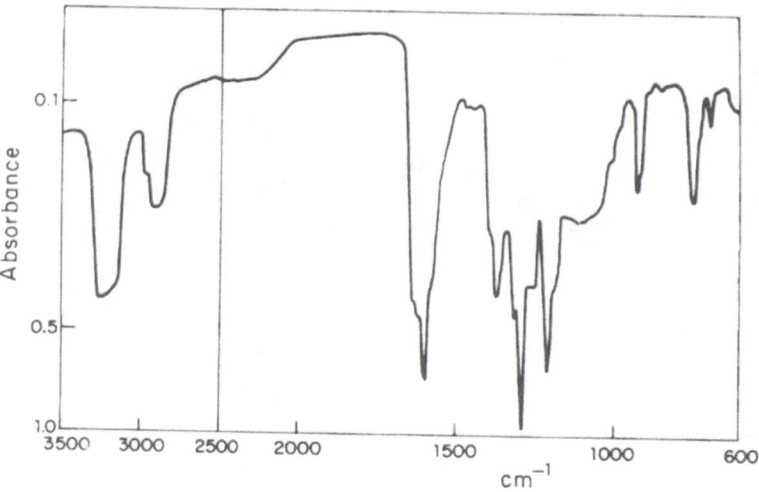

Fig. 5. IR spectrum (KBr) of crystalized perezone extracted from a transformed root culture derived from *P. cuernavacana*

et al. 1965; F. Vazquez, unpubl.). Whether transformed root cultures eventually become commercial sources of secondary metabolites will depend, among other factors, on the production of the compound at very high levels (Hilton et al. 1989). Although the yield of perezone from the transformed root culture analyzed was somewhat lower than the yield obtained from field-grown *P. cuernavacana* plants, the latter is very variable and depends on several environmental factors which cannot be controlled.

3 Summary and Conclusions

The establishment of transformed hairy root cultures from two *Perezia* species, *P. cuernavacana* and *P. thyrsoidea*, has been demonstrated. These cultures were obtained following the infection of nodal-internodal segments from both species with *A. rhizogenes*. Actively growing transformed root cultures from these species were established on hormone-free MS liquid culture medium. The study of their growth kinetics and the associated perezone production was performed only for the hairy root cultures derived from *P. cuernavacana*, because of their higher growth rate. The transformed nature of the tissue was confirmed by genomic integration and expression of the marker *gus* gene. The production of perezone by the transformed root culture was evidenced by IR spectroscopy. Perezone was produced at a significant level (1.3% dry weight) in the stationary growth phase.

Our results offer an alternative to enhanced production of perezone, and represent an advantage over its extraction from natural plant populations which present problems in their agronomic culture. Additionally, the hairy root cultures derived open the possibility for the study of the biosynthetic sesquiterpene quinone pathways of the *Perezia* species.

References

Arellano J, Vázquez F, Villegas T, Hernández G (1996) Establishment of transformed root cultures of *Perezia cuernavacana* producing the sesquiterpene quinone perezone. Plant Cell Rep 15:455–458

Bacigalupi R (1931) A monograph of the genus *Perezia*, section *Acourtia*. Contribution from the Gray Herbarium of Harvard University. XCVII The Gray Herbarium of Harvard University, Cambridge, Massachusetts, pp 1–163

Benfey NP, Ren L, Chua N (1989) The CaMV 35S enhancer contains at least two domains which can confer different developmental and tissue-specific expression patterns. EMBO J 8:2195–2202

Flores HE, Curtis WR (1992) Approaches to understanding and manipulating the biosynthetic potential of plant roots. In: Pedersen H, Mutharasan R, DiBiasio D (eds) Biochemical engineering VII. The NY Acad Sci, New York, pp 188–209

Hamill JD, Rounsley S, Spencer A, Todd G, Rhodes MJC (1991) The use of the polymerase chain reaction in plant transformation studies. Plant Cell Rep 10:221–224

Hansen J, Jorgensen J, Stougaard J, Marcker KA (1989) Hairy roots – a short cut to transgenic root nodules. Plant Cell Rep 8:12–15

Hilton MG, Wilson PDW, Robins RJ, Rhodes MJC (1989) Transformed root cultures – fermentation aspects. In: Robins RJ, Rhodes MJC (eds) Manipulating secondary metabolism in culture. Cambridge University Press, Cambridge, pp 239–245

Jefferson RA, Kavanagh TA, Bevan MW (1987) GUS fusion: beta-glucuronidase as a sensitive and versatile gene fusion marker in higher plants. EMBO J 6:3901–3907

Joseph-Nathan P, Santillan RL (1986) The chemistry of perezone and its consequences. In: Atta-Ul-Rahman (eds) Natural products, vol 1. Elsevier, Amsterdam, pp 763–813

Murashige T, Skoog F (1962) A revised medium for rapid growth and bio-assays with tobacco tissue cultures. Physiol Plant 15:473–497

Walls F, Salmon M, Padilla J, Joseph-Nathan P, Romo J (1965) La estructura de la perezona. Bol Inst Quim Univ Nac Auton Mex 17:3–15

XVI Genetic Transformation of *Pimpinella anisum* (Anise)

B.V. Charlwood[1,2] and K.M.S.A. Salem[1]

1 Introduction

1.1 Importance of the Plant

Pimpinella anisum L. (Apiaceae), known as anise or aniseed, is native to Egypt, Asia Minor, and Greece, but is now cultivated worldwide, particularly in Spain, France, Turkey, Syria, Chile, China, and USA.

The commercial and medicinal value of anise is due mainly to its essential oil, the major source of which is the ripe, dried fruits. The fruit oil contains around 94% of *trans*-anethole (**1**, Fig. 1), together with small amounts of other phenylpropanoids and various sesquiterpenoid lactones (Table 1). The leaves, shoots, and roots also produce significant amounts of oil (Reichling et al. 1985, 1986), but with much reduced levels of phenylpropanoids and concomitant increases in isoprenoids, both mono- and sesquiterpenoids, as indicated in Table 1.

As a medicinal plant, anise has been used as a traditional remedy for asthma, bronchitis, indigestion, nausea, nephritis, and epilepsy. The fruit itself is claimed to possess abortifacient, antispasmodic, carminative, dia- phoretic, diuretic, and expectorant properties, and can also have a tran- quilizing activity through a CNS-depressant activity (Ibrahem et al. 1988). Leaves of anise, which are rich in vitamin C, are used for salads and herbal tea, whilst extracts of the herb are used in aromatherapy for treating menopausal symptoms.

1.2 Tissue Culture and Secondary Metabolites

Callus and suspension cultures of *P. anisum* may be readily initiated from explants after incubation on either MS medium (Murashige and Skoog 1962) in the absence of plant growth regulators, or on B5 medium (Gamborg et al. 1968) supplemented with 2,4-D (Noma et al. 1979). As has

[1] Division of Life Sciences, King's College London, Campden Hill Road, London W8 7AH, UK
[2] Departamento de Química – CCEN, Universidade Federal de Alagoas, Campus Universitário, 57072-970 Maceió-AL, Brazil

Biotechnology in Agriculture and Forestry, Vol. 45
Transgenic Medicinal Plants (ed. by Y.P.S. Bajaj)
© Springer-Verlag Berlin Heidelberg 1999

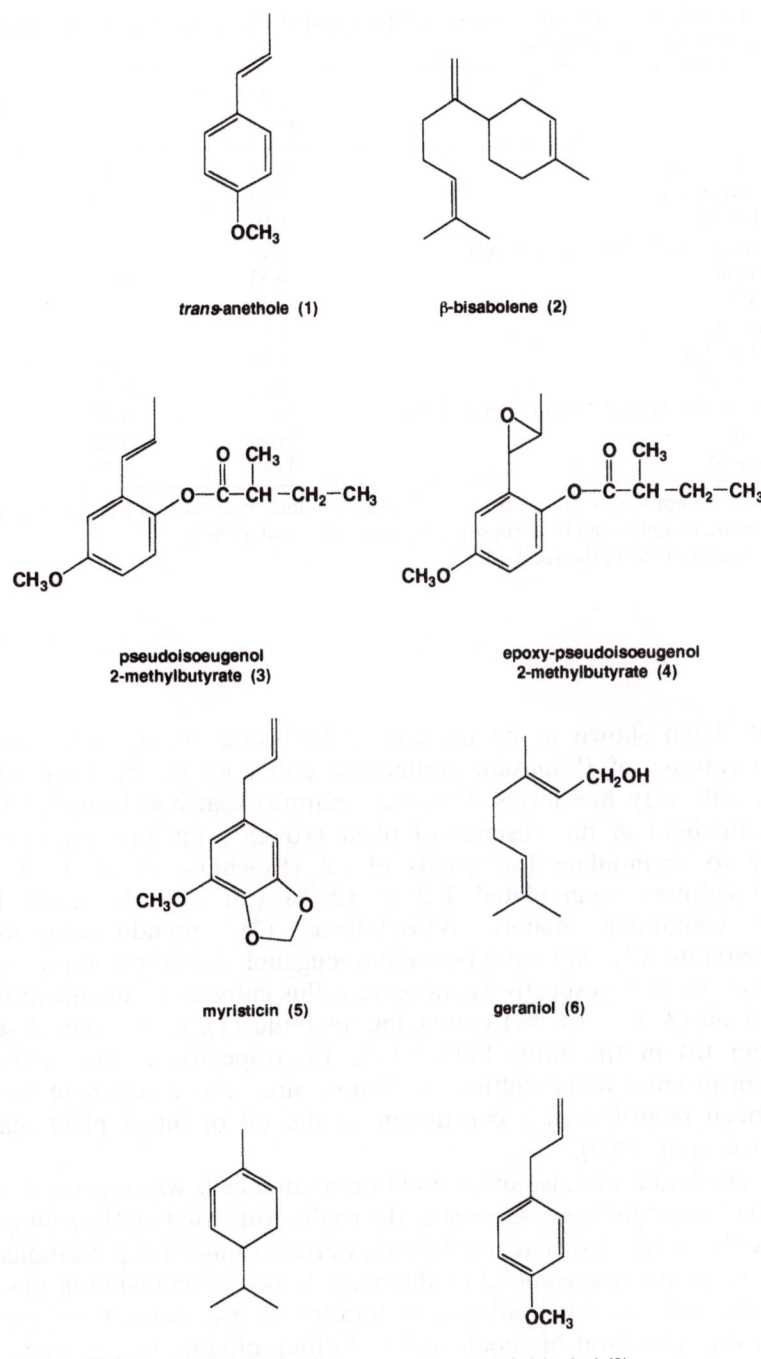

Fig. 1. Structures of some compounds identified in the essential oils of *P. anisum*

Table 1. Partial percentage compositions of the essential oils from fruits, herb, and roots of *P. anisum*. (Kubeczka et al. 1986)

Compound	Percentage composition[a]		
	Fruit	Herb	Root
trans-Anethole (**1**)	94.14	29.40	3.39
Methylchavicol (**8**)	1.99	1.55	trace[b]
γ-Himachalene	0.99	3.37	–
Pseudoisoeugenol 2-methylbutyrate (**3**)	0.97	13.13	2.28
Anisaldehyde	0.53	0.29	trace
cis-Anethole	0.22	1.18	–
α-Zingiberene	0.15	4.37	1.23
β-Bisabolene (**2**)	0.13	11.83	52.46
Germacrene D	0.12	14.75	0.29
Epoxy-pseudoisoeugenol 2-methylbutyrate (**4**)	Trace	4.37	6.22
Pregeijerene	Trace	1.62	12.78
ar-Curcumene	Trace	0.58	0.37

[a] The partial compositions (phenylpropanoids and terpenoids only) shown have been abstracted from the more complete analyses reported by Kubeczka et al. (1986).
[b] Not more than 0.1% of the total oil.

generally been shown to be the case (Charlwood 1993), callus and suspension cultures of *P. anisum* accumulate either no essential oils or produce at only very low levels. However, cultures that had been habituated on MS medium in the absence of plant growth regulators maintained a capacity to accumulate low yields of oil (Reichling et al. 1986). Suspension cultures accumulated $1–2 \times 10^{-3}\%$ (on a fresh weight basis) of oil containing mainly β-bisabolene (**2**), pseudoisoeugenol 2-methylbutyrate (**3**) and epoxy-pseudoisoeugenol 2-methylbutyrate (**4**) in the ratio 7.5:2.5:1, respectively, whereas callus cultures accumulated higher levels of oil ($3–4 \times 10^{-3}\%$) containing anethole (**1**), compounds **2–4**, and myristicin (**5**) in the ratio 150:20:1:43:10, respectively. The finding of myristicin in anise tissue cultures is of note since this component has only rarely been recorded as a constituent of the oil of intact plant material (Harborne et al. 1969).

Accumulation of anise oil in undifferentiated cells was increased by the addition of a lipophilic phase, such as the triglyceride Miglyol (Reichling et al. 1985), to the culture to act as an artificial external storage site: β-bisabolene (**2**) and epoxy-pseudoisoeugenol 2-methylbutyrate (**4**) accumulated in this layer during the culture. Alternatively, restoration of the natural storage sites through the formation of shoot- and leaf-differentiating tissues leads to an increase in oil production but with a profile different from that of the oil of the parent plant (Charlwood et al. 1990). Indeed, a leaf-differentiating culture has been obtained from a habituated suspension culture by Reichling et al. (1995) which accumulated up to 480 µg/g (on a fresh weight basis) of **4** during the last third of the culture.

2 Transformation

2.1 Previous Studies

P. anisum has been previously transformed with *A. tumefaciens* using a cocultivation technique (Komari 1989). Anise callus was induced from hypocotyls by incubation on solidified (0.9% agar) GB medium supplemented with 1 mg/l 2,4-D, resuspended in liquid medium, and incubated with *A. tumefaciens* (strain A281 harbouring pTOK119) containing the selectable marker gene *npt* II (coding for neomycin phosphotransferase which confers resistance to kanamycin) and the reporter gene firefly luciferase. Transformed cells were selected on solidified medium containing 250 mg/l cefotaxime and 300 mg/l kanamycin: proof of transformation was carried out by Southern hybridization (for *npt* II) and liquid scintillation assay (for luciferase activity). Plant regeneration from the transformed callus material was not reported. Our work (Salem 1994; Salem and Charlwood 1995) is summarized here.

2.2 Methodology

2.2.1 Production of Tissue Cultures of P. anisum

Seeds of *P. anisum* (gift of, and certified by, Dr. F. Afifi, Faculty of Pharmacy Botanic Garden, University of Jordan) were soaked in sterile water for 3 h, the seed coats were removed (except around the embryonic axis), and the seeds placed in a 7% sodium hypochlorite solution containing one or two drops of Tween 80 for 10 min (Salem 1994). Subsequently, the seeds were washed four times in sterile distilled water, the remaining portions of the seed coats removed, and the seeds were germinated on solidified (1% agar) MS medium (pH 5.6) containing 3% sucrose (medium MS0) in the dark at 25 °C for 3 days. After this time, seedlings were transferred to continuous light (21.5 μE/m^2/s; Philips cool, white fluorescent tubes) for 2–3 weeks. Stems and roots from sterile seedlings were cut into sections of 2 cm in length, divided longitudinally, and placed with their cut surfaces down onto MS0 medium containing 0.2 mg/l kinetin and 1 mg/l 2,4-D (medium MS1). Discs (1 cm diameter) derived from young shoots were obtained using a sterile cork borer and incubated on similar medium under the conditions above. After 5–7 days callus appeared at the wound sites and this was excised after 14 days and cultured on MS1 medium in continuous light. Callus was subcultured every 2 weeks for at least five passages before placement on MS0 medium, whereupon a high level of shoot differentiation occurred, giving rise to a shoot-differentiating callus (SDC) culture with a doubling time (t_D^3) of 4.8 days.

---

[3] $t_D = [\ln 2\,(t_1 - t_0)]/\ln(w_1 - w_0)$, where w_0 is the biomass at the inoculation of the culture (t_0) and w_1 is the biomass at t_1 days into the culture period

   The SDC culture remained essentially unchanged in morphology (Fig.
2a) over 3 years of subculturing, and consisted (typically) of shoot mass
(48% of total fresh weight) and callus (52%), with roots being formed only
rarely. The SDC could also be maintained in liquid MS0 medium with a similar
morphology and a slightly shorter t_D (4.1 days). When the SDC was incubated
under a photoperiod (16h light/8h dark), the shoot yield and t_D value (5.2
days) remained almost unchanged, but the amount of root tissue formed
increased (up to 5% of total fresh weight): such root tissue thickened upon
protracted culturing and callus formed on its surface and, eventually,
produced spontaneous shoots (Fig. 2b). Incubation of SDC on MS0 medium
supplemented with 5mg/l BAP and 1mg/l NAA gave a culture which
produced slightly higher amounts of shoot tissue (52% of total fresh weight)

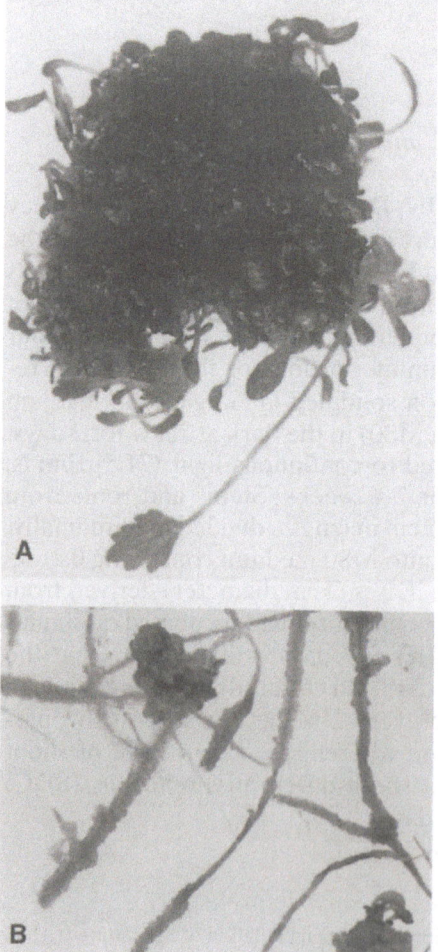

Fig. 2. A Shoot-differentiating callus (SDC)
culture of *P. anisum* grown for 3 weeks under
continuous light on solidified MS0 medium. **B**
Thickening roots, showing callus formation
and spontaneous shoot induction, produced
from SDC incubated on solidified MS0 medium under a 16-h light photoperiod. (Salem
1994)

but with increased root formation (ca. 2%). The ability of the SDC to maintain shoot differentiation diminished progressively as the cytokinin: auxin ratio in the incubation medium approached 1:1. The addition of zeatin (0.012 mg/l; medium MS6), as sole plant growth regulator, or of thiamine (1.5 mg/l; medium MS7) to MS0 medium also led to a decrease in the percentage of shoots formed within the SDC, yielding values of only 5 and 31%, respectively.

2.2.2 Transformation with A. tumefaciens

Transformation experiments with *A. tumefaciens* were carried out with strain T37 (a nopaline strain bearing pTiT37), with strain 337 (an octopine strain bearing the *tms* 1 mutant pTiA6NC::Tn5; Garfinkle et al. 1981) harbouring the binary vector pBI121.1 (containing *npt* II and the reporter gene *gus* coding for β-glucuronidase (GUS) activity), and with strain LBA4404 (a disarmed octopine strain bearing pTiAL4404; Ooms et al. 1982) harbouring the binary vector pMON9793 (containing *gus* and *npt* II genes). Binary vector plasmids were mobilized from *Escherichia coli* into the appropriate strain of *A. tumefaciens* using the triparental mating protocol of Ditta et al. (1980) employing *E. coli* bearing the helper plasmid pRK2013. For transformation experiments, *Agrobacterium* strains were cultured in YT broth (5 g/l yeast extract, 8 g/l tryptone, 5 g/l sodium chloride) for 24 h at 25 °C with continuous shaking at 120 rpm. The bacterial suspension was centrifuged (1000 g, 5 min) and washed twice with sterile water: the pellet was resuspended in MS0 medium to a final absorption value of ca. 0.6 at 600 nm.

Direct infection of 3-week-old sterile seedlings of anise with T37 was carried out by inoculating bacterial aliquots (50 μl) at different sites on the stem, blade, and petiole using a hypodermic syringe with a 25G-gauge needle. The wound sites were probed with the needle to ensure direct contact between bacterial cells and plant tissue. Similar wounds were made on control seedlings but with aliquots of sterile water instead of bacterial suspension. Wounded seedlings were incubated on MS0 medium at 25 °C under continuous light for 3 weeks. After this time, any galls which initially formed at the wound sites could be excised from the parent plant, transferred to solidified MS0 medium, and incubated under continuous light at 25 °C when shoots began to emerge from the tumourous mass. Alternatively, galls could be allowed to grow for up to 6 weeks at the wound sites on the parent plant, and during this time shooty outgrowths formed on the surfaces of the galls. Excised shoots were transferred to solidified MS0 medium containing 100 mg/l cefotaxime and 400 mg/l carbenicillin and subcultured every 15 days. Samples of the putatively transformed tissues were monitored for the presence of contaminating bacteria by squashing the shooty teratomas and culturing the mass in YT broth. When tissue samples were free of bacteria (typically after three passages), the shooty teratomas were transferred to liquid MS0 medium containing 75 mg/l cefotaxime for one passage and subcultured every 3 weeks thereafter on liquid MS0 medium for maintenance.

Cocultivation of 14-day-old callus cultures of anise with *A. tumefaciens* strains 337 (pBI121.1) and LBA4404 (pMON9793) was carried out on solidified MS0 medium. Aliquots (50–100 µl) of a 24-h bacterial suspension were placed on differentiating callus lumps and incubated for 30, 45, or 60 min. The plant tissue was blot-dried on sterile filter paper and cocultivated for 24 or 48 h under a photoperiod or continuous light, after which time the plant material was washed (2 × 5 min each) with liquid MS0 medium, blot-dried, washed with sterile water containing 400 mg/l carbenicillin, blot-dried, and incubated on solidified MS0 medium for 2–4 days. After this time, callus material was either subcultured on MS0 medium containing 400 mg/l carbenicillin (for gall formation) or transferred to MS medium supplemented with 0.2 mg/l BAP, 0.2 mg/l NAA, 400 mg/l carbenicillin, and 50 mg/l kanamycin (for transformant selection and subsequent shoot regeneration).

Following cocultivation protocols, callus and shoots showed significant contamination with *Agrobacterium* and had to be washed at subculture for 6 h with a cocktail of antibiotics containing, typically, carbenicillin (200 mg/l), ampicillin (200 mg/l), and cefotaxime (100 mg/l). Many of the shoots which formed were, however, sensitive to this mixture and an alternative, less effective, cocktail containing carbenicillin (100 mg/l), ampicillin (75 mg/l), and cefotaxime (25 mg/l) had to be used. Samples of the putatively transformed tissues were monitored for the presence of contaminating bacteria by squashing plant material and culturing the mass in YT broth. Treatment with an antibiotic cocktail was maintained until tissues could be demonstrated to be free from contaminating bacteria.

2.2.3 Analysis of Transformed Callus and Shoots

Confirmation of transformation of callus and shoots was obtained by polymerase chain reaction (PCR) analysis of plant DNA using the primers shown in Table 2. DNA was extracted from fresh plant tissue (5–7 g) using the methods of Dellaporta et al. (1983) and Mettler (1987), and from

Table 2. Primers used in PCR experiments to verify transformation of *P. anisum* tissues

Oligonucleotide primers	Gene segment	Size (kb)	Reference
GGTGGGAAAGCGCGTTACAAG GTTTACGCGTTGCTTCCGCCA	*gus* 400 → 420 1599 ← 1579	1.2	Jefferson et al. (1986)
GAGGCTATTCGGCTATGACTG ATCGGGAGCGGCGATACCGTA	*npt* II 201 → 222 900 ← 879	0.70	Beck et al. (1982)
ATGGATCTGCGTCTAATTTTC CTAATACATTCCGAATGGATG	*ipt (tmr)* 659 → 679 1381 ← 1361	0.72	Goldberg et al. (1984)
ATGTCGCAAGGACGTAAGCCCA GGAGTCTTTCAGCATGGAGCAA	*vir* D1	0.45	Hirayama et al. (1988)

Agrobacterium strains (5 ml of 24-h suspension cultures) using the method of Zhou et al. (1990). PCR analysis was carried out using the protocol described by Hamill et al. (1991) under the following conditions: initial denaturation – 92 °C for 3 min; *Taq* addition (1.5 U) 72 °C; amplification – 26 cycles of 55 °C for 2 min (annealing), 72 °C for 2 min (extension), 92 °C for 1 min (denaturation); final annealing – 55 °C for 2 min; final extension – 73 °C for 10 min; sample size – 3 μl containing 1.2–2.1 μg DNA in TE buffer (Sambrook et al. 1989); product analysis – electrophoresis on 1% agarose gel using 0.5 × TBE running buffer and staining for 20 min with 0.5 μg/ml ethidium bromide (Sambrook et al. 1989).

The localization of GUS activity was determined in callus and shoots by incubating samples of plant material for 1–24 h at room temperature in a solution prepared by dissolving 0.3 mg of 5-bromo-4-chloro-3-indolyl glucuronide (X-Gluc) in a small volume of N,N-dimethylformamide and diluted to 10 ml with 50 mM sodium phosphate buffer (pH 7). After incubation, plant material was washed with 70% aqueous ethanol, and either sectioned or mounted in 10% glycerol prior to microscopic examination.

2.2.4 Analysis of Essential Oil Content

Fresh shoot tissues (1–10 g) were cut into small segments and subjected to steam distillation with continuous extraction with dichloromethane (Chrompack microsteam distillation-extraction apparatus) for 2 h. Concentrated extracts were analyzed by gas chromatography-mass spectrometry (Hewlett Packard model HP5890 series 2 chromatograph connected to a Jeol model AX505W spectrometer) using a BP1 capillary column (25 m × 0.22 mm id: film thickness 0.25 μm). The chromatographic conditions were: *column temperature program* – 140 °C held for 4 min and then increased to 290 °C at a rate of 8 °C/min and held for 4 min; *carrier gas* – helium at a flow rate of 1.6 ml/min; *sample size* – 1 μl; *injector temperature* – 250 °C; *split ratio* – 1:15; *ion source temperature* – 200 °C; *ionizing potential* – 70 eV; *accelerating voltage* – 3 kV; *sensitivity* – 1.45; *scan speed* – 1 s; *scan range* – 33–600 amμ. Components were determined quantitatively using gas chromatography with flame ionization detection and calibration with appropriate standards.

2.3 Results and Discussion

2.3.1 Agrobacterium tumefaciens-Transformed Cultures

Small galls (spherical-shaped outgrowths of proliferating cells) initially formed at approximately 20% of the sites on seedlings which had been infected directly with *Agrobacterium* strain T37 after 3 weeks (Salem and Charlwood 1995). Galls which were separated from the mother plant at this time and incubated on solidified MS0 medium (continuous light; 25 °C) produced shooty outgrowths on the surface of the galls after about 10 days.

Galls which were allowed to remain on the mother plant for 4–5 weeks after the original infection also produced similar shooty teratomas, but such shoot tissues were much less contaminated with residual *Agrobacterium* than those produced by the former method. In control experiments, when seedlings were wounded in the absence of *Agrobacterium*, thin layers of callus tissue were formed at the wound sites within 10 days, but these tissues showed no capacity to proliferate either on the plant itself or when transferred to MS0 medium.

No galls were formed following cocultivation of callus with *Agrobacterium* strain 337; however, callus resistant to kanamycin could be selected following such transformation experiments with strains 337 and LBA4404. Shoots did not differentiate on the surface of the transformed callus, however, they were formed when the tissue was transferred to MS medium supplemented with 0.2 mg/l BAP and 2 mg/l NAA.

Shooty teratomas from which all contaminating bacteria had been removed were grown on for further studies. Proof that such lines were free of *Agrobacterium* was obtained by performing a PCR analysis of the total shoot DNA using primers for the *vir* D1 region of *A. tumefaciens* (Table 2). Figure 3, lane 1 shows DNA of *A. tumefaciens* strain T37 probed with the *vir* D1 primer (product band at 0.45 kb), whilst lane 2 shows that this product is not amplified in bacteria-free shooty teratomas. Shooty teratomas still contaminated with strain T37 were employed as positive controls (data not shown). That the shooty teratoma lines were actually transformed was demonstrated by PCR analysis using primers for the *ipt* gene of T37 (Tabel 2). Figure 3, lane 3 shows DNA of *A. tumefaciens* strain T37 probed with the *ipt* primer (product band at 0.72 kb), whilst lane 4 shows that this product was not amplified in nontransformed shoot cultures of *P. anisum*. A similar result (data not shown) was obtained when DNA from shoot material of an intact plant of *P. anisum* was probed with primers for *ipt*, demonstrating the validity of using an *ipt* sequence from *A. tumefaciens* as a marker for plant transformation. Lane 5

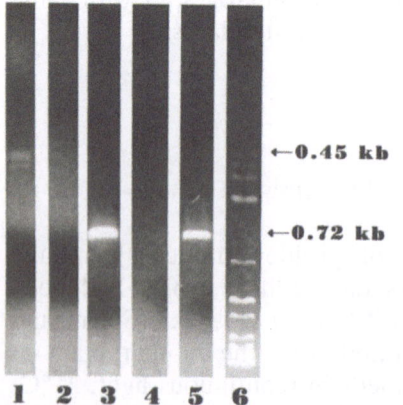

←0.45 kb

←0.72 kb

1 2 3 4 5 6

Fig. 3. PCR products obtained following amplification of the *vir* D1 and *ipt* genes in DNA extracted from *A. tumefaciens* (strain T37) and *P. anisum* cultures. *Lane 1 A. tumefaciens* T37 probed with primers for *vir* D1 (product at 0.45 kb); *lane 2* bacteria-free, transformed shoot cultures of *P. anisum* probed with primers for *vir* D1; *lane 3 A. tumefaciens* T37 probed with primers for *ipt* (product at 0.72 kb); *lane 4* nontransformed shoot cultures of *P. anisum* probed with primers for *ipt*; *lane 5* bacteria-free, transformed shoot cultures of *P. anisum* probed with primers for *ipt* (product at 0.72 kb); *lane 6* 1-kb ladder. (Salem 1994)

shows amplification of the *ipt* primer in a shooty teratoma line which had previously been demonstrated unambiguously not to be contaminated with bacterial strain T37.

An exactly similar PCR method was employed to demonstrate the transformation of callus material with strains 337 and LB4404, although the primers used here were for the *gus* and/or the *npt* II gene. A typical result is shown in Fig. 4, where amplification of the *npt* II gene (product band at 0.70 kb) was observed with bacteria-free, kanamycin-resistant callus (lane 3), whereas nontransformed callus showed no such amplification (lane 1). Furthermore, the permanent expression of the *gus* gene in kanamycin-resistant callus tissue could still be readily demonstrated after >ten subcultures of the callus, as shown in Fig. 5, which indicates that GUS activity is mainly present in cells associated with thickened vessels.

2.3.2 Analysis of Shoot Cultures

The morphologies of the transformed shoot culture and its nontransformed counterpart (SDC) were similar (Fig. 2a) except that the former was associated with a higher percentage (ca. 60%) of unorganized tissue which was harder, more compact, and contained less water (fresh wt./dry wt. ratio 12:1) than callus of the SDC (fresh wt./dry wt. 18:1). The leaf tissue of the transformed shoots was more coarsely serrated and slightly curlier and smaller in size than that of the SDC, but the density (15–20 hairs/cm^2) and morphology of the oil-bearing hairs on the leaf surfaces of both types of tissue were similar. The transformed shoot culture showed higher t_D values compared with the SDC when grown on solidified or in liquid MS0 medium (8.1 and 5.1 days, respectively).

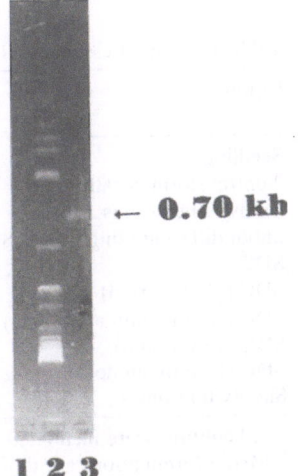

← **0.70 kb**

Fig. 4. PCR products obtained following amplification of the *npt* II gene in DNA extracted from callus of *P. anisum* transformed with *A. tumefaciens* (strain LBA4404 harboring the plasmid pMON9793). *Lane 1* nontransformed callus of *P. anisum* probed with primers for *npt* II; *lane 2* 1-kb ladder; *lane 3* bacteria-free, transformed callus of *P. anisum* probed with primers for *npt* II (product at 0.70 kb). (Salem 1994)

1 2 3

The total yields of essential oil from cultures of sterile seedlings, from nontransformed and transformed callus, from the SDC grown under a variety of conditions, and from shooty teratomas are shown in Table 3. Whilst the leaflets from the young seedling accumulated around 0.06% oil (on a fresh weight basis), the SDC cultured on solidified medium produced seven fold less than this when cultured under continuous light. It is of interest to note that the yield of essential oil increased 1.7 times when the SDC was cultured under a photoperiod, but this increase in yield was accompanied by a concomitant increase in root production. Although it has previously been reported that root formation enhances the accumulation of isoprenoids

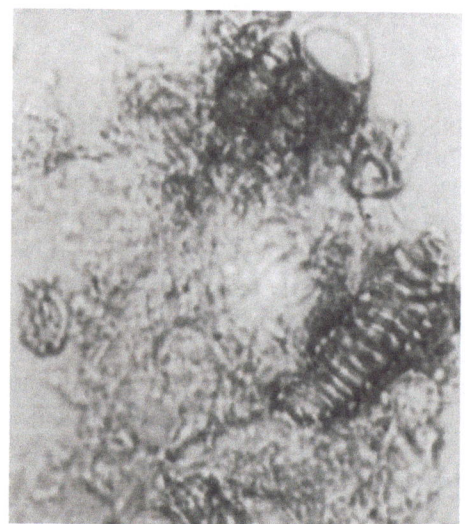

Fig. 5. Sectioned callus tissue (× 63) of *P. anisum* transformed with *A. tumefaciens* LBA4404 harboring the plasmid pMON9793 following incubation with X-Gluc for 2 h to show (*dark areas*) the localization of expression GUS activity. (Salem 1994)

Table 3. Yield of essential oil in cultures and seedlings of *P. anisum*. (Salem 1994)

Culture[a]	Growth conditions[b]	Total oil (µg/g fresh wt.)	Percentage (relative to seedling)
Seedling	MS0/solid	636	100
Nontransformed callus	MS0/solid	3.5	0.55
Transformed callus	MS0/solid	5.3	0.83
Shoot differentiating callus (SDC)	MS0/solid	88.3	13.9
SDC	MS0/liquid	22.1	3.5
SDC (photoperiod)	MS0/solid	148	23.3
SDC (with kinetin and 2,4-D)	MS1/solid	3.5	0.5
SDC (with zeatin)	MS6/solid	10.6	1.7
SDC (with thiamine)	MS7/solid	141	22.2
Shooty teratomas	MS0/solid	72.7	11.4

[a] All cultures were incubated under continuous white light unless indicated otherwise.
[b] Media formulations are described in the text.

in shoot-proliferating cultures of *Artemisia annua* (Ferreira and Janick 1996), in the present case, a photoperiod-induced activation of phenylalanine ammonia lyase might also be responsible for an increase in phenylpropanoid production.

A dramatic reduction in the oil content of the SDC to only 3% of that found in the seedling was observed when the culture was incubated in liquid medium, and this may be accounted for by the sharp decrease in glandular hair density (to around 3–5 hairs/cm^2) which occurred in submerged culture. It was observed that very low oil contents were associated with all cultures where undifferentiated tissue was the dominant morphological feature: thus minimal amounts of oil (in the order of 0.5×10^{-3}%; cf. Reichling et al. 1986) could be detected in non-transformed and transformed callus cultures, whilst SDC grown in MS1 or MS6 media showed dramatic reductions in yield compared with the SDC incubated in MS0 medium. Only in medium MS7 which contained thiamine, a vitamin often claimed to enhance redifferentiation and secondary compound accumulation, was the yield of oil enhanced in the absence of increased levels of shoot formation.

Transformed shoot cultures accumulated around 18% less essential oil than did their non-transformed counterparts when incubated under the same conditions. However, since the amounts of callus tissue associated with the transformed culture was 15% more than was present in the SDC, it may be considered that nontransformed and transformed shoot tissue accumulated almost identical amounts of essential oil pro rata.

The major components of the oils extracted from all cultured material (Table 4) were β-bisabolene (2), geraniol (6), pseudoisoeugenol 2-methylbutyrate (3) and *trans*-anethole (1), and these were unambiguously identified by comparison with published retention data, mass spectral patterns, and authentic standards as appropriate. L-phellandrene (7), methylchavicol (8) and myristicin (5) were also tentatively identified on the basis of retention times and mass spectral data only, although all of these compounds have been previously reported to occur in oils derived from anise. Isoprenoids formed the major components of the essential oils accumulated in all of the cultures.

Table 4. Percentage compositions of the essential oils from cultured seedlings and shoot cultures of *P. anisum*. (Salem 1994)

Culture	Percentage composition			
	Anethole	β-Bisabolene	Pseudoisoeugenol 2-methylbutyrate	Geraniol
	(1)	(2)	(3)	(6)
Seedling	5.7	49.5	12.8	32.0
Nontransformed shoot culture grown in continuous light	1.9	18.9	10.6	68.6
Nontransformed shoot culture grown in a photoperiod	2.5	22.3	24.0	51.2
Transformed shoot culture grown in continuous light	7.7	50.9	6.2	35.2

The seedling and the shooty cultures incubated in continuous light accumulated typically 85% isoprenoids and only 15% phenylpropanoids, but when the shooty cultures were transferred to a 16-h photoperiod, an increase in phenylpropanoid accumulation to around 26% occurred. The oils from the seedling and from the transformed shoots both accumulated β-bisabolene (2) as the major component, whereas geraniol (6) was the chief isoprenoid in oils derived from the nontransformed shoot cultures. Pseudoisoeugenol 2-methylbutyrate (3) was the principal phenylpropanoid in the oils of the seedling and the nontransformed shoot cultures, whilst the transformed shoot culture uniquely accumulated more *trans*-anethole (1) than 3. In this respect, our SDC contrasts with that developed by Reichling's group (Reichling et al. 1995; Kemmerer and Reichling 1996) which accumulated predominantly epoxy-pseudoisoeugenol 2-methylbutyrate (4) in large concentrations.

Transformed shooty teratomas have previously been obtained from species of, amongst others, *Nicotiana*, *Mentha*, *Artemisia*, and *Solanum* following infection with strains of *A. tumefaciens*. The amounts of secondary compounds reportedly accumulated by these transformed cultures varied. Whereas the shooty teratomas of *S. dulcamara* showed a glycoalkaloid content of 1% (on a dry weight basis; Ehmke et al. 1995), which is about five times higher than that found in the parent plant, the monoterpene contents of transformed shoot cultures of *M. citrata* and *M. piperita* were about the same as those of the parent tissues from which they were derived (Rhodes et al. 1992). In each case, however, there was a significant shift in the profile of the compounds accumulated in the transformed cultures compared with the intact plant. In contrast, there was little, if any, difference in sesquiterpene accumulation between transformed and nontransformed shoot cultures of *A. annua* (Paniego and Giulietti 1996) in that both accumulated minute amounts of artemisinin (in the order of 0.02–0.06% on a dry weight basis) compared with the leaves of the parent plant (which contain up to 1% of the isoprenoid).

Similar to the situation established for root cultures transformed with *A. rhizogenes*, shoot cultures transformed with *A. tumefaciens* remain genetically stable over a number of passages; for *P. anisum*, shooty teratomas remained unchanged with respect to morphology and oil accumulation for more than 2 years. Unlike transformed root cultures, however, the growth rates of transformed shooty cultures do not appear typically to be enhanced through *Agrobacterium* transformation.

3 Summary and Conclusions

It has been shown to be possible to produce transformed callus tissue and shoot cultures of *Pimpinella anisum* following stable insertion of T-DNA from *A. tumefaciens* strains 337 (harboring the pBI121.1 vector), LBA 4404 (harboring the pMON9793 vector) and T37, respectively, into the plant genome. Although both nontransformed and transformed callus cells accumulated very small (ca. $0.5 \times 10^{-3}\%$ on a fresh weight basis) amounts of essential oil, the levels in the latter were slightly higher than those of nontransformed tissue. Transformed shoot cultures grew more slowly than their nontransformed counterparts and showed levels of oil accumulation (ca. $8 \times 10^{-3}\%$) which were very similar to those found in normal shoot proliferation cultures. The oils from cultured seedlings and from transformed and non-transformed shoot cultures were similar in composition containing β-bisabolene, pseudoisoeugenol 2-methylbutyrate, *trans*-anethole, and geraniol. The yield of oil accumulated by nontransformed shoot cultures incubated in a photoperiod ($1.5 \times 10^{-2}\%$) was around fourfold lower than that found in the leaves of the cultured seedling.

References

Beck E, Ludwig G, Averswald EA, Reiss B, Schaller H (1982) Nucleotide sequence and exact localisation of the neomycin phosphotransferase gene from transposon TNS. Gene 19:327–336

Charlwood BV (1993) Recent advances in the production of aroma compounds in plant culture systems. In: van Beek TA, Breteler H (eds) Phytochemistry in agriculture. Oxford Science Publications, Oxford, pp 322–345

Charlwood BV, Charlwood KA, Molina-Torres J (1990) Accumulation of secondary compounds in organised plant cultures. In: Charlwood BV, Rhodes MJC (eds) Secondary products from plant tissue culture. Oxford Science Publications, Oxford, pp 167–200

Dellaporta SL, Woods J, Hicks JB (1983) A plant DNA minipreparation: version II. Plant Mol Biol Rep 1:19–21

Ditta G, Stanfield S, Corbin D, Helinski D (1980) Broad host range DNA cloning system for gram-negative bacteria: construction of a gene bank of *Rhizobium meliloti*. Proc Natl Acad Sci USA 77:7347–7351

Ehmke A, Ohmstede D, Eilert U (1995) Steroidal glycoalkaloids in cell and shoot teratoma cultures of *Solanum dulcamara*. Plant Cell Tissue Organ Cult 43:191–197

Ferreira JFS, Janick J (1996) Roots as an enhancing factor for the production of artemisinin in shoot cultures of *Artemisia annua*. Plant Cell Tissue Organ Cult 44:211–217

Gamborg OL, Miller RA, Ojima K (1968) Nutrient requirements of suspension cultures of soybean root cells. Exp Cell Res 50:151–158

Garfinkle DJ, Simpson R, Ream L, White F, Nester E (1981) Genetic analysis of crown gall: fine structure map of the T-DNA by site directed mutagenesis. Cell 27:143–153

Goldberg SB, Flick JS, Rogers SG (1984) Nucleotide-sequence of the *tmr* locus of *Agrobacterium tumefaciens* pTiT37 T-DNA. Nucleic Acids Res 12:4665–4677

Hamill JD, Rounsley S, Spencer A, Todd G, Rhodes MJC (1991) The use of polymerase chain reaction in plant transformation studies. Plant Cell Rep 10:221–224

Harborne JB, Heywood VH, Williams CA (1969) Distribution of myristicin in seeds of the Umbelliferae. Phytochemistry 8:1729–1732

Hirayama T, Muranaka T, Ohkawa H, Oka A (1988) Organisation and characterisation of the *vir* C and *vir* D genes from *Agrobacterium rhizogenes*. Mol Gen Genet 213:229–237

Ibrahem DK, Al-Maliki SJ, Hamad MN, Al-Khazraji SM (1988) Influence of *Pimpinella anisum* on behaviour of male mice. J Biol Sci Res 19:263–276

Jefferson RA, Burgess SM, Hirsh D (1986) β-Glucuronidase from *Escherichia coli* as a gene-fusion marker. Proc Natl Acad Sci USA 83:8447–8451

Kemmerer B, Reichling J (1996) S-Adenosyl-L-methionine: anol-O-methyltransferase activity in organ cultures of *Pimpinella anisum*. Phytochemistry 42:397–403

Komari T (1989) Transformation of callus cultures of nine plant species mediated by *Agrobacterium*. Plant Sci 60:223–230

Kubeczka K-H, Bohn I, Formacek V (1986) New constituents from the essential oils of *Pimpinella* species. In: Brunke E-J (ed) Progress in essential oil research. Walter de Gruyter, Berlin, pp 279–298

Mettler IJ (1987) A simple and rapid method for the minipreparation of DNA from tissue cultured plant cells. Plant Mol Biol Rep 5:346–349

Murashige T, Skoog F (1962) A revised medium for rapid growth and bio-assays with tobacco tissue cultures. Physiol Plant 15:473–497

Noma M, Huber J, Pharis RP (1979) Occurrence of gibberellin A_1 counterpart GA_1, GA_4, and GA_7 in somatic cell embryo cultures of carrot and anise. Agric Biol Chem 43:1793–1794

Ooms G, Hooykaas PJJ, Van Veen RJM, Van Beelen P, Regensburg-Tuink TJG (1982) Octopine Ti plasmid deletion mutants of *Agrobacterium tumefaciens* with emphasis on the right side of the T-region. Plasmid 7:15–29

Paniego NB, Giulietti AM (1996) Artemisinin production by *Artemisia annua* L.-transformed organ cultures. Enzyme Microb Technol 18:526–530

Reichling J, Becker H, Martin R, Burkhardt G (1985) Comparative studies on the production and accumulation of essential oil in the whole plant and in cell cultures of *Pimpinella anisum* L. Z Naturforsch 40C:465–468

Reichling J, Martin R, Burkhardt G, Becker H (1986) Comparative study on the production and accumulation of essential oil in the whole plant and in tissue cultures of *Pimpinella anisum*. In: Brunke E-J (ed) Progress in essential oil research. Walter de Gruyter, Berlin, pp 421–428

Reichling J, Martin R, Kemmerer B (1995) Biosynthesis of pseudoisoeugenol-derivatives in liquid tissue cultures of *Pimpinella anisum*. Plant Cell Tissue Organ Cult 43:131–136

Rhodes MJC, Spencer A, Hamill JD, Robins RJ (1992) Flavour improvement through plant cell culture. In: Patterson RLS, Charlwood BV, MacLeod G, Williams AA (eds) Bioformation of flavours. Royal Society of Chemistry, Cambridge, pp 42–64

Salem KMSA (1994) Biochemical studies of transformed cultures of *Pimpinella anisum* (Umbelliferae) and *Sinapis alba* (Cruciferae). PhD Thesis, University of London, London

Salem KMSA, Charlwood BV (1995) Accumulation of essential oils by *Agrobacterium tumefaciens*-transformed shoot cultures of *Pimpinella anisum*. Plant Cell Tissue Organ Cult 40:209–215

Sambrook J, Fritsch EF, Maniatis T (1989) Molecular cloning: a laboratory manual (2nd edn). Cold Spring Harbor Lab Press, New York

Zhou C, Yang Y, Jong A (1990) Mini-prep in ten minutes. BioTechniques 8:172–173

XVII Genetic Transformation of *Phyllanthus niruri* L. (*P. amarus*)

K. Ishimaru[1], K. Yoshimatsu[2], T. Yamakawa[3], H. Kamada[4], and K. Shimomura[2]

1 Introduction

1.1 Distribution and Importance of *Phyllanthus* Plants

The genus *Phyllanthus* (family Euphorbiaceae) consists of about 600 species of herbaceous plants, shrubs, nonsucculents, and trees. It is predominantly native to the tropics of the Old World but well represented in the floras of warmer parts in the world (particularly in the Americas). The name coming from the Greek *phyllon* (a leaf) and *anthos* (a flower), alludes to the brooms of some species on branches that have the form of leaves (Everett 1981).

Evergreen trees *P. acidus* (gooseberry tree) and *P. arborescens* (native to Java) which grow over 10~40 feet tall, and small shrub species *P. pulcher*, *P. angustifolius*, *P. epiphyllanthus*, etc. are popular as ornamentals for garden or potted plants and loved for their graceful foliage and attractively colored (yellow, pink, red, etc.) flowers.

Some herbal plants such as *P. niruri* L. (frequently known as *P. amarus*; paraparai mi), *P. sellowianus* Mueller Arg. (common name saradi blanco), *P. anisolobus*, *P. fraternus*, etc., also supply traditional folk medicines for various treatments of jaundice, asthma, hepatitis, flu, dropsy, urolitic disease (diabetes, kidney stone, etc.), and so on. Particularly on *P. niruri*, a small plant with originate in Central and South America (the regions around Peru and Ecuador), several chemical studies have been done.

1.2 Chemical Constituents and in Vitro Studies on *Phyllanthus*

P. niruri, one of the most important medicinal plants in the Euphorbiaceae, contains constituents such as lignans (Schneiders and Stevenson 1982;

[1] Department of Applied Biological Sciences, Faculty of Agriculture, Saga University, 1 Honjo, Saga, 840 Japan
[2] Tsukuba Medicinal Plant Research Station, National Institute of Health Sciences, 1 Hachimandai, Tsukuba, Ibaraki, 305 Japan
[3] Department of Global Agricultural Sciences, The University of Tokyo, Yayoi 1-chome, Bunkyo-ku, Tokyo, 113 Japan
[4] Gene Experiment Center, University of Tsukuba, Tsukuba, Ibaraki 305 Japan

Biotechnology in Agriculture and Forestry, Vol. 45
Transgenic Medicinal Plants (ed. by Y.P.S. Bajaj)
© Springer-Verlag Berlin Heidelberg 1999

Satyanarayana et al. 1988; Singh et al. 1989), flavonoids (Hnatyszyn et al. 1987), alkaloids (Mulchandani and Hassarajani 1984; Joshi et al. 1986), phthalic acid (Singh et al. 1986), and tannins (Foo and Wong 1992; Foo 1993, 1995). In other medicinal herbs, *P. sellowianus* (Tempesta et al. 1988) and *P. anisolobus* (Bachmann et al. 1993), similar secondary metabolites (lignans and alkaloid) have been isolated. As shown in biological studies on *P. niruri* (Thyagarajan et al. 1982; Syamasundar et al. 1985; Ueno et al. 1988), corresponding to antivirus (hepatitis) and inhibitory activity against angiotensin-converting enzyme, the most important constituent of this medicinal herb correlated to several pharmaceutical actions was presumed to be tannin (high-molecular polyphenols).

Compared to the numerous research works on the chemistry of secondary metabolites, not much work has been done on tissue culture, micro propagation, transformation, etc. of *Phyllanthus* plants. Caulogenesis from hypocotyl explants of *P. fraternus*, whose extract has been shown to possess antiviral properties against hepatitis viruses, cultured on modified B5 medium (Kumari and Saradhi 1992) supplemented with 2,4-D or NAA in the presence and absence of 6-benzylaminopurine (BA) was determined (Rajasubramaniam and Saradhi 1994). In the study, adventitious shoots differentiated from callus developed from the cut ends of 12.5% of the hypocotyl segments on the medium with 10^{-6}M BA in combination with 10^{-7}M 2,4-D or 10^{-6} M NAA. Profuse rooting also occurred from the hypocotyl explants on medium with 10^{-6}M BA and 10^{-6}M NAA. Addition of casein hydrolysate in the medium along with 10^{-6}M BA and 10^{-7}M 2,4-D enhanced the frequency of cultures with adventitious shoots up to 68.0%; this effect could partially be substituted by some amino acids (glutamine, glutamic acid, or proline). *P. fraternus* plantlets in vitro, after hardening, were successfully transferred to the soil, which presented a useful system for micropropagation of the plant.

Unander (1991, 1996) also obtained callus from four herbaceous species, i.e., *P. amarus* (*P. niruri*), *P. abnormis*, *P. urinaria*, and *P. caroliniensis*, on modified MS medium (Murashige and Skoog 1962). Optimum induction and growth of friable calli of *P. amarus* were observed when cultured on the medium with either 0.5 mg/l or 1 mg/l BA and 1 mg/l of either 2,4-D or indole-3-butyric acid (IBA). Although Khanna and Staba (1968) had satisfactory results on increasing and growing callus of the woody *P. emblica*, the experiments indicated that a cytokinin-containing medium is superior for the herbaceous *Phyllanthus*. Unander (1991) also reported that aqueous extracts from field-grown plants were more active in vitro against viral DNA polymerase and reverse transcriptase than those of the calli.

In this chapter, the successful establishment of transformed cell cultures (crown gall and hairy roots induced by *Agrobacteria*) of *P. niruri* and the determination of the production of useful pharmaceuticals (tannins) in the cells and organs under various culture conditions (Ishimaru et al. 1992) are discussed.

2 Transformation and Secondary Metabolism of *Phyllanthus niruri*

2.1 Methods for Transformation and Analysis of Secondary Metabolites

2.1.1 Plant Material and Bacterial Strain

Seeds of *P. niruri* collected in Peru were surface sterilized and germinated aseptically on hormone-free half-strength MS (1/2 MS) solid medium (Gelrite 2 g/l) under light (16 h light/day, 2000 lx). Axenic plants were subcultured and used as explants for *Agrobacterium* inoculation.

1. *Hairy Root Culture.* For the induction of hairy roots, two types of *Agrobacterium* strains were used, *A. rhizogenes* A4 harboring the root-inducing plasmid pRi A4b, and *A. tumefaciens* R-1000 + 121 having two plasmids, a root-inducing plasmid (pRi A4b) and a mini Ti plasmid, pBI 121, which contained chimeric genes encoding for NPT II (neomycin phosphotransferase II) and GUS (β-glucuronidase) on the T-DNA. These strains were individually inoculated via a needle onto the stems of axenic plants in vitro. After 2 to 3 weeks, several hairy roots appeared at the infected sites (Fig. 1). Hairy roots were cut off, transferred, and cultured in the dark on hormone-free MS solid medium containing an antibiotic (0.5 mg/ml Claforan) to eliminate bacteria. After removal of bacteria, axenic hairy roots were transferred and subcultured in hormone-free 1/2 MS liquid medium (in an Erlenmeyer flask) in the dark on a rotary shaker (100 rpm). Among some axenic hairy roots established, eight clones (HR-A4-1~5 induced by *A. rhizogenes* A4 and HR-121-1~3 by *A. tumefaciens* R-1000 + 121) which showed sufficient growth were used for the experiments.

2. *Crown Gall Culture.* Crown galls were induced from stem segments of axenic plants by direct infection with *A. tumefaciens* C58 harboring the mutant-type tumor-inducing plasmid pGV 2215 as described in (1) above. Crown galls, after elimination of bacteria on hormone-free MS solid medium containing Claforan, were subcultured on hormone-free MS solid medium in the light (the same condition as described in (1) above). In the culture, some crown galls (CG-A) turned green and were separated from galls (CG-B) which did not show green coloration.

3. *Adventitious Root Culture.* Nontransformed organ cultures (adventitious root) were also established and determined the secondary metabolism compared to those of the transformed cells and organs. Stem segments from axenic plants were cut off and cultured on 1/2 MS solid medium supplemented with 0.1 mg/l NAA. After 2 to 3 weeks of culture in the dark, adventitious roots appeared at the cut ends with some calli. The roots were cut off and maintained in 1/2 MS liquid medium containing 0.1 mg/l NAA on a rotary shaker under light.

Fig. 1. Induction of *Phyllanthus niruri* hairy roots by *Agrobacterium rhizogenes* A4 on hormone-free 1/2 MS medium. (Photo Kamada 1990)

2.1.2 Confirmation of the Transformation

Extraction and detection of opines (agropine and mannopine) was done by the method of Petit et al. (1983). Hairy roots (HR-A4-1~5 and HR-121-1~3) were macerated with 0.05 N HC1 for 1 h at room temperature. The supernatants were spotted on a whatman 3-MM paper. Paper electrophoresis was performed in formic acid/acetic acid/water at the ratio of 1:3:16 (v/v/v) (Otten and Schilperoort 1978) for 4 h at 20 V/cm. The electrophoretogram was visualized with alkaline silver nitrate reagent (Trevelyan et al. 1950). Silver nitrate-positive substances were identified by comparing their electrophoretic mobility with those of authentic samples (agropine and manopine).

In HR-121-1~3, GUS expression was also confirmed with X-Gluc (5-bromo-4-chloro-3-indoly1-β-D-glucronide) as a substrate (Jefferson et al. 1987).

2.1.3 Semilarge Scale Culture

1. *Hairy Root Culture.* HR-121-2, which showed the strongest GUS expression, was inoculated into hormone-free 1/2 MS liquid medium (2% sucrose) in a 6-1 air-lift-type fermenter equipped with a paddle (11 cm in diameter, 20 rpm) and supplied with air (0.05 VVM) in the dark (Fig. 2). Hairy roots were harvested after 7 weeks of culture.

2. *Shoot Culture.* Nodal segments of axenic plants were cut off and cultured on 1/2 MS solid medium supplemented with 1 mg/l IAA in the dark. In this culture adventitious shoots formed from the basal end of inoculated plants. Shoots were cut off and subcultured in hormone-free 1/2 MS liquid medium on a rotary shaker in light. Shoot segments (1 cm length, 30 pieces) were transferred to either a 6-1 air-lift type fermenter (mentioned above) or a 2-1 rotating-drum fermenter (3 rpm) (Fig. 3), both with hormone-free 1/2 MS liquid medium containing 2% sucrose. Shoots (Sh-A) cultured in the former were harvested after 7 weeks of culture in the light. Shoots (Sh-B) inoculated in the latter were cultured with air supplement (at 0.1 VVM) in the light and harvested after 10 weeks of culture.

Fig. 2. *Phyllanthus niruri* hairy roots (HR-121-2) cultured in hormone-free 1/2 MS liquid medium for 4 weeks in a 6-1 air-lift-type fermenter equipped with a paddle. (Photo Shimomura 1990)

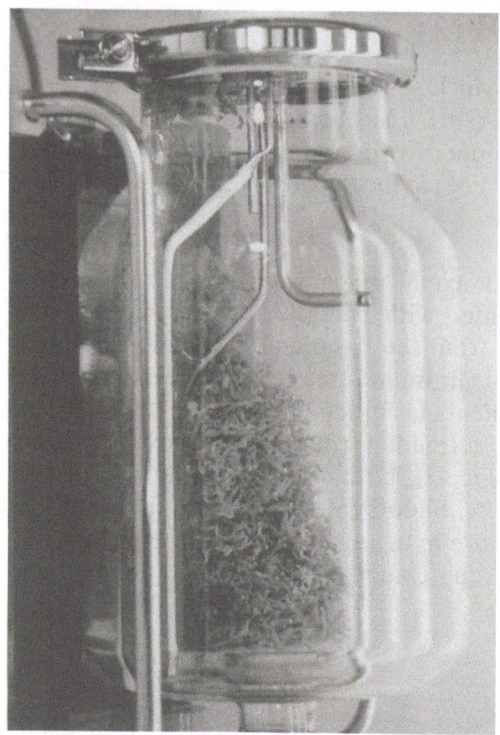

Fig. 3. *Phyllanthus niruri* shoots cultured in hormone-free 1/2 MS liquid media for 4 weeks in a 2-1 rotating-drum-type fermenter. (Photo Shimomura 1990)

2.1.4 Plant Cultivation Under Hydroponics

The plants of *P. niruri* were also cultivated in a hydroponic facilities and the growth and secondary metabolites were compared to those of the field grown plants, in vitro plants, cells and organs, transformed cells and organs. Axenic plantlets (in vitro) were transferred and cultivated under hydroponics at 20 °C in the light (14 h light/day). Culture conditions in hydroponic facilities were as follows; cultivation medium: balanced solution; EC 2-4 mS Ω/cm, relative humidity: 70 ± 10%, light intensity: 150 μE/sm^2.

2.1.5 Analysis of Secondary Metabolites

1. *Isolation of Compounds from Hairy Roots.* HR-121-2, harvested after 6 weeks of culture in hormone-free 1/2 MS liquid medium on a rotary shaker, were lyophilized and homogenized with MeOH. The MeOH extract was

evaporated to dryness under reduced pressure and subjected to Sephadex LH-20 and MCI-gel CHP-20P column chromatographies to afford six phenolic compounds, gallic acid (**1**), (−)-epicatechin (**2**), (+)-gallocatechin (**3**), (−)-epigallocatechin (**4**), (−)-epicatechin 3-O-gallate (**5**), and (−)-epigallocatechin 3-O-gallate (**6**) (Nonaka et al. 1983).

2. *Isolation of Compounds from Leaves and Stems of Parent Plants Collected in Peru.* Dried leaves and stems of the plants collected in Peru were mashed and extracted with MeOH. The extract was treated as above (1) and subjected to Sephadex LH-20, MCI-gel CHP-20P, and μ-Bondapak C18 Porasil B column chromatographies to afford **1**, brevifolin carboxylic acid (**7**), corilagin (Tanaka et al. 1985) (**8**), geraniin (Okuda et al. 1982) (**9**), 1,2-di-O-galloyl-3,6-(R)-hexahydroxydiphenoyl-β-D-glucose (Saijo et al. 1989) (**10**), and terchebin (Schmidt et al. 1967) (**11**).

3. *Isolation of Compounds from Plants Grown Under Hydroponic Conditions.* After 3 months of cultivation in hydroponic facilities, the mature plants were harvested. Leaves and stems were cut off, dried and extracted with 80% aqueous acetone twice. The extracts were concentrated and partitioned with $CHCl_3$. The aqueous layer which contained polar phenolic constituents was chromatographed as described above to give **1, 8, 9**, 1,2,4,6-tetra-O-galloyl-β-D-glucose (Haddock et al. 1982) (**12**), and phyllanthusiin D (Foo and Wong 1992) (**13**). Roots were also treated by the same procedure as described above (1) to afford **2–6**.

4. *HPLC Analysis of Tissue Cultures.* Samples were mashed and extracted with MeOH with a sonicator at room temperature. Each extract was filtered and evaporated to dryness, which was dissolved in 80% aqueous MeOH, filtered through a Milipore filter (0.45 μm) and injected into a HPLC system. In the system, **1~6** and related phenolic (+)-catechin (**14**) were analyzed. HPLC conditions were as follows; column: Nucleosil 100 $5C_{18}$ (4.6 mm id × 250 mm), mobile phase: MeCN-50 mM NaH_2PO_4 (1:9→2:3), column temperature: 40 °C, flow rate: 0.9 ml/min, detection: 280 nm, R_ts (min): **1** (5.3), **3** (12.5), **4** (16.1), **14** (17.3), **2** (20.0), **6** (20.6), and **5** (24.4).

2.2 Results and Discussion

2.2.1 Hairy Root Cultures

Two types of *Agrobacterium* strains (*A. rhizogenes* A4 and *A. tumefaciens* R-1000 + 121), used for transformation of *P. niruri*, showed no clear difference in the induction of the hairy roots. Among hairy roots induced by *A. rhizogenes* A4, five clones (HR-A4-1~5) which showed sufficient growth in hormone-free 1/2 MS medium were selected and analyzed for the production of phenolic compounds (**1~6** and **14**) (Table 1). The main constituents contained in these hairy roots was **6** (0.177~0.449% as dry wt). In these five

Table 1. Contents of compounds **1–6** and **14** (% as dry wt.) in different cultures of *Phyllanthus niruri*. (Ishimaru et al. 1992)

Culture	1	2	3	4	5	6	14
HR-A4-1	0.006	0.100	0.056	0.118	0.042	0.379	0.005
-2	0.004	0.065	0.054	0.132	0.014	0.306	0.045
-3	0.002	0.021	0.067	0.111	0.002	0.449	0.002
-4	0.003	0.039	0.024	0.110	0.014	0.267	0.008
-5	0.002	0.072	0.015	0.092	0.011	0.177	0.006
HR-121-1	0.006	0.054	0.040	0.084	0.008	0.314	0.010
-2	0.005	0.110	0.078	0.109	0.050	0.424	0.005
-3	0.004	0.238	0.074	0.106	0.066	0.525	0.020
NR	0.001	0.014	0.007	nd	trace	0.026	trace
CG-A, CG-B	nd	nd	nd	nd	nd	nd	nd
HR-121-2Fa	0.014	0.024	0.040	0.084	0.015	0.055	0.010
-2Fb	0.005	0.006	0.022	0.021	0.008	0.036	trace
Sh-A	trace	0.028	0.037	0.127	0.009	0.050	0.010
Sh-B	0.004	0.033	0.031	0.058	0.014	0.043	0.007

nd = not detected, trace = content below 0.001%.

clones, the ratio of the contents of these phenolic compounds did not differ very much. Although the contents of **14** (0.002%) and **5** (0.002%) in HR-A4-3 were lower than those of the other hairy roots, this clone produced the highest amount of **6** (0.449%).

Three clones of hairy roots (HR-121-1~3) induced by *A. tumefaciens* R-1000 + 121 also produced flavan-3-ols similar to those observed in HR-A4-1~5 (Table 1). There was no clonal variation in the production of these phenolic compounds among the eight clones induced by two types of *Agrobacterium*. In hairy roots, clone HR-121-3 produced the highest amounts of **2** (0.238%), **5** (0.066%), and **6** (0.525%).

2.2.2 Crown Gall and Adventious Root Cultures

Although a relatively high level of polyphenol products was observed in hairy root tissues, in the other transformed cells, crown galls (both CG-A and CG-B), no detectable level of polyphenol was yielded. This result implied the importance of organogenesis in *P. niruri* for the production of the phenolic constituents.

Adventitious roots (NR) were also derived and cultured in 1/2 MS liquid medium containing 0.1 mg/l NAA. In the cultures, low contents of flavan-3-ols were yielded (Table 1). In NR the amounts of **5** and **14** were fairly low and **4** was not detected. The content of **6** (0.026%) in NR was almost 1/9 to 1/20 less than those in hairy roots. Therefore, for the production of flavan-3-ols to succeed with the biosynthetic regulation, transformed organ cultures (hairy roots) seemed to be more useful than the normal root cultures.

2.2.3 Semilarge-Scale Cultures

Two types of semilarge-scale fermenters were tested, a 6-1 air-lift-type fermenter equipped with a paddle and a 2-1 rotating-drum-type fermenter (see Sect. 2.1.3).

Hairy root clone HR-121-2 was inoculated and cultured in a 6-1 air-lift-type fermenter in hormone-free 1/2 MS liquid medium. The hairy roots grew satisfactorily (Fig. 2) and the weight of the roots reached ca. 694 g fr.wt (inoculation 7.9 g fr.wt) after 7 weeks of culture. In this culture, some hairy roots (HR-121-2Fa) were cut by the paddle and formed a globular mass of ca. 3 cm in diameter, while the others (HR-121-2Fb) were not clumped. The polyphenol contents both in HR-121-2Fa and HR-121-2Fb were also examined (Table 1). The contents of **2**, **5**, and **6** in both hairy roots were almost 1/3 to 1/20 of that produced in HR-121-2 which was cultured in 100-ml flasks on a rotary shaker. Therefore, the semilarge-scale propagation of *P. niruri* hairy roots was effective for growth but not for polyphenol production, especially for galloyl derivatives of flavan 3-ols such as **5** and **6**.

Semilarge-scale cultures of *P. niruri* shoots were also tried using both types of fermenters. About 0.7 g (fr.wt) of shoots, subcultured in hormone-free 1/2 MS liquid medium, were transferred to each fermenter and cultured in the same medium in the light. In both cultures, shoots grew rapidly without aggregation until the fw reached ca. 686 g (Sh-A, in a 6-1 air-lift type, 7 weeks culture) and ca. 234 g (Sh-B, in a 2-1 rotating-drum type, 10 weeks culture), respectively. In both these cultures, only flavan 3-ols were produced (Table 1) and the hydrolyzable tannins (**8~13**), which were observed in rich amounts in leaves and stems of the parent plant, were not detected. This result indicated that, cultures in a fermenter in liquid medium condition, the biosynthesis of hydrolyzable tannins in *P. niruri* shoots was depressed (by decrease of esterification and/or increase of hydroxylation of galloyl groups) compared to that in the intact plants, which was fairly enhanced when the plants were cultured in the field and under hydroponic facilities. The contents of flavan-3-ols produced in Sh-A and Sh-B were almost similar to those of HR-121-2Fa and HR-121-2Fb, showing low contents of the galloyl derivatives **5** and **6**. For the production of hydrolyzable tannins using such liquid cultures (hairy root, adventitious root, and shoot), further investigation (arrangement of several culture conditions) will be required.

2.2.4 Hydroponic Cultivation

Plantlets were also transferred and cultivated in a hydroponic system. In about 3 months after transfer from in-vitro condition, the plants grew satisfactorily, showing the length of ca. 50 cm of aerial part and ca. 30 cm of root portion (Fig. 4). Furthermore, the leaves and stems of the plant, cultivated under hydroponic conditions, contained similar kinds of hydrolysable tannins to those produced in the parent plants collected in Peru, the original region of this plant [see Sect. 2.1.5: (1), (2)]. This result demonstrates the profitable use

Fig. 4. *Phyllanthus niruri* plantlet cultivated for 2 months under hydroponic facilities. (Photo Shimomura 1990)

of hydroponics for a good supply of the important medicinal constituents (tannins) as well as for the mass proliferation of this traditional medicinal herb.

2.2.5 Tissue-Specific Metabolites

Amongst phenolic compounds produced in aerial parts of *P. niruri*, the main constituent was hydrolyzable tannin, **9**. On the other hand, the root part of the plant contained mainly flavan-3-ols (**2~4**) and their galloyl derivatives (**5** and **6**) which are the structural elements of condensed tannins (proanthocyanidins) (see Sect. 2.1.5). The hydrolyzable tannins observed mainly in the aerial parts could not be isolated from the roots. This tissue-specific distribution of phenolic compounds (hydrolyzable tannins and flavan-3-ols) observed in *P. niruri* is interesting when considering the biosynthetic position of secondary metabolites in this plant. Concerning the production of flavan-3-ols contained in the roots of this plant, hairy root cultures were the most efficient method amongst those tested. The constitution of the flavan-3-ols found in the parent plant and hairy root and shoot cultures of *P. niruri* is very similar to that observed in leaves of

Thea sinensis (green tea) (Nonaka et al. 1983) which are expected to be one of the most promising materials as functional foods for human health (antitumor, -virus, -bacteria, oxidatives, etc.). Cultures of *P. niruri* (both transformed and nontransformed) also might have the same pharmacological effects as expected from tea leaves.

3 Summary and Conclusions

Transformation of *Phyllanthus niruri* was performed by the *Agrobacterium*-mediated method (with *A. rhizogenes* A4, *A. tumefaciens* R-1000 + 121, and *A. tumefaciens* C58) and succeeded in establishing the transformed organ (hairy root) and cell (crown gall) cultures. Eight clones of the hairy roots induced by two types of *Agrobacterium* showed no clear difference in growth and polyphenol production, mainly of flavan-3-ols and their galloyl derivatives. The crown gall cells yielded no detectable level of polyphenols.

Semilarge-scale cultures of hairy roots and shoot tissues (nontransformed) were successful by use of 6-1 air-lift-type and 2-1 rotating-drum-type fermenters. In both cultures, flavan-3-ols similar to those produced in roots of parent plants were formed.

For mass proliferation of this important medicinal plant, cultivation under hydroponic facilities was also examined. Under the hydroponic condition, *P. niruri* plantlets showed good growth, producing sufficient amounts of its medicinal constituents, hydrolyzable tannins such as corilagin, geraniin, galloylglucose, etc., in the aerial part, and flavan-3-ols (condensed tannins) in the root portion.

References

Bachmann TL, Ghia F, Torssell KBG (1993) Lignans and lactones from *Phyllanthus anisolobus*. Phytochemistry 33:189–191

Everett TH (1981) PHYLLANTHUS The New York Botanical Garden illustrated encyclopedia of horticulture, vol 8. Garland Publishing, New York, pp 2616–2617

Foo LY (1993) Amariin, a di-dehydrohexahydroxydiphenoyl hydrolysable tannin from *Phyllanthus amarus*. Phytochemistry 33:487–491

Foo LY (1995) Amarinic acid and related ellagitannins from *Phyllanthus amarus*. Phytochemistry 39:217–224

Foo LY, Wong H (1992) Phyllanthusiin D, an unusual hydrolysable tannin from *Phyllanthus amarus*. Phytochemistry 31:711–713

Haddock EA, Gupta RK, Al-Shafi SMK, Haslam E, Magnolato D (1982) The metabolism of gallic acid and hexahydroxydiphenic acid in plants. Part 1. Introduction. Naturally occurring galloyl esters. J Chem Soc Perkin Trans I:2515–2524

Hnatyszyn O, Ferraro G, Coussio JD (1987) A biflavonoid from *Phyllanthus sellowianus*. J Nat Prod 50:1156–1157

Ishimaru K, Yoshimatsu K, Yamakawa T, Kamada H, Shimomura K (1992) Phenolic constituents in tissue cultures of *Phyllanthus niruri*. Phytochemistry 31:2015–2018

Jefferson RA, Kavanagh TA, Bevan MW (1987) GUS fusions: β-glucuronidase as a sensitive and versatile gene fusion marker in higher plants. EMBO J 6:3901–3907

Joshi B, Gawad DH, Pelletier SW, Kartha G, Bhandary K (1986) Isolation and structure (X-ray analysis) of *ent*-norsecurinine, an alkaloid from *Phyllanthus niruri*. J Nat Prod 49:614–620

Khanna P, Staba EJ (1968) Antimicrobials from plant tissue cultures. Lloydia 31:180–189

Kumari N, Saradhi PP (1992) Regeneration of plants from callus cultures of *Origanum vulgare* L. Plant Cell Rep 11:476–479

Mulchandani NB, Hassarajani SA (1984) 4-methoxy-nor-securinine, a new alkaloid from *Phyllanthus niruri*. Planta Med 50:104–105

Murashige T, Skoog F (1962) A revised medium for rapid growth and bio-assays with tobacco tissue cultures. Physiol Plant 15:473–497

Nonaka G, Kawahara O, Nishioka I (1983) A new class of dimeric flavan-3-ol gallates, theasinensins A and B, and proanthocyanidin gallates from green tea leaf (1). Chem Pharm Bull 31:3906–3914

Okuda T, Yoshida T, Hatano T (1982) Constituents of *Geranium thunbergii* Sieb. *et* Zucc. Part 12. Hydrated stereostructure and equilibration of geraniin J Chem Soc Perkin Trans I:9–14

Otten LABM, Schilperoort RA (1978) A rapid micro scale method for the detection of lysopine and nopaline dehydrogenase activities. Biochem Biophys Acta 527:497–500

Petit A, David C, Dahl GA, Ellis JG, Guyon P, Casse-Delbart F, Tempé J (1983) Further extention of the opine concept: Plasmids in *Agrobacterium rhizogenes* cooperate for opine degradation. Mol Gen Genet 190:204–214

Rajasubramaniam S, Saradhi PP (1994) Organic nitrogen stimulates caulogenesis from hypocotyl callus of *Phyllanthus fraternus*. Plant Cell Rep 13:619–622

Saijo R, Nonaka G, Nishioka I (1989) Isolation and characterization of five new hydrolyzable tannins from the bark of *Mallotus japonicus*. Chem Pharm Bull 37:2063–2070

Satyanarayana P, Subrahmanyam P, Viswanatham KN, Ward RS (1988) New seco- and hydroxy-lignans from *Phyllanthus niruri*. J Nat Prod 51:44–49

Schmidt OT, Schulz J, Wurmb R (1967) Terchebin. Liebigs Ann Chem 706:169–179

Schneiders GE, Stevenson R (1982) Structure and synthesis of the aryltetralin lignans hypophyllanthin and nirtetralin. JC Soc Perkin I:999–1003

Singh B, Agrawal PK, Thakur RS (1986) Chemical constituents of *Phyllanthus niruri* Linn. Indian J chem 25B:600–602

Singh B, Agrawal PK, Thakur RS (1989) A new lignan and a new neolignan from *Phyllanthus niruri*. J Nat Prod 52:48–51

Syamasundar KV, Singh B, Thakur RS, Husain A, Kiso Y, Hikino H (1985) Antihepatotoxic principles of *Phyllanthus niruri* herbs. J Ethnopharmacol 14:41–44

Tanaka T, Nonaka G, Nishioka I (1985) Punicafolin, an ellagitannin from the leaves of *Punica granatum*. Phytochemistry 24:2075–2078

Tempesta MS, Corley DG, Beutler JA, Metral CJ, Yunes RA, Giacomozzi CA, Calixto JB (1988) Phyllanthimide, a new alkaloid from *Phyllanthus sellowiannus*. J Nat Prod 51:617–618

Thyagarajan SP, Thiruneelakantan K, Subramanian S, Sundaravelu (1982) *In vitro* inactivation of HBsAg by *Eclipta alba* Hassk and *Phyllanthus niruri* Linn. Indian J Med Res 76:124–130

Trevelyan WE, Procter DP, Harrison JS (1950) Detection of sugars on paper chromatograms. Nature 166:444–445

Ueno H, Horie S, Nishi Y, Shogawa H, Kawasaki M, Suzuki S, Hayashi T, Arisawa M, Shimizu M, Yoshizaki M, Morita N, Berganza LH, Ferro E, Basualdo I (1988) Chemical and pharmaceutical studies on medicinal plants in Paraguay. Geraniin, an angiotensin-converting enzyme inhibitor from Paraparai Mi, *Phyllanthus niruri*. J Nat Prod 51:357–359

Unander DW (1991) Callus induction in *Phyllanthus* species and inhibition of viral DNA polymerase and reverse transcriptase by callus extracts. Plant Cell Rep 10:461–466

Unander DW (1996) *Phyllanthus* species: In vitro culture and the production of secondary metabolites. In: Bajaj YPS (ed) Biotechnology in agriculture and forestry, vol 37. Medicinal and aromatic plants IX. Springer, Berlin Heidelberg New York, pp 304–318

XVIII Genetic Transformation of *Salvia miltiorrhiza*

Z.B. Hu[1], D. Liu[1], and A.W. Alfermann[2]

1 Introduction

In China, the genus *Salvia* consists of 78 species, 24 varieties, and 8 forms, and is distributed throughout, especially in areas of southwestern such as the Sichiuan, Yunnan, and Xizang provinces (Huang et al. 1981). Many *Salvia* species have been screened as possible sources of tanshinones, for example, *S. przewalskii* var. *mandarinorum*, *S. przenwalskii*, *S. miltiorrhiza*, *S. yunnanensis*, and *S. bowleyana* contain more tanshinone II-A, and *S. przewalskii*, *S. miltiorrhiza*, *S. yunnanensis*, *S. digitaloides*, and *S. kiangsiensis* more tanshinone I, in the ten *Salvia* species studied (Huang et al. 1980). Among them, *S. miltiorrhiza* Bunge is a well-known traditional Chinese medicinal plant. The roots of this plant, Tan-shen, are widely used in China for treatment of hematologic and cardiovascular disorders (Guo 1992).

Two types of biologically active compounds have been discovered in *S. miltiorrhiza*; one is liposoluble, such as a diterpenoid, and the other is water-soluble, such as a phenolic acid compound (Guo 1992). The majority of diterpenoids in *S. miltiorrhiza* are diterpenoid quinones, including cryptotanshinone, tanshinone I, II, etc. (Fig. 1). These compounds were considered as the main components for the treatment of cardiovascular diseases (Li et al. 1981). The other type of compound is mainly phenolic acids including rosmarinic acid, salvianolic acid, lithospermic acid, etc. (Fig. 2). Phenolic acid compounds are thought to be effective against hepatic fibrosis and other forms of hepatic injuries (Liu et al. 1993).

The high-yield production of tanshinone in untransformed root cultures has been reported, but the period of cultivation was relatively long (Shimomura et al. 1991). Studies on cell culture in *S. miltiorrhiza* have also been carried out; however, the main product was ferruginol, while the production of tanshinones was not stable and decreased gradually in successive subcultures (Miyasaki et al. 1986a,b).

[1] Laboratory of Biotechnology of Chinese Materia Medica, Shanghai University of Traditional Chinese Medicine, Shanghai 200032, China
[2] Institut für Entwicklungs-und Molekularbiologie der Pflanzen, Heinrich-Heine-Universität Düsseldorf, Universitätsstr. 1, D-40225 Düsseldorf, Germany

Biotechnology in Agriculture and Forestry, Vol. 45
Transgenic Medicinal Plants (ed. by Y.P.S. Bajaj)
© Springer-Verlag Berlin Heidelberg 1999

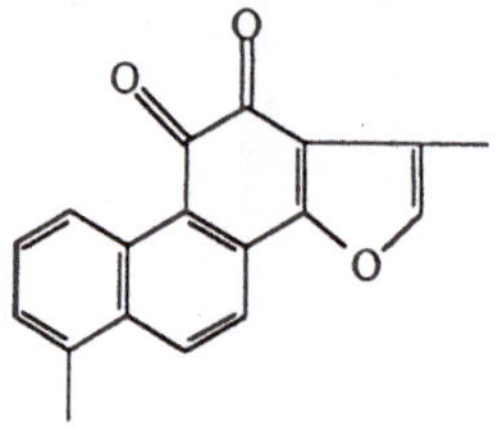

Tanshinone I

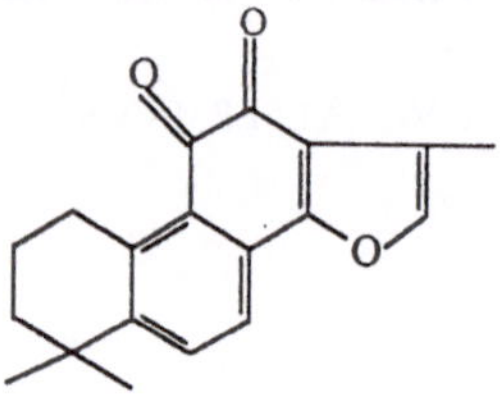

Tanshinone IIA

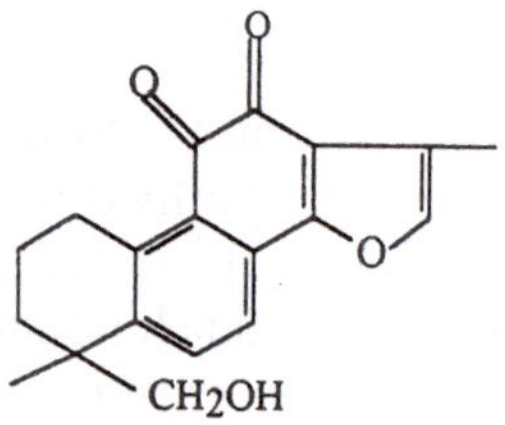

Tanshinone IIB

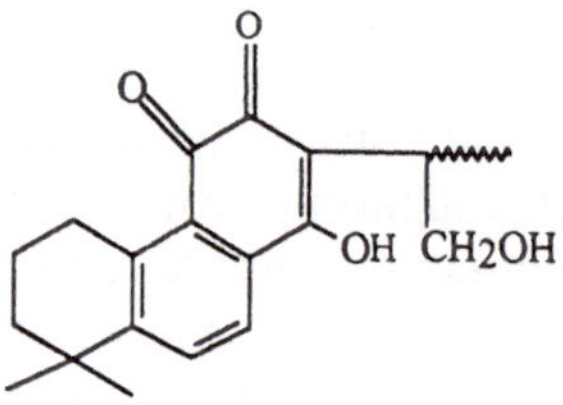

Tanshinone V

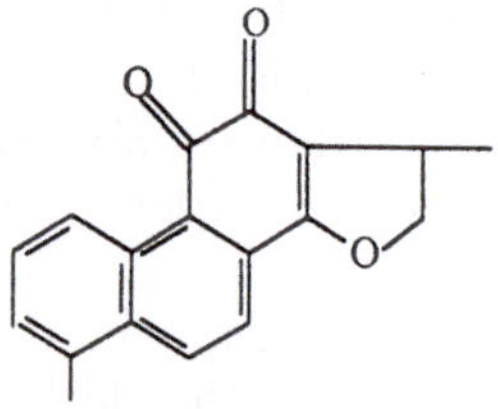

Dihydrotanshinone I

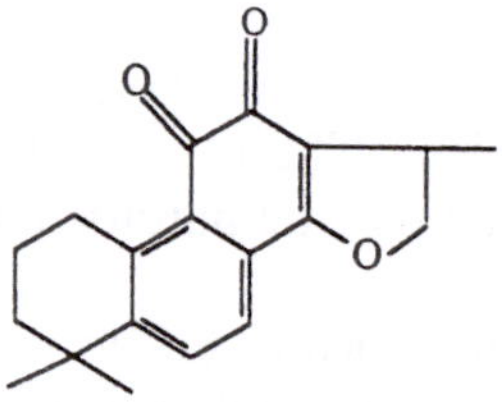

Cryptotanshinone

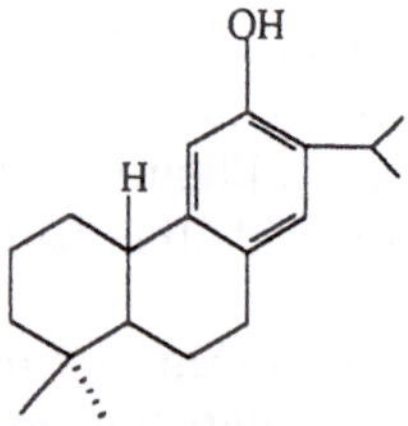

Ferruginol

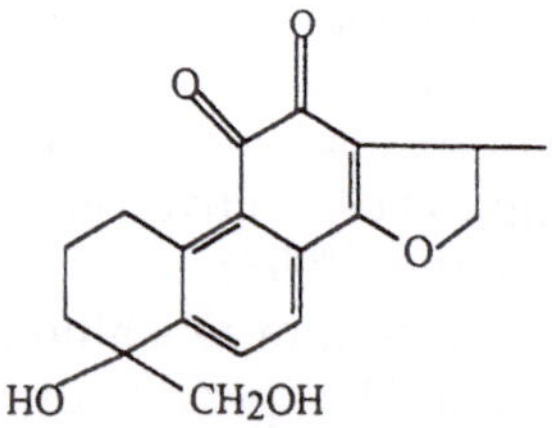

Tanshindiol A

Fig. 1. Structure of some diterpenoids in *S. miltiorrhiza*

2 Genetic Transformation

Transformation by *Agrobacteria* in *Salvia* species has been reported in various species and the results are summarized in Table 1.

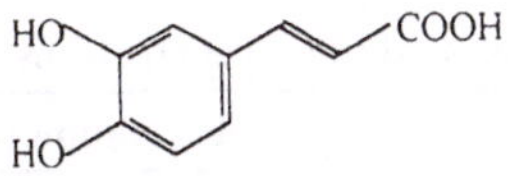

Caffeic acid 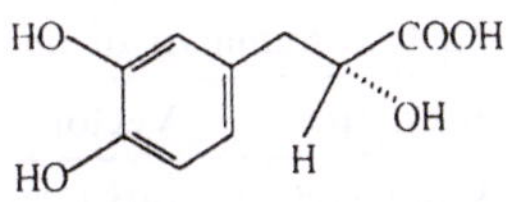   Danshensu

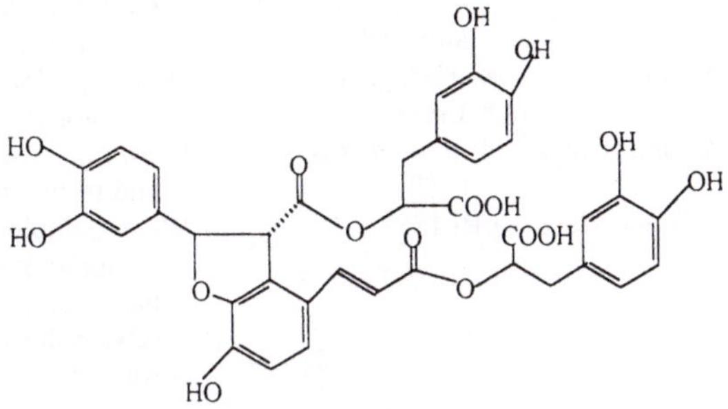

Salvianolic acid B

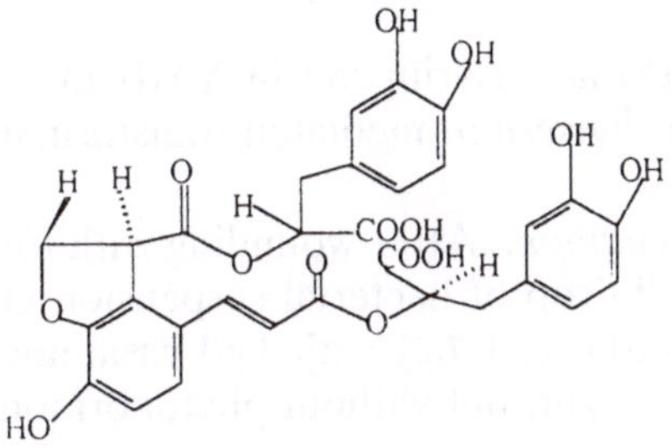

Lithospermic acid B

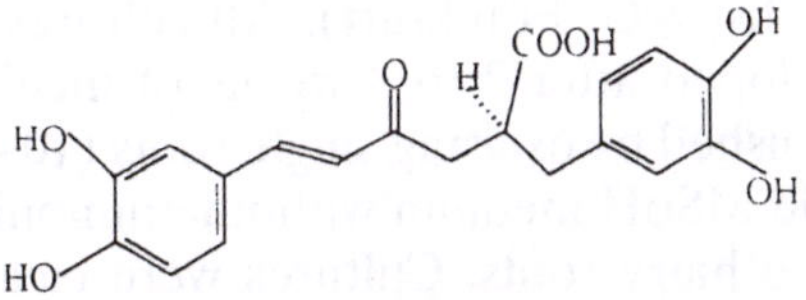

Rosmarinic acid

Fig. 2. Structure of some phenolic acid compounds in *S. miltiorrhiza*

2.1 Protocol

Plant Material. Leaf segments (about $10 \times 10\,mm$ in size) from sterile-grown plants, regenerated from untransformed callus of *S. miltiorrhiza*, were used as explants.

Bacteria. The bacterial strain used for obtaining the transformed roots was the virulent *A. rhizogenes* LBA 9402, the optimal strain for inducing the hairy roots in *S. miltiorrhiza*. The strain was incubated overnight in the presence of

Table 1. Summary of various studies on genetic transformation in *Salvia* species

Salvia sp.	Vector	Observation/remarks	Reference
S. miltiorrhiza	pRi 1855	Hairy roots were formed and diterpenoid was produced	Hu and Alfermann (1993)
S. chinensis	Binary vector CPAL 4404 and pBI 121 and C58 C1	Transformed shoots and teratoma-like buds were obtained. GUS was expressed	Xu and Yang (1993)
S. miltiorrhiza	*A. rhizogenes* 15834	Hairy roots were formed and tashinones produced	Zhang et al. (1995a)
S. miltiorrhiza	*A. tumefaciens* C 58	Crown gall tissue was formed and tashinone produced	Zhang et al. (1995b)
S. miltiorrhiza	pRi 1855	Water-soluble phenolic acid compounds were formed in hairy roots, among them, salvianolic acid A content was 2.17 times higher than that in dry roots	Huang et al. (1997)

$20\,\mu M$ acetosyringone in YMB medium (Hooykaas et al. 1977) to increase *Agrobacterium*-mediated transformation frequencies.

Inoculation. After wounding with forceps, leaf segments were infected with a small drop of bacterial suspension and cultured on MS medium (Murashige and Skoog 1962) with 1 g/l casamino acids, 2% sucrose, and solidified with 0.8% agar, but without phytohormones (MSoH).

Elimination of Bacteria. After 3 days, the explants were transferred to MSoH solid medium supplemented with 500 mg/l sodium Cefotaxin (Claforan, Hoechst AG, Frankfurt). All cultures were kept at 25 °C in the dark. Roots developed after 2 to 3 weeks of inculcation. Root suspension cultures were estabished by excising single roots (10–15 mm in length) and placing them into liquid MSoH medium without ammonium nitrate, which inhibited the growth of the hairy roots. Cultures were cleared of bacteria by several passages in medium containing 500 mg/l Claforan.

Hairy Root Culture. Root tips of 1–2 cm length and 0.2 g fr wt (about 10 mg dry wt) were inoculated in 25 ml medium in 100-ml flasks and cultivated on a gyratory shaker at 120 rpm in the dark at 25 °C for subculture at 2-week intervals.

Opine Assay. Opines in the hairy roots were extracted according to the procedure of Shaw et al. (1988). TLC was performed according to the protocol of Dräger and Schaal (1992), opines were detected with the reagent of Trevelyan et al. (1950). Agropine and mannopine were used as standards.

Product Analysis. The lyophilized roots were extracted according to the method of Okamura et al. (1991) for determining liposoluble compound contents. The quantitative analyses of tanshinones were performed with HPLC. A Nucleosil $100 C_{18}$ (5-μm porticle size) column with a ware length of 290 mm and an inner diameter of 4.6 mm was used. The two solvents were A: H_2O containing 5 mM KH_2Po_4 (pH 2.6) and B: acetonitrite-MeoH (2:1). The flow rate was 1.5 ml/min. The elution program was as follows: 0–12.5 min: linear gradient from 70–95% of B; 12.5–15 min: 95% B; 15–16 min: linear gradient from 95 to 70% B; 16–20 min: 70% B. The detector wavelength was 270 nm. Seven abietane-type diterpenoids and ferruginol in the hairy roots were successfully determined in one run using the HPLC system described earlier (Hu and Alfermann 1993).

The lyophilized roots were extracted according to the method of Huang et al. (1997) for determining phenolic acid compound contents. the quantitative analysis of salvianolic acid A, B, danshensu, and protocatechuic aldehyde were performed with HPLC. An ODS column (HP) with 100×4.6 mm (5 μm) was used. The two solvents were A: 0.2 M KH_2Po_4 (pH 3) and B: MeoH. The flow rate was 1.4 ml/min and linear gradient was from A:B = 100:0 to A:B = 50:50 after 50 min. The detector wavelength was 280 nm (Huang et al. 1997).

2.2 Results and Discussion

2.2.1 Response of Leaf Explants Infected with Various Bacterial Strains

Table 2 shows that different strains of *A. rhizogenes* exhibited various abilities to induce hairy roots on leaf explants of *S. miltiorrhiza*. The percentages of explants rooted were 0, 5, 10, 20, and 85% infected with bacterial strains A 41027, TR 105, R 1601, ATCC 15834, and LBA 9402, respectively. After adding acetosyringone (20 μM) to the growth medium of the bacteria, the values infected were increased, except in A 41027. The bacterial strain LBA 9402, which contains the plasmid p Ri 1855 (Mathews et al. 1990), is the best strain for inducing hairy roots in *S. miltiorrhiza*. Acetosyringone, an activator

Table 2. Response of *S. miltiorrhiza* roots produced on leaf explants with various strains of *A. rhizogenes* and the effects of acetosyringone. (Hu and Alfermann 1993)

Bacterial strain	Treatment of bacteria	No. of explants	No. of producing roots
A 41027	a	20	0 (0%)
	b	20	0 (0%)
TR 105	a	20	1 (5%)
	b	20	5 (25%)
R 1601	a	20	2 (10%)
	b	20	6 (33%)
ATCC 15834	a	20	4 (20%)
	b	20	14 (70%)
LBA 9402	a	20	17 (85%)
	b	20	20 (100%)

a, Without acetosyringone, b, activated with acetosyringone.
Time when observation recorded: 31 days after infection.

of the *vir* genes of *Agrobacterium*, could be used to increase transformation frequencies in *S. miltiorrhiza* as well, from 85 to 100%.

All the hairy roots tested were positive for agropine and mannopine, showing that the hairy roots contained the appropriate enzymes from the plasmids responsible for opine synthesis, but the roots of the parent did not synthesize opines.

2.2.2 Diterpene Production in Hairy Root Cultures

Using the HPLC method described above, the main abietane-type compounds in the hairy root cultures determined were tanshinone I (1, Rt 9.46 min), tanshinone II A (2, Rt 11.24 min), tanshinone II B (3, RT 5.34 min), tanshinone V (4, Rt 7.14 min), dihydrotanshinone (5, Rt 7.45 min), cryptotanshinone (6, Rt 9.03 min), tanshinone VI (7, Rt 5.94 min), and ferruginol (8, Rt 2.26 min). Okamura et al. (1991) used HPLC to determine several tanshinones and ferruginol; however, the determination was carried out under two different systems with normal phase HPLC procedure, and the polar tanshinone II B, V and VI could not be detected using normal phase HPLC.

2.2.3 Comparison of Diterpene Contents of Various Hairy Root Line

Suspension cultures were initiated from four different hairy root lines (S 9402, S 15834, S 1601, and S 105). These lines produced different levels of diterpenes in the roots (Fig. 3). The line S 9402 showed the highest production. An interesting result is that, in contrast to cell cultures (Miyasaki et al. 1986a,b), all these hairy root cultures produced tanshinone (**1–7**) concomitantly with ferruginol (**8**). The product levels of total tanshinone and ferruginol in different root lines varied between 0.5–1.9 and 0.02–0.07%, respectively. Cryptotanshinone, as the main compound, accounted for 40–60% of the total diterpene contents.

2.2.4 Effect of Nitrogen and Carbon Sources on Root Growth and Diterpene Production

The growth of the hairy root cultures was investigated in MS medium with or without ammonium nitrate over a time period of 21 days (Fig. 4). In both media similar growth curves were observed. After a short lag phase of 2 days, the hairy roots grew rapidly until day 16. However, the resulting root dry weight and the diterpene accumulation in the roots were higher in the absence of ammonium nitrate (ca. 220 mg dry roots in 25 ml medium and 19 mg diterpenes/g dry wt.) than in its presence (ca. 140 mg dry roots in 25 ml and 7 mg diterpenes/g dry wt.). The growth of the hairy root cultures and formation of diterpenes were impaired at high levels of nitrogen, and the roots became rather short and thick. This is similar to the results reported by Berlin et al.

Fig. 3. Diterpene production in various hairy root culture lines of *S. miltiorrhiza* in MS hormone-free medium without ammonium, detected on the 20th day

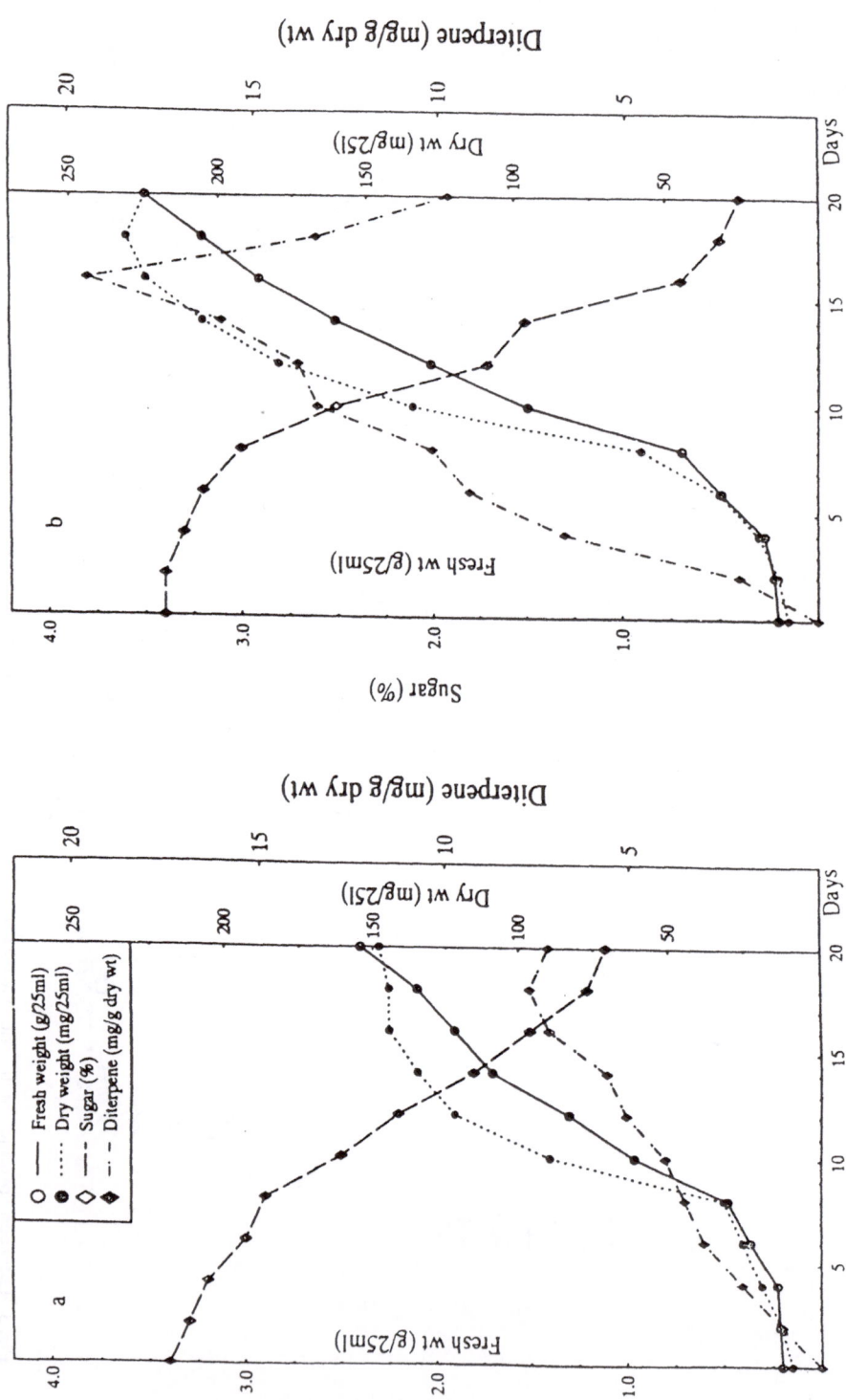

Fig. 4a,b. The time course of growth and diterpene formation by *S. miltiorrhiza* hairy root cultures (line S 9402) in the MS hormone-free medium with (**a**) or without (**b**) ammonium

(1990) for hairy roots from other plants. Furthermore, the hairy roots of *S. miltiorrhiza* released about 40% of the total diterpenes into the medium (Fig. 5).

2.2.5 Sugar Consumption in Culture Course of Hairy Roots

The measurement of refractive index indicated that the sugar in the medium was consumed until day 16. This coincided with the termination of growth. Therefore, on day 12 the medium was supplemented with sucrose up to 3%. By this treatment the final diterpene accumulation increased substantially (Fig. 5). Line S 9402 cultured in hormone-free MS medium without ammonium nitrate and with supplementation of sucrose on day 12 increased 22-fold in dry weight and produced 20 mg cryptotanshinone/g dry weight and a total tanshinone content of 43 mg/g dry weight within 20 days of cultivation. In comparison, untransformed root cultures increased 20-fold in dry weight, and produced 38 mg/g dry weight of cryptotanshinone and up to 80 mg/g dry weight of total tanshinone within 8 weeks (Okamura et al. 1991). The hairy root cultures of *S. miltiorrhiza*, therefore, could be an interesting system by which fast growth of the biomass as well as relatively high tanshinone production could be achieved, facilitating further studies on tanshinone biosynthesis.

2.2.6 Water-Soluble Phenolic Acid Compounds in Hairy Roots

Table 3 shows that salvianolic acid B content was 80% of four water-soluble phenolic acid compounds detected in the hairy roots and solvianolic acid A 0.15% dry weight, which was 2.17-fold more than that in the rough material, i.e., a Chinese herb which is not processed maintains its original constituents (Huang et al. 1997).

3 Summary and Conclusions

Transformed root cultures of *Salvia miltiorrhiza* have been established by infecting pieces of sterile-growing plants with *A. rhizogenes* strains LBA 9402. A promoting effect of acetosyringone on the *Agrobacterium*-mediated hairy root initiation was also observed. All transformed root cultures were assayed for opines by TLC, and were positive for agropine and mannopine. The quantitative HPLC procedures were developed for determining liposoluble and water-soluble phenolic acid compounds. Seven major tanshinones and ferruginol were found not only in the roots, but also in the liquid medium. The effects of culture conditions on the hairy root growth and the diterpene production have also been studied. The most apparent effect was seen when the ammonium nitrate level of the medium was modified.

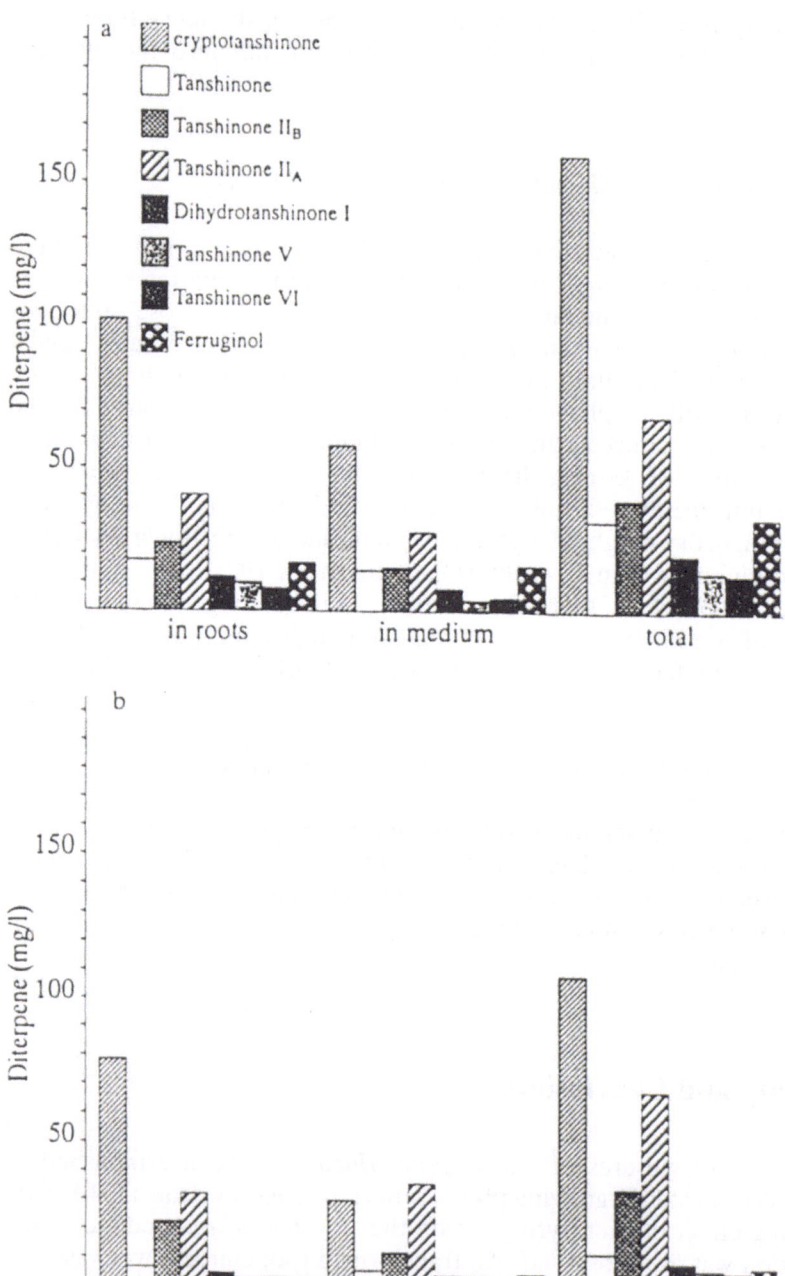

Fig. 5. Diterpene content of *S. miltiorrhiza* hairy root cultures (line S 9402) in MS hormone-free medium without ammonium nitrate with (a) or without (b) an additional sucrose supplementation (3%) on day 12. The diterpene content was measured on day 20 of cultivation

Table 3. Comparison of water-soluble phenolic acid compound contents in hairy root cultures of *S. miltiorrhiza*. (Huang et al. 1997)

Compound	Content (% dry wt.)	Relative content[a] (% dry wt.)
Danshensu	0	0
Protocatechuic aldehyde	0.5	4.85
Salvianolic acid A	0.15	14.56
Salvianolic acid B	0.83	80.58

[a] % of each compound in detected four phenolic compounds.

References

Berlin J, Mollenschott C, Greidziak N, Erdogan S, Kuzovkina I (1990) Affecting secondary product formation in suspension and hairy root cultures – a comparison. In: Nijkamp HJJ, Van der Plas LHW, van Aartrijk J (eds) Progress in plant cellular and molcular biology. Kluwer, Dordrecht, pp 763–768

Dräger B, Schaal A (1992) Pseudotropine formation and calystegins in *Atropa belladonna* root cultures. Int Symp Phytochem and Agricult, Wageningen, April 22–24, 1992

Guo JX (ed) (1992) Research and clinical application of Tan-shen. China Medicine and Pharmacy Science Press, Beijing

Hooykaas PJJ, Klapwijk PM, Nuti MP, Schelperoort RA, Rosch A (1977) Transfer of the *Agrobacterium tumefaciens* Ti plasmid to avirulent agribacteria and to *Rhizobium* ex planta. J Gen Microbiol 98:477–484

Hu ZB, Alfermann AW (1993) Diterpenoid production in hairy root cultures of *Salvia miltiorrhiza* Phytochemistry 32:699–703

Huang XL, Yang BJ, Haung HZ, Xu Y, Hu ZB (1980) Studies on the active principles of Dan-shen III. Searching for plant resources containing tanshinone II-A. Acta Bot Sin 22: 98–100

Huang XL, Yang BJ, Hu ZB (1981) Diterpene quinone of *Salvia* Linn. and their taxonomic signification. Acta Phytotaxon Sin 19:421–433

Huang LD, Hu ZB, Liu D (1997) Production of water-soluble phenolic acid compounds in hairy root of *Salvia miltiorrhiza*. Plant Physiol Commun 33:259–261

Li Y, Gu ML, Guo JX (1981) Studies on chinese drugs for coronary heart disease III. The tissue cultures of *salvia miltiorrhiza* and thin layer chromatography of its callus. Acta Acad Med Prim Shanghai 8:191–196

Liu P, Mizoguzhi Y, Morizawa S (1993) Anti liver fibrosis action of magnesium-salvianolic acid B on Ccl₄-induced rats. Traditional Chinese Med Acta 8 (Suppl):65–67

Mathews H, Bharatha N, Litz RE, Narayanan KR, Rao PS, Bhatia CR (1990) The promotion of *Agrobacterium* mediated transformation in *Atropa belladonna* L. by acetosyringone. J Plant Physiol 136:404–409

Miyasaka H, Nasu M, Yamamoto T, Endo Y, Yoneda K (1986a) Regulation of ferruginol and cryptotanshinone biosynthesis in cell suspension cultures of *Salvia miltiorrhiza*. Phytochemistry 25:637–640

Miyasaka H, Nasu M, Yamamoto T, Endo Y, Yoneda K (1986b) Production of cryptotanshinone and ferruginol by immobilized cultured cells of *Salvia miltiorrhiza*. Phytochemistry 25: 1621–1624

Murashige T, Skoog F (1962) A revised medium for rapid growth and bio-assays with tobacco tissue cultures. Physiol Plant 15:473–497

Okamura N, Kobayashi K, Yagi A, Kitazawa T, Shimomura K (1991) High-performance liquid chromatography of abietane-type compounds. J Chromatogr 542:317–326

Ooms G, Karp A, Burrell MM, Twell D, Roberts J (1985) Genetic modification of potato development using Ri T-DNA. Theor Appl Genet 70:440–446

Shaw ML, Conner AJ, Lancaster JE, Williams MK (1988) Quantitations of nopaline and octopine in plant tissue using Sakaguchi's reagent. Plant Mol Biol Rep 6:155–164

Shimomura K, Kitazawa T, Okamura N, Yagi A (1991) Tanshinone production in adventitious roots and regenerates of *Salvia miltiorrhiza*. J Nat Prod 54:1583–1587

Trevelyan WE, Procter DP, Harrison JS (1950) Detection of sugar on paper chromatograms. Nature 166:444–445

Xu ZH, Yang LJ (1993) Transformation in *Salvia chinensis*. In: Bajaj YPS (ed) Biotechnology in agriculture and forestry, vol 22. Plant protoplasts and genetic Engineering III. Springer, Berlin Heidelberg New York, pp 308–313

Zhang YL, Song JK, Lu KL, Liu HL (1995a) Establishment of hairy root cultures and production of tanshinones in *Salvia miltiorrhiza*. Chin Meteria Med Acta 20:269–271

Zhang YL, Song JY, Zhao BH, Liu HL (1995b) Crown gall tissue culture and production of tanshinone in *Salvia miltiorrhiza*. Chin J Biotechnol 11:150–152

XIX Genetic Transformation of *Scoparia dulcis* L.

T. Hayashi

1 Introduction

Scoparia dulcis L. (Scrophuraliaceae) is a medicinal plant widely distributed in the tropical and subtropical regions. The fresh or dried plant has been used as a crude folk drug for stomach disorders, bronchitis, diabetes, hypertension, hemorroids, and hepatosis, and as an analgesic and antipyretic (Satyanarayana 1969; Chow et al. 1974; Chiang Su New Medical College 1977; Perry 1980; Gonzales Torres 1986; Farias Freie et al. 1993).

From the roots of *S. dulcis*, 6-methoxybenzoxazolinone (MBOA) was isolated and found to exert hypotensive activity (Chen and Chen 1976). MBOA was later reported to have an anti-inflammatory effect (Otsuka et al. 1988). This compound has been shown to inhibit seed germination and root growth in the wild oat, *Avena fatua* (Pérez 1990), as well as auxin-inducing bending in the *Avena* curvature test and auxin-inducing elongation in the *Avena* coleoptile section test (Hasegawa et al. 1992). On analyses of the distribution of MBOA within the plant *S. dulcis*, it was detected not only in roots but also in leaves, stems, and fruit (T. Hayashi et al. 1994).

Our studies on biologically active substances from a Paraguayan collection of this plant led to the isolation of a tetracyclic diterpenoid named scopadulcic acid B (SDB) together with a labdane-type diterpenoid, scoparic acid A (SA) (T. Hayashi et al. 1987; Kawasaki et al. 1987). SDB showed many-sided biological activities such as inhibitory effects on replication of Herpes simplex virus type 1 (HSV-1) (K. Hayashi et al. 1988), gastric H^+,K^+-ATPase (Asano et al. 1990), and bone resorption stimulated by parathyroid hormone (Miyahara et al. 1996), as well as antitumor and antitumor-promoting activities (K. Hayashi et al. 1992; Nishino et al. 1993). On the other hand, SA was found to be a potent inhibitor of β-glucuronidase (T. Hayashi et al. 1992).

HPLC analysis of diterpenoids in individual plants of this collection revealed the presence of two chemotypes as the major components of diterpenoids, i.e., the SA and SDB types (T. Hayashi et al. 1991a). However, a chemotype (SDX type) different from that of the Paraguayan collection was

Faculty of Pharmaceutical Sciences, Toyama Medical and Pharmaceutical University, 2630 Sugitani, Toyama 930-0194, Japan

Biotechnology in Agriculture and Forestry, Vol. 45
Transgenic Medicinal Plants (ed. by Y.P.S. Bajaj)
© Springer-Verlag Berlin Heidelberg 1999

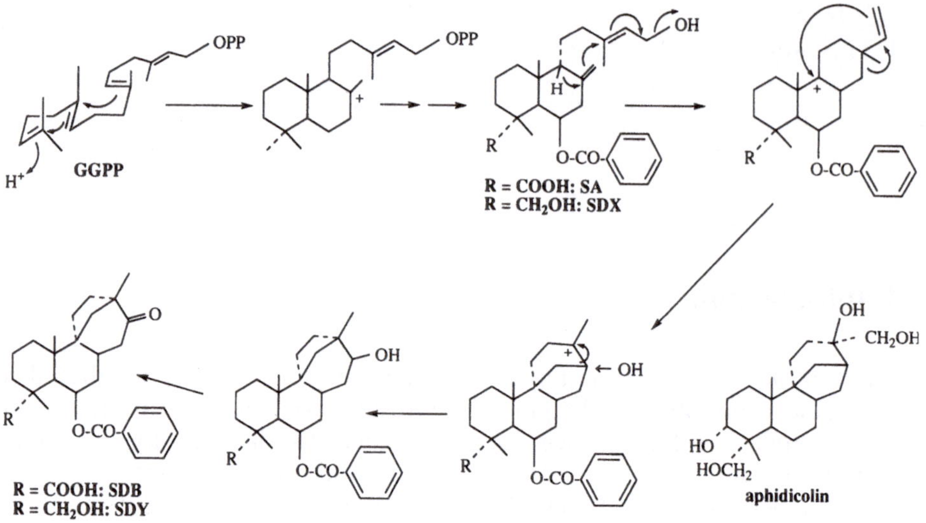

Fig. 1. Proposed biosynthetic pathway of scopadulcic acid B (SDB) and scopadulciol (SDY) in *Scoparia dulcis*. (Hayashi et al. 1991a)

observed in Asian collections which was characterized by the presence of scopadiol (SDX) and scopadulciol (SDY) (T. Hayashi et al. 1993). SDY exhibited inhibitory effects on gastric H^+,K^+-ATPase (T. Hayashi et al. 1991b) and bone resorption (Miyahara et al. 1996). In addition, it was found to be an inhibitor of HSV-1 and a potentiator of acyclovir (K. Hayashi and T. Hayashi 1996). When the distribution of these biologically active diterpenoids within the plant was examined by HPLC, leaves were found to contain the largest amount and no diterpenoid was detected in roots and seeds (T. Hayashi et al. 1991a).

A unique carbon skeleton of SDB and SDY was assumed to be biosynthesized through labdan via pimaran and aphidicolan from geranylgeranyldiphosphate (GGPP) because of their similarity to aphidicolin, a secondary metabolite of *Cephalosporium aphidicola* (Fig. 1). Aphidicolin has been proved to be biosynthesized through the mevalonate pathway (Ackland and Hanson 1984). Recently, Rohmer et al. (1993) proved the involvement of a nonmevalonate pathway in terpenoid biosynthesis. In higher plants, mono- and diterpenoids were proposed to be biosynthesized in chloroplasts through a nonmevalonate pathway from glucose (Romer et al. 1996). Therefore, it is interesting to examine the involvement of the non-mevalonate pathway in the biosynthesis of SDB and SDY. For further study to clarify the molecular basis for characterizing the regulatory mechanism of biosynthesis of these diterpenoids in each biogenetic chemotype, it is necessary to establish a method for the genetic transformation of this plant.

2 Genetic Transformation

By using *Agrobacterium rhizogenes* as a vector, a number of studies on the production of biologically active principles of plants have been performed. In addition, various efforts have been made to confer some agriculturally important traits, such as resistance to herbicide, viruses, and insects, to plants by using transgenic technology. *A. rhizogenes* causes the formation of transformed hairy roots by introducing T-DNA of the Ri plasmid into genomic DNA of the host plant cells. Hairy roots transformed with Ri plasmid have the advantage of increased growth rate without addition of plant hormones.

2.1 Establishment of Hairy Root Culture

Seeds of *S. dulcis* were sterilized and germinated aseptically on 1/2 MS (Murashige and Skoog 1962) solid medium. The hairy roots were induced in seedlings directly infected with *A. rhizogenes* (ATCC 15834). The roots appearing at the infected sites were excised and placed onto 1/2 MS solid medium containing 100 μg/ml antibiotic Claforan in the dark at 25 °C. After elimination of the bacteria by repeated subculture in the same medium, the hairy roots were transferred to 1/2 MS liquid medium and grown in the dark in a rotary shaker at 100 rpm (Fig. 2). The hairy roots thus obtained were found to contain opines such as mannopine and agropine characteristic of the Ri plasmid (ATCC 15834), indicating the transformation of T-DNA, bacterial plasmid DNA, to the plant nuclear genome (T. Hayashi et al. 1994).

2.2 Regeneration from Hairy Roots

Regeneration of the plants was observed when the hairy roots were incubated under continuous light irradiation in 1/2 MS solid medium (Fig. 2). To confirm the transformation, the total DNA of the regenerated plants was extracted by the CTAB method (Rogers and Bendich 1985). For PCR analysis of the extracted DNA, four oligonucleotide primers listed in Table 1 were selected on the basis of nucleotide sequences of *rol* B genes in Ri plasmids. The amplified bands were detected by the agarose electrophoretic method reported by Saiki et al. (1988). As shown in Fig. 3, the amplified band of *rol* B gene was detected in the regenerated plant as well as in *A. rhizogenes*, while no corresponding band was observed in the control plant. Furthermore, the DNA obtained from the amplified band of the regenerated plants was found to contain a nucleotide sequence completely consistent with that of *rol* B gene of Ri plasmid.

Fig. 2. A Hairy roots induced from seedling of *Scoparia dulcis* by inoculation of *Agrobacterium rhizogenes*. **B** Hairy roots incubated on 1/2 MS solid medium containing the antibiotic Claforan in the dark. **C** Hairy roots incubated in 1/2 MS liquid medium in the dark in a rotary shaker at 100 rpm. **D** Regenerated plants obtained by incubation of the hairy roots on 1/2 MS medium in the light. (Hayashi et al. 1994)

Table 1. Oligonucleotide primers used for PCR analysis

Name	Sequences (5' → 3')	Length (mer)
1861S	ATGGATCCCAAATTGCTATTCCTTCCA	27
2640A	TTAGGCTTCTTTCTTCAGGTTTA	23
TL-B	TGGCAGGATATATTGTGATGT	21
ORF 1	GATGACATCGCAGTCGATGA	20

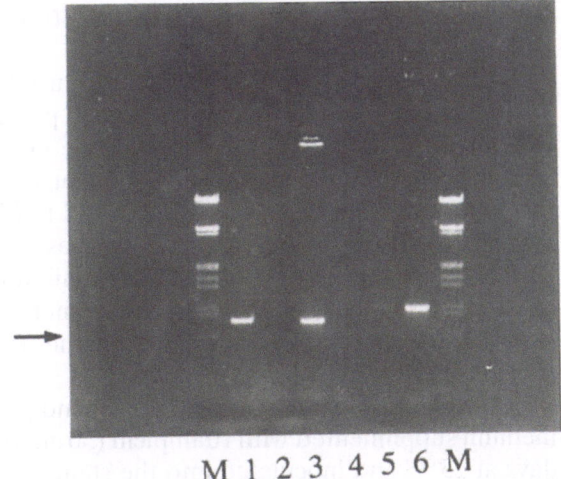

M 1 2 3 4 5 6 M

Fig. 3. Electrophoretic patterns of PCR products. *M* Molecular weight marker of λ DNA digested with Hind III and Eco RI; *lanes 1* and *4* hairy roots; *lanes 2* and *5* control; *lanes 3* and *6* ATCC 15834; *Primer: lanes 1–3* 1861S and 2640A; *lanes 4–6* TL-B and ORF 1. *Arrow* indicates the amplified bands of the *rol* B gene

Table 2. Content of MBOA in roots of parent plants, hairy roots, and roots of regenerated plants from hairy roots

Sample	MBOA (mg/g dry wt.)
Roots of parent plants	1.4 ± 0.3
Hairy roots	0.9 ± 0.1
Roots of regenerated plants	1.2 ± 0.1

Values are expressed as means ± SE (n = 3–6).

2.3 Production of MBOA and Diterpenoids by Hairy Roots

Generally, hairy roots obtained by transformation with Ri plasmids produce the same metabolites as those occurring in the roots of the original plants. In the hairy roots of *S. dulcis*, MBOA, but not diterpenoids, was detected. The content level of MBOA in the hairy roots was 64% that in the roots of parent plants, while the roots of the regenerated plants increased the content level of MBOA to 85% of the parent plants (Table 2; T. Hayashi et al. 1994). HPLC analysis of the leaves of the regenerated plants showed the same diterpene patterns as those of the parent plants, suggesting that the transformation with Ri plasmid does not affect genetic determination of chemotypes of *S. dulcis*.

2.4 Production of Transgenic Herbicide-Resistant Plants

A herbicide-resistant trait has been introduced into *S. dulcis* by using a binary vector system based on *Agrobacterium*-Ri plasmids (Yamazaki et al. 1996). In the binary vector, pARK5, the *bar* gene encoding phosphinothricin (PPT) acetyltransferase (PAT) was placed under the transcriptional control of the promoter for cauliflower mosaic virus 35S RNA, and flanked by the terminator of the *nos* gene for nopaline synthase. PAT inactivates the synthetic herbicide, PPT (Basta, Hoechst), an inhibitor of glutamine synthase, thus causing the death of plant cells. The chimeric neomycin phosphotransferase II (NPT-II) gene was also present in the T-DNA as a reporter gene for transformation.

Agrobacterium harboring pRi15834 and pARK5 was precultured in YEB medium supplimented with rifampicin (50 mg/l) and kanamycin (25 mg/l) for 2 days at 25 °C and inoculated into the stems of 3-week-old sterile *S. dulcis* by scratching with a needle. Within 2 weeks after inoculation, hairy roots appeared at the infected sites on the stems. The hairy roots were excised and transferred to B5 medium containing antibiotics (Claforan, 500 mg/l) and herbicide (Bialaphos, 5 mg/l). Hairy roots resistant to Bialaphos were selected and further cultured in the same medium for 1 month, resulting in regeneration of adventitious shoots. The regenerated shoots were transferred to A1 medium (1/2 MS medium containing 1% sucrose and 0.8% agar) for rooting. The rooted plantlets were grown for 3 weeks on vermiculite supplemented with 10% (v/v) A1 medium without agar and then to culture soil. Finally, four clones of transformed plants, clones B, C, E, and G2, were established. The integration of the T-DNAs of pRi15834 and pARK5 was analyzed by DNA-blot hybridization, indicating that all clones except for clone E contained the *bar* gene. The expression of the chimeric *bar* gene was confirmed by PAT assays. Positive activity was shown by clones C and G2, but not by clones B and E. It was speculated that the fragment copy of the *bar* gene was not functional and the integrated *bar* gene might have been rearranged in clone B during regeneration after selection with the herbicide. Clone E was presumed to have escaped Bialaphos selection. When an aqueous solution of herbicide was applied to the leaves of a regenerated plant of clone C and a control plant, the control leaves died 7 days after application, while the leaves of clone C showed resistance to the herbicide (Fig. 4).

To confirm inheritance of the transgenic trait by progenies, seeds of the offspring and control untransformed plants were germinated on A1 agar medium containing Bialaphos. Within 2 weeks, all the control seeds died, but 48 out of 67 seedlings of the S1 progenies could develop cotyledons. In addition, all the herbicide-resistant seedlings showed PAT activity and the same hybridization pattern with the *bar* probe as that of their parent plant. The segregation ratio of resistant and sensitive progenies was statistically significant being $3:1$ ($\chi^2 = 0.4$, $p > 0.05$). These results indicated that integrated *bar* genes in the parent clone C were linked and transmitted to the offspring as an apparent single dominant allele according to the Mendelian rule. DNA blot hybrydization of the progenies indicated that 10 out of 19 Bialaphos-resistant

Fig. 4. Resistance of transgenic *S. dulcis* expressing the *bar* gene to herbicide (Bialaphos). **a** Control plant derived from a hairy root transformed with pRi15834 only. The leaf treated with Herbiace (Bialaphos sodium salt content, 20%) died completely 7 days after application (indicated by a *red tag*). **b** Clone C 7 days after application. No visible change was observed in the leaf treated with Herbiace (indicated by a *red tag*). (Yamazaki et al. 1996)

Table 3. Content of scopadulcic acid B (SDB) in leaves of regenerated plants and S1 progenies of clone C

Clone	SDB content (% dry wt.)
B	0.44
C	0.44
E	0.32
G2	0.11
HR[a]	0.45
UT[b]	0.74
S1	0.52

[a] Transformed with only pRi 15834.
[b] Untransformed control. The diterpenoid fraction was extracted from freeze-dried leaves of each regenerated plant. SDB was determined by HPLC using p-toluene-sulfonic acid as an internal standard. Data are the means of duplicate determinations.

progenies (53%) and 16 out of 19 resistant progenies (84%) contained TL- and TR-DNA, respectively (unpubl. data). The clones carrying TL-DNA showed the phenotypes typical of Ri syndrome such as shorter internodes and less apical dominance compared with normal plants. In contrast, three clones carrying no TL- and TR-DNAs showed no such Ri syndrome. Thus, S1 progenies carrying only *bar* but not TL- and TR-DNAs causing Ri syndrome were obtained.

2.5 Production of Scopadulcic Acid B (SDB) by the Transformants

The content of SDB in leaves of transgenic clones and offspring of the clone C was analyzed by HPLC. As illustrated in Table 3, the leaves of transgenic clones accumulated a lesser amount (15–60%) of SDB compared with that in untransformed plants (Yamazaki et al. 1996). Particularly, clone G2, which showed marked phenotypic changes of the Ri syndrome, accumulated the lowest amount of SDB. On the other hand, the leaves of S1 progenies of the clone C produced SDB in almost the same amount as those of the parent plants (unpubl. data). These experimental results suggest that some physiological change caused by expression of the genes in the Ri plasmid might affect the production of diterpenoids.

3 Summary and Conclusions

Hairy root culture of *S. dulcis* was established by introducing T-DNA of the Ri plasmid. The genetic transformation of the hairy roots was confirmed by detection of mannopine and agropine. In addition, analyses of the agarose electrophoretic pattern and nucleotide sequence of the amplified DNA by PCR revealed DNA of the regenerated plants from the hairy roots to contain *rol* B gene of Ri plasmid. Hairy roots thus obtained produced MBOA but not diterpenoids, while the regenerated plants produced both MBOA and diterpenoids.

Transgenic herbicide-resistant *S. dulcis* plants were obtained by using *Agrobacterium* harboring pRi 15834 and pARK5. The transgenic state of the regenerated plants and progenies was confirmed by DNA-blot hybridization and assaying of PAT activity. The production of diterpenoid by transformed plants was less than that of untransformed plants. However, the capacity of the progenies for production of diterpenoid was almost the same as that of the parent plant.

So far, the biosynthetic mechanism of diterpenoids produced by *S. dulcis* has not been well clarified. Therefore, identification of the DNA sequences responsible for genetically determined chemotypes such as SA, SDB, and SDX types of *S. dulcis* might be useful to reveal the biosynthetic pathway of SDB and SDY. The protocol of genetic transformation described above is applicable to this study.

References

Ackland M, Hanson JR (1984) Studies in terpenoid biosynthesis, Part 30. The acetate and mevalonate labelling patterns of the diterpenoid, aphidicolin. J Chem Soc Perkin Trans I:2751–2754

Asano S, Mizutani M, Hayashi T, Morita N, Takeguchi N (1990) Reversible inhibitions of gastric H$^+$,K$^+$-ATPase by scopadulcic acid B and acetyl scopadol, New biological tools of H$^+$,K$^+$-ATPase. J Biol Chem 265:22167–22173

Chen C, Chen M (1976) 6-Methoxybenzoxazolinone and triterpenoids from roots of *Scoparia dulcis*. Phytochemistry 15:1997–1999

Chiang Su New Medical College (1977) Dictionary of Chinese Materia Medica (Zhong Yao Da Ci Dian), Jiangsu New Medical College, Shanghai Scientific and Technological Publisher, Shanghai, 2132 pp

Chow SY, Chen SM, Yang CM, Hsu H (1974) Pharmacological studies on Chinese herb (1) Hypotensive effect of 30 Chinese herbs. J Formosan Med Assoc 73:729–739

Farias Freie SM, Silva Emin JA, Lapa AJ, Souccar C, Brandao Torres LM (1993) Analgesic and anti-inflammatory properties of *Scoparia dulcis* L. extract and glutinol in rodents. Phytother Res 7:408–414

Gonzales Torres DM (1986) Catalogo de plantas medicinales (y Alimenticias y Otiles) Osada en Paraguay. Asuncion, Paraguay, 394 pp

Hasegawa K, Togo S, Urashirna M, Mizutani J, Kosemura S, Yamamura S (1992) An auxin-inhibiting substance from light-grown maize shoots. Phytochemistry 31:3673–3676

Hayashi K, Hayashi T (1996) Scopadulciol is an inhibitor of Herpes simplex virus type 1 and a potentiator of acyclovir. Antiviral Chem Chemother 7:79–85

Hayashi K, Hayashi T, Morita N (1992) Cytotoxic and antitumour activity of scopadulcic acid B from *Scoparia dulcis* L. Phytother Res 6:6–9

Hayashi K, Niwayama S, Hayashi T, Nago R, Ochiai H, Morita N (1988) *In vitro* and *in vivo* antiviral activity of scopadulcic acid B from *Scoparia dulcis*, Scrophulariaceae, against Herpes simplex virus type 1. Antiviral Res 9:345–354

Hayashi T, Kishi M, Kawasaki M, Shimizu M, Suzuki S, Yoshizaki M, Morita N, Tezuka Y, Kikuchi T, Berganza LH, Ferro E, Basualdo I (1987) Scopadulcic acid-A and -B, new diterpenoids with a novel skeleton, from a Paraguayan crude drug Typychá kuratû (*Scoparia dulcis* L.). Tetrahedron Lett 28:3693–3696

Hayashi T, Okamura K, Kakemi M, Asano S, Mizutani M, Takeguchi N, Kawasaki M, Tezuka Y, Kikuchi T, Morita N (1990) Scopadulcic acid B, a new tetracyclic diterpenoid from *Scoparia dulcis* L. Its structure, H$^+$,K$^+$-adenosine triphosphatase inhibitory activity and pharmacokinetic behaviour in rats. Chem Pharm Bull 38:2740–2745

Hayashi T, Okamura K, Kawasaki M, Morita N (1991a) Two chemotypes of *Scoparia duclis* in Paraguay. Phytochemistry 30:3617–3620

Hayashi T, Asano S, Mizutani M, Takeguchi N, Kojima T, Okamura K, Morita N (1991b) Scopadulciol, an inhibitor of gastric H$^+$,K$^+$-ATPase from *Scoparia dulcis* and its structure-activity relationships. J Nat Prod 54:802–809

Hayashi T, Kawasaki M, Okamura K, Tamada Y, Morita N, Tezuka Y, Kikuchi T, Miwa Y, Taga T (1992) Scoparic acid A, a β-glucuronidase inhibitor from *Scoparia dulcis*. J Nat Prod 55:1748–1755

Hayashi T, Okamura K, Tamada Y, Iida A, Fujita T, Morita N (1993) A new chemotype of *Scoparia dulcis*. Phytochemistry 32:349–352

Hayashi T, Gotoh K, Ohnishi K, Okamura K, Asamizu T (1994) Content of 6-methoxy-2-benzo-xazolinone (MBOA) in *Scoparia dulcis* and its production by cultured tissues. Phytochemistry 37:1611–1614

Kawasaki M, Hayashi T, Arisawa M, Shimizu M, Horie S, Ueno H, Syogawa H, Suzuki S, Yoshizaki M, Morita N, Tezuka Y, Kikuchi T, Berganza LH, Ferro E, Basualdo I (1987) Structure of scoparic acid A, a new labdane-type diterpenoid from Paraguayan crude drug Typychá kuratû (*Scoparia dulcis* L.). Chem Pharm Bull 35:3963–3966

Miyahara T, Hayashi T, Matsuda S, Yamada R, Ikeda K, Tonoyama H, Komiyama H, Matsumoto M, Nemoto N, Sankawa U (1996) Inhibitory effects of scopadulcic acid B and its derivative on bone resorption and osteoclast formation in vitro. Bioorg Med Chem Lett 6:1037–1042

Murashige T, Skoog F (1962) Revised medium for rapid growth and bio-assays with tabacco tissue cultures. Physiol Plant 15:473–497

Nishino H, Hayashi T, Arisawa M, Satomi Y, Iwashima A (1993) Antitumor-promoting activity of scopadulcic acid B from medicinal plant *Scoparia dulcis* L. Oncology 50:100–103

Otsuka H, Hirai Y, Nagao T, Yamazaki K (1988) Anti-inflammatory activity of benzoxazolinoids from roots of *Coix lachryma-jobi* var. Ma-yuen. J Nat Prod 51:74–79

Pérez FJ (1990) Allelopathic effect of hydroxamic acids from cereals on *Avena sativa* and *A. fatua*. Phytochemistry 29:773–776

Perry LM (1980) Medicinal plants of East and Southeast Asia: attributed properties and uses. MIT Press, Cambridge, Massachusetts, 385 pp

Rogers SO, Bendich AJ (1985) Extraction of DNA from miligram amounts of fresh, herbarium and mummifield plant tissue. Plant Mol Biol 5:69–76

Rohmer M, Knani M, Simonin P, Sutter B, Sahm H (1993) Isoprenoid biosynthesis in bacteria: a novel pathway for the early steps leading to isopentenyl diphosphate. Biochem J 295:517–524

Rohmer M, Seemann M, Horbach S, Bringer-Meyer S, Sahm H (1996) Glyceraldehyde 3-phosphate and pyruvate as precursors of isoprenic units in an alternative non-mevalonate pathway for terpenoid biosynthesis. J Am Chem Soc 118:2564–2566

Saiki RK, Gelfand DH, Stoffel S, Schalf SJ, Higuchi R, Horn GT, Mullis KB, Erlich HA (1988) Primer-directed enzymatic amplification of DNA with a thermostable DNA polymerase. Science 239:487–491

Satyanarayana K (1969) Chemical examination of *Scoparia dulcis* (Linn): part I. J Indian Chem Soc 46:765–766

Yamazaki M, Son L, Hayashi T, Morita N, Asamizu T, Murakoshi I, Saito K (1996) Transgenic fertile *Scoparia dulcis* L., a folk medicinal plant, conferred with a herbicide-resistant trait using an Ri binary vector. Plant Cell Rep 15:317–321

XX Genetic Transformation of *Scutellaria baicalensis*

M. Hirotani

1 Introduction

The distribution of the genus *Scutellaria* (family Labiatae) and the importance of *Scutellaria baicalensis* Georgi have been reviewed in detail (Yamamoto 1991). Wogon, Scutellariae Radix, is the dry root, excepting the exodermis, of *S. baicalensis*, which is collected in spring and fall, and is a very old and well-known drug in traditional Chinese medicine (Chiang Su New Medical College 1977; Tang and Eisenbrand 1992). It is officially listed in the Japanese Pharmacopeia JPXIII and Chinese Pharmacopeia, and is one of the most widely used crude drugs for the treatment of bronchitis, hepatitis, diarrhea, and tumors. It is also used frequently as an important medicine in Chinese clinical practice. Recent papers reported that flavonoids from the roots of this species have an inhibitory effect on human immunodeficiency virus (HIV-1) (Li et al. 1993), human T cell leukemia virus type I (HTLV-I) (Baylor et al. 1992), and mouse skin tumor promotion (Konoshima et al. 1992). The root of *S. baicalensis* is known to contain a number of flavone derivatives. The first flavone isolated from its root was wogonin; its structure was determined by Hattori (1930). Wogonin is present in only small amounts in the root; the flavone glycoside baicalin predominates by far (Shibata et al. 1923). Baicalin was also extracted from the roots and its acid hydrolysis yielded glucuronic acid and a flavone aglycone named baicalein. Further flavones and related compounds isolated from *S. baicalensis* are some 50 kinds of flavonoids (Tang and Eisenbrand 1992; Miyaichi and Tomimori 1994, 1995; Zhang et al. 1994; Ishimaru et al. 1995; Zhou et al. 1997).

In China, a number of *Scutellaria* species have been described as substitutes for *S. baicalensis*. They are identified as *S. amoena*, *S. hypericifolia*, *S. likiangensis*, *S. rehderiana*, *S. tenax*, and *S. viscidula* (Song 1981). The major components in all species are baicalin, baicalein, and wogonin. The constituents of the other *Scutellaria* species have been investigated, i.e., *S. rivularis* Wall (Chou 1978; Tomimori et al. 1984, 1986a, 1990), *S. discolor* Colebr (Tomimori et al. 1985a, 1986b), *S. indica* L (Chou and Lee 1986; Miyaichi et al. 1987, 1989), and *S. scandens* (Miyaichi et al. 1988a,b).

School of Pharmaceutical Sciences, Kitasato University, 9-1 Shirokane 5 Chome, Minato-ku, Tokyo, 108 Japan

Biotechnology in Agriculture and Forestry, Vol. 45
Transgenic Medicinal Plants (ed. by Y.P.S. Bajaj)
© Springer-Verlag Berlin Heidelberg 1999

2 Genetic Transformation

2.1 Methodology/Protocol

2.1.1 Establishment of Hairy Root Cultures

Plant Material. Sterile plants of *S. baicalensis* Georgi were germinated from seeds surface sterilized with 10% (w/v) chlorinated lime solution for 20 min followed by washing 3× with sterile H_2O. Subsequently, they were cultured on MS basal medium (Murashige and Skoog 1962) containing 3% sucrose and 0.9% agar at 25 °C in the dark.

Bacterial Strains. Agrobacterium rhizogenes pRi15834, *A. rhizogenes* Rif^R harboring pRi15834/pBI121 (chimeric CaMV 35S-GUS gene in pBI121 the mini-Ti plasmid containing *gus* and *neo*: Jefferson et al. 1987) and pRi15834/pGSGluc 1 (a binary vector containing chimeric *neo* and *gus* genes driven by the T-DNA TR1' and TR2' promoters, respectively: Saito et al. 1990) were used for hairy root induction by transformation. They were grown on YEB agar (Vervliet et al. 1975), YEB-B agar (YEB agar containing 50 mg/l rifampicin and 25 mg/l kanamycin) and YEB-A agar (YEB agar containing 50 mg/l rifampicin, 300 mg/l streptomycin and 100 mg/l spectinomycin), respectively, and maintained by subculturing onto the same media at 24 °C in the dark.

Induction of Hairy Roots. Hairy roots were induced by direct inoculation with three strains of *A. rhizogenes* on sterile seedlings 3 weeks after germination. The bacteria in the generated adventitious root tips were eliminated on MS agar medium containing 2.5 mg/l indolebutyric acid (IBA) and 80 mg/l claforan at first, then to 500 mg/l claforan, and to 5 mg/l IBA and 100 mg/l claforan in the third medium. Finally, adventitious roots were cultured on YEB agar medium to confirm elimination of bacteria. Three axenic adventitious root clones (S.b pBI121, S.b pG-1 and S.b pRi-1) were isolated (Fig. 1). Axenic adventitious roots were subcultured in hormone-free B5 liquid medium (Gamborg et al. 1968) every 5 weeks.

Detection of Opine. Extraction and detection of opine was carried out by the following method, described by Petit et al. (1983). Hairy roots (ca. 500 mg fr wt.) were added to 500 μl water and ground with sea sand using a glass pestle in a microtube and centrifuged at 10000 rpm for 5 min. The supernatant (40 μl) and standard samples (agropine and mannopine) were spotted on a Whatman 3 MM paper and subjected to electrophoresis (ca. 50 V/cm, 40 min) using Toyo High Voltage Electrophoresis Equipment model HPE-406. The buffer used was formic acid/acetic acid/water (30/60/910, v/v/v). The dried chromatogram was stained with an alkaline silver nitrate reagent (Trevelyan et al. 1950).

Amplification of Transformed DNA by Polymerase Chain Reaction (PCR). Hairy roots and untransformed callus tissue (fr wt 50 mg) were homogenized with 400 μl extraction buffer (20 mM Tris-HCl, pH 7.5, 250 mM

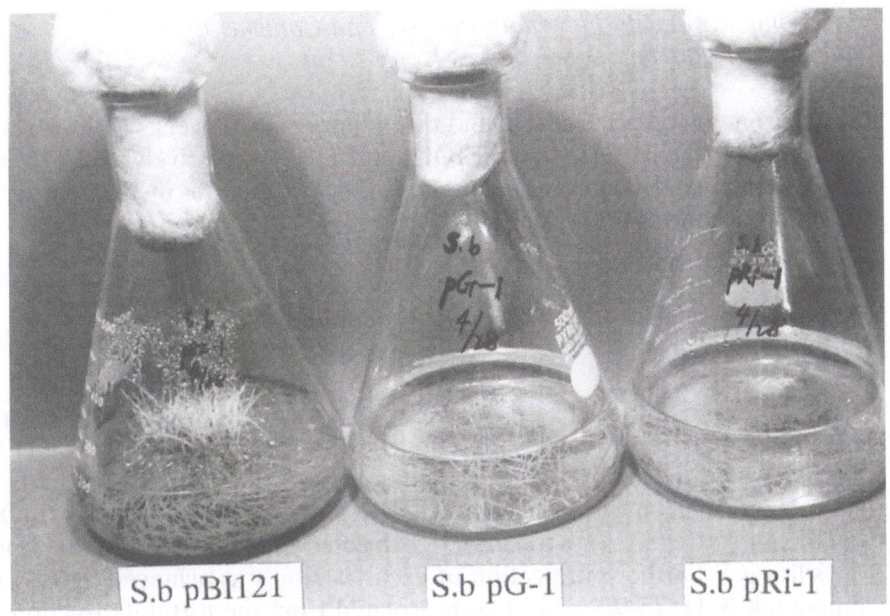

S.b pBI121   S.b pG-1   S.b pRi-1

Fig. 1. Three axenic adventitious root clones (*S.b. pBI121, S.b. pG-1, S.b pRi-1*), cultured for 5 weeks in B5 medium

NaCl, 25 mM EDTA, 0.5% SDS), and genomic DNAs were extracted with usual manner (Zhou et al. 1997). Five μl of DNA solution was used for 100 μl PCR reaction mixture.

The amplification reaction mixture contained 10 mM tris-HCl pH 8.3 at 25 °C, 50 mM KCl, 1.5 mM MgCl$_2$, 200 mM dNTPs, 20 pM oligonucleotide primers, and 2.5 U Taq DNA polymerase in 100 μl. PCR reactions were performed in a HOEI Thermocycler Model TC-100 using the following profile: 30 cycles of 94 °C for 1 min, 55 °C for 2 min, and 72 °C for 3 min, plus and extension step at 72 °C for 7 min. Amplified DNAs were analyzed by EtBr staining after 1.2% agarose gel electrophoresis. Six oligonucleotides, rol A-1, TR-1, AGS-2, rol B-2, TR-2, and AGS-1, were designed to identify the insertion of TL-DNA or TR-DNA of the Ri-plasmid into the plant genome by PCR amplification. Rol A-1 (5'-ATGGAA TTAGCCGGACTAAACG-3') or TR-1 (5'-GGCCCAAAAAAACATTC CCACC-3') and AGS-2 (5'-AATCGTTCAGAGAGCGTCCGAAGTT-3') correspond to the *N*-terminal-coding region of the TL-DNA rol A gene (Slightom et al. 1986) or the TR-DNA rol B^{TR} gene (Bouchez and Camilleri 1990) and the agropine synthase gene (Bouchez and Tourneur 1991), respectively. Also, Rol B-2 (5'-ATGGAT CCCAAATTGCTATTCC-3') or TR-2 (5'-CTAAGCGCTCGTCCGTGTCTCC-3') and AGS-1 (5'-CGGAAATTGTGGCTCGTTGTGGAC-3') correspond to the complementary sequence of the C-terminal-coding region of the TL-DNA rol B gene or the rol B^{TR} and the agropine synthase gene, respectively.

2.1.2 Comparison with Flavonoid Patterns and Contents Between Scutellariae Radix and S. baicalensis Hairy Roots

Powder from Scutellariae Radix and dried hairy root (5 weeks' culture) was refluxed with 50% EtOH, 100% EtOH, and H_2O, successively. All extracts were combined and evaporated to dryness. Each extract was dissolved in 50% EtOH and used for HPLC analysis.

2.1.3 Isolation and Identification of the Compounds from Hairy Root Cultures

After 5 weeks' culture, hairy roots were harvested and dried at 60°C for 1 week. Dried hairy roots (1 kg) were refluxed successively with EtOH and H_2O. Part of the EtOH extract (41.86 g, in total 207.6 g) was dissolved in H_2O (900 ml) and shaken successively with Et_2O and n-BuOH. The n-BuOH (6.17 g) and Et_2O (9.22 g) extracts were subjected to column chromatography over silica gel. Further purification of each fraction was achieved by HPLC. Compounds **8–11**, **15–20**, and **21** were isolated from the n-BuOH fraction of the EtOH extract. Compounds **1–6** and **7** were also isolated from the Et_2O fraction of the EtOH extract. At the same time, the H_2O extract was subjected to Diaion HP20 column chromatography, washed with H_2O, and then eluted with MeOH. The MeOH eluate gave compounds **12**, **13**, and **14**. The fraction obtained by column chromatography of n-BuOH extract was purified by HPLC. Compound **8**, pale yellow amorphous powder (5.9 mg), was analyzed for $C_{21}H_{21}O_{11}$ by FAB-MS. UV λ_{max}^{MeOH} nm (log ε): 257 (4.02), 298 (3.75); +AlCl$_3$: 267 (4.03), 312 (3.75); +AlCl$_3$/HCl: 267 (4.03), 312 (3.75); +NaOAc: 263 (4.04), 322 (3.68); +NaOAc/H_3BO_3: 257 (4.02), 298 (3.75).

HPLC Conditions. Analysis and isolation of compounds were run on an instrument fitted with an absorbance detector and a differential refractometer. The column (4.6 mm × 150 mm) packed with Hibar RP-18 (5 μ) for analysis and the column (19 mm × 150 mm) packed with μ-Bondasphere 5 μ C18-100 Å for isolation of compounds were used and the solvents used were as follows: eluent MeOH-0.06M H_3PO_4 (55:45), flow rate 0.8 ml/min for analysis, and MeOH-H_2O-HOAc solvent system, flow rate 6 ml/min for isolation of compound.

2.2 Results and Discussion

2.2.1 Establishment of Hairy Root Cultures

Three adventitious root clones of S. baicalensis were induced by direct inoculation on sterile seedlings (Fig. 1). In order to prove transformation, opine was detected in the usual way (Fig. 2) and direct confirmation of transferred DNA was tried by the PCR amplification method using three different parts in T-DNA as the primers. To determine the insertion of TL-DNA or TR-DNA of

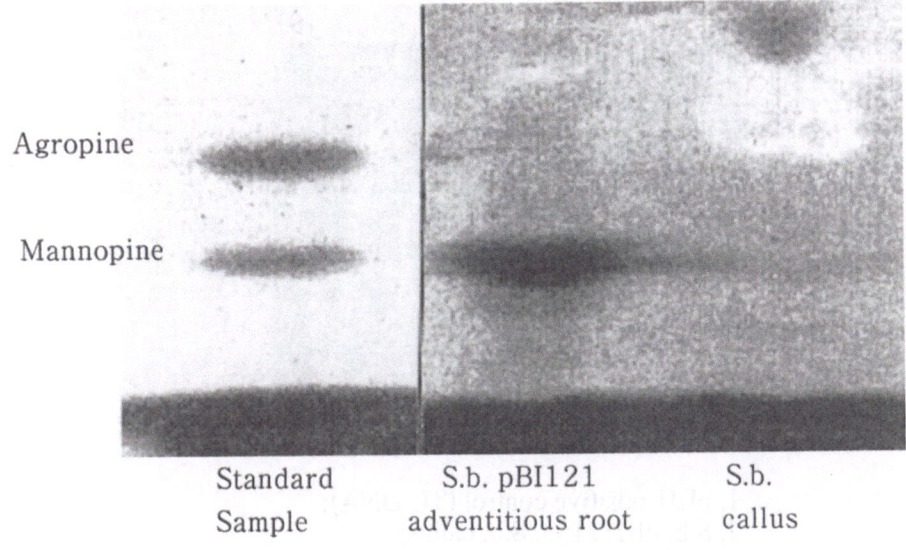

Agropine

Mannopine

| Standard | S.b. pBI121 | S.b. |
| Sample | adventitious root | callus |

Fig. 2. Detection of opine by paper electropharetic analysis

the Ri plasmid, PCR was performed using primers which can detect either TL-DNA rol A and B (Slightom et al. 1986) or the rol B^{TR} gene (Bouchez and Camilleri 1990) and the agropine synthase gene (Bouchez and Tourneur 1991). PCR amplification analysis of DNAs from transformed hairy roots gave the expected 1541-, 1672-, and 816-bp bands with the primer combination rol A-1 and rol B-2 (lane 2), AGS-2 and AGS-1 (lane 5), and TR-1 and TR-2 (lane 8) same bands (lanes 1, 4, and 7) that pLJ1 and pLJ85 yielded as positive controls (Fig. 3). PLJ1 and pLJ85 are cosmid clones containing entire TL-DNA and TR-DNA, respectively, of pRiHRI (Jouanin 1984). On the contrary, no visible band was obtained from untransformed callus (lanes 1, 3, and 6). These results show that the hairy root strain S.b. pBI121 contained both the TL-DNA and TR-DNA region of the Ri plasmid.

The most actively growing strain, S.b. pBI121, was isolated and used in the following experiments (Fig. 1). Growth of hairy roots of S.b. pBI121 was investigated in a time-course experiment by culturing in five different basal media, B5, MS, 1/2 MS, woody plant (Lloyd and McCown 1980), and NN (Nitsch and Nitsch 1969) liquid media. Growth was best in B5 liquid medium (Fig. 4). On the other hand, in MS medium, growth was inhibited, even at half-strength. Thus, B5 medium was used for following experiments.

2.2.2 HPLC Patterns and Contents of Baicalin and Baicalein in the Extracts of Scutellariae Radix and Hairy Root

To clarify the similarity of the flavonoid components between Scutellariae Radix and *S. baicalensis* hairy root, HPLC analysis was carried out. As shown

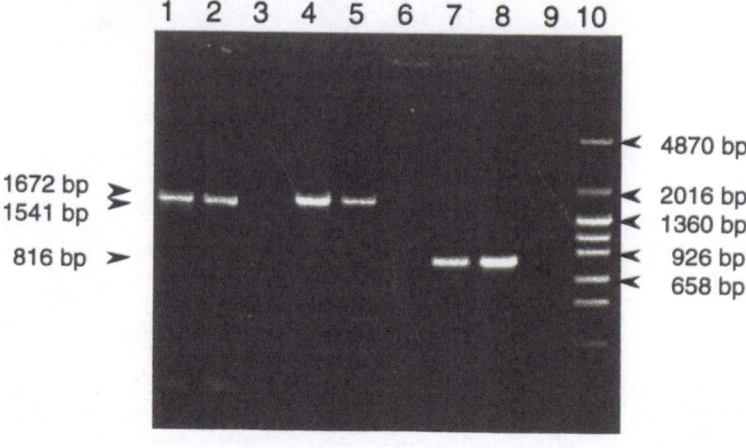

1. pLJ1 positive control (TL-DNA);
2. S.b. pBI121; 3. S.b. callus;
4. pLJ85 positive control (TR-DNA);
5. S.b. pBI121; 6. S.b. callus;
7. pLJ85 positive control (TR-DNA);
8. S.b. pBI121; 9. S.b. callus;
10. pHY marker DNA

Fig. 3. PCR amplifications from DNA of adventitious root and callus of *Scutellaria baicalensis*. (Zhou et al. 1997)

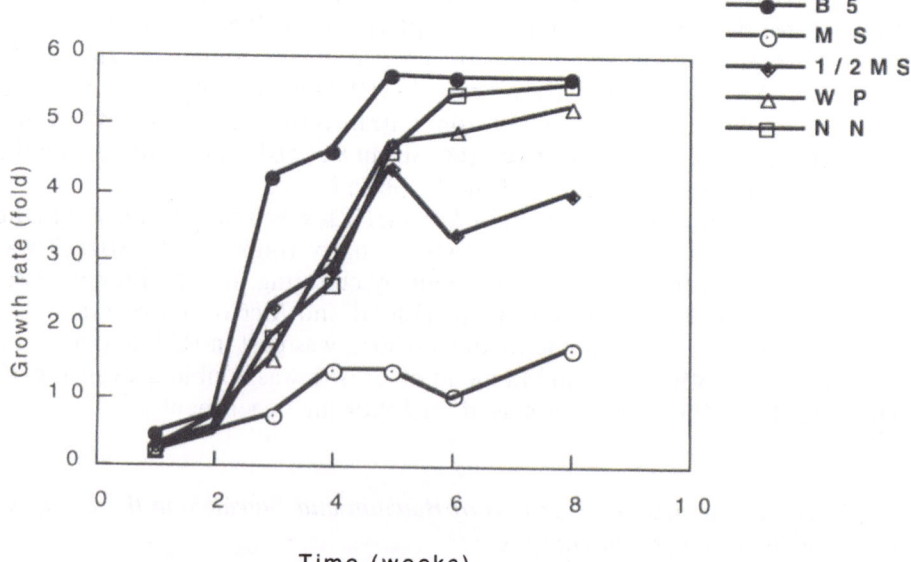

Fig. 4. Growth of *Scutellaria baicalensis* hairy roots (S.b pBI121 strain) cultured in five basal liquid media. (Zhou et al. 1997)

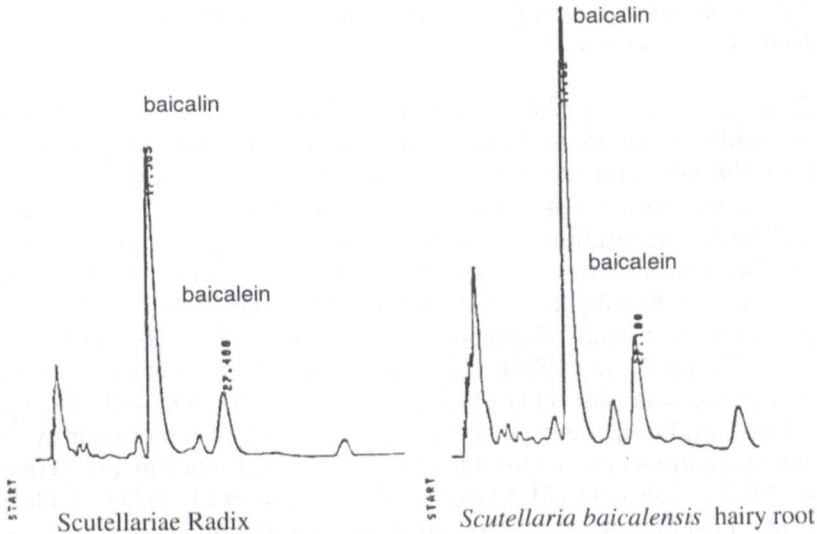

Fig. 5. HPLC patterns of the extracts of Scutellariae Radix and *Scutellaria baicalensis* hairy root. (unpubl.)

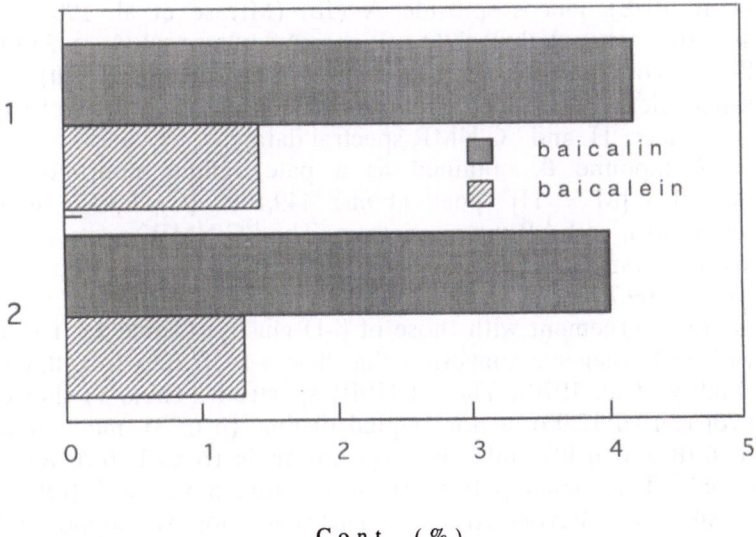

Fig. 6. Baicalin and baicalein contents in Scutellariae Radix and *Scutellaria baicalensis* hairy root. (unpubl.). *1* Scutellariae Radix; *2* hairy root

in Fig. 5, HPLC patterns of the extracts of Scutellariae Radix agreed with those of *Scutellaria baicalensis* hairy root. Furthermore, it was realized that the content of baicalin and baicalein in Scutellariae Radix and hairy root was almost the same, namely 4.2 and 4.0% for baicalin and 1.4 and 1.3% for baicalein (Fig. 6).

2.2.3 Isolation and Identification of the Compounds from Hairy Root Cultures

By a combination of silica gel column chromatography and HPLC, six flavone glycosides (8–11, 15, and 16) and five phenylethanoids (17–21) were isolated from the n-butanol extract and seven flavones (1–7) from the ether extract. The water extract was subjected to Diaion HP20 column chromatography and further separation was achieved by HPLC; three flavone glucuronides (12–14) were isolated. Among these, 20 were identified as the known compounds, 5,2',6'-trihydroxy-7, 8-dimethoxyflavone 2'-O-β-D-glucopyranoside (9) (Miyaichi and Tomimori 1995), viscidulin III-2'-O-β-D-glucoside (10) (Zhang et al. 1994), 5,2',6'-trihydroxy-6,7,8-trimethoxyflavone 2'-O-β-D-glucopyranoside (11) (Ishimaru et al. 1995), 6-C-β-D-glucopyranosyl-8-C-α-L-arabinopyranosylchrysin (15), 6-C-α-L-arabinopyranosyl-8-C-β-D-glucopyranosylchrysin (16) (Takagi et al. 1981), chrysin (2) (Tomimori et al. 1985b), baicalein (3), wogonin (1) (Shibata et al. 1923), skullcapflavone I (4) (Takido et al. 1979), rivularin (6) (Chou 1978), 5,2'-dihydroxy-6,7,8-trimethoxyflavone (5) (Tomimori et al. 1983), skullcapflavone II (7) (Takido et al. 1975), baicalin (12) (Shibata et al. 1923), wogonoside (13) (Liu and Gao 1982), 5,2'-dihydroxy-7,8,6'-trimethoxyflavone 2'-O-β-D-glucuronopyranoside (14) (Tomimori et al. 1990), martynoside (17) (Sasaki et al. 1978), leucosceptoside A (18) (Miyase et al. 1982), 2-(3-hydroxy-4-methoxyphenylethyl) 1-O-α-L-rhamno pyranosyl-(1 → 3)-(4-O-caffeoyl)-β-D-glucopyranoside (19) (Sasaki et al. 1989), acteoside (20), and 4-hydroxy-β-phenylethyl-β-D-glucopyranoside (21) (Birkofer et al. 1968), by analyses of their mass, ^{1}H, and ^{13}C NMR spectral data.

Compound 8, obtained as a pale yellow amorphous powder, exhibited a $[M + H]^+$ peak at m/z 449.1081 ($C_{21}H_{21}O_{11}$) in its high resolution positive FAB mass spectrum. The ^{13}C NMR spectrum showed signals arising from one flavone (15 carbons, C-2-10 and 1'–6') and a hexose (δ 60.8, 69.7, 73.3, 76.8, 77.2, and 100.5), whose chemical shift values were in good agreement with those of β-D-glucose (Table 1). The UV spectrum with shift reagents suggested that 8 was a 5,7-dihydroxyflavone derivative (Mabry et al. 1970). The ^{1}H NMR spectrum (Table 1) showed a chelated hydroxyl (δ 12.90), a noncoupled olefinic (δ 6.23), meta-coupled aromatic (δ 6.18 and 6.30) and ABC-type aromatic (δ 6.61, 6.70 and 7.25) proton signals. This signal pattern suggested that 8 had a 5 (OH), 7 (OH), 2', 6'-substituted flavone structure. This suggestion was supported by data from the HMBC spectrum (Table 2). In the ^{1}H NMR spectrum, the coupling constant (J = 8 Hz) of the anomeric proton signal at δ 4.87 indicated the presence of the β-configuration of the glucopyranoside moiety. The position of attachment of the glucose in 8 was decided by a HMBC experiment. The anomeric proton signal at δ 4.87 (H-1″) showed long-range correlation with the carbon at 156.3 (C-2'), indicating that the glucose moiety is linked to the C-2' hydroxyl group of the B-ring (Table 2). From the combined spectral data, 8 was concluded to be 5,7,2',6'-tetrahydroxyflavone 2'-O-β-D-glucopyranoside.

Table 1. ^{13}C and ^1H NMR spectral data of compound 8 (DMSO-d_6)

Atom	C	H
2	161.3	
3	112.3	6.23 1H, s
4	181.9	
5	161.6	
6	98.7	6.18 1H, d(2)
7	164.2	
8	94.0	6.30 1H, d(2)
9	158.4	
10	104.1	
1'	110.3	
2'	156.3	
3'	105.6	6.70 1H, d(8)
4'	132.2	7.25 1H, dd(8,8)
5'	109.6	6.61 1H, d(8)
6'	156.5	
Glucose		
1"	100.5	4.87 1H, d(8)
2"	73.3	3.04 1H, dd(9,8)
3"	76.8	3.19 1H, dd(9,9)
4"	67.7	3.08 1H, dd(9,9)
5"	77.2	3.27 1H, m
6"	60.8	3.40 1H, brd(12)
		3.66 1H, brd(12)
OH		12.90 1H, s
		10.10 1H, s

Table 2. ^1H-^{13}C Long-range correlation by HMBC of compound **8**. (Zhou et al. 1997)

H	Correlated C			
4.87 (H-1")	156.3 (C-2')			
6.18 (H-6)	161.6 (C-5)	164.2 (C-7)	94.0 (C-8)	104.1 (C-10)
6.23 (H-3)	104.1 (C-10)	181.9 (C-4)	161.3 (C-2)	110.3 (C-1')
6.30 (H-8)	164.2 (C-7)	98.7 (C-6)	104.1 (C-10)	158.4 (C-9)
6.61 (H-5')	110.3 (C-1')	105.6 (C-3')	156.5(C-6')	
6.70 (H-3')	110.3 (C-1')	156.3 (C-2')	109.6 (C-5')	
7.25 (H-4')	156.3 (C-2')	156.5 (C-6')		
12.90 (HO-C-5)	161.6 (C-5)	98.7 (C-6)	104.1 (C-10)	

Of the 21 compounds, **8** was isolated from hairy root cultures of *S. baicalensis* for the first time as a new compound. Although compound **14** in *S. rivularis*, compounds **18** and **20** in *S. rivularis* and *S. prostrata*, compound **19** in *Rehmannia glutinosa*, and **21** in *Syringa vulgaris* had already been identified (Birkofer et al. 1968; Miyase et al. 1982; Sasaki et al. 1989; Tomimori et al. 1990), these compounds were isolated from *S. baicalensis* for the first time (Fig. 7).

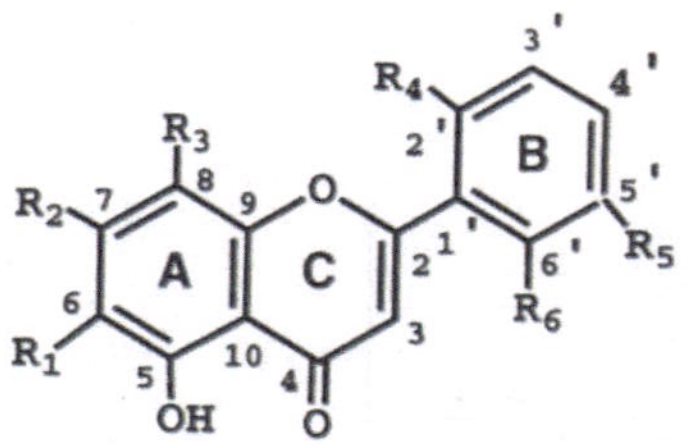

Phenylethanoids

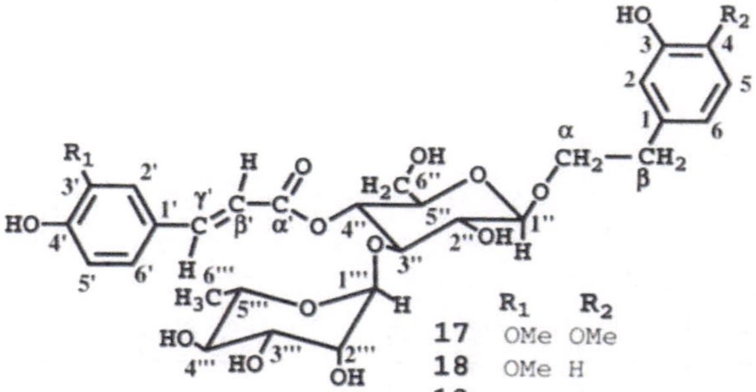

	R_1	R_2
17	OMe	OMe
18	OMe	H
19	H	OMe
20	H	H

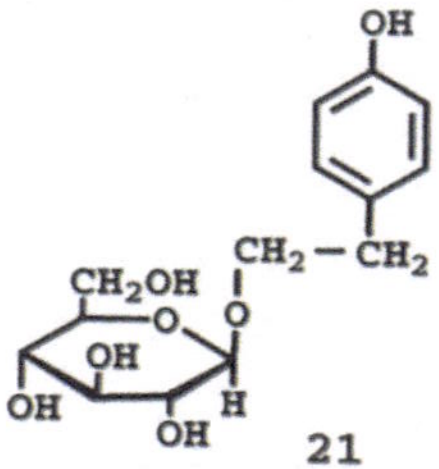

21

17: martynoside, **18**: leucosceptoside A, **19**: 2-(3-hydroxy-4-methoxyphenylethyl) 1-O-α-L-rhamnopyranosyl-(1→3)-(4-O-caffeoyl)-β-D-glucopyranoside, **20**: acteoside, **21**: 2-(4-hydroxyphenyl)ethyl O-β-D-glucopyranoside

Flavonoids

	R_1	R_2	R_3	R_4	R_5	R_6
1	H	OH	OMe	H	H	H
2	H	OH	H	H	H	H
3	OH	OH	H	H	H	H
4	H	OMe	OMe	OH	H	H
5	OMe	OMe	OMe	OH	H	H
6	H	OMe	OMe	OH	H	OMe
7	OMe	OMe	OMe	OH	H	OMe
***8**	H	OH	H	O-Glc	H	OH
9	H	OMe	OMe	O-Glc	H	OH
10	H	OH	OMe	O-Glc	OH	OMe
11	OMe	OMe	OMe	O-Glc	H	OH
12	OH	O-GlcA	H	H	H	H
13	H	O-GlcA	OMe	H	H	H
14	H	OMe	OMe	O-GlcA	H	OMe
15	Glc	OH	Ara	H	H	H
16	Ara	OH	Glc	H	H	H

1: wogonin, **2**: chrysin, **3**: baicalein, **4**: skullcapflavone I, **5**: 5, 2′-dihydroxy-6, 7, 8-trimethoxyflavone, **6**: rivularin, **7**: skullcapflavone II, ***8**: 5, 7, 2′, 6′-tetrahydroxyflavone 2′-O-β-D-glucopyranoside, **9**: 5, 2′, 6′-trihydroxy-7, 8-dimethoxyflavone 2′-O-β-D-glucopyranoside, **10**: viscidulin III 2′-O-β-D-glucoside, **11**: 5, 2′, 6′-trihydroxy-6, 7, 8-trimethoxyflavone 2′-O-β-D-glucopyranoside, **12**: baicalin, **13**: wogonoside, **14**: 5, 2′-dihydroxy-7, 8, 6′-trimethoxyflavone 2′-O-β-D-glucuronopyranoside, **15**: 6-C-β-D-glucopyranosyl-8-C-α-L-arabinopyranosylchrysin, **16**: 6-C-α-L-arabinopyranosyl-8-C-β-D-glucopyranosylchrysin

Fig. 7. Structures of compounds isolated from *S. baicalensis* hairy root cultures. (unpubl.)

3 Summary and Conclusions

Three strains of adventitious roots cultures of *Scutellaria baicalensis* were established by direct inoculation of sterile seedlings with three strains of *A. rhizogenes* pRi15834, *A. rhizogenes* Rif^R harboring pRi15834/pBI121 and pRi15834/pGSglucl. In one of the three adventitious roots, strain pBI121, transformation was proved by an opine assay and direct detection of the inserted T-DNA by the PCR. To determine the optimal medium for growth of hairy roots, the effects of five basal media were investigated and growth was best in B5 liquid medium. Flavonoid patterns on HPLC between Scutellaria Radix and *S. baicalensis* hairy root were almost identical. Furthermore, baicalin and baicalein contents of Scutellaria Radix completely agreed with that of *S. baicalensis* hairy root. These results suggest that it is possible for hairy root cultures of *S. baicalensis* to be used in herbal medicine as a substitute for Scutellaria Radix.

A new flavone glucoside, 5,7,2',6'-tetrahydroxyflavone 2'-O-β-D-glucopyranoside, 15 known flavonoids, and 5 known phenylethanoids were isolated from the hairy root cultures of *S. baicalensis* for the first time. Their structures were elucidated by various spectroscopic data.

References

Baylor NW, Fu T, Yan Y-D, Ruscetti FW (1992) Inhibition of human T cell leukemia virus by the plant flavonoid baicalin (7-glucuronic acid, 5,6-dihydroxyflavone). J Infect Dis 165:433–437

Birkofer L, Kaiser C, Thomas U (1968) Acteoside and neoacteoside, sugar esters from *Syringa vulgaris*. Z Naturforsch 23:1051–1058

Bouchez D, Camilleri C (1990) Identification of a putative rol B gene on the TR-DNA of the *Agrobacterium rhizogenes* A4 Ri plasmid. Plant Mol Biol 14:617–619

Bouchez D, Tourneur J (1991) Organization of the agropine synthase region of the T-DNA of the Ri plasmid from *Agrobacterium rhizogenes*. Plasmid 25:27–39

Chiang Su New Medical College (ed) (1977) Dictionary of Chinese crude drugs. Shanghai Scientific Technological Publishers, Shanghai, pp 2017–2021

Chou C-J (1978) Rivularin, a new flavone from *Scutellaria rivularis*. J Taiwan Pharm Assoc 30:36–43

Chou C-J, Lee S-Y (1986) Studies on the constituents of *Scutellaria indica* root (I). J Taiwan Pharm Assoc 38:107–118

Gamborg OL, Miller RA, Ojima K (1968) Nutrient requirements of suspension cultures of soybean root cells. Exp Cell Res 50:151–158

Hattori S (1930) Spectrography of the flavone series. III. The constitution of wogonin. Acta Phytochim 5:99–116

Ishimaru K, Nishikawa K, Omoto T, Asai I, Yoshihira K, Shimomura K (1995) Two flavone 2'-glucosides from *Scutellaria baicalensis*. Phytochemistry 40:279–281

Jefferson RA, Kavanagh TA, Bevan MW (1987) GUS fusions: β-glucuronidase as a sensitive and versatile gene fusion marker in higher plants. EMBO J 6:3901–3907

Jouanin L (1984) Restriction map of an agropine-type Ri plasmid and its homologies with Ti plasmids. Plasmid 12:91–102

Konoshima T, Kokumai M, Kozuka M, Iinuma M, Mizuno M, Tanaka T, Tokuda H, Nishino H, Iwashima A (1992) Studies on inhibitors of skin tumor promotion. XI. Inhibitory effects of flavonoids from *Scutellaria baicalensis* on Epstein-Barr virus activation and their anti-tumor-promoting activities. Chem Pharm Bull 40:531–533.

Li B-Q, Fu T, Yan Y-D, Baylor NW, Ruscetti FW, Kung H-F (1993) Inhbition of HIV infection by baicalin – a flavonoid compound purified from Chinese herbal medicine. Cell Mol Biol Res 39:119–124

Liu ML, Gao W (1982) Analysis of flavonoids in *Scutellaria baicalensis* G. Chin J Pharm Anal 2:134–140

Lloyd G, McCown B (1980) Commercially feasible micropropagation of mountain laurel, *Kalmia latifolia*, by use of shoot-tip culture. Combined Proc Int Plant Propagators' Soc, vol 30, pp 421–427

Mabry TJ, Markham KR, Thomas MB (1970) The systematic identification of flavonoids. Springer, Berlin Heidelberg New York, pp 33–60

Miyaichi Y, Tomimori T (1994) Studies on the constituents of *Scutellaria* species XVI. On the phenol glycosides of the root of *Scutellaria baicalensis* Georgi. Nat Med 48:215–218

Miyaichi Y, Tomimori T (1995) Studies on the constituents of *Scutellaria* species XVII. Phenol glycosides of the root of *Scutellaria baicalensis* Georgi (2). Nat Med 49:350–353

Miyaichi Y, Imoto Y, Tomimori T, Lin C-C (1987) Studies on the constituents of *Scutellaria* species XI. On the flavonoid constituents of the aerial parts of *Scutellaria indica* L. Chem Pharm Bull 35:3720–3725

Miyaichi Y, Imoto Y, Kizu H, Tomimori T (1988a) Studies on Nepalese crude drugs. IX. On the flavonoid constituents of the root of *Scutellaria scadens* Buch.-Ham. D. Don. Chem Pharm Bull 36:2371–2376

Miyaichi Y, Imoto Y, Kizu H, Tomimori T (1988b) Studies on Nepalese crude drugs. (X) On the flavonoid and stilbene constituents of the leaves of *Scutellaria scadens* Buch.-Ham. D. Don. Shoyakugaku Zasshi 42:204–207

Miyaichi Y, Imoto Y, Tomimori T, Lin C-C (1989) Studies on the constituents of *Scutellaria* species IX. On the flavonoid constituents of the root of *Scutellaria indica* L. Chem Pharm Bull 37:794–797

Miyase T, Koizumi A, Ueno A, Noro T, Kuroyanagi M, Fukushima S, Akiyama Y, Takemoto T (1982) Studies on the acyl glycosides from *Leucoseptrum japonicum* (MIQ.) Kitamura *et* Murata. Chem Pharm Bull 30:2732–2737

Murashige T, Skoog F (1962) A revised medium for rapid growth and bio-assays with tobacco tissue cultures. Physiol Plant 15:473–497

Nitsch JP, Nitsch C (1969) Haploid plants from pollen grains. Science 163:85–87

Petit A, David C, Dahl GA, Ellis JG, Guyon P, Casse-Debart F, Tempe J (1983) Further extension of the opine concept: plasmids in *Agrobacterium rhizogenes* cooperate for opine degradation. Mol Gen Genet 190:204–214

Saito K, Kaneko H, Yamazaki M, Yoshida M, Murakoshi I (1990) Stable transfer and expression of chimeric genes in licorice (*Glycyrrhiza ularensis*) using an Ri plasmid binary vector. Plant Cell Rep 8:718–721

Sasaki H, Taguchi H, Endo T, Yoshioka I, Higashiyama K, Otomasu H (1978) The glycosides of *Martynia louisiana* Mill. Chem Pharm Bull 26:2111–2121

Sasaki H, Nishimura H, Chin M, Mitsuhashi H (1989) Hydroxycinnamic acid esters of phenethylalcohol glycosides from *Rehmannia glutinosa* var. *purpurea*. Phytochemistry 28:875–879

Shibata K, Iwata S, Nakamura M (1923) Baicalin, a new flavone-glucuronic acid compound from the roots of *Scutellaria baicalensis*. Acta Phytochim 1:105–139

Slightom J, Durand-Tardif M, Jouanin L, Tepfer D (1986) Nucleotide sequence analysis of *Agrobacterium rhizogenes* agropine type plasmid. J Biol Chem 261:108–121

Song WZ (1981) Studies on the resource of the Chinese herb *Scutellaria baicalensis* Georgi. Acta Pharm Sin 16:139–145

Takagi S, Yamaki M, Inoue K (1981) Flavone di-C-glycosides from *Scutellaria baicalensis*. Phytochemistry 20:2443–2444

Takido M, Aimi M, Takahasi S, Yamanouchi S, Torii H, Dohi H (1975) Studies on the constituents in the water extracts of crude drugs. I. On the roots of *Scutellaria baicalensis* Georgi (Wogon) (1). Yakugaku Zasshi 95:108–113

Takido M, Yasukawa K, Matsuura S, Iinuma M (1979) On the revised structure of skullcapflavone I, a flavone compound in the roots of *Scutellaria baicalensis* Georgi (Wogon). Yakugaku Zasshi 99:443-444

Tang W, Eisenbrand G (1992) Chinese drugs of plant origin. Springer, Berlin Heidelberg New York, pp 919-929

Tomimori T, Miyaichi Y, Imoto Y, Kizu H, Tanabe Y (1983) Studies on the constituents of *Scutellaria* species. II. On the flavonoid constituents of the root of *Scutellaria baicalensis* Georgi (2). Yakugaku Zasshi 103:607-611

Tomimori T, Miyaichi Y, Imoto Y, Kizu H (1984) Studies on the constituents of *Scutellaria* species. V. On the flavonoid constituents of Ban Zhi Lian, the whole herb of *Scutellaria rivularis* Wall (1). Shoyakugaku Zasshi 38:249-252

Tomimori T, Miyaichi Y, Imoto Y, Kizu H, Namba T (1985a) Studies on Nepalese crude drugs. V. On the flavonoid constituents of the root of *Scutellaria discolor* COLEBR. (1). Chem Pharm Bull 33:4457-4463

Tomimori T, Jin H, Miyaichi Y, Namba T (1985b) Studies on the constituents of *Scutellaria* species. VI. On the flavonoid constituents of the root of *Scutellaria baicalensis* Georgi (5) Quantitative analysis of flavonoids in *Scutellaria* roots by high-performance liquid chromatography. Yakugaku Zasshi 105:148-155

Tomimori T, Miyaichi Y, Imoto Y, Kizu H (1986a) Studies on the constituents of *Scutellaria* species (VIII). On the flavonoid constituents of Ban Zhi Lian, the whole herb of *Scutellaria rivularis* Wall (2). Shoyakugaku Zasshi 40:432-433

Tomimori T, Miyaichi Y, Imoto Y, Kizu H, Namba T (1986b) Studies on Nepalese crude drugs. VI. On the flavonoid constituents of the root of *Scutellaria discolor* COLEBR. (2). Chem Pharm Bull 34:406-408

Tomimori T, Imoto Y, Miyaichi Y (1990) Studies on the constituents of *Scutellaria* species. XIII. On the flavonoid constituents of the root of *Scutellaria rivularis* Wall. Chem Pharm Bull 38:3488-3490

Trevelyan WE, Procter DP, Harrison JS (1950) Detection of sugars on paper chromatograms. Nature 166:444-445

Vervliet G, Holsters M, Teuchy H, Montagu M, Schell J (1975) Characterization of different plaque-forming and defective temperate pharges in *Agrobacterium* strains. J Gen Virol 26: 33-48

Yamamoto H (1991) *Scutellaria baicalensis* Georgi: in vitro culture and the production of flavonoids. In: Bajaj YPS (ed) Biotechnology in agriculture and forestry vol 15. Medicinal and aromatic plants III. Springer, Berlin Heidelberg New York, pp 398-418

Zhang Y-Y, Guo Y-Z, Onda M, Hashimoto K, Ikeya Y, Okada M, Maruno M (1994) Four flavonoids from *Scutellaria baicalensis*. Phytochemistry 35:511-514

Zhou Y, Hirotani M, Yoshikawa T, Furuya T (1997) Flavonoids and phenylethanoids from hairy root cultures of *Scutellaria baicalensis*. Phytochemistry 44:83-87

XXI Genetic Transformation of *Serratula tinctoria* (Dyer's Savory) for Ecdysteroid Production

M.-F. Corio-Costet[1], L. Chapuis[1], J.-P. Delbecque[2], and K. Ustache[1]

1 Introduction

Serratula tinctoria is a perennial plant of the Compositae family with medium-sized, serrated leaves and purple flowers (Loste 1937). This plant, also known as dyer's savory, is widespread in Europe but with an irregular distribution. Inflorescences (capitula) are purple and are usually unisexual, staminate, or pistillate. In Europe, the flowering period extends from July to September. More than 40 species have been described in Europe, North Africa, and Asia. The plants produce large amounts of secondary metabolites, in particular ecdysteroids at very high concentration in roots (up to 2% dry wt.), in flowers, and in leaves (Bathori et al. 1986; Rudel et al. 1992; Corio-Costet et al. 1993b). Three major ecdysteroids are present in *S. tinctoria* (Fig. 1): 20-hydroxyecdysone (20E), its corresponding 3-acetate (20E3Ac), and polypodine B (Pol B), together with numerous minor ecdysteroids such as diacetates of 20E, rubrosterone, poststerone, their corresponding 3-epimers and compounds with a 22-oxo group (Rudel et al. 1992).

 S. tinctoria thus represents a convenient source of ecdysteroids, which may be a powerful defense for the plant, and can be used against some pests (Mondy et al. 1997). However, it is obvious from our observations in the greenhouse that several insect species that live on such plants have adapted their metabolism to resist this excess of molting hormones. The question also arises whether a few insect species may even benefit from this hormonal source to regulate their populations.

 The in vitro production of ecdysteroids from various plant systems has been considered by several investigators as a valuable tool for the study of the roles and the biosynthetic pathways of ecdysteroids (see Corio-Costet et al. 1996). Therefore, the potential of hairy roots for the production of ecdysteroids has been recently explored, since they generally allow a higher level of productivity, in addition to better genetic and biochemical stabilities: hairy root lines producing phytoecdysteroids have been obtained with *Ajuga reptans*, *Achyranthes fauriei*, *Pfaffia iresinoides*, and *Vitex stickeri* (Matsumoto and Tanaka 1991). We established hairy root cultures of *Serratula tinctoria*, in

[1] INRA-Bordeaux, URIV-Phytopharmacie, BP 81, 33883 Villenave d'Ornon, France
[2] Université de Bourgogne, CNRS, UMR 5548, Faculté des Sciences Gabriel, 21000 Dijon, France

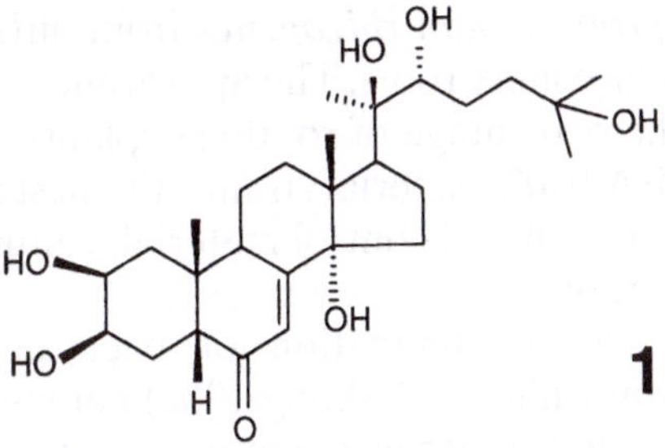

Fig. 1. Structures of the major ecdysteroids found in *Serratula tinctoria*. (Rudel et al. 1992). **1** 20-Hydroxyecdysone; **2** 3-acetate of 20-hydroxyecdysone; **3** 5β,20-dihydroxyecdysone (polypodine B). (Corio-Costet et al. 1996)

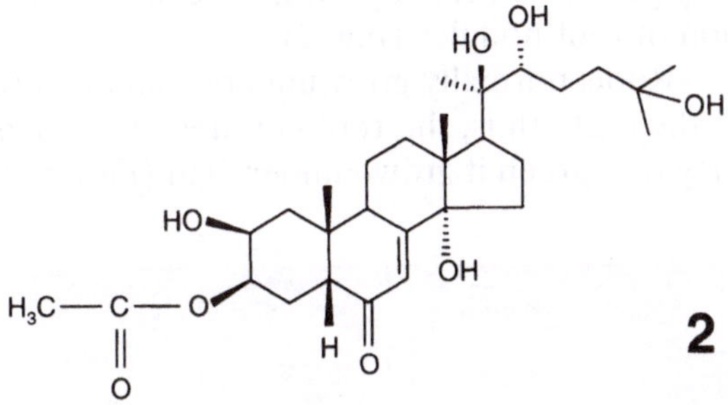

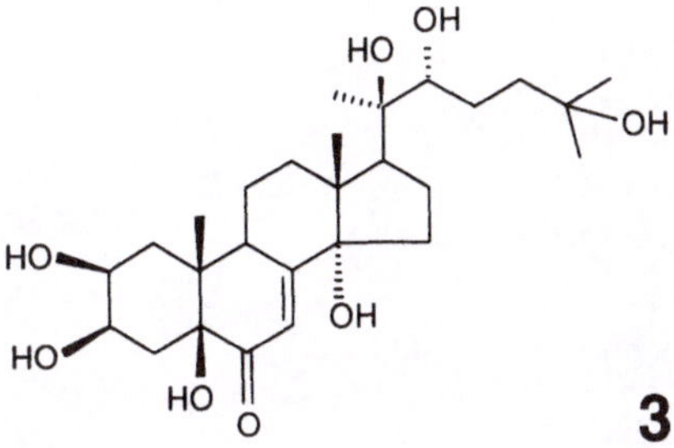

solid and liquid media, which appeared capable of producing high amounts of ecdysteroids, after transformation with *Agrobacterium rhizogenes* (Corio-Costet et al. 1994, 1996).

2 Genetic Transformation

2.1 Establishment of Hairy Root Cultures

Hairy root cultures of *S. tinctoria* were initiated from seedlings obtained under sterile conditions by germinating seeds collected from plants in the greenhouse, and sterilized appropriately (see protocol below). Freshly cut stem segments from sterile seedlings were then inoculated with varying amounts of

Agrobacterium rhizogenes from different strains, showing various abilities to induce hairy roots. The appearance of roots was 2 to 6 weeks after inoculation. The percentage of rooting explants was 0 to 45%, when infected with A_4 and LBA 9402 bacterial strains, the best results being obtained with the A_4 strain. Uninoculated control material, cultured under the same conditions, produced no roots.

After elimination of bacteria by several transfers in MS medium (Murashige and Skoog 1962) containing antibiotics, where they grew slowly, selected clones were cultured on the same medium without antibiotics, where they grew more rapidly, with little lateral branching and characteristic formation of root nodules (Fig. 2).

Roots normally grow underground and root cultures are usually cultured in the dark; thus, the root cultures of *S. tinctoria*, as well as of other species, may turn green if grown under light (Flores et al. 1993). Generally, roots grew

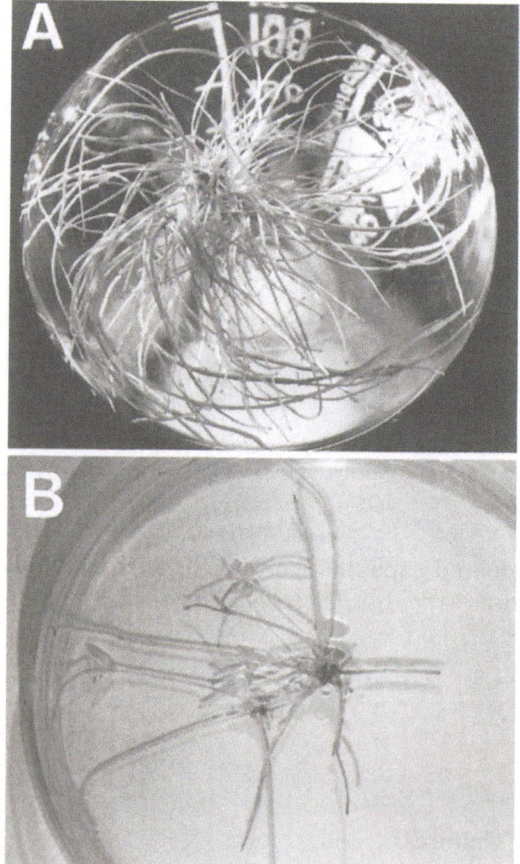

Fig. 2A,B. Hairy roots cultured **A** in the light (70-day-old culture) in liquid medium, and **B** in solid medium in the dark (10-day-old culture). (M.F. Corio-Costet and L. Chapuis 1992, unpubl.)

0.5–2.8 mm per day without extensive branching. For the most part, the obtained clones grew ca. 1.5–2.5 mm per day (Corio-Costet et al. 1996).

After successive culture for more than 8 months, the clones which grew more rapidly were selected: in liquid medium, the elongation was 2–4.5 mm per day. The selected hairy roots were subcultured on MS medium at 25 °C in the dark. It is important to stress that all the results were obtained without any addition of hormones.

2.2 Molecular Analysis of Hairy Roots

After transformation, various lines showed the characteristic phenotype of hairy roots, such as lack of geotropism. Moreover, in contrast to other tissue cultures in *S. tinctoria*, no growth regulator was necessary for maintaining such hairy roots, confirming that they were most likely transformed. However, this transformation was definitely demonstrated at the molecular level by PCR (polymerase chain reaction) with specific rol A (ORF10) primers, specific to T_L-DNA of the A_4 strain of *A. rhizogene*: positive, left-hand primer (ATGGAATTAGCCGGACTAAACGT), and negative right hand primer (TTGTTTGGATGCCCTAATA) were used for PCR amplification of DNA extracted from hairy roots and from plant roots. The results (Fig. 3) clearly show that *S. tinctoria* hairy roots contain the rol A-specific fragment, demonstrating that they have been transformed by bacteria, whereas plant roots gave no positive response, employing from 0.01 to 0.05 µg of DNA.

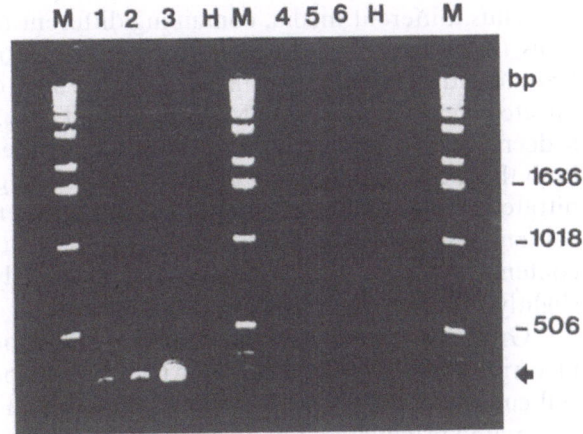

Fig. 3. PCR amplification products from genomic DNA extracted from hairy roots, using primers rolA. Lanes: *M* molecular weight marker (1 kb DNA); *1* and *2*: dilutions of DNA from hairy roots of *S. tinctoria* (1/200 and 1/500); *3* 2 µl of T_LDNA cosmid pLJ₁ (positive control); *H* H₂O negative control (no DNA); *4*, *5*, and *6* dilutions of DNA from roots of *S. tinctoria* (1/200, 1/500 and 1/1000). *Arrow* shows amplified fragments. (L. Chapuis and M.F. Corio-Costet 1996, unpubl.)

2.3 Phytoecdysteroid Production in Hairy Roots

Important amounts of ecdysteroids were found in hairy root cultures compared to other tissue or cell cultures, and did not decrease with time. The content in hairy roots was ca. 1.4 to 4 mg/g dry weight, which is less that of original roots (12–14.9 mg/g) but far more than calli and cell cultures (Corio-Costet et al. 1993b, 1996). Two major phytoecdysteroids were detected in hairy roots, namely 20E and its 3-monoacetate (20E3Ac) (Fig. 4). In roots, 20E3Ac generally was the major phytoecdysteroid, whereas it was 20E in hairy root and cell cultures (Corio-Costet et al. 1996).

The effects of changing medium composition was examined for ecdysteroid production in hairy roots. MS medium was used as basal medium. First, the addition of different concentrations of auxin have been checked for ecdysteroid production: 2,4-D (2,4-dichlorophenoxy acetic acid) was added at various concentrations, giving media named MS1 to MS6 (Table 1). Phytoecdysteroid production revealed a negative dose-dependent relationship for 2,4-D with a decrease in ecdysteroid concentrations from 36 to 91%. Besides, hairy root growth was slowed on media containing from 0.05 to 0.5 mg/l of 2,4-D and addition of 1 mg/l of 2,4-D even stopped growth by tissue necrosis (Table 1).

Hairy root growth is also known to be influenced by factors such as pH, or sugar, phosphate, nitrate, and ammonia concentrations. A decrease in both phosphate and nitrate concentrations produced an increase in the synthesis of hyoscyamine in hairy roots of *Datura stramonium* (Payne et al. 1987). Similar results have been obtained for the accumulation of secondary metabolites in *Catharanthus roseus* hairy roots (Parr et al. 1988). More recently, an increase in 20-hydroxyecdysone production in *Ajuga* hairy roots has been obtained after phosphate depletion in culture media (Uozumi et al. 1995).

Thus, different media, containing different nitrogen forms or concentrations (MS7 to MS9) have been tried in our model. The highest content of ecdysteroid was obtained with a decrease in 50% of ammonium and potassium nitrate concentrations, with MS9 medium (Table 2). These results suggest that a decrease in ammonium concentrations in the culture medium stimulates both the increase in ecdysteroid production and growth (MS7). If ammonium nitrate is replaced by sodium nitrate, a supplementary increase in ecdysteroid content and growth is observed (MS8), but if the ammonium and the nitrate contents decrease simultanously, the best ecdysteroid production with a slightly slower growth (MS9) was obtained.

Growth and secondary metabolite production are often inversely related in various other models: factors which limit exponential growth of callus or cell cultures induce stages of slower growth and usually stimulate secondary compound production. For example, the addition of growth regulators such as 2,4-D decreased the secondary metabolites in *Duboisia myoporoides* and *Ajuga* hairy roots (Deno et al. 1987; Uozumi et al. 1995).

Moreover, previous studies have also demonstrated that the highest ecdysteroid concentrations were found in hairy roots as well as in plant roots, in the first centimeter of root tip, which corresponds to the growing zone

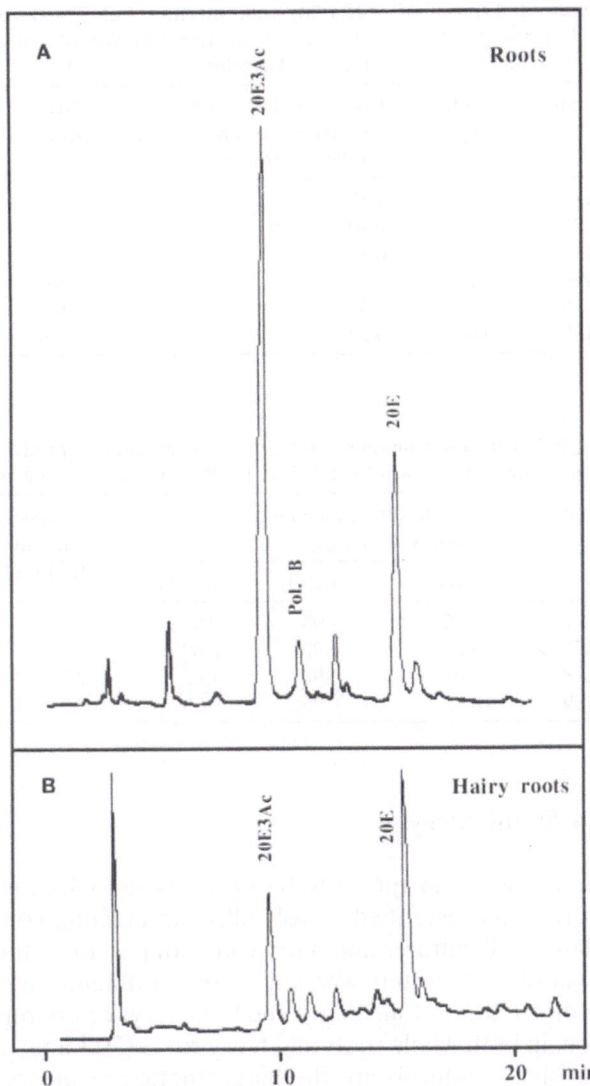

Fig. 4A,B. HPLC profile of ecdysteroids from *Serratula tinctoria* roots (**A**) and hairy roots (**B**). NP-HPLC conditions: Merck Lichrospher 100 Diol 25 cm column, linear gradient from 5 to 20% ethanol in dichloromethane in 20 min at 1 ml/min, UV detection at 244 nm. (Corio-Costet et al. 1993b)

(Delbecque et al. 1995): thus ecdysteroid production seems to be associated with root growth in our model. Furthermore, during experiments on MS7 and MS8 (without ammonium), we also observed a modification in root geotropism, i.e., hairy roots coiled around themselves (Fig. 5). Such a phenomenon, whereby some clones were more affected than others, is difficult to interpret.

Table 1. Effects of different concentrations of 2,4-D on ecdysteroid production and growth in *Serratula tinctoria* hairy roots. (M.F. Corio-Costet and K. Ustache 1996, unpubl.)

Media	2,4-D (mg/l)	Ecdysteroid contents EIA measurements (µg Eq of 20E/mg dry wt.)	Growth (mm/day)
MS1	0	1.17	0.93
MS2	1	0.10	0.00
MS3	0.5	0.17	0.35
MS4	0.1	0.30	0.37
MS5	0.05	0.44	0.62
MS6	0.01	0.59	0.98

Table 2. Effect of ammonium and nitrate on ecdysteroid production and growth of hairy roots of *S. tinctoria*. (M.F. Corio-Costet, J.-P. Delbecque and K. Ustache 1996, unpubl.)

Media	Modification in medium (concentration g/l)			Ecdysteroid contents EIA measurements (µgEq of 20E/ mg dry wt.)	Growth (mm/day)
	NH_4NO_3	KNO_3	$NaNO_3$		
MS1	1.65	1.90	0.00	1.17	0.9
MS7	0.00	1.90	0.00	1.82	1.47
MS8	0.00	1.90	1.65	2.55	1.62
MS9	0.83	0.95	0.00	3.85	0.8

2.4 Sterol Analyses

Hairy roots also appear to be a promising tool for metabolic studies. Analyses of free and esterified sterols allow interesting comparisons between rooted plants, cell cultures and hairy roots. In previous studies, it was shown that the level of steryl esters was important to obtain a high content of ecdysteroids (Corio-Costet et al. 1996). Indeed, steryl esters represent an important fraction, in both whole roots and hairy roots (70.1 and 42.9%, respectively, of total sterols). As sterols are the biosynthetic precursors of ecdysteroids in insects (Karlson and Hoffmeister 1963; Svoboda et al. 1978), it is presumed that they serve a similar function in plants. It was thus interesting to observe the distribution of the three sterol classes: 4,4-dimethylsterols, 4α-methylsterols, and 4-desmethylsterols.

The most important difference was the high accumulation of 4,4-dimethylsterols (essentially amyrins) in esterified sterols of roots (Corio-Costet et al. 1993b), whereas hairy roots accumulated 4-desmethylsterols. Another interesting point is the cholesterol level, which represented up to 11% of total sterols in roots but only 1.6% in cell cultures. In 1-month-old hairy roots, cholesterol represented 3.5 to 9.2% of total sterols, which is very interesting to notice, because cholesterol is probably the privileged precursor of ecdysteroid biosynthesis: indeed, its level appears higher in tissues able to synthetize a high quantity of ecdysteroids.

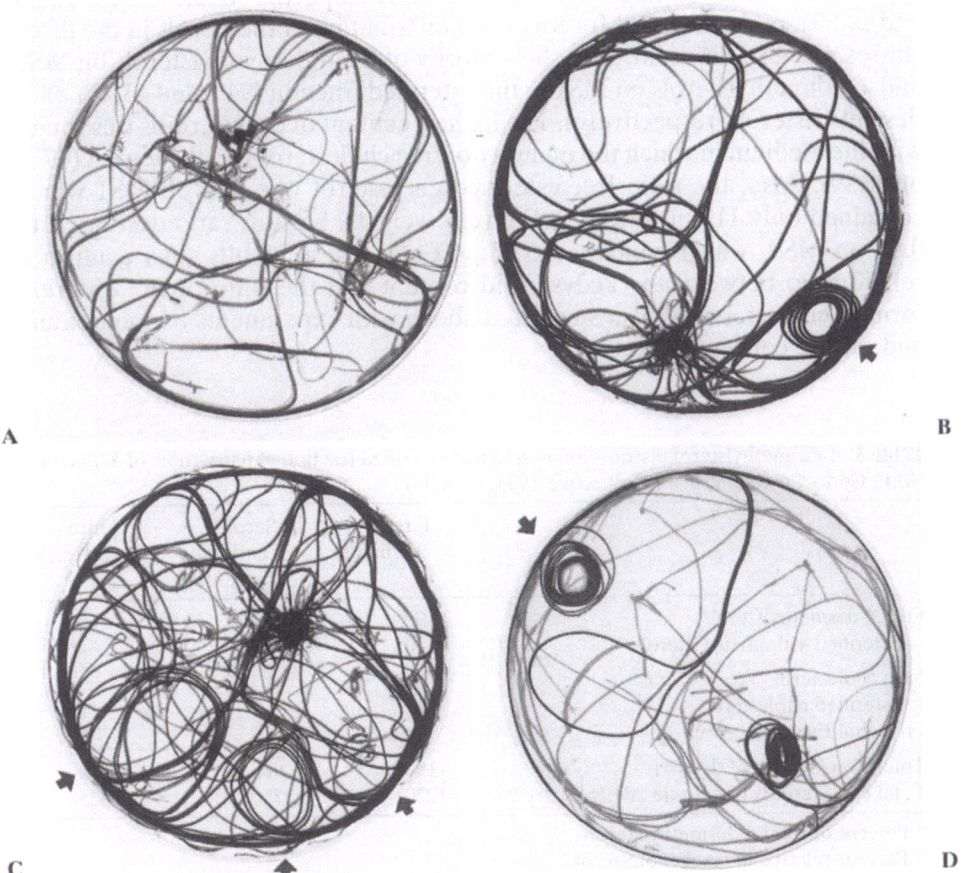

Fig. 5A–D. Hairy roots growing on MS1 (**A**), MS7 (**B**), MS8 (**C** and **D**) media. *Arrows* show the rolling up of the roots. (M.F. Corio-Costet and K. Ustache, unpubl.)

As indicated above, the root tip contains higher ecdysteroid quantities than the other parts (Delbecque et al. 1995; Corio-Costet et al. 1996). We thus also examined the 4-desmethylsterol distribution in root tip centimeter by centimeter (Table 3). The esterified sterol fraction was more important in the second and third centimeter from the apex than in the first. Inversely, the relative percentage of cholesterol was higher in the esterified fraction than in the free fraction of the apex. These results suggest that the level of esterified cholesterol could be a good indicator of the potentiality of ecdysteroid biosynthesis, and so the ratio free cholesterol/esterified cholesterol could be used as a significant value. It appears that the highest ecdysteroid production was obtained for the lowest ratio (in the first centimeter of the hairy root).

The sterol distribution of the three classes of sterols in hairy roots was also compared after 1 month of growth on media MS1 and MS9 (Table 4). The content of esterified sterols on MS9 was more important than on MS1 (25.5 and 19.5%, respectively). Moreover, the distribution of the sterols in the three classes was also different, with the majority of the 4-desmethylsterols on MS9 and 4,4-dimethylsterols on MS1 in the esterified fraction (94.5 and 43.7% of 4-desmethylsterols, respectively). The highest content of ecdysteroids was found with the medium in which the quantity of free cholesterol was the lowest (47% on MS9). Also, the ratio free/esterified cholesterol was 3.12 on MS1 which contained only 1170 µg-Eq of 20E/g (dry weight), but this ratio decreased to 0.86 on MS9, which contained 3789 µg-Eq/g. These results thus confirm a relationship between the ecdysteroid content and the ratio of the different form of cholesterol, as also suggested above with experiments on ammonium and nitrate concentrations.

Table 3. 4-desmethylsterol distribution in free and esterified fraction of hairy roots of *S. tinctoria*. (M.F. Corio-Costet and J.-P. Delbecque 1993, unpubl.)

	First centimeter	Second centimeter Percentage	Third centimeter
Free 4-desmethylsterols	90[a]	64[a]	66[a]
Esterified 4-desmethylsterols	10[a]	36[a]	34[a]
Free cholesterol	82.3[b]	93[b]	89[b]
Esterified cholesterol	17.7[b]	7[b]	11[b]
Free chol/Esterified chol[c]	4.6	13.3	8
Total sterols (µg/g of dry wt)	1190	989	1090
Total ecdysteroids (µg-Eq de 20 E/g of dry wt)	1220	780	625

[a] Percent of total 4-desmethylsterols.
[b] Percent relative to total cholesterol.
[c] Ratio of free cholesterol on esterified cholesterol.

Table 4. Analysis of sterol distribution of three classes in free and esterified fractions in hairy root of *Serratula tinctoria* after growing on MS1 and MS9. (M.F. Corio-Costet 1997, unpubl.)

	Medium MS1		Medium MS9	
	Free sterols	Esterified sterols	Free sterols	Esterified sterols
4-Desmethylsterols	93.9[a]	43.7[b]	93.6[a]	94.5[b]
4α-Methylsterols	1.9[a]	8.7[b]	5.0[a]	4[b]
4,4-Dimethylsterols	4.2[a]	47.6[b]	1.4[a]	1.5b[b]
Total sterols	80.5	19.5	74.5	25.5
Cholesterol	76	24	47	53
Free cholesterol/ Esterified cholesterol	3.12		0.86	
Total sterols (µg/g of dry wt)	820		788	
Total ecdysteroids (µg-Eq de 20 E/g of dry wt.)	1170		3789	

[a] Percentage relative to total free sterols.
[b] Percentage relative to total esterified sterols.

2.5 In Vitro Incorporation of Radiolabeled Cholesterol in Hairy Roots

Ecdysteroid biosynthesis was investigated in the hairy root by incubation of radiolabeled precursors. Tritiated cholesterol was first administered in liquid medium (10 µCi/flask) of hairy roots during up to 15 days. Then, roots were extracted as described below and analyzed using RP-HPLC. A significant incorporation of radiolabeled cholesterol into ecdysteroids was obtained, as radioactive 20E reached more than 0.1% of the incorporated cholesterol after 15 days of culture, as shown by high performance liquid chromatography analysis (Fig. 6). Though the chemical nature of the different steroids was not completed, it appeared that the major steroid was 20E (comigration with both RP and NP-HPLC), whereas other minor steroids were presumed to be, respectively, polypodine B and 20E3Ac. This study thus demonstrates that

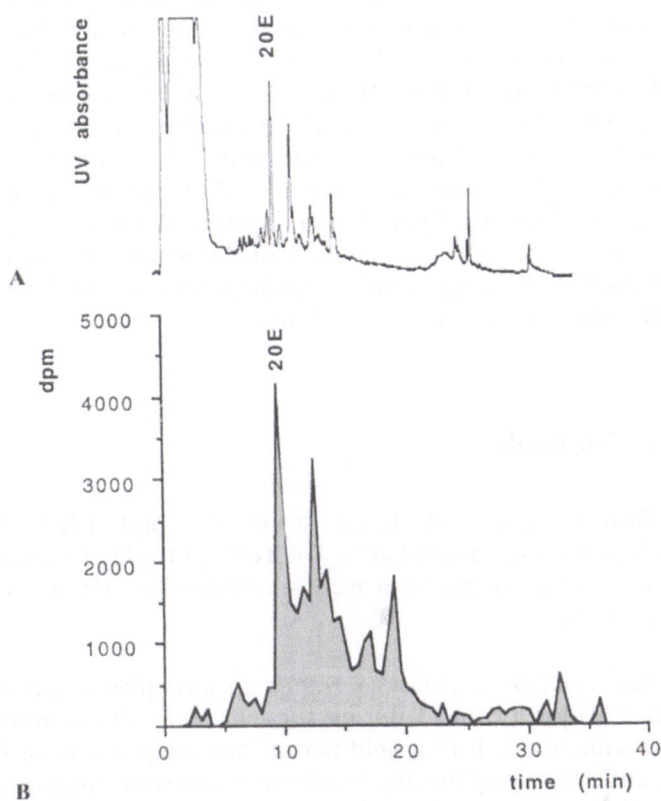

Fig. 6A,B. RP-HPLC analysis of steroids from *Serratula tinctoria* hairy roots, after incorporation of [³H] cholesterol during 15 days: (top) UV trace at 244 nm and (*bottom*) radioactivity measurements (dpm) of collected fractions. RP-HPLC conditions: Merck Lichrospher 100-RP18 column, with linear gradient elution from 16 to 40% acetonitrile in water at 1 ml/min. (Corio-Costet et al. 1996)

several ecdysteroids, in particular 20E, are really synthetized from tritiated cholesterol in hairy root cultures, suggesting that plant root itself is a synthetizing organ. Though the incorporation rates are rather low using cholesterol, they can be improved in the plant with more polar precursors, such as mevalonic acid. Indeed, experiments have recently been performed using radioactive mevalonic acid, giving a much higher efficiency (Delbecque et al. 1995). However, radioactive 25-hydroxycholesterol into could not be incorporated 20E.

3 Summary and Conclusions

Hairy root cultures of *Serratula tinctoria* appear very attractive with their high production of ecdysteroid, their easy cultivation methods, and the possibility of incorporating different precursors of ecdysteroid biosynthesis. Hairy root cultures will also probably be a good model for the study of enzymes involved in the biosynthetic pathway of phytoecdysteroids. Ecdysteroids are biosynthesized de novo from acetate, mevalonate, or cholesterol (Nagakari et al. 1994; Delbecque et al. 1995) and the role of cytochrome P450 in the hydroxylation reaction was demonstrated, but most of the biochemical pathways, including the first step of ecdysteroid biosynthesis and conjugation, remain to be investigated. Hairy roots of *Serratula tinctoria* thus can be potentially useful for the elucidation of the metabolism of ecdysteroids, with the possibility of using inhibitors or activators of cytochrome P450, as well as of the roles of ecdysteroids in plants.

4 Protocols

Bacterial Strain. Bacterial strain A4 and LBA 9402, *Agrobacterium rhizogenes* was provided by Dr. Tepfer (Tepfer 1984) and was used for hairy root initiation. Bacteria were maintained at 4°C on 15 g/l of nutrient agar medium.

Plants and Hairy Roots. *Serratula tinctoria* plants were collected in Burgundy near Dijon (France) during summer 1989. Plants were grown in pots in a greenhouse with 16-h light period, at a temperature of 18°C during the night and 25°C during the day. Seeds were collected regularly, and surface sterilized in 0.05% mercuric chloride for 10 min; then they were thoroughly washed with distilled water and germinated on MS medium, supplemented with sucrose and solidified with 0.8% agar. Cultures were incubated at 22 ± 2°C and exposed to a photoperiod of 16h, without hormones. Stem segments from these cultures were then inoculated with various amounts of *Agrobacterium rhizogenes*, making an incision with a disposable scalpel, they were then allowed to grow overnight in either liquid or solid media. Inoculated tissues were kept at 23°C and high humidity until roots appeared. Roots were excised from the inoculation site and rinsed four times with sterile water before explantation onto the appropriate medium containing antibiotics (Cefotaxime 500 µg/l) and agar (15 g/l). After several transfers in order to eliminate bacteria, hairy roots were cultured on the same medium without antibiotics at 25°C, in the dark. A culture derived from a single root, comprising an organ clone, was established from each root tip. Liquid culture was pursued as follows: 20–30 mg of fresh hairy roots were incubated in 250-ml flasks containing 100 ml of MS medium and placed on a rotary shaker (125 rpm) at 25°C, in dark or light.

Molecular Analysis. Total DNA was isolated from 100 mg of hairy roots and 200 mg of roots of whole plant. DNA extractions were performed using 0.8 ml DNA Now monophasic reagent (Biogentex, 1993, patent pending) per sample and following the manufacter's instructions. After final air drying, DNA was dissolved in sterilized distilled water at a concentration of approximately 2 µg/ml and kept at −20°C. DNA solutions were used to perform PCR assays with the following reaction mixture in a total volume of 20 µl: 70 mM tris-HCl; 2 mM $MgCl_2$, 17 mM $(NH_4)_2 SO_4$, 10 mM β-mercaptoethanol, 0.05% (w:v) polyoxyethylene-ether W1 (Sigma), 0.2 mg/ml bovine serum albumin, 200 µM each dATP, dCTP, dGTP, and dTTP, 2 µl of primers, approximately 10 ng of template DNA and 0.8 unit Red-Goldstar DNA polymerase. The solution was overlaid with 20 µl of paraffin oil and submitted to 37 cycles each of 30 s melting at 95°C, 1 min annealing at 65°C, and 1 min 30 s extension at 72°C on a thermal cycler. The PCR products were electrophoresed at 90 V for 3 h on ethidium bromide-stained, 10% (w/v) agarose gel run in TBE buffer (borate EDTA).

Isolation and Analysis of Ecdysteroids. Plant material was extracted as described previously (Corio-Costet et al. 1993a). The eluted fractions of ecdysteroids were evaporated and submitted to HPLC (high performance liquid chromatography in reverse or normal phases, i.e., RP or NP) and/or EIA (enzyme immunoassay) as described by Corio-Costet et al. (1993b).

Isolation and Analysis of Sterols. The sterol composition of plant organs and cell cultures was determined as previously described (Benveniste et al. 1984). The various sterol fractions were acetylated at room temperature overnight as previously described (Costet-Corio and Benveniste 1988) and identified by capillary gas-liquid chromatography and gas-liquid chromatography-mass spectrometry.

Acknowledgments. The authors thank the Regional Council of Burgundy for financial support by a state-region grant (Contrat Etat-Region Bourgogne), N. Pitoizet for EIA measurements, M. Corio and M. Chaignaud for photographs and Dr. D. Tepfer for the generous supply of *A. rhizogenes* strains and PCR primers.

References

Bathori M, Szendrei K, Herke I (1986) Chromatography of ecdysteroids originated from *Serratula tinctoria*. Chromatographia 21:234–238

Benveniste P, Bladocha M, Costet MF, Ehrard A (1984) Use of inhibitors of sterol biosynthesis to study plasmalemma structure and functions. In: Boudet AM, Alibert G, Marigo J, Lea J (eds) Membrane and compartmentation in the regulation of plant functions vol 24. Proc Phytochemical Soc Eur. Clarendon Press, Oxford, pp 283–300

Corio-Costet MF, Chapuis L, Scalla R, Delbecque JP (1993a) Analysis of sterols in plants and cell cultures producing ecdysteroids. I. *Chenopodium album*. Plant Sci 91:23–33

Corio-Costet MF, Chapuis L, Mouillet JF, Delbecque JP (1993b) Sterol and ecdysteroid profiles of *Serratula tinctoria* (L.): plant and cell cultures producing steroids. Insect Biochem Mol Biol 23:175–180

Corio-Costet MF, Beydon P, Chapuis L, Malosse C, Delbecque JP (1994) Sterol and ecdysteroid biosynthesis in *Serratula tinctoria*: in vitro incorporation of radiolabelled cholesterol and mevalonic acid in hairy roots cultures. In: de Kouchkovsky Y, Larher F (eds) Plant sciences 1994, Saint-Malo (France), imprimerie Univ Rennes I, p 152

Corio-Costet MF, Chapuis L, Delbecque JP (1996) *Serratula tinctoria*(L.) (Dyer's savory): In vitro culture and the production of ecdysteroids and other secondary metabolites. In: Bajaj YPS (ed) Biotechnology in agriculture and forestry vol 37. Medicinal and aromatic plants IX. Springer, Berlin Heidelberg New York, pp 384–401

Costet-Corio MF, Benveniste P (1988) Sterol metabolism in wheat treated by N-substituted morpholines. Pestic Sci 22:243–247

Delbecque JP, Beydon P, Chapuis L, MF Corio-Costet (1995) In vitro incorporation of radiolabelled cholesterol and mevalonic acid into ecdysteroids by hairy root cultures of a plant, *Serratula tinctoria*. Eur J Entomol 92:301–307

Deno H, Yamagata H, Emoto T, Yoshioka T, Yamada Y, Fujita Y (1987) Scopolamine production by hairy root cultures of *Duboisia myoporoides*. II. Establishment of a hairy root culture by infection with *Agrobacterium rhizogenes*. J Plant Physiol 131:315–323

Flores HE, Yao-Rem D, Cuello JL, Maldonado-Mendoza IE, Loyola-Vargas VM (1993) Green roots: photosynthesis and photoautotrophy in an underground plant organ. Plant Physiol 101:363–371

Karlson P, Hoffmeister M (1963) Zur Biogenese des Ecdysons. I. Umwandlung von Cholesterin in Ecdyson. Z Physiol Chem 331:298–300

Loste H (1937) Flore descriptive et illustrée de la France, tome II, Librairie Sciences et arts, Paris, pp 400–401

Matsumoto T, Tanaka N (1991) Prodution of phytoecdysteroids by hairy root cultures of *Ajuga reptans* var. *atropurpurea*. Agric Biol Chem 55:1019–1025

Mondy N, Caissa C, Pitoizet N, Delbecque JP, Corio-Costet MF (1997) Effects of the ingestion of *Serratula tinctoria* extracts, a plant containing phytoecdysteroids on the developement of the vineyard pest *Lobesia botrana*. Arch Insect Biochem Physiol 35:227–235

Murashige T, Skoog F (1962) A revised medium for rapid growth and bio-assays with tobacco tissue cultures Physiol Plant 15:473–797

Nagakari M, Kushiro T, Matsumoto T, Tanaka N, Kakinuma K, Fujimoto Y (1994) Incorporation of acetate and cholesterol into 20-hydroxyecdysone by hairy root clone of *Ajuga reptans* var. *atropurpurea*. Phytochemistry 36:907–910

Parr AJ, Peerless ACJ, Hamill JD, Walton NJ, Robins RJ, Rhodes MJC (1988) Alkaloid production by transformed root cultures of *Catharanthus roseus*. Plant Cell Rep 7:309–312

Payne J, Hamill JD, Robins, RJ, Rhodes MJC (1987) Production of tropane alkaloids by hairy root cultures of *Scopolia japonica*. Agric Biol Chem 50:2715–2722

Rudel D, Bathori M, Gharbi J, Girault JP, Racz I, Melis K, Szendrei K, Lafont R (1992) New ecdysteroids from *Serratula tinctoria*. Planta Med 58:358–364

Svoboda JA, Thompson MJ, Robbins WE, Kaplanis JN (1978) Insect steroid metabolism. Lipids 13:742–753

Tepfer D (1984) Transformation of several species of higher plants by *Agrobacterium rhizogenes*; sexual transmission of the transformed genotype and phenotype. Cell 37:959–967

Uozumi N, Makino S, Kobayashi T (1995) 20-hydroxyecdysone in *Ajuga* hairy root controlling intracellular phosphate content based on kinetic model. J Ferment Bioeng 80: 362–368

XXII Genetic Transformation of *Solanum aculeatissimum*

T. IKENAGA[1] and T. MURANAKA[2]

1 Introduction

Steroid saponins, aculeatiside A and aculeatiside B (Fig. 1), were isolated from the root of *Solanum aculeatissimum* Jacq. (Solanaceae) (Saijo et al. 1983). Natural steroid sapogenin (diosgenin), from tubers of *Dioscorea* species is used in the commercial production of steroid hormones, female steroid hormones, and steroidal pharmaceuticals (Ikegami 1981; Imada et al. 1981; Tal et al. 1984; Rokem et al. 1985). Since nuatigenin and isonuatigenin obtained by hydrolysis of aculeatiside A and aculeatiside B (Fig. 1) can be transformed into pregdienolone (Tschesche et al. 1964; Imada et al. 1981; Saijo et al. 1983), it is conceivable that *S. aculeatissimum* may be a more useful source of pregnane derivatives than the *Dioscorea* species (Imada et al. 1981). This plant also contains these steroid saponins at a high level; however, it had never been cultivated, and the properties and seasonal variations of growth and steroid saponin production, etc. were also not known. Results of an earlier experiment (Ikenaga et al. 1988a) showed that the content of steroid saponins per dry weight, aculeatiside A and B, increased as the plant grew, reached a peak in November, and decreased slightly in December when the aerial parts of the plant died. The yield of steroid saponins per plant increased in accordance with the rise in root dry weight until November, and then decreased by slow degrees (Table 1).

It was therefore suggested that November was the appropriate time for harvest from the standpoint of steroid saponin production. The total content of steroid saponins was more than 10% in November. Since this content is more than or the same as the steroid saponin content in *Dioscorea* species (Imada et al. 1981), *S. aculeatissimum* would be profitable as a source plant for steroid hormone material.

[1] School of Environmental Sciences, Nagasaki University, 1-14 Bunkyo-machi, Nagasaki, 852 Japan
[2] Biotechnology Laboratory, Sumitomo Chemical Co., Ltd., 2-1,4-chome Takatsukasa, Takarazuka, Hyogo, 665 Japan

Biotechnology in Agriculture and Forestry, Vol. 45
Transgenic Medicinal Plants (ed. by Y.P.S. Bajaj)
© Springer-Verlag Berlin Heidelberg 1999

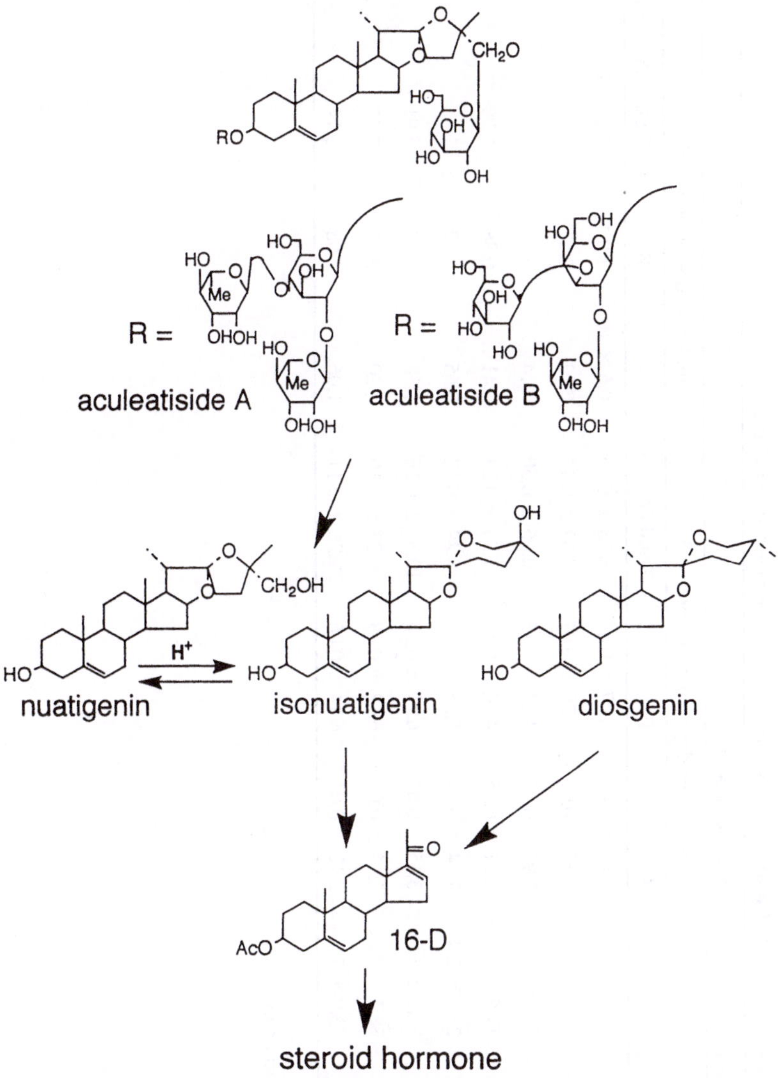

Fig. 1. Schematic representation of steroidal hormone synthesis. (Ikenaga et al. 1988a)

2 Genetic Transformation

Recently, transformation of *Solanum* species via *Agrobacterium* spp. has been described in several reports. Three different accessions of *S. commersonii* were tested for transformability with three *A. tumefaciens* strains (Cardi et al. 1992). Cultural conditions as well as bacterial and plant genotypes had significant effects on transformation frequencies. The transgenic nature of regenerants was demonstrated by Southern blot analysis, callus growth on medium selective for the transgene (*npt*II), and expression of a cotransformed

T. Ikenaga and T. Muranaka

Table 1. Growth and production of steroid saponins in *Solanum aculeatissimum* during one vegetation period

| Harvest time, 1997 | Length of main stem (cm) | Dry weight (g) | | Steroid saponins | | | | | |
| | | Aerial part | Root | Content per dry weight of root (%) | | | Yield per root of plant (g) | | |
				A*	B*	Total (A + B)	A*	B*	Total (A + B)
June	8.0 ± 0.5	0.44 ± 0.05	0.22 ± 0.03	0.26 ± 0.01	0.47 ± 0.01	0.73 ± 0.06	0.6×10^{-3}	1.0×10^{-3}	1.6×10^{-3}
July	29.7 ± 1.1	14.9 ± 0.0	2.2 ± 0.1	1.52 ± 0.05	3.29 ± 0.07	4.82 ± 0.11	0.03	0.07	0.11
August	55.3 ± 1.8	49.9 ± 5.5	7.5 ± 0.8	1.94 ± 0.17	4.22 ± 0.13	6.16 ± 0.50	0.14	0.32	0.46
September	87.1 ± 5.3	156.4 ± 20.1	17.9 ± 2.0	2.31 ± 0.15	4.94 ± 0.12	7.25 ± 0.24	0.41	0.88	1.29
October	94.0 ± 7.7	213.3 ± 75.2	23.8 ± 2.4	2.74 ± 0.04	5.55 ± 0.08	8.29 ± 0.12	0.65	1.32	1.98
November	104.3 ± 4.5	292.2 ± 49.2	37.7 ± 3.9	3.65 ± 0.07	6.71 ± 0.10	10.36 ± 0.16	1.38	2.53	3.91
December	113.1 ± 5.9	318.1 ± 30.8	38.7 ± 2.1	3.11 ± 0.03	6.12 ± 0.08	9.23 ± 0.10	1.20	2.37	3.57
January	–	–	33.9 ± 2.7	3.12 ± 0.04	6.14 ± 0.06	9.26 ± 0.11	1.06	2.08	3.14

* A = aculeatiside A, B = aculeatiside B.
Each value is mean of ten plants ± standard error.

gus gene. Inheritance patterns for one of the transgenes were determined by sexual crosses and showed Mendelian segregation for single and multiple integrated copies of the *neo* marker.

In shake cultures of *S. aviculare* hairy roots, the maximum alkaloid yield was comparable with levels found in the aerial parts of intact plants. Taking into account both growth and alkaloid yield, accumulation of steroidal alkaloids was improved by about 40% at gibberellic acid concentrations of 10 and 100 µg/l (Subroto and Droran 1994).

Hairy root cultures of *S. mauritianum* grew rapidly and could be maintained using a MS medium (Murashige and Shoog 1962) supplemented with 0.1 g/l myo-inositol and 3% sucrose, as either a solid or liquid culture. Under these conditions, the solasodine level was much higher in the hairy roots than in the roots of an intact plant (Drewes and van Staden 1995).

We used grafting to determine where steroid saponins are synthesized in *S. aculeatissimum*, which contains aculeatiside A and aculeatiside B at high levels in its roots (Ikenaga et al. 1988a). The results suggest that the steroid saponins aculeatiside A and aculeatiside B were synthesized in the root (Ikenaga et al. 1990). Moreover, soft callus produced less steroidal saponins, but were recognized in hard callus which was regenerating roots (unpubl.).

From these observations, it was necessary to culture roots to produce steroidal saponins in vitro. However, it was very difficult to increase their roots by root culture; therefore, hairy roots of *S. aculeatissimum* were cultured.

Although numerous attempts have been made to produce pharmaceuticals by hairy root cultures (Flores et al. 1987; Signs and Flores 1990). There are no reports in which hairy root cultures have been used to produce steroidal saponins. In this chapter our work on the production of the steroidal saponins aculeatiside A and aculeatiside B in *S. aculeatissimum* hairy root cultures transformed with *Agrobacterium rhizogenes* (strain 15834) is described. Culture conditions such as basal media and auxins, as well as the effects of light on root growth and steroidal saponin production were also investigated.

2.1 Materials and Methods

2.1.1 Hairy Root Induction and Culture

Hairy roots of *Solanum aculeatissimum* were initiated with sterilized plantlets inoculated with *Agrobacterium rhizogenes* strain ATCC 15834. Excess bacteria were eliminated on agar-solidified MS medium supplemented with 0.5 mg/ml Claforan (Hoechst Japan co., Tokyo). After 2 weeks of culture, the tips (2 to 3 cm in length) of newly formed roots (Fig. 2) were excised and placed on fresh medium of the same composition. Two weeks later, the grown root was cut off and placed on MS solid medium without Claforan. Axenic roots, 2 to 3 cm long, were transferred to 40 ml of B5 liquid medium (Gamborg et al. 1968) in 100-ml Erlenmeyer flasks. The roots were cultured at 25 °C in the dark, at 80 rpm (Circular Motion Shaker, Tomy Seiki Co., Tokyo). Hairy roots

Fig. 2. Hairy roots induced by the infection of aseptically grown plantlets of *S. aculeatissimum* with *A. rhizogenes* 4 weeks after infection. (Ikenaga et al. 1995)

were subcultured at 3-week intervals. One root inoculum (one single tip approx. 2 cm length and about 0.5 mg dry weight) was used for all studies in 100-ml Erlenmeyer flasks (five replicates) and was cultured under continuous fluorescent white light (ca. $60\,\mu E/m^2/s$).

2.1.2 Determination of Steroidal Saponins

Hairy roots were dried at 40 °C for 48 h and then powdered mechanically and dried again at 40 °C for 2 h. Crude extracts for quantitative analysis were obtained by adding 0.5 g of this hairy root powder to 100 ml ethanol. After filtration, the residue was extracted once again in the same manner. The whole filtrate was evaporated to less than 10 ml, transferred into a 10-ml volumetric flask, and diluted with methanol to the mark. After filtration on a Millipore filter (0.22 μm), the filtrate was used as a sample for HPLC. Aculeatiside A and aculeatiside B in the extract were identified by HPLC using a C-8 phase column with 30% acetonitril as the mobile phase, as described previously (Ikenaga et al. 1988b).

2.1.3 PCR Analysis of Insertion of the T-DNA

Four oligonucleotides, ROLB-TL1, ROLB-TR1, ROLB-TL3R, and ROLB-TR3R, were designed to identify the insertion of the TL-DNA or TR-DNA of the Ri plasmid into the plant genome by PCR amplification. ROLB-TL1 (5'-TAGCCGTGACTATAGCAAACCCCTCC-3') orROLB-TR1(5'-CAACC AGTCCTTTCGTGGTGAGATAA-3') corresponds to the N-terminal coding region of TL-DNA *rol* B gene (Slightom et al. 1986) or *rol* B^{TR} (Bouchez and Camilleri 1990), respectively, and ROLB-TL3R (5'-GGCTTCTTT CTTCAGGTTTACTGCAG-3')orROLB-TR3R(5'-AGCGCTC GTCCGTGTCTCCCCTGGCC-3') correspond to the complementary sequence of the C-terminal coding region of TL-DNA *rol* B gene or *roll* B^{TR}, respectively. An aliquot of genomic DNA isolated from three independent hairy root clones and untransformed root tissues was amplified in 50 µl reaction volume with 1 µM oligonucleotide primers, 200 µM of dNTPs and 0.5 µ TaKaRa Taq polymerase (Takara Shuzo, Kyoto, Japan) in a buffer containing 10 mM Tris HCl (pH 8.3), 50 mM KCl, 1.5 mM MgCl$_2$, and 0.001% gelatin. PCR amplification was performed in a programmed temperature control system (model PC-800, Astec, Tokyo, Japan) using the following parameters: 5 min at 95 °C, 2 min at 55 °C, and 2 min at 72 °C for 1 cycle, 1 min at 95 °C, 1 min at 55 °C, and 1 min at 72 °C for 29 cycles followed by an extra cycle with a 10-min extension step at 72 °C. Amplified DNAs were analyzed by EtBr staining after 0.8% agarose gel electrophoresis.

2.2 Results and Discussion

2.2.1 Establishment of Hairy Root Cultures

Each hairy root clone generated from seedlings by infection with *A. rhizogenes* was isolated and cultured in liquid B5 medium. To determine the insertion of TL-DNA or TR-DNA of the Ri plasmid, PCR was performed by using primers which can detect either TL-DNA *rol* B (Slightom et al. 1986) or the *rol* B^{TR} gene (Bouchez and Camilleri 1990). As shown in Fig. 3, PCR amplification analysis of DNAs from the hairy root clones yielded the expected 670-bp band with the primer combination ROLB-TL1 and ROLB-TL3R (lanes 2–4), although no visible band was obtained from untransformed root (lane 1). Also, DNAs from the hairy root clones yielded the expected 673-bp band with the primer combination ROLB-TR1 and ROLB-TR3R (lanes 6–8), and no band was obtained from untransformed root (lane 5). From these results, it was revealed that all hairy root clones examined contain both the TL-DNA and TR-DNA regions of the Ri plasmid. The primer specificity was confirmed that no band was observed with the primer combination ROLB-TL1 (TL-DNA *rol* B-specific sequence) and ROLB-TR3 (*rol* B^{TR}-specific sequence; lanes 10–12).

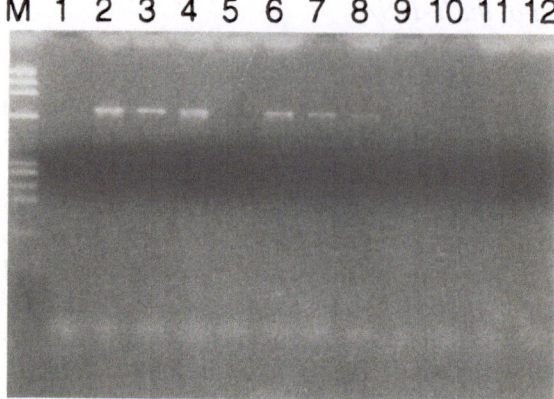

Fig. 3. PCR amplification of the T-DNA region from genomic DNA isolated from untransformed root (*lanes 1, 5, 9*), hairy root clone #1 (*lanes 2, 6, 10*), hairy root clone #2 (*lanes 3, 7, 11*) and hairy root clone #3 (*lanes 4, 8, 12*). Primer combinations are ROLB-TL1/ROLB-TL3R to detect the TL-DNA *rol* B region (*lanes 1–4*), ROLB-TR1/ROLB-TR3R for the *rol* B^{TR} region (*lanes 5–8*), and ROLB-TL1/ROLB-TR3R (*lanes 9–12*). M φ × 174 *Hae*III *DNA* size marker (500 ng). (Ikenaga et al. 1995)

2.2.2 *Effect of Light on Growth and Steroidal Saponin Production*

The roots of seedlings not inoculated with *A. rhizogenes* showed no growth either in the light condition (Fig. 4) or in the dark; opines were also not contained in them.

The influence of light on growth and steroidal saponin production was investigated by culturing hairy roots in hormone-free B5 liquid medium supplemented with 3% sucrose for 21 days. The appearance of transformed roots was different in dark- and light-grown cultures. Dark-grown roots showed a pale brown color, whereas light-grown roots showed a greenish color (Fig. 4). From an initial inoculum of 0.5 mg (one root tip), the dry weight increased more than 300 times in light-grown hairy root cultures (Fig. 5). In contrast, dark-grown hairy roots increased in dry weight about 200 times. The light-grown hairy roots contained the steroidal saponins aculeatiside A (0.020% dry weight) and aculeatiside B (0.019% dry weight; Fig. 3). These levels were lower than in the parent plant, their content per dry weight of roots was aculeatiside A 3.65% and aculeatiside B 6.71% (Ikenaga et al. 1995).

Steroidal saponins were not detected in dark-grown hairy roots. The fact that the hairy roots produce the secondary metabolite only under light conditions is in agreement with the results reported on *Lippia dulcis* (Sauerwein et al. 1991) and *Digitalis lanata* (Yoshimatsu et al. 1990).

Transformed roots were incubated in B5 liquid medium supplemented with 3% sucrose during a 10-week period. At 1-week intervals, five flasks were harvested and both growth and steroidal saponin production were measured. From an initial inoculum of about 0.5 mg of root tip, the dry weight increased

Fig. 4. Hairy roots of *S. aculeatissimum* cultured in hormone-free B5 liquid medium supplemented with 3% sucrose for 21 days in the light (*left*) and in the dark (*center*), and nontranformed roots (*right*) from seedlings cultured in the same medium for 21 days in the light. (Ikenaga et al. 1995)

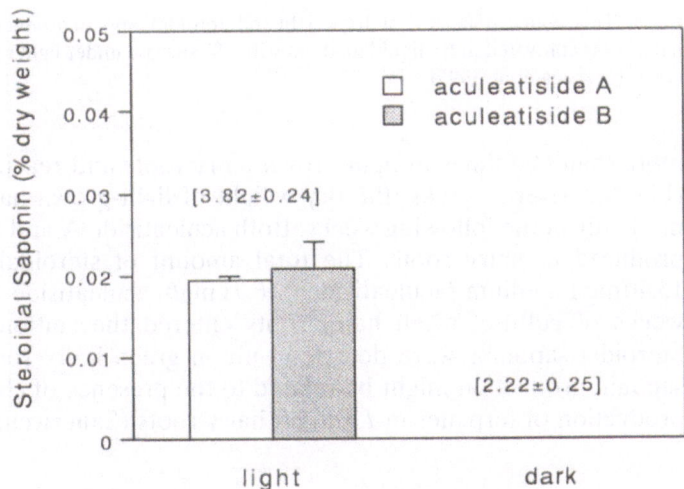

Fig. 5. Steroidal saponin content in *S. aculeatissimum* hairy root cultured in B5 liquid medium with 3% sucrose under light and dark conditions. *Numbers in brackets* show the dry weight/11 medium (g). Values are the mean of five observations and the standard error. (Ikenaga et al. 1995)

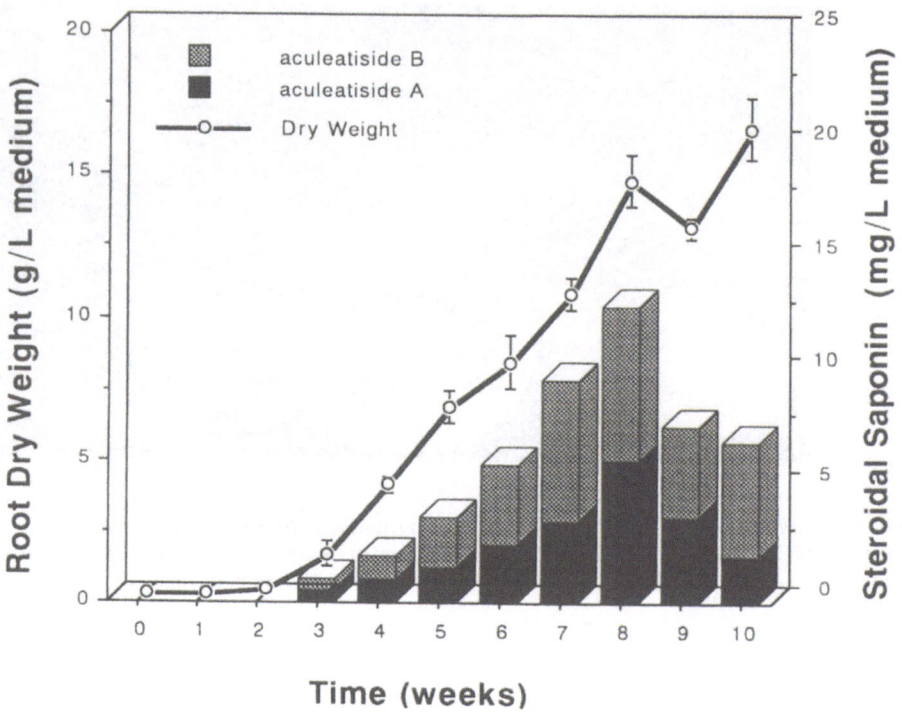

Fig. 6. Time course of hairy root dry weight and steroidal saponin production in *S. aculeatissimum* hairy roots cultivated in B5 liquid medium with 3% sucrose under light conditions. *Bars* standard error. (Ikenaga et al. 1995)

more than 600 times in light-grown hairy roots and reached 17.4 g/l medium (Fig. 6). After 2 weeks, the dry weight of light-grown hairy roots increased markedly in the following weeks. Both aculeatiside A and aculeatiside B were produced in hairy roots. The total amount of steroidal saponins reached 13.40 mg/l medium (aculeatiside A 6.71 mg/l, aculeatiside B 6.39 mg/l) after 8 weeks of culture, when hairy roots entered the stationary growth phase. Steroidal saponins were detected only in green hairy roots. Thus, steroidal saponin production might be related to the presence of chloroplasts as in the production of terpenes in *L. dulcis* hairy roots (Sauerwein et al. 1991).

2.2.3 Effect of Basal Medium on Growth and Steroidal Saponin Production

The effect of various culture media containing 3% sucrose on growth and steroidal saponin content was examined after a culture period of 28 days under light conditions. The media tested were MS, B5, and LS (Linsmaier and Skoog 1965). Green hairy roots showed fastest growth in B5 medium, while growth in LS medium was poor (Fig. 5). Aculeatiside A and aculeatiside B were detected in hairy roots cultured in MS and B5 medium. Those in B5 medium showed

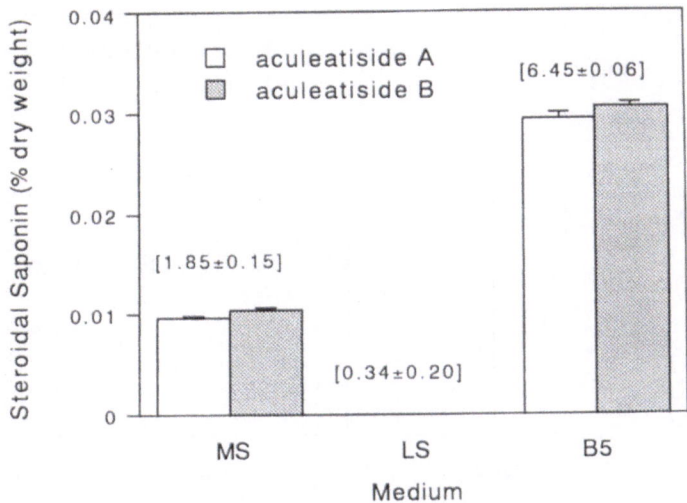

Fig. 7. Steroidal saponin content in *S. aculeatissimum* hairy roots cultured in different media with 3% sucrose under light conditions. *Numbers in brackets* show the dry weight/1l medium (g). Values are the mean of five observations and the standard error. (Ikenaga et al. 1995)

a higher steroidal saponin content (aculeatiside A 0.030% dry weight; aculeatiside B 0.028% dry weight). Saponins were not observed in transformed roots cultured with LS medium (Fig. 7).

The hairy root cultures may grow rapidly using MS medium supplemented with myo-inositol, as Drewes and van Staden (1995) reported in *S. mauritianum.*

2.2.4 Effect of Auxin on Growth and Steroidal Saponin Production

Auxin influenced the growth of dark-grown hairy roots; these grew best in B5 medium containing 100 μg/l NAA (unpubl.). In view of this result, auxin (NAA, IAA, or 2,4-D) was added to B5 medium at 100 μg/l, in order to increase growth of transformed roots under light conditions. Although the addition of NAA and IAA had no effect on growth when compared to hormone-free growth under light conditions, the production of steroidal saponins in hairy roots tended to increase (Fig. 8). The steroidal saponins yield per 1l of medium (aculeatiside A 2.75 mg/l; aculeatiside B 3.50 mg/l) was highest when 100 μg/l NAA was added to the medium. The addition of 2,4-D inhibited both hairy root growth and steroidal saponin production. The effect of auxins on root growth and the production of steroidal saponins in the hairy roots of *S. aculeatissimum* was similar to that on root growth and hernandulcin production in *L. dulcis* hairy roots, although the auxin concentrations were different (Sauerwein et al. 1991).

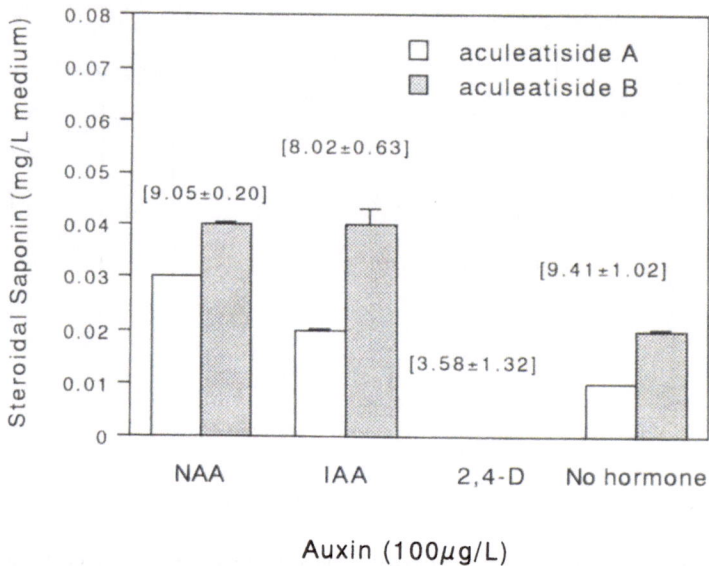

Fig. 8. Steroidal saponin product in light-grown *S. aculeatissimum* hairy roots cultured in B5 liquid media with 3% sucrose with and without various auxins under light conditions. *Numbers in brackets* show the dry weight/1l medium (g). Values are the mean of five observations and the standard error. (Ikenaga et al. 1995)

The results obtained in this study indicate that transformed roots can be used to produce steroidal saponins as a source for steroidal hormone materials. Further experiments will be required to investigate the relationship between culture conditions and biosynthesis in order to increase steroidal saponin production in hairy roots.

For example, addition of gibberellic acid to the medium may promote growth of hairy roots in *S. aculeatissimum*, because growth in hairy roots of *S. aviculare* was enhanced when supplemented with it (Subroto and Droran 1994).

2.2.5 Regeneration of Plants from Hairy Roots

Upon transfer from dark to light hairy roots cultured without phytohormones showed extensive regeneration of adventitious buds within 2 weeks throughout the surface of the roots. Their buds developed into plantlets which upon excision could be easily grown into whole plants.

Often exclusively associated with the transgenic regenerants from hairy roots are various phenotypic alterations such as wrinkled leaves, rounded or indented leaf margins, shortened internode distances, reduced apical dominance, and reduced fertility (White and Sinkar 1987). Wrinkled leaves and shortened internode distances were also observed in regenerants of *S. aculeatissimum* hairy roots induced with *A. rhizogenes* (Fig. 9).

Fig. 9. Transformed plant (*left*) regenerated from hairy root and nontransformed plant (*right*) generated from seed in *S. aculeatissimum*. (Ikenaga et al. 1995)

The regenerated shoots from hairy roots were proved by PCR amplification analysis to be transgenic, which gave the same band pattern of the hairy root as shown in Fig. 1.

The regenerated plants were aseptically cultivated for 3 months in vitro using soil at 25 °C under 12 h light condition. Seedlings grown from seed were used as control. In the results of the experiment, steroidal saponin production in roots of transgenic plant was at the same level as in the control, although their contents were low. A few plants are now growing in pots.

3 Summary and Conclusions

Hairy root cultures of *Solanum aculeatissimum* were established by transformation using *Agrobacterium rhizogenes* strain 15834. Root growth and production of steroidal saponins were investigated under various culture conditions. Transformed roots grew better in B5 medium containing 3% sucrose under continuous light than in the dark. Also, the roots turned light green when cultured under continuous light. Green hairy roots produced aculeatiside A (6.71 mg/l) and aculeatiside B (6.39 mg/l) after 8 weeks of culture, while no steroidal saponins were detected in hairy roots cultured in the dark. Of the three culture media tested, B5 medium was superior for growth and steroidal saponin production. Growth and steroidal saponin production were enhanced when 100 μg/l auxin, except for 2,4-D, was added to the medium; the addition of 2,4-D inhibited growth. Production of steroidal saponins was highest with NAA. Transformed roots used in this experiment confirmed that the hairy roots examined contain both the TL-DNA and TR-DNA regions of the Ri plasmid by PCR amplification analysis of DNA.

Shoots regenerated from the hairy roots contained both DNA regions of the Ri plasmid. Their roots had a steroidal saponin content at the same level as roots of untransformed plants when they were cultured at 25 °C under 12-h light.

References

Bouchez D, Camilleri C (1990) Identification of a putative *rol* B gene on the TR-DNA of the *Agrobacterium rhizogenes* A4 Ri plasmid. Plant Mol Biol 14:617–619

Cardi T, Iannamico V, D'Ambrosio F, Filippone E, Lurquin F (1992) *Agrobacterium*-mediated genetic transformation of *Solanum commersonii* Dun. Plant Sci 87:179–189

Drewes FE, Staden J van (1995) Initiation of and solasodine production in hairy root cultures of *Solanum mauritianum* Scop. Plant Growth Regul 17:27–31

Flores HE, Hoy MW, Pickard JJ (1987) Secondary metabolites from root cultures. *TIBTECH* 5:64–69

Gamborg OL, Miller RA, Ojima K (1968) Nutrient requirements of suspension cultures of soybean root cells. Exp Cell Res 50:151–158

Ikegami N (1981) Steroid production in industry. J Syn Org Chem 39:433

Ikenaga T, Kikuta S, Itimura K, Nakashima K, Matubara T (1988a) Growth and production of steroid saponin in *Solanum aculeatissimum* during one vegetation period. Planta Med 140–142

Ikenaga T, Kikuta S, Nakashima K (1988b) High-performance liquid chromatographic determination of steroidal sapnin, aculeatiside A and B in *Solanum aculeatissimum*. JACQ. Chem Pharm Bull 36:416–418

Ikenaga T, Kikuta S, Kistuki M, Yamada M, Nakashima K (1990) Production of steroid saponin in grafts: *Solanum aculeatissimum* and *Lycopersicon esculentum*. Hort Science 25:1657–1658

Ikenaga T, Oyama T, Muranaka T (1995) Growth and steroidal saponin production in hairy root cultures of *Solanum aculeatissimum*. Plant Cell Rep 14:418–422

Imada Y, Ishikawa H, Nishikawa D (1981) Development of new steroid fermentation. Nippon Nogeikagaku Kaishi 55:713–721

Linsmaier EM, Skoog F (1965) Organic growth factor requirements for tobacco tissue culture. Physiol Plant 18:100–127

Murashige T, Skoog F (1962) A revised medium for rapid growth and bio-assays with tobacco tissue cultures. Physiol Plant 15:473–497

Rokem JS, Tal B, Goldberg I (1985) Methods for increasing diosgenin production by *Dioscorea* cells in suspension cultures. J Nat Prod 48:210–222

Saijo R, Fuke C, Murakami K, Nohara T, Tomimastu T (1983) Two steroidal glycosides, aculeatiside A and B from *Solanum aculeatissimum*. Phytochemistry 22:733–736

Sauerwein M, Yamazaki T, Shimomura K (1991) Hernandulcin in hairy root cultures of *Lippia dulcis*. Plant Cell Rep 9:579–581

Signs MW, Flores HE (1990) The biosynthesis potential of plant roots. BioEssays 12:7–13

Slightom JL, Durand-Tardif M, Jouanin L, Tepfer D (1986) Nucleotide sequence analysis of TL-DNA of *Agrobacterium rhizogenes* agropine type 7 plasmid: identification of open reading frames. J Biol Chem 261:108–121

Subroto MA, Droran PM (1994) Production of steroidal alkaloids by hairy roots of *Solanum aviculare* and the effect of gibberellic acid. Plant Cell Tissue Organ Cult 38:93–102

Tal B, Rokem JS, Gressel J, Goldberg I (1984) The effect of chlorophyll-bleaching herbicides on growth, cartenoid and diosgenin levels in cell suspension cultures of *Digitalis deltoidea*. Phytochemistry 23:1333–1335

Tschesche R, Richert KH (1964) Über Saponine der Spirostanoleihe-xi Nuatigenin, ein Cholegenin-Analogon des Pflanzenreiches. Tetrahedron 20:387–398

White FF, Sinkar VP (1987) Molecular analysis of root induction by *Agrobacterium rhizogenes*. In: Hohn T, Schell J (eds) Plant DNA infection agents. Springer, Berlin Heidelberg New York, pp 140–177

Yoshimatsu K, Satake M, Shimomura K, Sawada J, Terao T (1990) Determination of cardenolides in hairy root cultures of *Digitalis lantana* by enzyme-linked immunosorbent assay. J Nat Prod 53:1498–1502

XXIII Genetic Transformation of *Solanum commersonii* Dun.

T. CARDI

1 Introduction

Solanum commersonii (Cmm) is a tuber-bearing wild relative of the common potato (*S. tuberosum*). It is widespread in Paraguay and along the coasts of Argentina, Uruguay, and southern Brazil, from sea level up to about 400 m (Correll 1962; Hawkes 1990). Typically, it is a small, bushy plant, with flowers (deep mauve to purple on the outer surface, paler within) arranged in compact inflorescences. Generally, it is diploid ($2n = 2x = 24$) with an endosperm balance number (EBN) equal to 1, but autotriploid forms ($2n = 3x = 36$) are also known (Hanneman and Bamberg 1986).

Since the early 1970s, *S. commersonii* has been extensively studied for its response to cold stress, being considered both highly frost-resistant and able to cold-harden (Chen and Li 1980). Several other interesting characteristics are also present in this species, including resistance to biotic and abiotic stresses, plant and tuber quality-related traits, and meiotic mutations (Table 1). However, due to the EBN value, crossability of *S. commersonii* with diploid 2EBN and tetraploid 4EBN *Solanum* genotypes is poor, and its utilization in potato breeding has been limited so far. Only recently has gene transfer from this species been accomplished through manipulation of ploidy or somatic fusion of protoplasts (Ehlenfeldt and Hanneman 1988; Cardi et al. 1993a; Carputo et al. 1995). Further, it can be foreseen that in the near future genes cloned from Cmm (Table 1) will be transferred to cultivated potato and other species by genetic transformation.

Genetic transformation of *S. commersonii* can play a significant role in genetic and breeding studies involving this species. Selectable marker genes, such as those encoding resistance to antibiotics or herbicides, can be transferred into fusion partners for the selection of somatic hybrids or for monitoring partial genome transfer, allowing the exploitation of various Cmm accessions (Cardi et al. 1992). Moreover, isolated genes or their promoters can be used in chimeric constructs and inserted into Cmm in either sense or antisense orientation to study their function and expression patterns (Zhu et al. 1995a, 1996). Finally, transformation technology can be used to produce

CNR-IMOF, Research Institute for Vegetable and Ornamental Plant Breeding, Via Universita' 133, 80055 Portici, Italy

Biotechnology in Agriculture and Forestry, Vol. 45
Transgenic Medicinal Plants (ed. by Y.P.S. Bajaj)
© Springer-Verlag Berlin Heidelberg 1999

Table 1. Some useful agronomic traits present in *Solanum commersonii*[a] and genes with putative relevance cloned from this species

Trait	Pest	Relevant genes cloned	Reference	
			Traits	Genes
Biotic stress resistance				
Virus	PVX, PVM	–	Tozzini et al. (1991), Bamberg et al. (1994)	–
Bacteria	*P. solanacearum, E. carotovora*	–	Bamberg et al. (1994)	–
Fungi	*A. solani, Verticillium* spp.	Osmotin-like proteins	Alam (1985), Bamberg et al. (1994)	Zhu et al. (1993, 1995a,b)
Insects	*P. operculella M. euphorbiae*	–	Chavez et al. (1988), Bamberg et al. (1994)	–
Abiotic stress resistance				
Freezing tolerance	–	Stearoyl-ACP- and oleoyl-desaturases, dehydrin	Chen and Li (1980), Bamberg et al. (1994)	Grillo et al. (1996), Baudo et al. (1996)
Ability to cold harden	–	"	Chen and Li (1980)	"
Heat tolerance	–	"	Bamberg (1995)	"
Plant and tuber quality				
High dry matter	–	–	Hanneman and Bamberg (1986)	–
Flowering vigor	–	–	Bamberg et al. (1994)	–
Meiotic mutations				
Asynapsis	–	–	Johnston et al. (1986)	–

[a] A more comprehensive list can be found in Hanneman and Bamberg (1986), Hawkes (1990), and Bamberg et al. (1994).

mutants by insertional mutagenesis or to tag and eventually isolate target genes, using either T-DNA (Fraley et al. 1985; De Block 1993) or, provided that their mobility in Cmm is demonstrated, heterologous transposable elements (Knapp et al. 1988).

Genetic transformation could also be aimed at in vitro production of secondary metabolites. In fact, various alkaloids are produced in different species of the Solanaceae family, and some of them are reported to have anticancer properties (Valkonen et al. 1996). Five glycoalkaloids, namely commersonine, demissine, tomatine, dehydrocommersonine, and the hitherto unknown dehydrodemissine, are mainly found in *S. commersonii* (Osman et al. 1976; Deahl et al. 1993; Vázquez et al. 1997). Hairy root cultures obtained by transformation with *Agrobacterium rhizogenes* could be used as a reproducible and continuous source of these compounds to study not only their properties against potato pests, but also their toxicity to humans and animals, and their potential pharmaceutical uses (Valkonen et al. 1996). In other *Solanum* species, such as *S. aviculare* (Subroto and Doran 1994), *S. mauritianum* (Drewes and van Staden 1995), and *S. aculeatissimum* (Ikenaga et al. 1995), it has been demonstrated that transgenic root cultures produce steroidal compounds at levels significantly higher than those obtained from disorganized callus or suspension cultures.

2 Genetic Transformation

2.1 Purposes

Hitherto, genetic transformation in *S. commersonii* has been applied basically for two purposes: the introduction of marker genes and gene expression studies (Tables 2, 3).

In order to produce parental lines carrying marker genes to be used for the selection of hybrid cell lines after somatic fusion experiments, *nptII* and *uidA* genes were introduced into three accessions of *S. commersonii*. Factors affecting transformation efficiency were studied and optimal conditions for each genotype were devised. The presence and expression of the transgenes were analyzed in the primary transformants and in their progenies (Cardi et al. 1992).

Baertlein et al. (1992) introduced the ice nucleation gene *inaZ* from *Pseudomonas syringae* into *S. commersonii* and *Nicotiana tabacum* to study its expression in a freezing-tolerant and a freezing-sensitive species, respectively.

Table 2. Summary of published transformation experiments on *Solanum commersonii*

Aim of the experiment	Genes introduced	Traits analyzed in the transgenic plants	Reference
Introduction of marker genes	*nptII, uidA*	Kanamycin resistance, GUS expression	Cardi et al. (1992)
Gene expression studies	*inaZ* (from *P. syringae*, sense and antisense orientation), *nptII*	Ice nucleation activity, kanamycin resistance	Baertlein et al. (1992)
"	*uidA* (w/Cmm osmotin-like protein gene promoters), *nptII*	GUS expression, kanamycin resistance	Zhu et al. (1995a)
"	pA13 cDNA (osmotin-like protein gene from Cmm, sense and antisense orientation), *nptII, uidA*	Resistance to *Phytophthora infestans* and freezing stress, kanamycin resistance	Zhu et al. (1996)

Table 3. Genetic transformation in various accessions of *Solanum commersonii*

Accession	Explant	*Agrobacterium* strain[a]	Plasmid vector (type)	Selection of transgenic explants (mg/l)	Transgenes Copies[b] (no.)	segregation[c] (kan^r:kan^s)	Reference
PI 243503	Leaf segments	GV2260	p35SGUSINT (binary)	Kanamycin (25, 50)	ns[d]	ns	Cardi et al. (1992)
PI 472833	"	EHA105	p35SGUSINT (binary)	"	1,3	1.1:1, 3.2:1	"
PI 472834	"	GV2260	p35SGUSINT (binary)	"	2	1.1:1	"
PI 458817	Leaf disks	"	pGSJ280, pGSH160 (cointegration)	Kanamycin (100)	ns	ns	Baertlein et al. (1992)
ns	"	LBA4404	pBI101.1 (binary)	Kanamycin (ns)	1–6	ns	Zhu et al. (1995a)
PI 458817	"	"	pBI121.1 (binary)	Kanamycin (ns)	1–3	ns	Zhu et al. (1996)

[a] Only the strains that gave the best results are reported.
[b] Based on Southern analysis.
[c] After cross with control (kan^s) *S. commersonii*.
[d] Not specified.

Transgenic plants containing the bacterial gene in sense and antisense orientation were produced and submitted to ice nucleation assays. In both species, transformants with sense constructs showed increased ice nucleation activity over untransformed controls, whereas plants transformed with antisense constructs were undistinguishable from the untransformed controls.

Osmotin-like proteins are encoded by a multigene family in *S. commersonii*; the deduced amino acid sequence of one cDNA clone isolated from Cmm (pA13) shows 89 and 91% identity with tobacco osmotin and tomato NP24 proteins, respectively (Zhu et al. 1993, 1995b). To study the transcriptional activation of two osmotin-like protein gene promoters isolated from the same species, Cmm plants were transformed with chimeric constructs containing the 5' flanking DNA sequences of the two genes fused to the coding region of the *uidA* gene. Transgenic plants were challenged with abscissic acid, low temperature, sodium chloride, salicylic acid, wounding, or fungal infection, and then assayed by enzymatic and histochemical analysis for GUS activity (Zhu et al. 1995a). In a second series of experiments, to further study the role of osmotin-like proteins in *S. commersonii*, Zhu et al. (1996) produced transgenic Cmm plants constitutively expressing sense or antisense RNAs from chimeric gene constructs consisting of the cauliflower mosaic virus 35S promoter and the pA13 cDNA. The transgenic plants were evaluated for fungal and freezing resistance. Analyses of transformants obtained in both studies indicated that the 5' flanking regions of osmotin-like protein genes are involved in regulating gene expression at the transcriptional level after treatments with ABA, SA, or NaCl, wounding, and fungal infection, but not after exposure to low temperature. Further, pA13 osmotin-like protein produced by *S. commersonii* has a strong antifungal activity, but is not a major determinant of freezing tolerance.

Studies on the manipulation of secondary metabolites such as commersonine and demissine in transgenic plants would be rewarding.

2.2 Methodology

All transformation procedures employed so far in *S. commersonii* were based on the regeneration and cocultivation of leaf explants with *Agrobacterium tumefaciens*. Several protocols for in vitro culture and shoot regeneration, however, have been established in different genotypes of Cmm, and they could be used for either transient or stable transformation.

Four accessions, i.e. PI 243503, PI 458317, PI 472833, and PI 472834, maintained at the Inter-Regional Introduction Station, Sturgeon Bay, WI, USA, have been used in both regeneration and transformation studies (Tables 3, 4).

From the above accessions, leaf protoplasts as well as stem and leaf explants were used in regeneration studies (Cardi et al. 1990, 1993b; Iapichino et al. 1991). Explants derived from in vitro shoot cultures, MS (Murashige and Skoog 1962) basal medium, and combinations of growth regulators comprising NAA and IAA as auxins, ZEA and BAP as cytokinins, and GA$_3$, were used in

Table 4. Invitro shoot regeneration in various accessions of *Solanum commersonii*

Accession	Explant[a]	Basal medium[b]	Growth regulators (μM)	Regeneration frequency (%)	Shoots/ explant (no.)	Diploid regenerants (%)	Reference
PI 243503	Mesophyll protoplasts	MS	NAA (0.05)/ZEA (4.6) → BAP (1.1)/GA$_3$ (0.3)	0	–	–	Cardi et al. (1990)
"	Leaves	"	NAA (1)/BAP (10) → BAP (10)/GA$_3$ (14.3)	71	7.7	29	Cardi et al. (1993b)
PI 472833	Mesophyll protoplasts	"	NAA (0.05)/ZEA (4.6) → BAP (1.1)/GA$_3$ (0.3)	35	2.2	14	Cardi et al. (1990)
"	Leaves	"	NAA (1)/BAP (10) → BAP (10)/GA$_3$ (14.3)	52	6.1	60	Cardi et al. (1993b)
PI 472834	Mesophyll protoplasts	"	NAA (0.05)/ZEA (4.6) → BAP (1.1)/GA$_3$ (0.3)	66	2.0	24	Cardi et al. (1990)
"	Leaves	"	NAA (1)/BAP (10) → BAP (10)/GA$_3$ (14.3)	75	6.5	44	Cardi et al. (1993b)
PI 458317	Leaves, stems	MS-Co	IAA (11.4)/ZEA (22.8)	95–100	9.9–11.7	ns[c]	Iapichino et al. (1991)

[a] In all cases, in vitro shoot cultures were used as source of explants.
[b] MS = Murashige and Skoog (1962); MS-Co = MS + 1 g/l casein hydrolysate +40 mg/l adenine sulphate.
[c] Not specified.

all protocols (Table 4). Somaclonal variation was estimated by checking the number of chloroplasts in the stomata guard cells and of chromosomes in root tip cells of regenerated clones (Cardi et al. 1993b).

Agrobacterium strains GV2260, EHA105, and LBA4404 with either binary (p35SGUSINT, pBI101.1, and pBI121.1) or cointegration (pGSJ280, pGSH160) vectors were employed in cocultivation with leaf explants (Table 3). GV2260 has the C58C1 chromosomal background and the disarmed helper plasmid pGV2260 derived from the octopine Ti plamid pTiB6S3, EHA105 has the A281 chromosomal background and the pEHA105 helper plasmid derived from the leucinopine Ti plasmid pTiBo542, while LBA4404 has the Ach5 chromosomal background and the helper plasmid pAL4404 derived from the octopine Ti plasmid pTiAch5 (Hoekema et al. 1983; Hood et al. 1986; Deblaere et al. 1987; Melchers and Hooykaas 1987; Leone et al. 1993). All three binary vectors previously mentioned are pBin19 derivative (Jefferson 1987; Vancaynnet et al. 1990), whereas the two cointegrative vectors derived from the prototype cointegration vector pGV1500 (Deblaere et al. 1987). Besides the transgenes of interest, all vectors carried the *nptII* gene and hence the transgenic explants were selected in the presence of 25–100 mg/l of kanamycin. The three binary vectors also carry the *uidA* reporter gene; in p35SGUSINT the latter is interrupted by the potato ST-LS1 intron (Vancaynnet et al. 1990). In some experiments, factors potentially influencing transformation efficiency were tested. They included: the use of a feeder layer of potato cells, the postponement of kanamycin selection, the use of acetosyringone during growth of bacteria and cocultivation of explants (Fig. 2). To confirm the presence and expression of transgenes, transgenic plants were checked by Southern, Northern, and Western blot analysis, callus growth and GUS assays, and germination of progenies on selective medium (Baertlein et al. 1992; Cardi et al. 1992; Zhu et al. 1995a, 1996).

2.3 Results and Discussion

Several in vitro systems for the transformation of potato and related wild species are available (Mitten et al. 1990; Visser 1991; Fehér et al. 1991; Kumar et al. 1995; Liu et al. 1995). In *S. commersonii*, leaves are the preferred explants in transformation experiments with *Agrobacterium tumefaciens*; they are relatively easy to obtain and cultivate in vitro, by contrast with tuber disks or other explants, while the variability observed after in vitro culture is lower in comparison with other systems, such as callus or protoplast cultures (Ramulu 1987). Protoplasts with good transformability and regenerability, however, can be used for direct gene transfer of naked DNA in experiments to study transient expression of specific constructs as well as in transformation experiments with large and/or relatively uncharacterized sequences, such as in "shotgun"-type cloning or complementation analyses with YAC vectors (Fraley et al. 1985; Davey et al. 1989; Fehér et al. 1991). Direct gene transfer using protoplasts, however, has not yet been demonstrated with this technique.

Before starting any transformation activity, it is important to have some information about the regeneration procedures available and the degree of genetic variability induced in vitro. Although differences due to genotype, explant, and/or protocol are evident, in vitro regeneration of shoots in *S. commersonii* was accomplished in all accessions tested so far (Table 4). When the same protocol was used (Cardi et al. 1990, 1993b), PI 472834 gave the best results both from protoplasts and tissue explants. Protoplast-derived calli from PI 243503 did not regenerate any shoots in initial experiments (Cardi et al. 1990), but regeneration was achieved in later experiments with a clone selected within the seed population of the same accession (Bastia et al. 1995). Variability for in vitro response among clones was also demonstrated for regeneration from leaf explants (Cardi et al. 1993b). In protoplast-derived calli, shoot induction was obtained in the presence of the auxin NAA and the cytokinin ZEA in a molar ratio close to 1:10. For shoot elongation, calli were transferred to a medium containing BAP and GA_3. In experiments with tissue explants, a combination of NAA and BAP (1:10) was the best for achieving a slight induction of callus and shoot regeneration after transfer to a medium containing BAP and GA_3 (Cardi et al. 1993b) (Fig. 1). Good results were also obtained with a one-step regeneration procedure in which IAA, an auxin weaker than NAA, and ZEA was used at a ratio of 1:2 (Iapichino et al. 1991). However, since in addition to a different protocol a different accession was also used, it is not possible to attribute differences in regeneration frequencies to either the genotype or the cultural conditions. About two shoots were obtained per protoplast-derived callus, while 6–12 shoots were regenerated from each leaf or internode explant (Table 4).

As expected, a proportion of regenerated shoots doubled their chromosome complement during in vitro culture. Although there were some differences due to genotype and regeneration procedure (Cardi et al. 1993b), the proportion of regenerants that remained diploid was higher after regeneration from tissue explants (29–60%) than from protoplast-derived calli (14–24%;

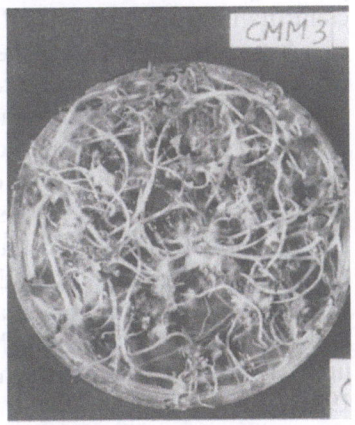

Fig. 1. In vitro shoot regeneration of *Solanum commersonii*, accession PI 472834

Table 4). Similar results were obtained in other *Solanum* genotypes (Ramulu 1987).

The effect of various factors influencing transformation frequency is reported in Fig. 2. A striking effect of plant genotype was evident: PI 243503 did not regenerate any transgenic shoots under a range of experimental conditions, in contrast with PI 472833 and PI 472834, which, on average, regenerated shoots in 5 and 3.2% of explants, respectively (Fig. 2A–E). In the absence of kanamycin, regeneration from leaf explants of PI 243503 was better than that of PI 472833 and similar to that of PI 472834 (Table 3). Thus, a good regeneration ability is a necessary, but not sufficient, requisite for transformation efficiency in Cmm. Similarly, in diploid and tetraploid *S. tuberosum* genotypes a 5-to-100-fold drop in regeneration efficiency after transformation was reported (Visser 1991; El-Kharbotly et al. 1995), while only 35% of the variation in transformation efficiency between varieties was attributable to variation in regeneration efficiency (Dale and Hampson 1995). Therefore, competence for regeneration and transformation in potato may be due to different genetic factors (El-Kharbotly et al. 1995). Various endogenous and exogenous factors, however, influenced the response to transformation of *S. commersonii* and for each accession tested optimal conditions should be devised.

Agrobacterium strain EHA105 (binary vector p35SGUSINT) gave better results than GV2260 (p35SGUSINT) both in PI 472833 and PI 472834, but the effect was much more evident in the former, where transformation frequencies equal to 13.9 and 5.3% were obtained with EHA105 and GV2260, respectively (Fig. 2A). EHA105 has a helper plasmid derived from the wild-type plasmid pTiBo542, from which various supervirulent strains for solanaceous crops were derived (Cardi et al. 1992). GV2260 was used also by Baertlein et al. (1992), whereas LBA4404 was used in later experiments by Zhu et al. (1995a, 1996). In our experiments, LBA4404 (pGA492) produced transgenic shoots only in PI 472834 with a transformation frequency of 8%, using a feeder layer of potato cells during cocultivation (Cardi et al. 1992). However, since in our experiments GV2260 and LBA4404 carried a different binary vector, no conclusion may be drawn about their relative efficiency with Cmm. In the wild potato species *S. brevidens*, however, GV2260 (p35SGUSINT) was more efficient than LBA4404 (p35SGUSINT) (Liu et al. 1995).

A positive effect of a feeder layer on transformation of PI 472834 was evident also with *Agrobacterium* strain GV2260, since 10 and 2.5% of transformation frequency were obtained with and without feeder layer, respectively. By contrast, under the same conditions it had an adverse effect in PI 472833 (Fig. 2B). The preculture of explants and a delayed start of selection had a negative effect on both accessions (Fig. 2C,D).

As previously stated, accession PI 243503 was recalcitrant to any treatment. Only with the addition of acetosyringone were transgenic shoots of this genotype obtained: average transformation frequencies were 11.1 and 12.9% with 20 and 100 μM of acetosyringone, respectively (Fig. 2F), but up to 31.3% of explants regenerated shoots in some experiments (Cardi et al. 1992).

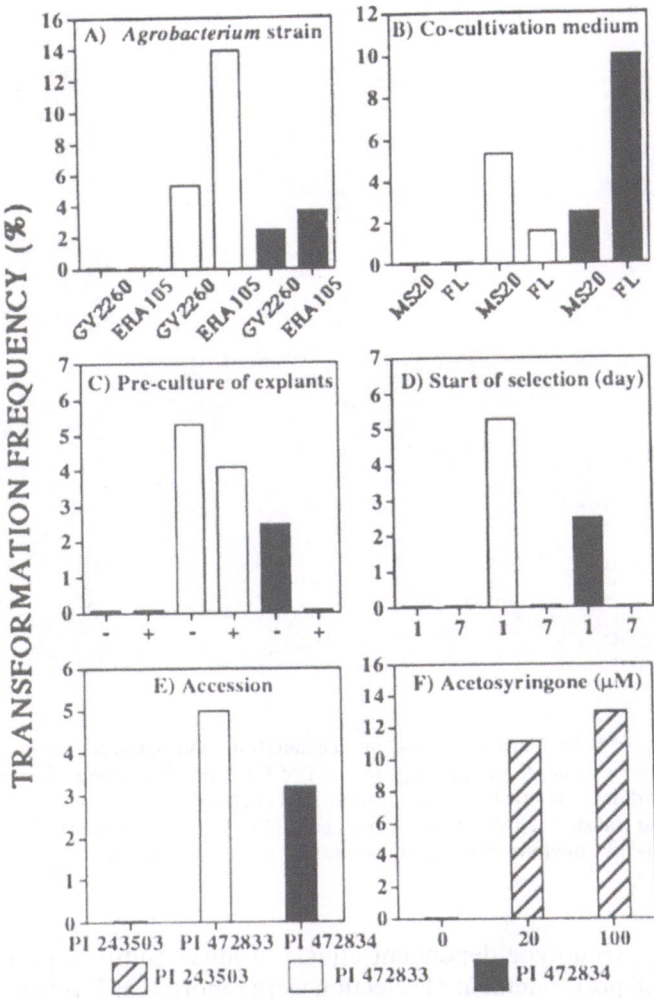

Fig. 2A–F. Effect of various factors on transformation frequency in three accessions of *Solanum commersonii*. (After Cardi et al. 1992). Standard conditions were: *A. tumefaciens* strain GV2260, p35SGUSINT as binary vector, MS20 as cocultivation medium, no preculture of explants, kanamycin added soon after cocultivation. **A** Both GV2260 and EHA105 strains contained the p35SGUSINT plasmid as binary vector. **B** *MS20* MS medium plus 20 g/l sucrose; *FL* feeder layer of potato cells in UM medium plus 30 g/l sucrose, 2 g/l casein hydrolysate, 5 mg/l 2,4D, and 0.25 mg/l kinetin. **C** Preincubation of explants for 24 h on the same medium used for cocultivation. **D** Kanamycin added either at the 1st or the 7th day of culture of the explants. **E** Data for accessions averaged from **A** through **D**. **F** Acetosyringone added both to *Agrobacterium* culture medium and to cocultivation medium

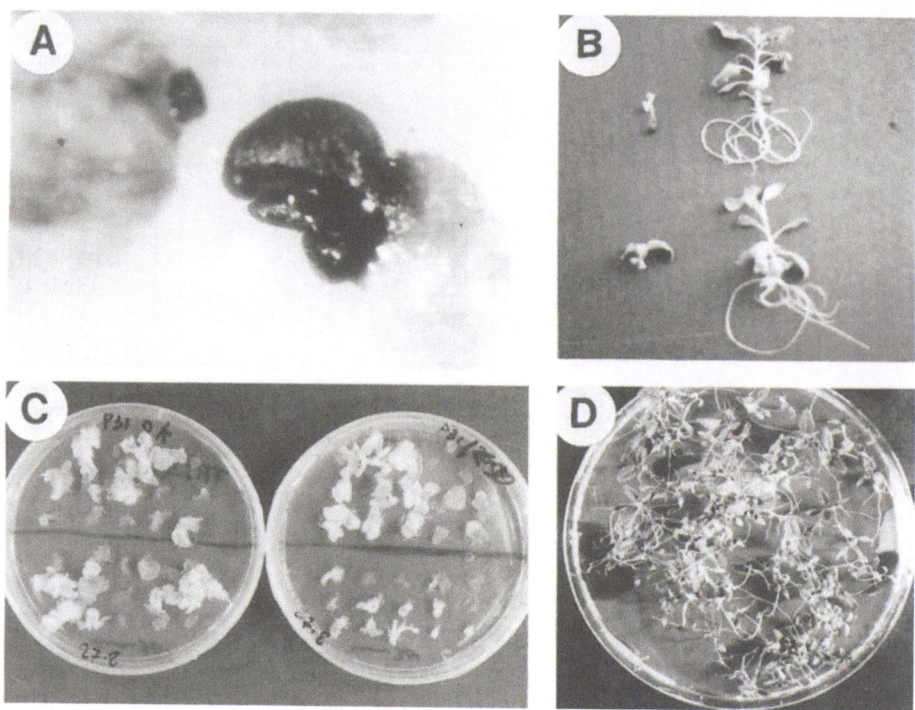

Fig. 3A–D. Various stages of regeneration and characterization of transgenic *Solanum commersonii*. **A** In vitro regeneration of GUS-positive shoots. **B** In vitro rooting on selective medium of control (*left*) and transformed (*right*) shoots. **C** Growth of calli from control (*below*) and transformed shoots (*above*) on control (*left*) and selective (*right*) media. **D** Germination on selective medium of seeds obtained from transformed plants

Genotype-dependent effects of nurse cultures, pretreatment of explant, and postponement of selection were reported in *S. tuberosum* and *S. brevidens* (Mitten et al. 1990; Visser 1991; Liu et al. 1995), while the use of aceto-syringone was suggested in some potato and wild genotypes (Kumar et al. 1995).

Transgenic shoots were able to root on kanamycin, and were positive for callus growth and GUS assays (Fig. 3). However, both the level of kanamycin resistance and the expression of the *uidA* gene varied among regenerants (Cardi et al. 1992). Similarly, different primary transformants with the *inaZ* gene showed large differences in the amount of ice nuclei produced (Baertlein et al. 1992). Zhu et al. (1995a, 1996) reported variation in GUS activity among primary transformants with chimeric constructs with two osmotin-like protein gene promoters and variation in RNA and protein levels both within sense and antisense transgenic plants for an osmotin-like protein gene. Such variation was attributed to the location of transgene insertion in the host genome (Baertlein et al. 1992; Zhu et al. 1995a). Similarly, variation in gene expression among primary transformants was commonly found also in

S. tuberosum and other species (De Block 1993; Dale and Hampson 1995; Liu et al. 1995).

Based on Southern analysis, the number of integration sites in different papers ranged from 1 to 6 (Table 3). However, when the progenies obtained by test cross with untransformed control were genetically analyzed (Cardi et al. 1992), the number of gene integrations estimated by DNA gel blot and segregation analyses was consistent in the case of single integrations, but divergent in the case of multiple copies (Table 3). This means either that multiple copies were closely linked, and hence cosegregated, or that not all the copies evidenced by molecular analysis were correctly expressed, probably due to transgene rearrangements or silencing (Fraley et al. 1985; De Block 1993).

3 Summary and Conclusions

Plant and bacterial genotype were very important to obtain transformation and regeneration of *Solanum commersonii* (Cardi et al. 1992, 1993b). The transformation ability of accessions tested, however, was only partially correlated with their regeneration ability. Various exogenous factors influenced the transformation efficiency, and optimized conditions were set up for each accession. A feeder layer of potato cells was beneficial in some cases, while the use of acetosyringone was absolutely necessary with one genotype. The preculture of explants or the postponement of selection was detrimental. In the case of multiple gene copies, molecular and genetic analyses gave different results as far as the number of transgene integrations is concerned, probably due either to gene rearrangement or cosuppression of transgene expression.

So far, transgenic *S. commersonii* shoots with marker genes (*nptII*, *uidA*) or genes involved in response to stress have been obtained by cocultivation of leaf explants with *Agrobacterium tumefaciens* (Baertlein et al. 1992; Cardi et al. 1992; Zhu et al. 1995a, 1996). By contrast, no data are available on transformability of this species with *A. rhizogenes*; however, studies on induced hairy roots for manipulation of secondary metabolites of medicinal importance may be undertaken.

The possibility of using transformation technology to insert molecular markers into *S. commersonii* should facilitate the use of this reproductively isolated wild species in genetics and breeding of potato and other related crops, while transformation experiments with genes putatively involved in important metabolic processes should speed up gene expression and gene isolation studies.

Acknowledgments. Thanks are due to Drs S. Grillo (CNR-IMOF, Portici, Italy), L. Monti and E. Filippone (Univ. of Naples Federico II, Portici, Italy), and P.F. Lurquin (Washington State Univ., Pullman, USA) for the critical review of the manuscript. Contribution No. 151 from CNR-IMOF.

References

Alam Z (1985) Screening of *Solanum* species against *Alternaria solani*. Pak J Agric Res 6:180–182

Baertlein DA, Lindow SE, Panopoulos NJ, Lee SP, Mindrinos MN, Chen THH (1992) Expression of a bacterial ice nucleation gene in plants. Plant Physiol 100:1730–1736

Bamberg JB (1995) Screening potato (*Solanum*) species for male fertility under heat stress. Am Potato J 72:23–34

Bamberg JB, Martin MW, Schartner JJ (1994) Elite selections of tuber-bearing *Solanum* species germplasm. Inter-Regional Potato Introduction Station, NRSP-6, 56 pp

Bastia T, Hasani A, Carotenuto N, Pelosi M (1995) Raddoppiamento somatico in vitro del livello di ploidia in specie selvatiche diploidi (2n = 2x = 24, EBN = 1) del genere *Solanum*. Proc XXXIX Annu Meet Italian Society of Agricultural Genetics (SIGA), 27–30 Sept 1995, Vasto Marina, Italy, pp 167–168

Baudo MM, Meza-Zepeda LA, Palva TP, Heino P (1996) Induction of homologous and ABA-responsive genes in frost-resistant (*Solanum commersonii*) and frost-sensitive (*Solanum tuberosum* cv. Bintje) potato species. Plant Mol Biol 30:331–336

Cardi T, Puite KJ, Ramulu KS (1990) Plant regeneration from mesophyll protoplasts of *Solanum commersonii* Dun. Plant Sci 70:215–221

Cardi T, Iannamico V, D'Ambrosio F, Filippone E, Lurquin PF (1992) *Agrobacterium*-mediated genetic transformation of *Solanum commersonii* Dun. Plant Sci 87:179–189

Cardi T, Puite KJ, Ramulu KS, D'Ambrosio F, Frusciante L (1993a) Production of somatic hybrids between frost-tolerant *Solanum commersonii* and *S. tuberosum*: protoplast fusion, regeneration and isozyme analysis. Am Potato J 70:753–764

Cardi T, Iannamico V, D'Ambrosio F, Filippone E, Lurquin PF (1993b) In vitro regeneration and cytological characterization of shoots from leaf explants of three accessions of *Solanum commersonii*. Plant Cell tissue Organ Cult 34:107–114

Carputo D, Cardi T, Frusciante L, Peloquin SJ (1995) Male fertility and cytology of triploid hybrids between tetraploid *Solanum commersonii* (2n = 4x = 48, 2EBN) and Phureja-Tuberosum haploid hybrids (2n = 2x = 24, 2EBN). Euphytica 80:123–129

Chavez R, Schmiediche PE, Jackson MT, Raman KV (1988) The breeding potential of wild potato species resistant to the potato tuber moth *Phythorimaea operculella* Zeller. Euphytica 39: 123–132

Chen HH, Li PH (1980) Characteristics of cold acclimation and deacclimation in tuber-bearing *Solanum* species. Plant Physiol 65:1146–1148

Correll DS (1962) The potato and its wild relatives. Texas Research Foundation, Renner, Texas, 606 pp

Dale PJ, Hampson KK (1995) An assessment of morphogenic and transformation efficiency in a range of varieties of potato (*Solanum tuberosum* L.). Euphytica 85:101–108

Davey MR, Rech EL, Mulligan BJ (1989) Direct DNA transfer to plant cells. Plant Mol Biol 13:273–285

Deahl KL, Sinden SL, Young RJ (1993) Evaluation of wild tuber-bearing *Solanum* accessions for foliar glycoalkaloid level and composition. Am Potato J 70:61–69

Deblaere R, Reynaerts A, Höfte H, Hernalsteens JP, Leemans J, Van Montagu M (1987) Vectors for cloning in plant cells. Methods Enzymol 153:277–292

De Block M (1993) The cell biology of plant transformation: current state, problems, prospects and the implications for the plant breeding. Euphytica 71:1–14

Drewes FE, van Staden J (1995) Initiation of and solasodine production in hairy root cultures of *Solanum mauritianum* Scop. Plant Growth Regul 17:27–31

Ehlenfeldt MK, Hanneman RE Jr (1988) The transfer of the synaptic gene (sy-2) from 1EBN *Solanum commersonii* Dun. to 2EBN germplasm. Euphytica 37:181–187

El-Kharbotly A, Jacobsen E, Stiekema WJ, Pereira A (1995) Genetic localisation of transformation competence in diploid potato. Theor Appl Genet 91:557–562

Fehér A, Ferföldi K, Preiszner J, Dudits D (1991) PEG-mediated transformation of leaf protoplasts of *Solanum tuberosum* L. cultivars. Plant Cell Tissue Organ Cult 27:105–114

Fraley RT, Rogers SG, Horsch RB (1985) Genetic transformation in higher plants. CRC Crit Rev Plant Sci 4:1–46

Grillo S, Costa A, Tucci M, Amatruda MR, Consiglio F, Vigh L, Leone A (1996) Regulation of gene expression during cellular adaption to water stress. In: Grillo S, Leone A (eds) Physical stresses in plants. genes and their products for tolerance. Springer, Berlin Heidelberg New York, pp 163–169

Hanneman RE Jr, Bamberg JB (1986) Inventory of tuber-bearing *Solanum* species. Wis Agric Exp Stn Bull 533, 216 pp

Hawkes JG (1990) The potato. Evolution, biodiversity and genetic resources. Belhaven Press, London, 259 pp

Hoekema A, Hirsch PR, Hooykaas PJJ, Schilperoort RA (1983) A binary vector strategy based on separation of *vir-* and T-region of the *Agrobacterium tumefaciens* Ti-plasmid. Nature 303:179–180

Hood EE, Helmer GL, Fraley RT, Chilton MD (1986) The hypervirulence of *Agrobacterium tumefaciens* A281 is encoded in a region of pTiBo542 outside of T-DNA. J Bacteriol 168:1291–1301

Iapichino G, Lee SP, Chen THH, Fuchigami LH (1991) In vitro plant regeneration in *Solanum commersonii*. J Plant Physiol 137:734–738

Ikenaga T, Oyama T, Muranaka T (1995) Growth and steroidal saponin production in hairy root cultures of *Solanum aculeatissimum*. Plant Cell Rep 14:413–417

Jefferson RA (1987) Assaying chimeric genes in plants: the GUS gene fusion system. Plant Mol Biol Rep 5:387–405

Johnston SA, Ruhde RW, Ehlenfeldt MK, Hanneman RE Jr (1986) Inheritance and microsporogenesis of a synaptic mutant *sy-2* from *Solanum commersonii*. Can J Genet Cytol 28:520–524

Knapp S, Coupland G, Uhrig H, Starlinger P, Salamini F (1988) Transposition of the maize transposable element *Ac* in *Solanum tuberosum*. Mol Gen Genet 213:285–290

Kumar A, Miller M, Whitty P, Lyon J, Davie P (1995) *Agrobacterium*-mediated transformation of five wild *Solanum* species using in vitro microtubers. Plant Cell Rep 14:324–328

Leone M, Filippone E, Lurquin PF (1993) Transformation in *Solanum melongena* L. (Eggplant). In: Bajaj YPS (ed) Biotechnology in agriculture and forestry, vol 22. Plant protoplasts and genetic engineering III. Springer Berlin Heidelberg New York, pp 320–328

Liu THA, Stephens LC, Hannapel DJ (1995) Transformation of *Solanum brevidens* using *Agrobacterium tumefaciens*. Plant Cell Rep 15:196–199

Melchers LS, Hooykaas PJJ (1987) Virulence in *Agrobacterium*. Oxford Surv Plant Mol Cell Biol 4:167–220

Mitten DH, Horn M, Burrell MM, Blundy KS (1990) Strategies for potato transformation and regeneration. In: Vayda ME, Park WD (eds) The molecular and cellular biology of the potato. Redwood Press, Melkshom, pp 181–191

Murashige T, Skoog F (1962) A revised medium for rapid growth and bio-assays with tobacco tissue cultures. Physiol Plant 15:473–497

Osman SF, Herb SF, Fitzpatrick TJ, Sinden SL (1976) Commersonine, a new glycoalkaloid from two *Solanum* species. Phytochemistry 15:1064–1067

Ramulu KS (1987) Genetic instability during plant regeneration in potato: origin and implications. Plant Physiol (Life Sci Adv) 6:211–218

Subroto AM, Doran PM (1994) Production of steroidal alkaloids by hairy roots of *Solanum aviculare* and the effect of gibberellic acid. Plant Cell Tissue Organ Cult 38:93–102

Tozzini AC, Ceriani MF, Saladrigas MV, Hopp HE (1991) Extreme resistance to infection by potato virus X in genotypes of wild tuber-bearing *Solanum* species. Potato Res 34:317–324

Valkonen JPT, Keskitalo M, Vasara T, Pietilä L (1996) Potato glycoalkaloids: a burden or a blessing? Crit Rev Plant Sci 15:1–20

Vancaynnet G, Schmidt R, O'Connor-Sanchez A, Willmitzer L, Rocha-Sosa M (1990) Construction of an intron-containing marker gene: splicing of the intron in transgenic plants and its use in monitoring early events in *Agrobacterium*-mediated transformation. Mol Gen Genet 220:245–250

Vázquez A, González G, Ferreira F, Moyna P, Kenne L (1997) Glycoalkaloids of *Solanum commersonii* Dun. ex Poir. Euphytica 95:195–201

Visser RG (1991) Regeneration and transformation of potato by *Agrobacterium tumefaciens*. In: Lindsey K (ed) Plant tissue culture manual. Kluwer, Dordrecht, B5, pp 1–9

Zhu BL, Chen THH, Li PH (1993) Expression of an ABA-responsive osmotin-like gene during the induction of freezing tolerance in *Solanum commersonii*. Plant Mol Biol 21:729–735

Zhu B, Chen THH, Li PH (1995a) Activation of two osmotin-like protein genes by abiotic stimuli and fungal pathogen in transgenic potato plants. Plant Physiol 108:929–937

Zhu B, Chen THH, Li PH (1995b) Expression of three osmotin-like protein genes in response to osmotic stress and fungal infection in potato. Plant Mol Biol 28:17–26

Zhu B, Chen THH, Li PH (1996) Analysis of late-blight disease resistance and freezing tolerance in transgenic potato plants expressing sense and antisense genes for an osmotin-like protein. Planta 198:70–77

XXIV Genetic Transformation of *Swainsona galegifolia* (Darling Pea)

T.M. ERMAYANTI

1 Introduction

Swainsona galegifolia (Andr.) R. Br. (syn. *S. coronillifolia* Salisb.), called Darling pea or Smooth Darling pea, grows in Queensland and New South Wales (Anonymous 1981; Everist 1981). It is a shrubby perennial species which has small pinnate leaves about 10 cm long and in November has white, or pink to purple flowers. *S. galegifolia* is placed in the tribe Galegeae, of the family Fabaceae (Pohill and Raven 1981).

Some species of *Swainsona* were investigated initially because of their toxicity to animals (Dorling et al. 1980, 1987a,b), but later it was also found that the toxic compound swainsonine could be used for a number of medicinal purposes (Fellows 1989; Yagita et al. 1992). The chemical structure of swainsonine is close to that of castanospermine, which is produced by *Cantanospermum australe*, and which is being investigated for possible use as an anti-AIDS drug (Fellow 1989). Some species of *Swainsona* have been reported to produce a high level of swainsonine (Dorling et al. 1987a), and swainsonine was also analyzed in vitro (Bobek et al. 1989; Robinson et al. 1990). It was detected in tissue culture of *S. canescens*, but there are no reports on production of swainsonine from root cultures.

The objective of genetic transformation in *S. galegifolia* was to establish transformed root cultures in order to enhance the level of swainsonine production in vitro. In the present chapter our published work (Ermayanti et al. 1993, 1994a,b) is summarized.

2 Genetic Transformation and Swainsonine Production

2.1 Methodology

2.1.1 Bacteria and Plants

Agrobacterium rhizogenes strains A4 and LBA 9402 were grown overnight on liquid LB medium (Lichtenstein and Draper 1985). Seeds of *Swainsona*

Research and Development Centre for Biotechnology, Indonesian Institute of Sciences, Jalan Raya Bogor Km 46, Cibinong 16911, Indonesia

Biotechnology in Agriculture and Forestry, Vol. 45
Transgenic Medicinal Plants (ed. by Y.P.S. Bajaj)
© Springer-Verlag Berlin Heidelberg 1999

galegifolia were surface sterilized using 10% sodium hypochlorite for 15 min and rinsed three times in sterile water. The sterile seeds were maintained in the dark at 25 °C. Germinated seeds provided root tips for initiating root cultures and some seedlings were transferred to MS medium (Murashige and Skoog 1962) without hormones for further growth. These aseptic seedlings provided the stem material for callus induction and for transformation using *A. rhizogenes*.

2.1.2 Establishment of Root Cultures from Seedlings, Roots (untransformed roots), and Transformed Roots

Root tips of 1 cm length were excised from 5-day-old aseptic seedlings and placed in a 125-ml flask containing 20 ml medium with 1 mg/l IBA and medium without auxin. Flasks were maintained on a rotary shaker with 80 rpm in the dark at 25 °C.

Agrobacterium rhizogenes strains A4 and LBA 9402 were used for transformation after 2 days' growth in LB medium. Sterile young stems of *S. galegifolia* (about 1 cm long) on MS medium were inoculated by making a short slit in the stem with a sharp syringe needle which had been dipped into the bacterial broth, then stems were incubated at 25 °C in darkness. Three to 5 weeks after inoculation, transformed roots emerged from the infected site. After being freed from the bacteria using 100 mg/l Cefotaxime, transformed root tips were cut at 1 cm and cultured in MS liquid medium without exogenous hormones. Where different transformation events on a single genotype were studied, roots were isolated from separate explants of the same plant genotype.

2.1.3 Enhancement of Swainsonine Level in Untransformed and Transformed Root Cultures

Untransformed roots (0.3 g fresh weight) from stock cultures were transferred to 15 ml MS or White's liquid media with IBA or NAA. After 30 days culture, dry weight and swainsonine level were analyzed. For sugar treatments, 0.3 g untransformed and transformed roots were cultured in 15 ml MS medium with 30 g/l sucrose, glucose, or fructose. Dry weight and swainsonine production were determined after 30 days of culture.

Induction of polyploidy was attempted by addition of 0.05% colchicine to transformed roots and shoots grown in vitro, and regeneration of roots from callus or transformation of cotyledons with some strains of *A. rhizogenes* in MS medium with no hormones.

Stimulation of synthesis and release of swainsonine from transformed roots was conducted by addition of $CuSO_4$ to the medium, reduction of medium pH, and addition of pipecolic acid and malonic acid. $CuSO_4$ at 0.5–2 mM was added for 1–4 days before harvest on day 30. Reduction of medium pH was conducted by transferring roots from the medium with pH 5.7 to pH 4.62, 3.71 or 2.72 for 1–2 days before harvest. Similar procedures were used for addition of pipecolic acid and malonic acid at 0.01–2 mM for 1–12 days before harvest on day 30.

2.1.4 Swainsonine Extraction, Identification, and Quantification

Swainsonine was extracted from plants growing in the glasshouse and in vitro, from untransformed and transformed root cultures, and from medium according to methods of Dorling et al. (1987a). Identification was carried out using GC-MS (Perkin-Elmer Series 8500, USA) by comparing the spectra of identified swainsonine standard (Sigma, USA), or using GC (Shimadsu GC-8A, Japan) by spiking samples with swainsonine standard. The enlarged peak was confirmed as swainsonine. Swainsonine was quantified using GC by comparing the area of the swainsonine peak with the area of an internal standard castanospermine peak (Sigma, USA).

3 Results and Discussion

3.1 Swainsonine in Plants, Untransformed and Transformed Root Cultures, and in Callus

Swainsonine production in root cultures of *S. galegifolia* reached a maximum 30–35 days after subculture (Ermayanti et al. 1994a). The peak concentration of swainsonine in transformed roots was higher than in untransformed roots (62 cf. 24 µg/g dry wt.). These levels were higher than those for shoots (10.3 µg/g dry wt.), or roots (8.7 µg/g dry wt.) of plants grown in a glasshouse, or shoots in vitro (8.7 µg/g dry wt.). Roots and shoots of *S. galegifolia* from glasshouse-grown plants also gave swainsonine concentrations similar to some root lines derived from callus (Ermayanti et al. 1994a). The chromosome numbers of roots initiated from callus was variable with a modal frequency of 18 which is close to the haploid number of 16 for the species (Ermayanti et al. 1993). Similar higher levels of alkaloid produced by transformed roots in comparison with untransformed roots or with intact plants have been reported from other species such as *Duboisia leichhardtii* (Mano et al. 1989) and a number of solanaceous species (Shimomura et al. 1991). Production of swainsonine produced by transformed roots reached a maximum at the stationary phase. This suggests that two-stage culture is needed for maximum swainsonine production.

3.2 Swainsonine in Root Cultures Treated with Auxins, Sugar, and Colchicine, and in Roots from Different Transformation Events

Untransformed roots grown in MS liquid medium gave higher dry weight and swainsonine production than roots cultured in White's medium (Ermayanti et al. 1994a). Addition of IBA in MS liquid medium significantly increased biomass production, but did not significantly increase the concentration of swainsonine. Roots grew slowly in both MS and White's media with NAA and swainsonine concentration was also low.

Glucose gave highest biomass production of untransformed roots (0.99 g dry wt.) after 30 days in culture compared with addition of sucrose (0.51 g) and fructose (0.65 g). However, swainsonine production was not significantly different between roots grown in sucrose (18.13 µg/g dry wt.), glucose (21.35 µg/g) or fructose (15.72 µg/g) (Ermayanti et al. 1994a). Glucose and fructose inhibited growth of transformed roots by giving very low production of dry weight. Swainsonine production in transformed roots was not significantly different in media with different sugars, but concentration in transformed roots was much higher than in untransformed roots grown in the same medium (Ermayanti et al. 1994a).

Untransformed and transformed roots treated with 0.05% colchicine for 1–3 days grew very slowly in MS liquid medium. After 30 days in culture following colchicine treatment, biomass production of untransformed and transformed treated roots was reduced compared with the controls. Swainsonine production in colchicine-treated roots was comparable with that in untreated roots, but there was a significant difference in swainsonine level between transformed and untransformed roots. Only very few cells had aneuploid chromosome numbers, and more than 82% had the normal chromosome number (Ermayanti et al. 1994a). It is possible that the variability in chromosome number was adversely affecting swainsonine synthesis. The roots derived from callus showed chromosome numbers around the haploid number in most lines (Ermayanti et al. 1993), and a significant drop in swainsonine level was observed (Ermayanti et al. 1994a).

Roots induced from transformation of cotyledons gave swainsonine levels ranging from 28.5 to 49.5 µg/g dry wt. Chromosome numbers in these roots was 87% of cells had the diploid numbers of $2n = 32$, 6.7% of cells had chromosome numbers below the diploid, and 6.3% of cells had chromosome numbers above the diploid.

Different transformation events using *A. rhizogenes* strain LBA 9402, and three plant genotype gave roots which, when cultured, showed significantly different growth and swainsonine contents (Table 1). This could be the effects of the *rol*B gene and the (*aux*) genes which promote the formation of lateral roots by stimulating auxin synthesis by the plant, or increasing sensitivity to auxin (Shen et al. 1988, 1990).

3.3 Stimulation of Synthesis and Release of Swainsonine from Transformed Roots with Copper Sulfate, Reduction of Medium pH, and Addition of Pipecolic and Malonic acids

Addition of copper sulfate at 0.5–2 mM for 1–4 days caused some small changes in root dry weight but had a great effect on swainsonine production in both roots and medium (Table 2). Addition of copper sulfate at the same level

Table 1. Biomass and swainsonine concentration in transformed roots after eight different transformation events using three different plant genotypes transformed with *A. rhizogenes* strain LBA 9402. (Ermayanti et al. 1994a)

Plant	Transformation	Dry weight (g)	Swainsonine concentration (µg/g dry wt)
1	A	1.43 ± 0.024 c	59.41 ± 2.533 bc
	B	1.04 ± 0.015 a	57.54 ± 4.033 b
	C	1.24 ± 0.085 b	36.96 ± 2.162 a
2	D	1.42 ± 0.039 c	71.33 ± 2.871 d
	E	1.32 ± 0.065 bc	59.76 ± 3.572 bc
	F	1.45 ± 0.052 c	67.69 ± 3.784 cd
3	G	1.30 ± 0.039 bc	53.28 ± 1.912 b
	H	1.43 ± 0.068 c	71.30 ± 3.184 d

Different letters in each column indicate values significantly different at $P < 0.05$.

Table 2. Biomass and swainsonine concentration of transformed roots treated with copper sulfate for 0 to 4 days before harvest at day 30. (Ermayanti et al. 1994b)

Copper sulfate (mM)	Days	Dry weight (g)	Swainsonine concentration	
			Roots (µg/g dry wt.)	Medium (µg/50 ml)
0	–	1.1 ± 0.06 bc	69.5 ± 2.85 bc	trace a
0.5	1	1.1 ± 0.07 bc	75.1 ± 5.14 bcd	8.8 ± 0.83 b
	2	1.2 ± 0.13 cd	85.3 ± 8.50 cde	10.4 ± 0.69 bc
	3	1.3 ± 0.06 d	87.3 ± 9.20 de	12.0 ± 1.09 bc
	4	0.8 ± 0.06 a	84.6 ± 5.83 cde	16.8 ± 0.54 de
1.0	1	1.3 ± 0.11 d	88.1 ± 6.44 de	12.7 ± 0.88 bcd
	2	1.3 ± 0.06 d	107.9 ± 2.93 f	14.3 ± 1.43 cde
	3	1.0 ± 0.04 ab	99.3 ± 7.55 ef	22.8 ± 3.07 g
	4	1.1 ± 0.05 bc	84.0 ± 4.23 cde	28.7 ± 1.60 h
2.0	1	1.3 ± 0.07 d	92.8 ± 6.76 ef	14.4 ± 1.20 cde
	2	1.2 ± 0.05 cd	89.0 ± 7.11 de	17.9 ± 0.96 ef
	3	1.0 ± 0.03 ab	61.3 ± 5.08 ab	21.4 ± 0.17 fg
	4	0.9 ± 0.05 a	44.1 ± 4.80 a	23.6 ± 1.69 g

Different letters in each column indicate values significantly different at $P < 0.05$.

for 5–6 days killed the roots. When swainsonine from roots and medium was combined, the highest yield of swainsonine was found in roots cultured in medium with 1 mM copper sulfate for 2 days. The increase in yield was also reported for transformed roots of *Hyoscyamus albus* (Christen et al. 1992) and *Datura stramonium* (Rhodes et al. 1989).

Reduction of medium pH from 5.7 to 2.7 for 1 day had a significant effect on swainsonine production of *S. galegifolia* transformed root cultures (Table 3). The pH of the medium also affected the intracellular and extracellular alkaloid level of *Nicotiana glauca* transformed root cultures (Green and Thomas 1992). In *Beta vulgaris* transformed root cultures, however, increasing the medium pH did not stimulate the release of betanin (Buitelaar et al. 1992). The mechanism of enhancement and release of secondary metabolites in transformed roots may be similar to the endogenous mechanism by which root

Table 3. Biomass and swainsonine concentration of transformed roots after reduction of medium pH (pH after autoclaving is shown). (Ermayanti et al. 1994b)

Days	pH	Dry weight (g)	Swainsonine concentration	
			Roots (μg/g dry wt.)	Medium (μg/50 ml)
0	5.68	1.2 ± 0.06 b	73.1 ± 4.35 cd	Trace a
1	5.67	1.3 ± 0.01 b	65.8 ± 2.14 bc	Trace a
	4.62	1.3 ± 0.01 b	70.7 ± 7.48 cd	32.6 ± 0.66 b
	3.71	1.2 ± 0.02 b	81.9 ± 6.27 de	39.4 ± 1.39 b
	2.72	1.2 ± 0.01 b	93.3 ± 6.91 e	47.1 ± 3.40 c
2	5.67	1.3 ± 0.01 b	70.9 ± 6.14 cd	Trace a
	4.62	1.3 ± 0.01 d	60.6 ± 4.60 bc	37.5 ± 2.77 b
	3.71	1.3 ± 0.01 b	52.1 ± 2.29 b	32.9 ± 2.80 b
	2.72	0.9 ± 0.04 a	29.7 ± 2.28 a	34.6 ± 2.08 b

Different letters in each column indicate values significantly different at $P < 0.05$.

Table 4. Biomass and swainsonine concentration of transformed roots treated with pipecolic acid for 0–12 days before harvest on day 30. (Ermayanti et al. 1994b)

Days	Pipecolic acid (mM)	Dry weight (g)	Swainsonine concentration	
			Roots (μg/g dry wt.)	Medium (μg/50 ml)
0	0	1.1 ± 0.03 bc	64.4 ± 5.16 a	Trace a
1	0.01	1.1 ± 0.08 b	72.3 ± 5.66 ab	17.6 ± 1.42 b
	0.1	1.1 ± 0.06 b	98.7 ± 4.38 bc	20.2 ± 1.63 b
	1.0	1.1 ± 0.04 b	110.3 ± 6.78 cd	36.3 ± 3.62 cde
	2.0	1.2 ± 0.02 b	120.5 ± 6.51 d	37.8 ± 4.44 ef
2	0.01	1.0 ± 0.06 ab	96.0 ± 7.88 bc	30.3 ± 4.94 cd
	0.1	0.8 ± 0.03 c	110.5 ± 9.87 cd	32.6 ± 4.51 cd
	1.0	1.6 ± 0.03 c	117.7 ± 9.45 cd	40.2 ± 3.59 ef
	2.0	1.6 ± 0.05 c	120.7 ± 10.83 d	45.6 ± 3.78 f
6	0.01	1.2 ± 0.05 b	92.6 ± 11.26 bc	26.0 ± 0.56 bc
	0.1	1.1 ± 0.12 b	111.8 ± 7.39 cd	32.2 ± 3.11 cd
	1.0	1.2 ± 0.05 b	111.9 ± 4.77 cd	36.5 ± 4.04 cde
	2.0	1.2 ± 0.06 b	147.4 ± 12.34 e	43.3 ± 4.27 ef
12	0.01	1.2 ± 0.06 b	86.9 ± 7.36 ab	27.2 ± 1.73 bc
	0.1	1.2 ± 0.07 b	103.7 ± 4.21 cd	29.2 ± 4.52 cd
	1.0	1.3 ± 0.04 bc	120.4 ± 11.9 d	35.5 ± 2.98 cde
	2.0	1.3 ± 0.04 bc	122.9 ± 9.74 d	32.6 ± 2.98 cd

Different letters in each column indicate values significantly different at $P < 0.05$.

cells increase biosynthesis in response to the presence of elicitors or changes in membrane permeability (Furze et al. 1991).

Pipecolic acid at 1–2 mM for 2 days increased root biomass (Table 4). Even at low concentration, pipecolic acid enhanced the level of swainsonine in roots, and stimulated release of the product into the culture medium. Maximum swainsonine was detected in roots with 2 mM pipecolic acid for 6 days. No morphological changes were found in roots on addition of pipecolic acid, whereas the roots treated with copper sulfate or exposed to a pH drop were brown at the end of their treatments.

Table 5. Biomass and swainsonine concentration of transformed roots treated with malonic acid for 0–12 days before harvest on day 30. (Ermayanti et al. 1994b)

Days	Malonic acid (mM)	Dry weight (g)	Swainsonine concentration	
			Roots (µg/g dry wt.)	Medium (µg/50 ml)
0	0	1.1 ± 0.05 b	66.9 ± 5.21 a	Trace a
1	0.01	1.0 ± 0.04 ab	67.6 ± 6.77 a	12.2 ± 1.76 bc
	0.1	0.9 ± 0.06 a	93.5 ± 7.46 bc	15.7 ± 1.49 bc
	1.0	1.1 ± 0.08 b	96.3 ± 8.10 cd	21.0 ± 1.72 de
	2.0	1.2 ± 0.05 bc	100.1 ± 7.30 cd	24.2 ± 2.86 de
2	0.01	0.9 ± 0.04 a	77.4 ± 4.25 ab	13.9 ± 1.44 bc
	0.1	0.9 ± 0.06 a	99.8 ± 4.01 cd	15.2 ± 1.60 bc
	1.0	1.3 ± 0.03 c	112.9 ± 7.46 de	19.4 ± 0.75 cd
	2.0	1.3 ± 0.72 c	115.0 ± 6.16 de	24.6 ± 3.78 de
6	0.01	1.2 ± 0.09 bc	83.4 ± 6.50 ab	18.4 ± 2.70 cd
	0.1	1.1 ± 0.02 b	92.3 ± 6.28 bc	21.5 ± 2.99 de
	1.0	1.2 ± 0.10 bc	106.8 ± 7.53 cd	22.0 ± 1.86 de
	2.0	1.3 ± 0.14 c	114.9 ± 8.25 de	28.2 ± 1.14 ef
12	0.01	1.2 ± 0.04 bc	92.7 ± 4.27 bc	19.7 ± 1.85 cd
	0.1	1.3 ± 0.04 c	105.9 ± 5.82 cd	23.6 ± 2.92 de
	1.0	1.3 ± 0.04 bc	111.5 ± 10.48 de	28.8 ± 4.11 ef
	2.0	1.3 ± 0.04 bc	118.7 ± 9.28 e	33.6 ± 3.60 f

Different letters in each column indicate values significantly different at $P < 0.05$.

Malonic acid added to the medium for 1 day reduced or had little effect on root dry weight, but significantly increased the swainsonine level (Table 5). Successive additions of malonic acid to give an increase of 1 mM, on each of days 1–4, however, did not significantly increase root dry weight (Ermayanti et al. 1994b). Sequential addition of malonic did not give greater swainsonine yield in the roots than the addition of 2 mM malonic acid on day 2, 6 or 12 (Table 5). The increase in swainsonine level after addition of pipecolic acid and malonic acid may be as a result of their utilization by the swainsonine biosynthetic pathway.

Maximum total production of swainsonine by transformed root cultures of *S. galegifolia* after treatment with pipecolic acid was higher than that for malonic acid (220 µg cf. 187 µg per culture). However, malonic acid gave more advantages economically as the cost of pipecolic acid is more than 300× higher than the cost of malonic acid.

4 Summary and Conclusions

Root cultures of *Swainsona galegifolia* were established either from seedlings roots or after infection of plant organs with *Agrobacterum rhizogenes*. Swainsonine produced by root cultures reached a maximum 30–35 days after subculture, and the identity of the compound was confirmed by its GC retention time and mass spectrum in comparison to swainsonine from intact plants and the swainsonine standard. Roots and shoots from glasshouse-grown plants had similar swainsonine concentration to shoots of grown in vitro. The concentration of swainsonine in transformed roots was higher than in untransformed roots. Tests of the effect of plant genotype and *A. rhizogenes* strain LBA 9402 showed that plant genotype influences swainsonine level in transformed roots, but that different transformation events in one genotype resulted in equally large differences in the level of swainsonine. This chapter adds the evidence that roots transformed with *A. rhizogenes* show high and stable levels of secondary metabolite production and swainsonine production almost ten fold that measured in intact plants grown under glasshouse condition. The addition of precursors increased the level to almost 20-fold.

Acknowledgements. The author would like to thank A. Prof. Jen McComb and Dr. P.A. O'Brien for supervising the work, A. Prof P.R. Dorling and Dr. S.M. Colegate for their help on swainsonine analysis and for providing seeds and standards.

References

Anonymous (1981) *Swainsona galegifolia.* Growing native plants no 2. National Botanic Garden Canberra Australia 43 pp

Bobek GA, McFarlane IJ, Barrow KD (1989) Chemical capacity of tissue cultures from Australian plants. Proc 8th Australian Biochem Conf University of New South Wales Sydney, Australia 475 pp

Buitelaar RM, Cesario MT, Tramper J (1992) Elicitation of thiophene production by hairy roots of *Tagetes patula.* Enzyme Microb Technol 14:2–7

Christen P, Aoki T, Shimomura K (1992) Characteristics of growth and tropane alkaloid production in *Hyoscyamus albus* hairy roots transformed with *Agrobacterium rhizogenes* A4. Plant Cell Rep 11:597–600

Dorling PR, Huxtable CR, Colegate SM (1980) Inhibition of lysosomal α-mannosidase by swainsonine, an indolizidine alkaloid isolated from *Swainsona canescens.* Biochem J 191:649–651

Dorling PR, Colegate SM, Huxtable CR (1987a) Toxic species of the plant genus *Swainsona.* In: James LF, Elbein AD, Molyneux RJ, Warren CD (eds) Swainsonine and related glycosidase inhibitors. Symp Logan Utah pp 14–22

Dorling PR, Huxtable CR, Colegate SM (1987b) Isolation of swainsonine from *Swainsona canescens.* In: James LF, Elbein AD, Molyneux RJ, Warren CD (eds) Swainsonine and related glycosidase inhibitors. Symp Logan Utah pp 83–90

Ermayanti TM, McComb JA, O'Brien PA (1993) Cytological analysis of seedling roots, transformed root cultures and roots regenerated from callus of *Swainsona galegifolia* (Andr.) R Br. J Exp Bot 44:375–380

Ermayanti TM, McComb JA, O'Brien PA (1994a) Growth and swainsonine production of *Swainsona galegifolia* (Andr.) R Br. untransformed and transformed root cultures. J Exp Bot 45:633–639

Ermayanti TM, McComb JA, O'Brien PA (1994b) Stimulation of synthesis and release of swainsonine from transformed roots of *Swainsona galegifolia.* Phytochemistry 36:313–317

Everist SL (1981) Poisonous plants of Australia. Angus and Robertson Sydney pp 283–285

Fellows L (1989) Botany breaks into the candy store. New Sci August: 29–32

Furze JM, Rhodes MJC, Parr AJ, Robins RJ, Withehead IM, Threlfall DR (1991) Abiotic factors elicit sesquiterpenoid phytoalexin production but not alkaloid in transformed root cultures of *Datura stramonium.* Plant Cell Rep 10:111–114

Green KD, Thomas NH (1992) Product enhancement and recovery from transformed cultures of *Nicotiana glauca* roots. Biotechnol Bioeng 39:195–202

Lichtenstein C, Draper J (1985) Genetic engineering of plants. In: Glower DM (ed) DNA Cloning vol II. IRL Press, Oxford pp 67–119

Mano Y, Ohkawa H, Yamada Y (1989) Production of tropane alkaloids by root cultures of *Duboisia leichhardtii* transformed by *Agrobacterium rhizogenes*. Plant Sci 59:191–201

Murashige T, Skoog F (1962) A revised medium for rapid growth and bio-assays with tobacco tissue cultures. Physiol Plant 15:473–497

Pohill RM, Raven PH (1981) Advances in legume systematics I. Royal Botanic Gardens Kew England pp 357–360

Rhodes MJC, Robins RJ, Lindsay E, Aird ELH, Payne J, Parr AJ, Walton NJ (1989) Regulation of secondary metabolism in transformed root cultures. In: Kurz A (ed) Primary and secondary metabolism of plant cell cultures. Springer, Berlin Heidelberg New York, pp 58–72

Robinson C, Browne C, McFarlane IJ, Barrow KD (1990) Isolation and identification of swainsonine from seed extracts and plant tissue cultures of *Swainsona* species. Proc Aust Biochem Soc 22:78

Shen WH, Petit A, Guern J, Tempé J (1988) Hairy roots are more sensitive to auxin than normal roots. Proc Natl Acad Sci USA 85:3417–3421

Shen WH, Davioud E, David C, Barnier-Brygoo H, Tempé J, Guem J (1990) High sensitivity to auxin is a common feature of hairy root. Plant Physiol 94:554–560

Shimomura K, Sauerwein M, Ishimaru K (1991) Tropane alkaloids in the adventitious and hairy root cultures of solanaceous plants. Phytochemistry 30:2275–2278

Yagita M, Noda I, Maehara M, Fujieda S, Inoue Y, Hoshino T, Saksels E (1992) The presence of concanavalin-A (con-A)-like molecules on natural-killer (NK)-sensitive target cells: their possible role in swainsonine-augmented human NK cytotoxicity. Int J Cancer 52:664–672

XXV Transgenic Tobacco: Gene Expression and Applications

B.L.A. Miki[1], S.G. McTtugh[1], H. Labbe[1], T. Ouellet[1], J.H. Tolman[2], and J.E. Brandle[2]

1 Introduction

Tobacco is native to the southern reaches of North America and it is cultivated throughout the world. The major subtypes are flue-cured, burley, dark, oriental, pipe, and cigar. The major producers in the world are China, USA, Brazil, India, and Zimbabwe, and the value of this production is in the multibillion dollar range.

Tobacco has been instrumental in the development of many transformation technologies that are commonly used for producing transgenic plants and, consequently, studies on the mechanisms that govern gene expression in higher plants and the functional analysis of cloned genes. It has been used as a model system because of the high transformation frequencies and the ease with which large numbers of transgenic lines can be produced. The size of the plant and the uniformity of growth habit and morphology also provide researchers with large amounts of biological material for biochemical analysis and permit them to observe subtle phenotypic changes that might be related to the expression of the cloned genes inserted into the plant. Promoter trap and gene tagging experiments have been facilitated by the ease of the transformation process, especially using *Agrobacterium*-mediated protocols. In these studies, the predictability of the T-DNA structure after insertion into the tobacco genome has allowed the construction of T-DNA vectors designed to probe for genetic elements at the insertion sites, resulting in the cloning of novel promoters (Fobert et al. 1994) and genes (Hayashi et al. 1992). Unfortunately, there is no tobacco genetic map available, either classical or molecular, for map-based cloning of genes that might be interesting or useful.

As a model, transgenic tobacco has been pivotal to the development of agronomically useful gene technologies. It continues to be a favorite plant for generating the scientific knowledge that is a prerequisite for applications in crop species that are more difficult to manipulate. Some examples include the development of herbicide resistance (Comai et al. 1985; De Block et al. 1987), insect-pest resistance using insecticidal proteins (Barton et al. 1987), or resist-

[1] Eastern Cereal and Oilseed Research Centre, Agriculture and AgriFood Canada, Ottawa, Ontario, Canada K1A 0C6
[2] Southern Crap Protection and Food Research Centre, Agriculture and AgriFood Canada, 1391 Sandford Street, London, Ontario, Canada N5V 4T3

Biotechnology in Agriculture and Forestry, Vol. 45
Transgenic Medicinal Plants (ed. by Y.P.S. Bajaj)
© Springer-Verlag Berlin Heidelberg 1999

ance to viruses using coat protein genes (Powell et al. 1986). Furthermore, the development and understanding of key promoters such as that associated with the Cauliflower Mosaic Virus 35S transcript depended heavily on transgenic tobacco (Odell et al. 1985).

Although tobacco is a model system used to develop transformation technology, the benefits to tobacco as a crop have not been fully exploited. Tobacco is a high-value, but low-acreage crop and therefore not a target for the large biotechnology interests which have pioneered the commercialization of transgenic crops. Studies have also been conducted on the alkaloids nicotine, cadaverine, and anabasine in transgenic tobacco cultures (Saito et al. 1989; Fecker et al. 1993). In China, almost 1 million ha of transgenic tabacco was under cultivation in 1994; such plants showed resistance to TMV, and yielded 5–7% more leaves (Dale 1995).

2 Genetic Transformation

2.1 General Account/Review (Table 1)

For the recovery of transgenic plants, it is essential to have efficient plant regeneration systems in addition to efficient DNA-transfer systems. Tobacco can be regenerated from a diversity of cell and tissue types ranging from single cells such as protoplasts or microspores to explants from complex organs such as leaf. The relative ease of manipulating these processes in culture has allowed researchers to focus more attention on the details of DNA transfer and integration in tobacco. Consequently, a range of technologies have been developed (Table 1). Of these, *Agrobacterium*-mediated transfer of T-DNA-derived vectors is most commonly used and permits individual scientists to produce hundreds of transgenic plants in a relatively short period of time (Horsch et al. 1985). Direct DNA transfer to protoplasts using a variety of procedures such as DNA uptake (Paszkowski et al. 1984) or microinjection (Crossway et al. 1986) has also been demonstrated, but these have now been largely superseded by methods that do not depend on regenerable protoplast systems. For example, direct DNA transfer to a variety of tissue types by particle bombardment has yielded transgenic plants (Klein et al. 1988), and also provides a convenient method for transient expression analysis in a variety of organs. The great advantage of the *Agrobacterium*-mediated approach over the other methods is that the T-DNA structure after insertion is more predictable. This eliminates some of the experimental variables when comparing related gene constructs; for example, when investigating the influence of fine structural changes on expression patterns in a promoter deletion experiment.

2.2 Methodology

The leaf disk transformation protocol developed by Horsch et al. (1985) is the standard for producing transgenic tobacco. Tobacco is susceptible to infection

Table 1. Examples of methods developed for the genetic transformation of tobacco

Methodology	Comments	Sample reference
Direct DNA uptake	This method is limited to protoplasts but like particle bombardment is not restricted biologically to certain species or vectors	Paszkowski et al. (1984)
Agrobacterium-mediated	This technology is easy and cost-efficient to implement and produces large numbers of transgenic plants from diverse organs with predictable T-DNA inserts	Horsch et al. (1985)
Electroporation	This is a refinement of the direct DNA uptake method	Shillito et al. (1985)
Liposome-mediated	This has been successfully demonstrated but does not provide general advantages over the above methods	Deshayes et al. (1985)
Microinjection	This method has not advanced beyond the demonstration of transformation because of the emergence of particle bombardment as an easier technology to implement. It may find specialized uses	Crossway et al. (1986)
Particle bombardment	This approach provides a simple transient expression system for diverse organs and generates transgenic plants from diverse tissue sources	Klein et al. (1988)

by a wide range of *Agrobacterium* strains and Ti plasmid classes. Vector systems have been reviewed by Miki et al. (1993). A wide range of selectable marker genes have also been used successfully in the recovery of transgenic tobacco and have also been reviewed by Miki et al. (1993). Kanamycin resistance conferred by the bacterial *npt II* gene from *Tn 05* remains the most widely used.

Plants grown aseptically in vitro are very convenient to maintain and use as starting material. This practice also reduces losses due to contamination compared with leaves from greenhouse-grown plants. To maintain plantlets in vitro we first sterilize seeds in 70% ethanol for 1 min then in 6% sodium hypochlorite (Javex, Bristol-Myers) for 25 min followed by extensive washing with sterile distilled water. This is conveniently achieved by placing the seeds in envelopes made from Whatman No. 2 filter paper and staple-closed. Once the seeds have air-dried in a laminar air-flow unit, they are germinated and maintained on solid medium consisting of MS salts (Murashige and Skoog 1962), B5 vitamins (Gamborg et al. 1978), and 2% sucrose. Plantlets can be propagated vegetatively to maintain a supply.

For transformation, the leaf material with the midvein excised is cut into 1-cm^2 fragments using a scalpel or hole punch and precultured with the epidermal side down for 2 days on solid media consisting of MS salts, B5 vitamins, 3% sucrose, 1 mg/l BAA (6-benzylaminopurine), and 0.1 mg/l NAA (alpha-naphthalene acetic acid).

Infection with *Agrobacterium* strains is achieved using overnight cultures grown in LB medium and diluted by about 50%. The leaf explants are submerged quickly (about 30s) and blotted on Whatman No.2 filter paper to remove excess bacteria. They are then returned to the previous medium to coculture with *Agrobacterium* for 2–3 days. To subsequently inhibit bacterial growth and initiate the selection of transformed tissues, the explants are transferred to media with 500 µg/ml Cefotaxime (Claforan, Rousel Canada Inc. Montreal) or Timentin (Smith Klein and Beechem, Oakville) and the selective agent which is typically 100 µg/ml kanamycin (Sigma, St. Louis). The explants are transferred to fresh media every 2–3 weeks and shoots begin to emerge within 3–5 weeks.

Once well-defined stems have developed, the shoots are excised and transferred to Magenta boxes with solid media consisting of MS salts, B5 vitamins, 3% sucrose, 500 µg/ml Cefotaxime or Timentin and 100 µg/ml kanamycin. The plants can be transferred to the greenhouse once the roots are well developed. To ensure that each plant has arisen from independent transformation events, the explants are divided into separate fragments early in the procedure and only one shoot is selected from each fragment.

For self-fertilization, the flowers are covered with bags prior to anthesis. Once the seed capsules have dried and the seeds have been collected, they are germinated on solid media with 100 µg/ml kanamycin or other selective agents to assess the approximate number of T-DNA loci that are segregating among the progeny of the transgenic plants. Several biochemical approaches can be used to assess the transgenic status of the plants (Register 1997). To establish T-DNA copy number, it is essential to perform Southern blot analysis using enzymes that span the T-DNA borders and probes that are unique for the T-DNA. Vectors derived from pBIN 19 (Bevan 1984) frequently generate plants with multiple insertions, whereas others may produce predominantly single insertions. Typically, between 15 and 20% of transformants have multiple inserts. Since multiple inserts can lead to unpredictable phenomena like transgene silencing, they are best avoided (Finnegan and McElroy 1994; Brandle et al. 1995). We generally produce 20–50 plants with each vector to make sure that we have a sufficient number of plants to have a range of levels and/or patterns of expression of the introduced transgene for our subsequent studies. Our experience indicates that there are few, if any, differences in transformation efficiency between tobacco varieties. Studies conducted by Daub and coworkers (1994) demonstrated that there are some minor differences between cultivars but nothing of practical significance.

Following segregation analysis using kanamycin resistance, eight seedlings (n = 8 gives a 95% chance of detecting at least one homozygote) are selected from the plates showing single gene segregation. They are grown to flowering, selfed, and progeny-tested. One homozygote from each primary transformant is then selected for further analysis by taking the required number of seedlings from the progeny testing plates. There is typically no difference between homozygotes selected from the same primary transformant, so that a single individual is adequate for most purposes. The homozygotes are then transplanted to pots or cells where they can be tested under controlled-environment conditions for resistance to a target pest, for example. Lines surviving controlled-environment testing can then be used to form the basis of field trials. Our experience indicates that between 10 and 15 individuals are sufficient to ensure both high levels of transgene expression and the recovery of the parental phenotype. The time between transformation and homozygous lines is about 12 months, so if field trials are planned, the transformations must be completed a year before plant-bed seeding time. The transgene typically has no direct effect on the agronomic phenotype, although this depends on the gene being expressed. Most of the apparent phenotypic variation among lines is somaclonal in nature; however, traits like cherry red, that are unstable, need particular attention (J.E. Brandle unpubl.).

The field trials themselves must be designed to meet both scientific and regulatory requirements. The experimental design will depend on the trait to be evaluated and the nature of the experiment and does not lend itself to generalization, except that there is a firm requirement to compare the pheno-type with the original parent to ensure that it has been fully recovered. In most countries, regulations covering the small-scale release of genetically modified organisms were developed based on the OECD model. It is usually required that the breeder disclose the details of the constructs used, method of trans-formation, trait conferred, stability of the trait, and any potential ecological risk.

2.3 Results and Discussion

2.3.1 Expression of Marker Genes During the Development of Primary Transformants

Marker genes, such as GUS (beta-glucuronidase, *E. coli uid A*; Jefferson 1987) have been useful for studying the patterns of transgene expression in plants. This gene, fused to the 35S promoter, is widely available in vectors such as pBI 121 (Clontech, Palo Alto), which is derived from pBIN19 and has been used extensively in studies of transgenic tobacco. Although it is tempting to evalu-ate shoots or young primary transformants while still in culture, it is important to treat such data cautiously because GUS activity is often sectored at this stage of development. We have found that expression is stabilized in leaf tissues after transfer to soil and after a period of growth in the greenhouse. Quantitative measurements of GUS activity prior to this period can be ex-tremely variable and may yield erratic results. This was not always apparent in shoots of other species such as canola or alfalfa.

The range of expression among plants can vary widely reflecting the influence of the surrounding insertion sites on the expression of the genes

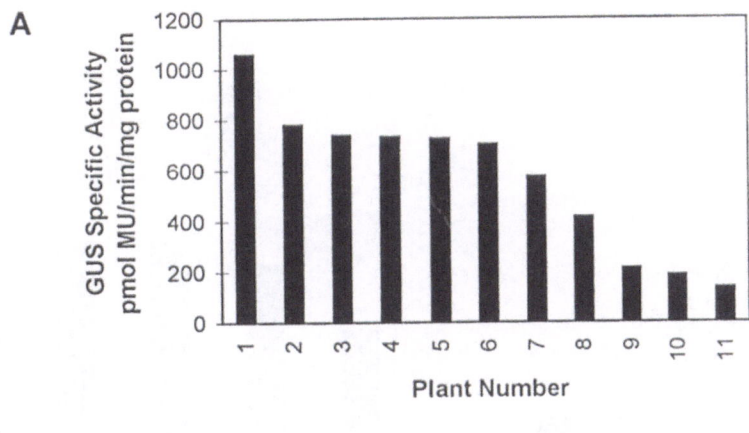

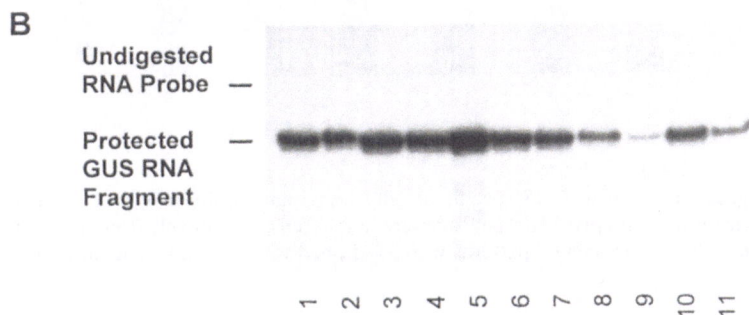

Fig. 1A,B. Variation in GUS-specific activity and mRNA levels in leaves of independent primary transgenic lines of tobacco transformed with the vector pBI 121. The GUS gene is under the control of the 35S promoter. The plants were grown in the greenhouse after transfer from tissue culture and the RNA levels were assessed by RNase protection assays using a GUS-specific RNA probe

within the inserted T-DNA. This is often referred to as position effects. Figure 1 shows data on leaves of independent primary transgenic plants generated under identical conditions and quantitatively assessed for GUS activity under the control of the 35S promoter. It is clear that the range of GUS activity may be very wide and generally reflects the level of GUS mRNA that has accumulated in the leaves. A number of developmental processes also influence expression levels. For instance, expression measured as GUS specific activity differs among the leaves on individual plants. This is illustrated in Fig. 2. The highest activity is found in the older leaves at the bottom of the plant and is progressively lower towards the younger top leaves. Much to our surprise, we have also found that expression is influenced by the developmental history of the plantlets in culture. For instance, a set of plantlets propagated in

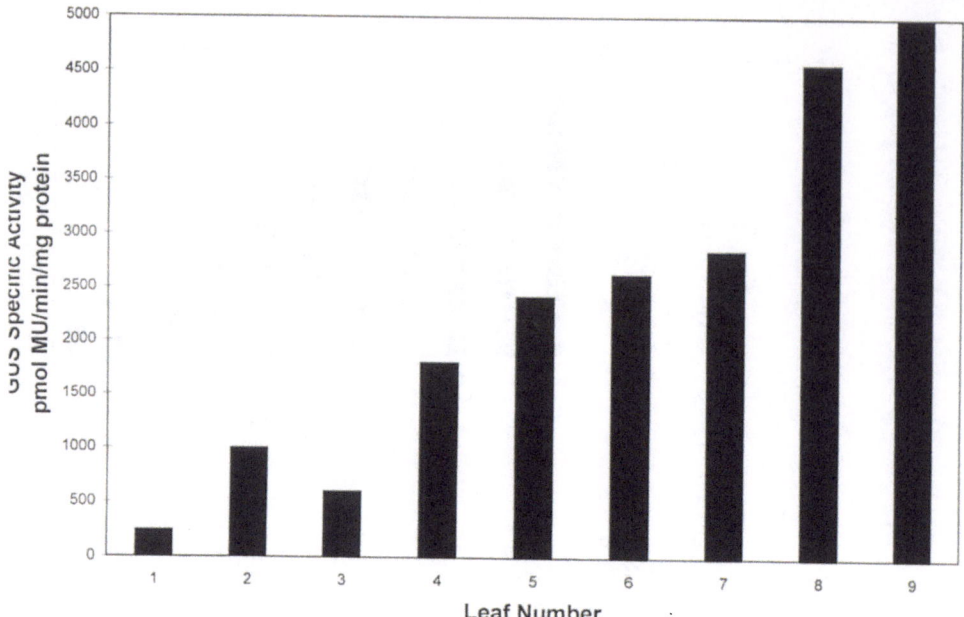

Fig. 2. Variation in GUS specific activity levels among the leaves of a single transgenic line transformed with pBI 121. The leaves are numbered consecutively from the top (*1*) to the bottom (*10*). The developmental patterns were confirmed with four independent transgenic lines

culture for a 3-month period after selection displayed much lower levels of GUS specific activity after transfer to the greenhouse than another set of plants generated in the same experiment and immediately transferred to the greenhouse (Fig. 3). It is not uncommon to find that clones of a single primary transgenic plant will yield different levels of expression. These studies teach us that a number of developmental and environmental factors have a profound influence on transgene expression and must be carefully controlled when evaluating levels of gene expression particularly in primary transgenic plants. Well-characterized progeny lines may be preferable sources of material for transgenic studies, as the growth conditions can be more carefully regulated.

2.3.2 Instability of Transgene Expression

The introduction of multiple T-DNA copies may result in unstable transgene expression and this instability may be inherited in an unpredictable way (reviewed by Matzke and Matzke 1993; Flavell 1994). If the transgene is homologous to other genes resident within the plant genome, coordinate inactivation of all of the genes may occur, a phenomenon known as cosuppression (Jorgenson 1990). Examples in tobacco include the

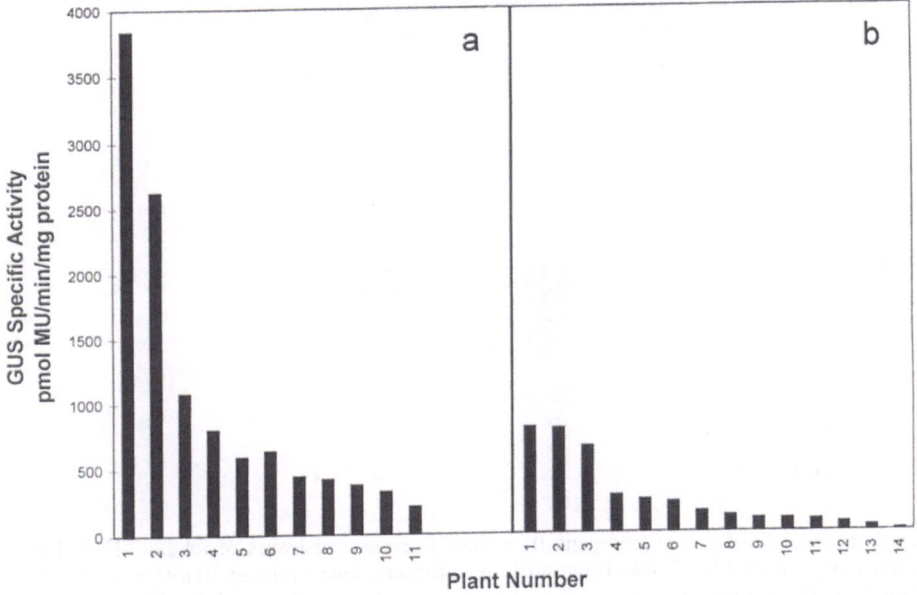

Fig. 3a,b. Range of GUS activities among two groups of plants that were transformed with pBI 121 at the same time and transferred to the greenhouse immediately (**a**) or propagated in culture through cuttings and transferred to the greenhouse 3 months later (**b**)

cosuppression of beta-1, 3-glucanase genes (de Carvalho et al. 1992), pyruvate kinase genes (Gottlob-McHugh et al. 1992), nitrate reductase genes (Dorlhac de Borne et al. 1994), and acetohydroxyacid synthase (Brandle et al. 1995) genes.

We observed the cosuppression of tobacco pyruvate kinase genes following the introduction of a transgene into tobacco containing a sequence coding for potato cytosolic pyruvate kinase, PKc, fused to the constitutive CaMV 35S promoter. This resulted in the recovery of two independent transgenic lines that had specifically eliminated cytosolic pyruvate kinase from the leaves (Gottlob-McHugh et al. 1992). These plants were able to grow despite the absence in the leaves of what is thought to be a key glycolytic enzyme. This phenotype was transmitted to progeny although in one transgenic line this phenotype was not transmitted in a mendelian fashion. In this line, many progeny plants had wild-type levels of pyruvate kinase in the leaves which was associated with a significant increase in potato transcript levels (Fig. 4). The lack of cytosolic pyruvate kinase activity was a tissue-specific phenomenon. The degree of reduction in potato transgene expression was significantly less in roots and was not associated with a reduction in PKc activity (Fig. 4). The mechanism explaining the difference in gene silencing observed between roots and leaves is unknown but may be related to tissue-specific isozymes. Tissue-specific gene silencing provides the unique opportunity to study the impact of selectively inactivating gene expression.

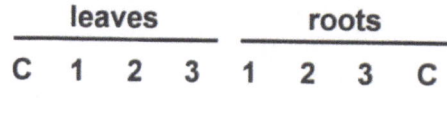

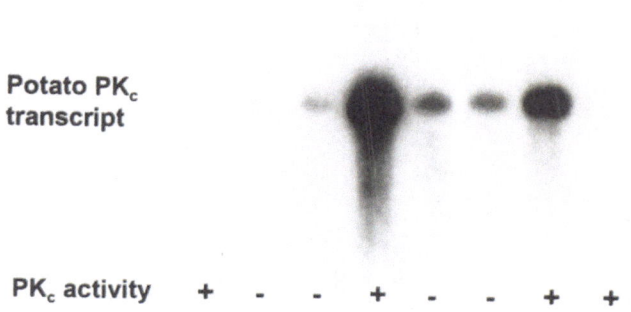

Fig. 4. Expression of potato transgene transcripts in roots and leaves of tobacco. Each lane contains 10 µg of total RNA; *lane C* untransformed tobacco; *lane 1* primary transformant (1-121-35) with no PKc activity in the leaves; *lane 2* progeny plant from selfing of 1-121-35 with no PKc activity in the leaves; *lane 3* progeny plant from selfing of 1-121-35 with wild-type levels of PKc activity in the leaves

The cosuppression of mutant acetohydroxyacid synthase transgenes resulted in a striking phenotype in field-grown plants (Fig. 5). The mutant AHAS genes confer resistance to a number of classes of herbicides, but up to 60% of homozygous field-grown plants showed damage ranging from mild to severe after herbicide application (Fig. 5). Interestingly, cosuppression was not seen in analyses performed in the greenhouse. The phenomenon was triggered by the transplantation of greenhouse-grown seedlings to the field, and therefore was only recognized at the stage of field evaluation. These studies emphasize the need to carefully select transgenic lines used for further studies and applications. If the phenotype of an inactivated transgene is less visible than herbicide resistance, for example, it may go unnoticed and has the potential to generate results that are difficult to interpret. By careful analysis of the plants, transgenic lines can be selected that do not exhibit gene inactivation phenomena.

2.3.3 Agronomic Properties and Field Evaluation

The current focus in transgenic tobacco field trials is crop protection, similar to that of the other major crops such as maize, canola and potato (Ahl-Goy and Duesing 1995). Trials include resistance to insects, viruses, fungal diseases, and herbicides. Tobacco was the model system used to develop a large part of this technology, but despite this pioneering role, much of the application of this

Fig. 5. Induction of cosuppression by transplantation to the field (*left row*). The plants express a mutant *Arabidopsis* acetohydroxyacid synthase gene that conferred resistance to sulfonylurea herbicides under greenhouse conditions, but showed sensitivity to herbicides under field conditions. (Brandle et al. 1995)

technology is now occurring in other crops. Ahl Goy and Duesing (1995) recently reported that tobacco accounted for 12.5% of all transgenic field trials conducted world-wide in 1994, down from 25% only a few years earlier. While the authors concluded that there was a significant trend away from tobacco, it is more probable that the reduction reflects a more mature research focus, where trials are now limited to those with a more direct impact on tobacco production. A time-course study of the ratio of trials conducted with commercial tobacco varieties compared with those done with SR1 and similar lab varieties would shed light on the real nature of this trend. There are now a large number of genes that could be used to solve tobacco improvement problems, such as blue-mold and insect resistance, that have resisted conventional efforts for years.

Herbicide Resistance. Brandle et al. (1992) introduced the *Arabidopsis csr 1-1* gene for sulfonylurea herbicide resistance into a Canadian flue-cured variety, Delgold, in an attempt to take advantage of the positive properties of sulfonylureas, to allow post-plant herbicide application, and to broaden the weed control spectrum. A number of primary transformants were screened and two lines with high levels of resistance in greenhouse trials were selected for extensive evaluation in field trials (Brandle and Miki 1993). Adequate levels of resistance were observed in one of the selections, but agronomic performance of the two transgenic lines was poor and resistance levels were not high enough to include a significant margin of safety. Poor agronomic performance was attributed to either somaclonal variation, insertional mutagenesis or retrotransposon induced mutations in the transformants. It was concluded that a larger population of transgenic lines would allow breeders to select around performance problems. Later field testing with homozygous lines derived from the two original primary transformants revealed that the line showing the best resistance in greenhouse trials suffered from transgene inactivation (Fig. 5). A line hemizygous for the same transgene was not affected and showed uniform, but slightly inadequate, resistance in field trials.

In order to improve the system, a new mutant gene derived from *Brassica napus* was introduced. It had resistance to a larger number of the sulfonylurea family members, and it was hoped that the problems with transgene inactivation and agronomic performance could be solved (Brandle et al. 1994). The gene was again introduced into Delgold and the three most resistant lines were selected in greenhouse trials and then evaluated in plant bed trials and normal yield trials. In field trials, the performance of the untreated lines was equal to untransformed controls but the transplanted homozygous lines in the yield trial again suffered from transgene inactivation (unpubl.). It is possible that the selection for high levels of resistance in greenhouse trials leads to expression levels in excess of what can be tolerated by the plant (Lindbo et al. 1993b). If this is indeed the case, it will be difficult to engineer herbicide resistance based on herbicide-insensitive mutant genes or direct overexpression.

Bromoxynil resistance is a good example of the successful application of the metabolic degradation approach to herbicide resistance. Bromoxynil herbicide is a photosynthetic inhibitor in plants that acts by binding a part of the quinone-binding complex of photosystem II. A soil bacterium was discovered that has the ability, via the action of a nitrilase enzyme, to degrade bromoxynil into its herbicidally inactive primary metabolite (Stalker et al. 1988). The bacterial plasmid carrying the gene (bxn) responsible for this nitrilase activity was isolated, subsequently subcloned into a plant expression vector, and then transformed to tobacco. The gene was under the control of a light-regulated tissue-specific promoter (RuBisCO) and resistance at eight times normal field rates was obtained. Subsequent commercial development has occurred in air-cured tobacco in France and agronomic performance of transgenic lines is equal to untransformed controls and herbicide resistance is at commercially acceptable levels (Delon et al. 1994). The transgenic cultivar has received approval for release without further need for confined testing, and is set for commercial production in 1995 (Freyssinet and Derose 1994).

A gene (bar) coding for an enzyme (phosphinothricin acetyl transferase, PAT) that detoxifies the herbicide phosphinothricin has been isolated from the bacteria *Streptomyces hygroscopicus*. Transformation experiments and subsequent field experiments with the bar gene under the control of the 35S promoter demonstrated that tobacco was tolerant to up to ten times normal field rates (DeBlock et al. 1987). There was no agronomic evaluation of the transgenic tobaccos included in this experiment

Resistance to Insect Pests. Recently, we tested four transgenic tobacco lines carrying a codon-optimized delta-endotoxin gene from *Bacillus thuringiensis* under control of the CaMV 35S promoter (Sardana et al. 1996). Four lines were selected from a group of ten for field tests following screening for single-copy inserts and for toxicity to tobacco hornworms (THW, *Manduca sexta*) in greenhouse trials. The trial was based on a randomized complete block design with four replications and normal production practices for tobacco research trials (e.g., Brandle and Miki 1993). Four plants were chosen at random from each row and one second instar THW larvae was placed on the third leaf from the top, which was then covered with a polyester mesh bag to prevent bird predation. Larvae were allowed to feed for 5 days. The leaf area consumed by each insect and the weight of each insect were recorded. Along with the efficacy studies, standard agronomic evaluation of the lines, relative to the untransformed parental lines used in the original transformations, were also conducted. In this case, the results of the trials demonstrated that selection for insect toxicity in the greenhouse was a valid screening strategy for transgene expression in that the toxicity of each line was high and there were no significant differences among lines. Similar toxicity results were obtained by Warren et al. (1992). There was, however, a great deal of variability in agronomic performance for yield, ground sucker weights, topping height, leaf numbers, and eighth and tip leaf area (Table 2). Despite this variability, it was still possible to select an individual line (e.g., T1-7) whose performance is very similar to that of the untransformed control (Delgold).

Resistance to Pathogens. Pathogen-derived resistance induced by the pathogen's own genetic material has been reported numerous times for plant viruses in transgenic plants (see the reviews by Beachy et al. 1990; Fitchen and

Table 2. Agronomic performance and insect resistance in transgenic (T1-7, T3-2, T4-1, T6-3, T10-3) and nontransgenic (Delgold) tobacco lines tested in field trials at the Delhi Research Station, Ontario in 1996

Line	Suckers (kg/ha)	Days to top	Leaf no.	8th leaf area (cm^2)	Tip leaf area (cm^2)	Yield (kg/ha)	Grade index ($/kg)	Leaf area consumed (cm^2)
Delgold	17.3	63	18.0	1419	999	3066	4.05	16.5
T1-7	36.8	62	17.2	1447	1054	2987	4.02	0.3
T3-2	49.5	68	19.8	1386	995	2862	3.95	0.2
T4-1	9.8	68	17.9	1196	865	2461	4.04	0.2
T6-3	55.2	63	17.7	1384	1108	2754	3.93	0.3
T10-3	77.1	60	16.9	1597	1132	3103	4.01	0.2
SE	14.1	0.7	0.3	52	42	82	0.03	1.0

Beachy 1993; Scholthof et al. 1993; Wilson 1993). So far, three main types of pathogen-derived resistance have been observed in virus-resistant transgenic tobacco.

The first type, the coat protein-mediated resistance, involves the expression and accumulation of a viral coat protein in the transgenic plants. Early examples included tobacco mosaic virus (Powell et al. 1986), tobacco rattle virus (van Dun et al. 1987), and alfalfa mosaic virus (Loesch-Fries et al. 1987; Tumer et al. 1987; van Dun et al. 1987). Many more viruses have been added to the list since then. In the coat protein-protected plants, the level of resistance correlates with the level of transgene product (Powell et al. 1990; Angenent et al. 1990). Those plants may also show a low but significant degree of protection against some other related viruses (Anderson et al. 1989; Nejidat and Beachy 1990). The coat protein-mediated resistance bears a striking resemblance to the classical cross-protection in which a plant infected with a mild strain of a virus is protected against superinfection by a related strain of that virus (Palukaitis and Zaitlin 1984). Detailed studies have suggested that the coat protein-mediated resistance involves interference with the initial infection, including uncoating of the viral particles (Wu et al. 1990), and the systemic spread of the virus (Wisniewski et al. 1990).

The second type of pathogen-derived resistance, the homology-dependent resistance, often involves the expression of a nonfunctional segment of the viral genome, such as part of the replicase gene (potato virus X, Braun and Hemenway 1992; cucumber mosaic virus, Anderson et al. 1992; tobacco mosaic virus, Donson et al. 1993), an untranslatable mutant form of the coat protein gene (tobacco etch potyvirus, Lindbo and Dougherty 1992) or the replicase (potato virus X, Longstaff et al. 1993). The transgene-derived RNA may also be translatable. This type of resistance is generally virus-specific. In the homology-dependent resistant plants, there is no apparent relationship between the level of transgene transcript or product accumulation and the degree of resistance (see Lindbo et al. 1993a for a review of the cases involving potyviruses). Plants displaying this virus-specific resistant phenotype have been shown to transcribe the transgene at a high rate yet accumulates the transgene transcript at low levels (Dougherty et al. 1994; Smith et al. 1994; Mueller et al. 1995). It is proposed that the homology-dependent resistance is mediated at the cellular level by a cytoplasmic activity that targets specific RNA sequences for inactivation (Lindbo et al. 1993b). Mueller et al. (1995) have suggested that the virus-specific resistance and the homology-dependent gene silencing observed in transgenic plants are due to the same RNA-based mechanism. It is also suggested that a threshold level of transgene-derived transcript must be exceeded to activate the cytoplasmic post-transcriptional RNA degradation process and elicit virus resistance (Smith et al. 1994). Data from Goodwin et al. (1996) support this suggestion. It also indicates that the post-transcriptional RNA degradation process is initiated via cleavage of specific sites within the target RNA sequence.

A third type of broad-spectrum pathogen-derived resistance has been achieved in transgenic plants by expression of dominant negative mutant forms of the movement proteins which are necessary for cell-to-cell movement of the virions (white clover mosaic virus, Beck et al. 1994; tobacco mosaic virus, Cooper et al. 1995).

Yie et al. (1995) developed and field-tested transgenic tobaccos expressing the satellite RNA and the coat protein gene from cucmber mosiac virus. Disease incidence was substantially reduced under both natural infection and mechanical inoculation conditions in the transgenic plants and yield and grade quality of the transgenic lines was significantly higher than controls under disease pressure. Performance of the transgenic lines was equal to controls under disease-free conditions.

Transgenes have been used in attempts to control other pathogens and pests of tobacco. The main strategies involve the expression of (1) detoxification genes for the toxins which play an important role in pathogenesis, like an acetyltransferase from *Pseudomonas syringae* that detoxifies tabtoxin produced in wildfire of tobacco (Anzai et al. 1989); (2) pathogenicity-related proteins from plants, such as a bean chitinase that increases the resistance of tobacco to the fungal pathogen *Rhizoctonia solani* (Broglie et al. 1991); (3) antifungal peptides, like the small cystein-rich antifungal proteins from radish which confer resistance in tobacco to the foliar fungal pathogen *Alternaria longipes* (Terras et al. 1995).

2.3.4 Production of Pharmaceutical Proteins and Industrial Enzymes in Tobacco

The use of tobacco for the production of pharmaceutical proteins or industrial enzymes is clearly feasible. Examples include the production of human protein C, a serum protease and human glucocerebrosidase for treatment of Gaucher disease (Cramer et al. 1995). It has also been shown that the four chains of a secretory immunoglobin were properly expressed and assembled in plants and that a fully functional antibody could be produced by intercrossing lines expressing each of the subunit chains (Ma et al. 1995). These experiments indicate that tobacco may eventually be used to produce large amounts of high-valued therapeutic proteins as a safe and efficient alternative to the current practices of isolating them from human tissues or mammalian production systems. Recently, it has also been shown that a bacterial antigen produced in transgenic plants could effectively immunize mice when the crude protein extracts from the transgenic plant tissue were administered orally (Haq et al. 1995). This may be developed into a cost-effective and easily administered process for vaccinations. An interesting study in animal nutrition revealed that expression of *Aspergillus niger* phytase genes in tobacco seeds produced a feed supplement that was very effective in raising the phosphate content in meal derived from other seed sources (Pen et al. 1993). New work has extended the list of examples to viral antigens and human haemoglobin (Mason et al. 1996; Diercyk et al. 1997).

Tobacco leaves are capable of producing high levels (8–10%) of soluble protein (fraction 1 protein, F1P) (Woodleif et al. 1981) and pilot systems have been developed to purify this fraction for use as a high-protein dietary supplement (Montanari et al. 1993). We have adapted the F1P system to transgenic tobacco expressing the sea raven type II antifreeze protein and defined the agricultural conditions that will maximize transgenic protein production (Kenward 1995). Any potential nicotine contamination in the final product was eliminated by using a locally adapted tobacco mutant (81V9) that has only a limited capacity to synthesize tobacco alkaloids (Chaplin 1977). This is especially relevant when considering the potential for oral delivery of a transgene protein as edible plant tissue.

As the production of plant recombinant proteins moves from the bench to pilot plant, regulatory issues must be addressed (Miele 1997). From both a regulatory and public safety standpoint, tobacco is an ideal species for the production of biologically active proteins. Tobacco is a nonfood crop and therefore the risk of accidental leakage of transgenic plant material expressing genes for biologically active proteins into the human food chain is near zero. Other plant bioreactor systems do not offer this advantage. Producing the tobacco in Canada, where there are no naturally occurring wild *Nicotiana* species, further minimizes the risk of gene leakage to the local flora. With the tobacco system, protein production is based on leaves, not seed or tubers. When coupled with the fact that the leaves are harvested before flowering, there is virtually no risk of uncontrolled bioreactor plants occurring in future crop seasons. Tobacco does not overwinter in Canada and there is subsequently no rattooning in the following season. The background genotype and the system used for transgene protein production addresses a number of potential regulatory concerns and begins to deal with the issue of biologically active secondary metabolites that may impair product quality.

3 Summary and Conclusions

Transgenic tobacco has played a prominent role in the advancement of plant science. It has been used as a valuable tool in gaining fundamental knowledge in diverse areas of plant biology ranging from gene expression to plant breeding. Tobacco provided one of the first transgenic model systems for higher eukaryotes. This remarkable achievement in plant biotechnology has been used to generate some of the first genetically modified plants with resistance to viruses, insects, fungi and herbicides. The adaptation of these technologies has been successfully applied to other plant species. Some of these species now provide alternatives to tobacco for the study and application of molecular genetic processes that confer significant agronomic and commercial value to crops. This is reflected in a decrease in the proportion of field trials that are performed each year with transgenic tobacco; however, a review of the scientific literature indicates that transgenic tobacco is still important for the generation of fundamental knowledge and new technologies. For example, applications are being developed that will improve production. As the acceptance of smoking tobacco in North America is on the decline, the development of tobacco as a valuable crop in the production of pharmaceuticals and industrial enzymes is becoming a reality. Examples have been provided above that can fundamentally change tobacco and create new crops through the use of biotechnology.

Acknowledgments. The authors are grateful to many colleagues who have contributed to the research and knowledge generated in our labs over the years, and to Radhey Pandeya and Liz Foster for reviewing the manuscript. ECORC contribution number 981305.

References

Ahl Goy P, Duesing JH (1995) From pots to plots: genetically modified plants on trial. Bio/Technology 13:454–459

Anderson EJ, Stark DM, Nelson RS, Powell PA, Tumer NE, Beachy RN (1989) Transgenic plants that express the coat protein genes of tobacco mosaic virus or alfalfa mosaic virus interfere with disease development of some nonrelated viruses. Phytopathology 79:1284–1290

Anderson JM, Palukaitis P, Zaitlin M (1992) A defective replicase gene induces resistance to cucumber mosaic virus in transgenic tobacco plants. Proc Natl Acad Sci USA 89:8759–8763

Angenent GC, van den Ouweland JMW, Bol JF (1990) Susceptibility to virus infection of transgenic tobacco plants expressing structural and nonstructural genes of tobacco rattle virus. Virology 175:191–198

Anzai H, Yoneyama K, Yamaguchi I (1989) Transgenic tobacco resistant to a bacterial disease by the detoxification of a pathogenic toxin. Mol Gen Genet 219:492–494

Barton KA, Whitely HR, Yang NS (1987) *Bacillus thuringiensis* delta-endotoxin expressed in transgenic *Nicotiana tabacum*. Plant Physiol 85:1103–1109

Beachy RN, Loesch-Fries S, Tumer NE (1990) Coat protein mediated resistance against virus infection. Annu Rev Phytopathol 28:451–474

Beck DL, Van Dolleweerd CJ, Lough TJ, Balmori E, Voot DM, Andersen MT, O'Brien IEW, Forster RLS (1994) Disruption of virus movement confers broad-spectrum resistance against systemic infection by plant viruses with a triple gene block. Proc Natl Acad Sci USA 91:10310–10314

Bevan M (1984) Binary *Agrobacterium* vectors for plant transformation. Nucleic Acids Res 12:8711–8721

Brandle JE, Miki BL (1993) Agronomic performance of flue-cured tobacco grown under field conditions. Crop Sci 33:847–852

Brandle JE, Labbe H, Zilkey BF, Miki BL (1992) Resistance to the sulfonylurea herbicides chlorsulfuron, amidosulfuron and DPX-R9674 in transgenic flue-cured tobacco. Crop Sci 32:1443–1445

Brandle JE, Morrison MJ, Hattori J, Miki BL (1994) A comparison of two genes for sulfonylurea herbicide resistance in transgenic tobacco seedlings. Crop Sci 34:226–229

Brandle JE, McHugh SG, James L, Labbe H, Miki BL (1995) Instability of transgene expression in field-grown tobacco carrying the *csr1-1* gene for sulfonylurea herbicide resistence. Bio/Technology 13:994–998

Braun CJ, Hemenway CL (1992) Expression of amino-terminal portions or full-length viral replicase genes in transgenic plants confers resistance to potato virus X infection. Plant Cell 4:735–745

Broglie K, Chet I, Holliday M, Cressman R, Biddle P, Knowlton S, Mauvais CJ, Broglie R (1991) Transgenic plants with enhanced resistance to the fungal pathogen *Rhizoctonia solani*. Science 254:1194–1197

Chaplin JF (1977) Breeding for varying levels of nicotine in tobacco. In: Recent advances in the chemical composition of tobacco and tobacco smoke. Proc 173rd Am Chem Soc Meet, Agriculture and Food, New Orleans, Louisiana, pp 328–339

Comai L, Facciotti D, Hiatt WR, Thompson G, Rose RE, Stalker DM (1985) Expression in plants of a mutant *aro*A gene from *Salmonella typhimurium* confers tolerance to glyphosate. Nature 317:741–744

Cooper B, Lapidot M, Heick JA, Dodds JA, Beachy RN (1995) A defective movement protein of TMV in transgenic plants confers resistance to multiple viruses whereas the functional analog increases susceptibility. Virology 206:307–313

Cramer L, Weissenborn DL, Oishi KK, Grabau EA, Bennet S, Ponce E, Grabowski GA, Radin DN (1995) Bioproduction of human enzymes in transgenic tobacco. In: Program and Abstr Int Symp on Engineering Plants for Commercial Products and Applications, October 1–4, 1995, Lexington, Kentucky

Crossway A, Oakes JV, Irvine JM, Ward B, Knauf VC, Shewmaker CK (1986) Integration of foreign DNA following microinjection of tobacco mesophyll protoplasts. Mol Gen Genet 202:179–185

Dale PZ (1995) R&D regulations and field trialing of transgenic crops. Tibtectt 13:398–403

Daub ME, Jenns AE, Urban LA, Brintle SC (1994) Transformation frequency and foreign gene expression in burley and flue-cured cultivars of tobacco. Tob Sci 38:51–54

De Block M, Botterman J, Vandewiele M, Dockx J, Thoen C, Gossele V, Rao Movva N, Thompson C, Van Montagu M, Leemans J (1987) Engineering herbicide resistance in plants by expression of a detoxifying enzyme. EMBO J 6:2513–2518

de Carvalho F, Gheyson G, Kushnir S, Van Montague M, Inzé D, Castresana C (1992) Suppression of beta-1,3-glucanase transgene expression in homozygous plants. EMBO J 11:2595–2602

Delon R, Pelissier B, San LH, Borrod G, Freyssinet G (1993) Transgenic tobaccos resistant to herbicides of the oxynil family: five years of field experiments. Ann Tab (Sect 2) 25:17–26

Deshayes AL, Herrer-Estrella L, Caboche M (1985) Lysosome-mediated transformation of tobacco mesophyll protoplasts by an *Escherichia coli* plasmid. EMBO J 4:2731–2737

Dieryck W, Pagnier J, Poyart C, Marden MC, Gruber V, Bournat P, Baudino S, Merot B (1997) Human haemoglobin from transgenic tobacco. Nature 386:29–30

Donson J, Kearney CM, Turpen TH, Khan IA, Kurath G, Turpen AM, Jones GE, Dawson WO, Lewandowski DJ (1993) Broad resistance to tobamoviruses is mediated by a modified tobacco mosaic virus replicase transgene. Mol Plant-Microbe Interact 6:635–642

Dorlhac de Borne F, Vincentz M, Chupeau Y, Vaucheret H (1994) Cosuppression of nitrate reductase host genes and transgenes in transgenic tobacco plants. Mol Gen Genet 243:613–621

Dougherty WG, Lindbo JA, Smith HA, Parks TD, Swaney S, Proebsting WM (1994) RNA-mediated virus resistance in transgenic plants: exploitation of a cellular pathway possibly involved in RNA degradation. Mol Plant-Microbe Interact 7:544–552

Fecker LF, Rügenhagen C, Berlin J (1993) Increased production of cadaverine and anabasine in hairy root cultures of *Nicotiana tabacum* expressing a bacterial lysine decarboxylase gene. Plant Mol Biol 23:11–21

Finnegan J, McElroy D (1994) Transgene inactivation: plants fight back. Bio/Technology 12:883–888

Fitchen JH, Beachy RN (1993) Genetically engineered protection against viruses in transgenic plants. Annu Rev Microbiol 47:739–763

Flavell RB (1994) Inactivation of gene expression in plants as a consequence of specific sequence duplication. Proc Natl Acad Sci USA 91:3490–3496

Fobert PR, Labbe H, Cosmopoulos J, Gottlob-McHugh S, Ouellet T, Hattori J, Sunohara G, Iyer VN, Miki BL (1994) T-DNA tagging of a seed coat-specific promoter in tobacco. Plant J 6:567–577

Freyssinet G, Derose RT (1994) Development of genetically modifed crops resistant to herbicides and pests. Agro-Food Industry Hi-Tech 5:3–7

Gamborg OL, Miller RA, Ojima K (1978) Nutrient requirements of suspension cultures of soybean root cells. Exp Cell Res 50:151–158

Goodwin J, Chapman K, Swaney S, Parks TD, Wernsman EA, Dougherty WG (1996) Genetic and biochemical dissection of transgenic RNA-mediated virus resistance. Plant Cell 8:95–105

Gottlob-McHugh SG, Sangwan RS, Blakely SD, Vanlerberghe GC, Ko K, Turpin DH, Plaxton WC, Miki BL, Dennis DT (1992) Normal growth of transgenic tobacco plants in the absence of cytosolic pyruvate kinase. Plant Physiol 100:820–825

Haq TA, Mason HS, Clements JD, Arntzen CJ (1995) Oral immunization with a recombinant bacterial antigen produced in plants. Science 268:714–716

Hayashi H, Czaja I, Lubenow H, Schell J, Walden R (1992) Activation of a plant gene by T-DNA tagging: auxin-independent growth in vitro. Science 258:1350–1353

Horsch RB, Fry JE, Hoffman NL, Eichholtz D, Rogers SG, Fraley RT (1985) A simple and general method for transferring genes into plants. Science 227:1229–1231

Jefferson RA (1987) Assaying for chimeric genes in plants; the GUS gene fusion system. Plant Mol Biol Rep 5:387–405

Jorgenson R (1990) Altered gene expression in plants due to *trans* interactions between homologous genes. Trends Biotechnol 8:340–344

Kenward K (1995) Expression of fish antifreeze protein genes in transgenic tobacco for increased freezing resistance and as a model for molecular farming. PhD Thesis, Queens University, Kingston

Klein TM, Harper E, Svab Z, Sanford JC, Fromm ME, Maliga P (1988) Stable genetic transformation of intact *Nicotiana* cells be the particle bombardment process. Proc Natl Acad Sci USA 85:8502–8505

Lindbo JA, Dougherty WG (1992) Untranslatable transcripts of the tobacco etch virus coat protein gene sequence can interfere with tobacco etch virus replication in transgenic plants and protoplasts. Virology 189:725-733

Lindbo JA, Silba-Rosales L, Dougerthy WG (1993a) Pathogen-derived resistance pt potyviruses: working, but why? Semin Virol 4:369–379

Lindbo J, Silva-Rosales L, Proebsting WM, Dougherty WG (1993b) Induction of a highly specific antiviral state in transgenic plants: implications for regulation of gene expression and virus resistance. Plant Cell 5:1749–1759

Loesch-Fries LS, Merlo D, Zinnen T, Burhop L, Hill K, Krahn K, Jarvis N, Nelson S, Halk E (1987) Expression of alfalfa mosaic virus RNA4 in transgenic plants confers virus resistance. EMBO J 6:1845–1851

Longstaff M, Brigneti G, Boccard F, Chapman S, Baulcombe DC (1993) Extreme resistance to potato virus X infection in plants expressing a modified component of the putative viral replicase. EMBO J 12:379–386

Ma JKC, Hiatt A, Hein M, Vine ND, Wang F, Stabila P, van Dolleweerd C, Mostov K, Lehner T (1995) Generation and assembly of secretory antibodies in plants. Science 268:716–719

Mason HS, Ball JM, Shi JJ, Jiang X, Estes MK, Arntzen CJ (1996) Expression of Norwalk virus capsid protein in transgenic tobacco and potato and its oral immunogenicity in mice. Proc Natl Acad Sci USA 93:5335–5340

Matzke M, Matzke AJM (1993) Genomic imprinting in plants: parental effects and *trans*-inactivation phenomena. Annu Rev Plant Physiol Plant Mol Biol 44:53–76

Miele L (1997) Plant bioreactors as hosts for biopharmaceuticals: regulatory considerations. TIBTECH 15:45–50

Miki BL, Fobert PF, Charest PJ, Iyer VN (1993) Procedures for introducing foreign DNA into plants. In: Glick BR, Thompson JE (eds) Methods in plant molecular biology and biotechnology. CRC Press, Boca Raton, pp 67–88

Montanari L, Fantozzi P, Pedone S (1993) Tobacco fraction 1 (F1P) utilization for oral and enteral feeding of patients. 1. Heavy metal evaluation. Food Sci Technol 26:259–263

Mueller E, Gilbert J, Davenport G, Brigneti G, Baulcombe DC (1995) Homology-dependent resistance: transgenic virus resistance in plants related to homology-dependent gene silencing. Plant J 7:1001–1013

Murashige T, Skoog F (1962) A revised medium for rapid growth and bio-assays with tobacco tissue cultures. Physiol Plant 15:473–497

Nejidat A, Beachy RN (1990) Transgenic tobacco plants expressing a coat protein gene of tobacco mosaic virus are resistant to some other tobamoviruses. Mol Plant-Microbe Interact 3:247–251

Odell JT, Nagy F, Chua N-H (1985) Identification of DNA sequences required for activity of the cauliflower mosaic virus 35S promoter. Nature 313:810–812

Palukaitis P, Zaitlin M (1984) A model to explain the "cross-protection" phenomenon shown by plant viruses and virioids. In: Kosuge T, Nester EW (eds) Plant-microbe interaction: molecular and genetic perspectives. Macmillan, New York, pp 420–429

Paszkowski J, Shillito RD, Saul M, Mandak V, Hohn T, Hohn B, Potrykus I (1984) Direct gene transfer to plants. EMBO J 3:2717–2722

Pen J, Verwoerd TC, Van Paridon PA, Beudeker RF, Van den Elzen PJM, Gearse K, Van der Klis JD, Versteegh HAJ, Van Ooyen AJJ, Hoekema A (1993) Phytase-containing transgenic seeds as a novel feed additive for improved phosphorous utilization. Bio/Technology 11:811–814

Powell AP, Nelson RS, De B, Hoffman N, Rogers SG, Fraley RT, Beachy RN (1986) Delay of disease development in transgenic plants that express the tobacco mosaic virus coat protein gene. Science 232:738–743

Powell PA, Sanders PR, Tumer N, Fraley RT, Beachy RN (1990) Protection against tobacco mosaic virus infection in transgenic plants requires accumulation of coat protein rather than coat protein RNA sequences. Virology 175:124–130

Register JC (1997) Approaches to evaluating the transgenic status of transformed plants. TIBTECH 15:141–146

Saito K, Murakoshi I, Inze D, Van Montagu M (1989) Biotransformation of nicotine alkaloids by tobacco shooty teratomas induced by Ti plasmid mutant. Plant Cell Rep 7:607–610

Sardana R, Dukiandjiev S, Giband M, Cheng X, Cowan K, Sauder C, Altosaar I (1996) Construction and rapid testing of synthetic and modified toxin gene sequences CryIA (b&c) by expression in maize endosperm culture. Plant Cell Rep 15:677–681

Scholthof K-BG, Scholthof HB, Jackson AO (1993) Control of plant virus disease by pathogen-derived resistance in transgenic plants. Plant Physiol 102:7–12

Shillito RD, Saul MW, Paszkowski M, Muller M, Potrykus I (1985) High-efficiency direct gene transfer to plants. Bio/Technology 3:1099–1103

Smith HA, Swaney SL, Parks TD, Wersnman EA, Dougherty WG (1994) Transgenic plant virus resistance mediated by untranslatable sense RNAs: expression, regulation, and fate of nonessential RNAs. Plant Cell 6:1441–1453

Stalker DM, McBride KE, Malyj LD (1988) Herbicide resistance in transgenic plants expressing a bacterial detoxification gene. Science 242:419–422

Terras FRG, Eggermont K, Kovalera V, Raikhel NV, Osborn RW, Kester A, Rees SB, Torrekens S, Van-Leuven F, Vanderleyden J, Cammue BPA, Broekaert WF (1995) Small cystein-rich antifungal proteins from radish: Their role in host defense. Plant Cell 7:573–588

Tumer NE, O'Connel KM, Nelson RS, Sanders PR, Beachy RN (1987) Expression of alfalfa mosaic virus coat protein gene confers cross-protection in transgenic tobacco and tomato plants. EMBO J 6:1181–1188

van Dun CMP, Bol JF, van Vloten-Doting L (1987) Exrpession of alfalfa mosaic virus and tobacco rattle virus coat protein genes in transgenic tobacco plants. Virology 159:299–305

Warren GW, Carozzi NB, Desai N, Koziel MG (1992) Field evaluation of transgenic tobacco containing a Bacillus thuringiensis insecticidal protein gene. J Econ Entomol 85:1651–1659

Wilson TMA (1993) Strategies to protect crop plants against viruses: pathogen-derived resistance blossoms. Proc Natl Acad Sci USA 90:3134–3141

Wisniewski LA, Powell PA, Nelson RS, Beachy RN (1990) Local and systemic spread of tobacco mosaic virus in transgenic tobacco. Plant Cell 2:559–568

Woodleif WG, Chaplin JF, Campbell CR, DeJong DW (1981) Effect of variety and harvest treatments on protein yield of close-grown tobacco. Tob Sci 25:83–86

Wu X, Beachy RN, Wilson TMA, Shaw JG (1990) Inhibition of uncoating of tobacco mosaic virus particles in protoplast form transgenic tobacco plants that express the viral coat protein gene. Virology 179:893–895

Yie Y, Wu ZX, Wang SY, Zhao SZ, Zhang TQ, Yao GY, Tien P (1995) Rapid production and field testing of homozygous transgenic tobacco lines with virus resistance conferred by expression of satellite RNA and coat protein of cucumber mosaic virus. Trans Res 4:256–263

XXVI Genetic Transformation of *Vinca minor* L.

N. Tanaka

1 Introduction

The lesser periwinkle (*Vinca minor* L.), a horticultural plant (family Apocynaceae), has lilac-blue flowers that bloom from spring to summer and is mainly cultivated in Europe as ground cover (Fig. 1A). Since the plant also produces vincamine (Fig. 1C), an important alkaloid that accumulates mainly in the leaves, it has also been utilized as a medicinal herb by Europeans for centuries. Indeed, this alkaloid has a powerful hypotensive action; it regulates cerebral circulation, enhances the usage of oxygen by the brain, and is very well tolerated by the human body. Moreover, vinpocetine (Fig. 1C), a derivative of vincamine (Oniscu et al. 1985), is a cerebral vasodilator that is marketed by some pharmaceutical companies in Europe and Japan.

In the field, the growth of *V. minor* plants is relatively slow. Moreover, since the plant is usually propagated from cuttings rather than from seeds, the number of plants that can be propagated at any one time is very limited. Thus, a method for improving propagation is needed. Stapfer and Heuser (1985) reported that culture in vitro, with a cytokinin such as benzyladenine or kinetin, resulted in an increase in shoot number; however, no similar effect was observed by Tanaka et al. (1995) when they attempted to reproduce these results using these growth regulators.

Some attempts have been made to establish callus cultures that produce vincamine in large amounts (Petiard and Demarly 1972; Garnier et al. 1975). However, stable, high rates of production of vincamine have not been achieved in such unorganized cell cultures. Thus, organ cultures, and, in particular, multiple-shoot cultures, are expected to be the best source of vincamine because the alkaloid is present mainly in the leaves. In the case of another related plant, *Catharanthus roseus (Vinca rosea)*, which produces important indole alkaloids such as ajmalicine, serpentine, catharanthine, vindoline, vinblastine, and vincristine, Hirata et al. (1987) have already reported the production of several alkaloids in multiple-shoot cultures. As noted above, it is difficult to induce multiple-shoot cultures of *V. minor* by addition of a cytokinin. Thus, the most likely way to enhance the proliferative capability of *V. minor* would seem to involve genetic manipulation to increase alkaloids.

Center for Gene Science, Hiroshima University, 1-4-2 Kagamiyama, Higashi-Hiroshima, Hiroshima 739, Japan

Biotechnology in Agriculture and Forestry, Vol. 45
Transgenic Medicinal Plants (ed. by Y.P.S. Bajaj)
© Springer-Verlag Berlin Heidelberg 1999

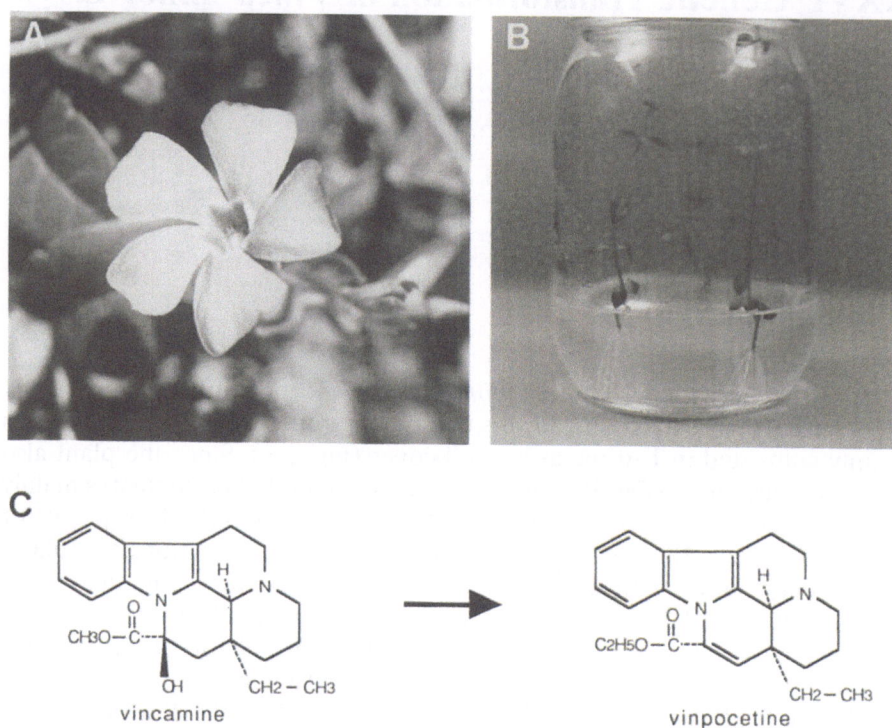

Fig. 1. A *Vinca minor* plant cultivated in the field. **B** Plantlets cultured on MS medium for 1 month. **C** Vincamine and its derivative, vinpocetine

Hairy roots can be cultured and shoots regenerate spontaneously, or upon artificial stimulation, from the roots, which also contain T-DNA (a pRi-transformed regenerant). The pRi-transformed regenerants frequently have decreased internode distances, reduced apical dominance, and altered geotropism of roots, the results of what is known as the hairy root syndrome. Moreover, the pRi-transformed regenerants cultured in vitro usually grow rapidly because of both increased formation of lateral buds and rapid development of leaves (Tanaka and Matsumoto 1993). This trait prompted us to use this technique to improve the proliferative capacity of *V. minor*.

2 Genetic Transformation

Many attempts have been made to produce the important indole alkaloids of *C. roseus* by applying tissue and organ culture (Payne et al. 1988; Scragg et al. 1989, Garnier et al. 1996a,b). Successful transformation with *Agrobacterium* has been achieved and some hairy root lines producing a high level of indole alkaloid have been established (Parr et al. 1988; Toivonen et al. 1989; Bhadra et al. 1993). Recently, suspension-cultured cells of *C. roseus* were transformed by infection with *A. tumefaciens* harboring a plasmid that contained a gene of isopentenyl transferase (Garnier et al. 1996a,b). Moreover, by particle bombardment, suspension-cultured cells were stably transformed with a plasmid containing a gene of *gusA* controlled by CaMV 35S and FMV 34S promoter derivatives (van der Fits and Memelink 1997).

By contrast, no attempts have been made, to our knowledge, to generate transformants of *V. minor*. In this chapter a new protocol for transformation of *V. minor* by induction of hairy roots is described.

3 Initiation and Culture of Hairy Roots

Plantlets cultured in vitro are suitable for inoculation with *A. rhizogenes* since they can be prepared aseptically to avoid contamination by other bacteria and fungi (Fig. 1B). Such cultured plantlets also contain vincamine in their leaves (Fig. 2B), but the level is only one third of that in leaves of plants cultivated for at least 3 years in the field.

Leaf disks and/or stem segments with a node (ca. 2 cm in length) were excised from the 30-day-cultured plantlets and cocultured with *A. rhizogenes* strain DC-AR2 (Tanaka et al. 1993), which harbored the mikimopine-type Ri plasmid pRi1724 (Tanaka and Oka 1994), as shown in Fig. 3, on water that had been solidified with 1% agar, at 25 °C for 3 days under light. Subsequently, the inoculated plant materials were cultured on 0.2% Gelrite-solidified MS medium (Murashige and Skoog 1962) with or without naphthaleneacetic acid (NAA), plus antibiotics to eliminate *Agrobacterium*, at 25 °C in the dark. About 20 days after inoculation, adventitious roots emerged from the edges and nodes of inoculated stem segments that has been cultured on MS medium plus 1 mg/l NAA (Fig. 4A), whereas no roots were generated on hormone-free medium or from leaf disks (Tanaka et al. 1994). However, since continuous culture on medium that contained a high level of auxin suppressed the elongation of adventitious root and induced them to dedifferentiate, even though new roots emerged one by one (Fig. 4B), the concentration of NAA was fixed between 0.01 and 0.05 mg/l after test culture at various concentrations of NAA (Fig. 4C). It appeared that a low concentration of NAA allowed growth as well as the initiation of hairy roots on *V. minor*. As shown in Fig. 3, pRi1724 is a weakly virulent Ri plasmid that contains no gene for biosynthesis of auxin (Tanaka and Oka 1994). Thus, an auxin is required for induction of hairy roots when plant tissues inoculated with *A. rhizogenes* fail to respond appropriately.

Since mikimopine, an opine produced by transformed cells, was detected in the adventitious roots, they were assumed to have been transformed. Results of Southern blot analysis also strongly supported this evidence of transformation since a signal that corresponded to a 7.6-kb fragment of T-DNA was detected (Fig. 5A).

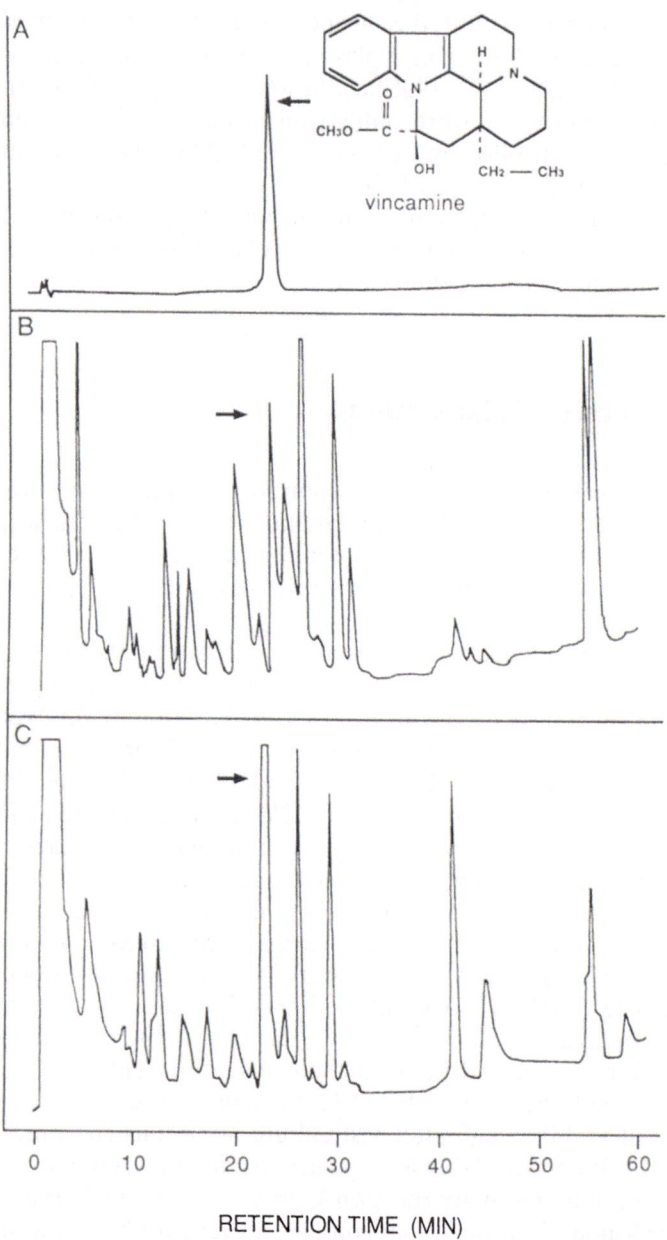

vincamine

Fig. 2A–C. HPLC profiles of methanol extracts from an original plant and a cultured plantlet. Chromatograms of methanol extracts detected at 283 nm. **A** Authentic vincamine. **B** An extract from the leaves of original plant. **C** An extract from leaves of the cultured plantlet. *Arrows* indicate peaks of vincamine. The structure of vincamine is also shown in **A**

RETENTION TIME (MIN)

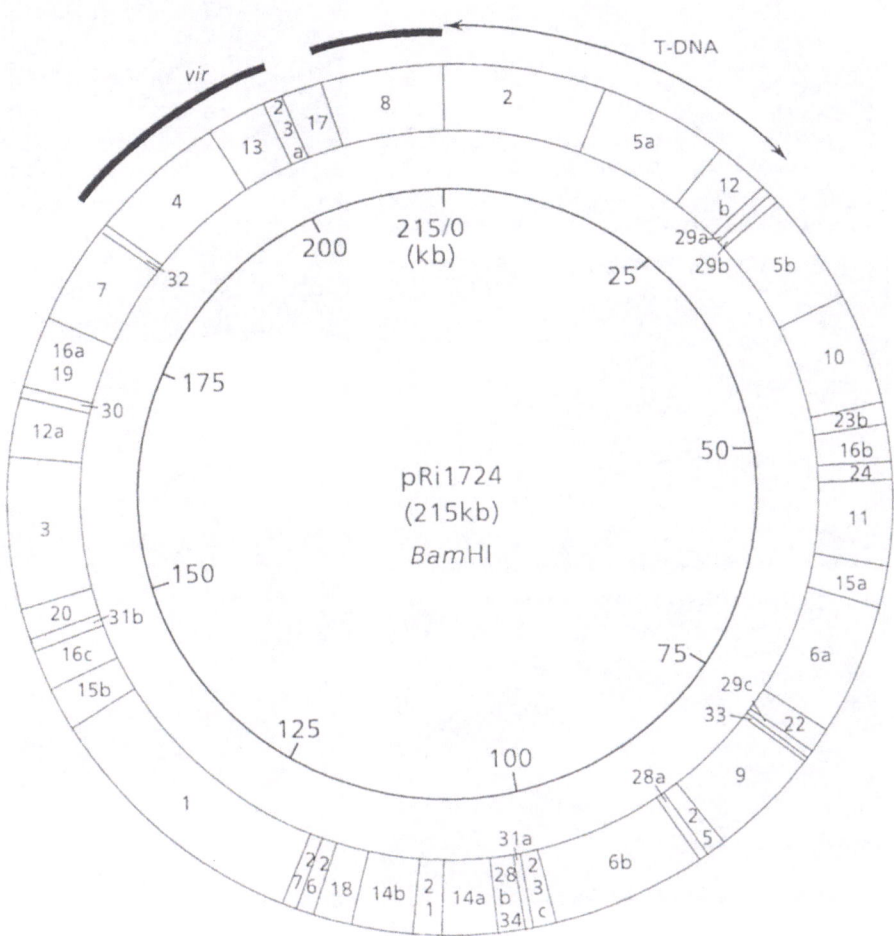

Fig. 3. Circular arrangement map of pRi1724

When *A. rhizogenes* harboring both pRi1724 and the binary vector pBI121 (Jefferson et al. 1987) that include genes for neomycin phospho-transferase II (NPTII) and β-glucuronidase (GUS) was used to inoculate stems cut from a cultured plant of *V. minor*, kanamycin-resistant adventitious roots were formed. These adventitious roots were transformed with the T-DNAs of both pRi1724 and pBI121, simultaneously. Southern blot analysis also revealed that most of the hairy roots that exhibited kanamycin resistance contained both T-DNAs (Fig. 5B), whereas some contained no T-DNA from pRi1724 and could be regarded as normal roots (Table 1).

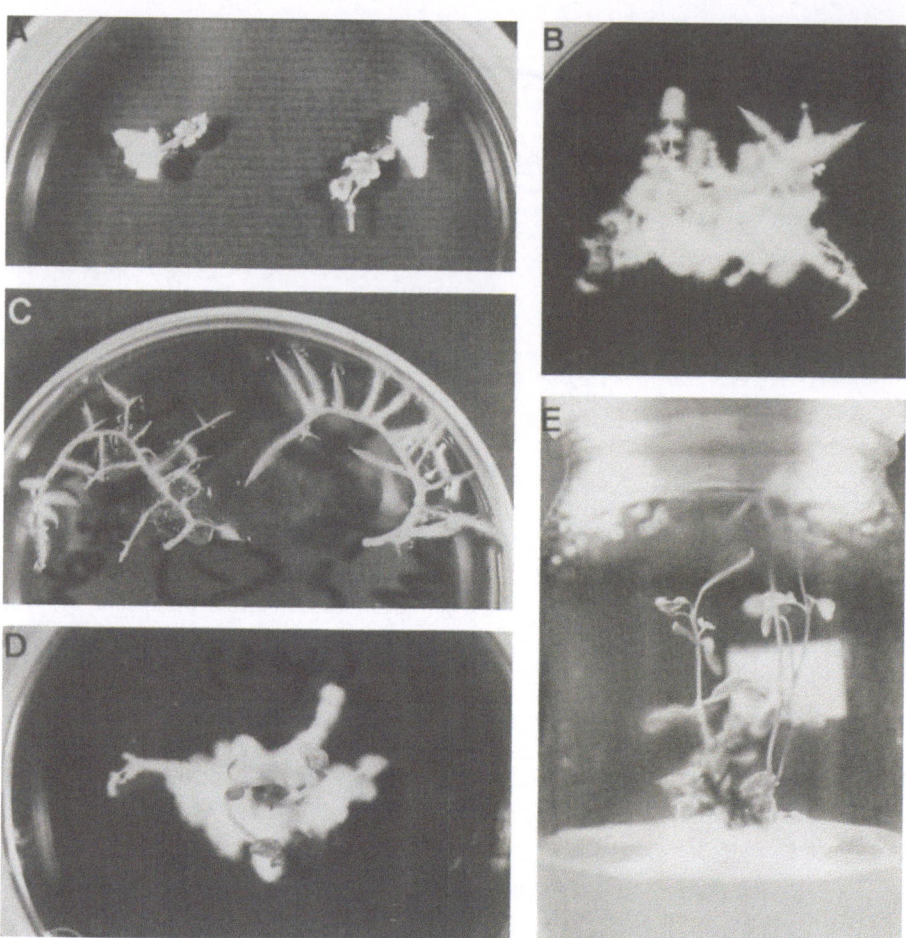

Fig. 4A–E. Induction and culture of hairy roots and their regenerants of *V. minor*. **A** Hairy roots induced by inoculation with *A. rhizogenes*. **B,C** Hairy roots cultured on MS medium supplemented with 1 mg/l (**B**) and 0.1 mg/l (**C**) NAA. **D,E** Shoots regenerated from hairy roots in the dark (**D**) and continuously cultured plantlets (**E**). (Tanaka et al. 1994)

Table 1. Characteristics of pBI121-transformed hairy roots of *Vinca* minor. (Tanaka et al. 1994)

Roots	Kanamycin resistance[a]	Mikimopine[b]	GUS activity[c]
pRi1724-transformed	−	+	−
pBI121-transformed			
1	+	+	+
2	+	+	+
3	+	+	+
4	+	+	+
5	+	−	+

[a] Cultured on MS medium containing 50 µg/ml kanamycin monophosphate.
[b] Analyzed by paper electrophoresis.
[c] Observed under UV light at 360 nm.

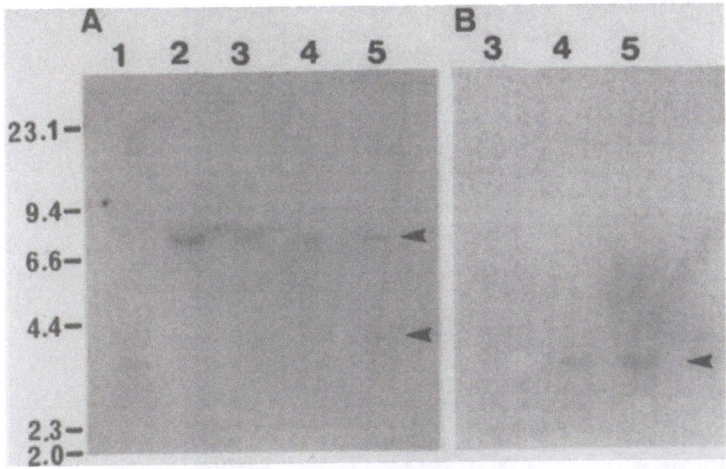

Fig. 5A,B. Detection of T-DNA and GUS gene of pBI121 in the pRi-transformed regenerants by Southern blot hybridization analysis. Two μg of each NDA was digested with *Eco*RI (**A**) or *Hind*III-*Eco*RI (**B**). **A** A 7.6-kb fragment, in which *rolABC* genes of pRi1724 exist, isolated from *Eco*RI digests of pRTE7.6, was used as a probe. An *upper arrowhead* and a *lower one* indicate the hybridization signal bands corresponding to 7.6 kb of *Eco*RI fragment and 3.5 kb of unknown fragment, detected in both transformed and untransformed plants, respectively. **B** A 3.0-kb fragment, in which the GUS gene exists, isolated from *Hind*III-*Eco*RI digests of pBI221 was used as a probe. An *arrowhead* indicates the hybridization signal band corresponding to 3.0 kb of *Hind*III-*Eco*RI fragment. *Lane 1* Untransformed plant; *lanes 2 and 3* pRi-transformed regenerants Vm-101 and Vm-102; *lanes 4 and 5* pRi- and pBI121-transformed regenerants Vm-201 and Vm-203. The *numerals on the left* are DNA sizes in kb referred from λ/*Hind*III digests. (Tanaka et al. 1994)

4 Characteristics of Transformed Regenerants Cultured in Vitro

Shoots were spontaneously regenerated from hairy roots during subsequent culture on MS medium that contained 1 mg/l NAA in the dark (Fig. 4D). The shoots cut from the hairy roots developed firm, green leaves; they rooted and grew vigorously during culture on hormone-free MS medium under light (Fig. 4E).

In addition to typical symptoms of the hairy root syndrome (Fig. 6), such as decreased internode distance, reduced apical dominance, and altered geotropism of roots, the various lines of regenerants displayed some differences (Tanaka et al. 1994). With variations in the severity of hairy root syndrome, in particular, the decrease in the internode distance and the reduction in apical dominance, the growth of regenerants also differed among lines. The mean numbers of shoots of two pRi-transformed regenerants, Vm-101 and Vm-102, were 11.1 and 8.8, whereas that of a cultured untransformed plantlet was 1 (Table 2). Benzyladenine has no effect on the number of shoots on pRi-transformed regenerants, which was similar to untransformed plantlets.

Fig. 6. A representative pRi1724-transformed regenerant of clone Vm-102. The plant was cultured for 2 weeks. (Tanaka et al. 1994)

Table 2. Effect of BA or sucrose on shoot number, vincamine content, and growth index on Vm-101 plants. (Tanaka et al. 1995, with kind permission from Kluwer Academic Publishers)

Treatment	Shoot number	Vincamine content (% in dry weight basis)	Growth index
BA (μM)[a]			
0	11.1 ± 2.3	–	10.4 ± 3.3
2.2	15.3 ± 4.3	–	10.2 ± 4.8
4.4	13.4 ± 3.1	–	5.8 ± 1.7
6.6	12.9 ± 2.5	–	6.7 ± 1.2
8.8	11.6 ± 2.1	–	4.9 ± 0.9
Sucrose (%)[b]			
3	–	0.39 ± 0.04	9.9 ± 2.9
4	–	0.38 ± 0.03	10.9 ± 3.7
5	–	0.50 ± 0.10	8.1 ± 2.6
6	–	0.31 ± 0.01	5.2 ± 1.7

[a] The data (mean value ± standard deviation) were calculated from the measurement of eight plants.
[b] The data (mean value ± standard deviation) were calculated from the measurement of four plants.

Moreover, Vm-101 has a growth index (defined as the quotient of fresh weight of the harvested plantlet divided by the fresh weight of the implanted shoot tip) of approximately 10 as compared with that of approximately 4 for the untransformed plantlets (Table 2). Thus, it appeared that the regenerants were plant materials with highly proliferative capacity that were suitable for micropropagation without a requirement for BA.

5 Production of Vincamine in Multiple-Shoot Cultures of pRi-Transformed Regenerants of *V. minor*

One superior cultured line, Vm-101, with high proliferative ability, also contained twice as much vincamine in its leaves as in leaves of untransformed plants (Table 3; Tanaka et al. 1995). The vincamine content of pRi-transformed regenerants was variable and, moreover, the high level in Vm-101 was not a general characteristic of pRi-transformed regenerants.

In the case of untransformed in vitro plantlets, culture on MS medium supplemented with 5–6% sucrose resulted in an increase in the growth index and vincamine content (Tanaka et al., unpubl.). When shoot tips of Vm-101 were cultured on MS medium supplemented from 3 to 6% sucrose, the maximum level of vincamine, 0.5% on a dry weight basis, was detected in shoots grown on the medium with 5% sucrose, while the growth index at 5% sucrose did not differ from that at 3%. Clearly, the pRi-transformed plant had a response to sucrose similar to that of the untransformed one although the extent of the response was small.

Table 3. Content of vincamine in untransformed and pRi-transformed regenerants of *V. minor*. (Tanaka et al. 1995, with kind permission from Kluwer Academic Publishers)

Plant	Source of tissue	Vincamine content in leaf[a] (% on dry weight basis)
Untransformed[b]	Tissue culture	0.24 ± 0.06
pRi-transformed[b]	Tissue culture	
Vm-101		0.44 ± 0.10
-102		0.33 ± 0.06
-103		0.30 ± 0.12
-104		0.18 ± 0.10
Untransformed[c]	Greenhouse	0.20 ± 0.10
pRi-transformed[c]	Greenhouse	
Vm-101		0.42 ± 0.11

[a] The data (mean value ± standard deviation) were calculated from the measurement of three plants.
[b] Cultured at 25 °C with 12-h photoperiod for 30 days.
[c] Cultivated for 3 months after 1 month acclimatization in a greenhouse.

6 Characteristics of Transformed Regenerants Cultivated in the Greenhouse

After acclimation, plants of the Vm-101 line were cultivated in the greenhouse for 3 months. These cultivated pRi-transformed plants had short internodes and reduced apical dominance. These plants show the same morphology as 5 years earlier. Thus, the rapid growth observed with the shoot cultures might have been a property that was displayed only in vitro (Fig. 6).

The vincamine content of leaves of pRi-transformed regenerants was twice that in the untransformed plants (Table 3), and the relative values were similar to those in the plant materials in tissue culture (Tanaka et al. 1995). Thus, the results indicated that the cultured line could be stably maintained during cultivation in a greenhouse. By contrast, the vincamine content of leaves of untransformed plants was only about 25% of that of leaves of the original untransformed plants that had been cultivated for at least 3 years in the field (Table 3). The difference in vincamine content between the two types of plant might have been attributable to the age of the leaves and/or the duration of cultivation. The environmental stress associated with production in culture of *V. minor* plants (effects of temperature and irradiation) might also have been a factor. Thus, when the pRi-transformed line Vm-101 is cultivated in the field, it might have a higher vincamine content than that of the original untransformed plants. This possibility remains to be examined.

7 Summary and Conclusions

Transformed regenerants of *Vinca minor* were obtained from hairy roots, induced by infection with *A. rhizogenes* strain DC-AR2 that harbored pRi1724 and/or the binary vector pBI121, after culture on MS medium supplemented with 1 mg/l NAA. The pRi-transformed regenerants that had been cultured in vitro displayed typical symptoms of hairy root syndrome with improved proliferative ability under culture conditions without cytokinin. Among the pRi-transformed regenerant lines, one superior line, Vm-101, not only had high proliferative ability but also produced almost double the amount of vincamine than untransformed cultured plantlets. An increase in the concentration of sucrose in the medium had no effects on the increased accumulation of vincamine in pRi-transformed regenerants but did alter it in untransformed plantlets. Vm-101 plants cultivated in a greenhouse accumulated high levels of vincamine in their leaves but did not have a high proliferative ability. Thus, it seems that propagation in vitro of pRi-transformed regenerants with both high proliferative ability and high vincamine productivity should permit the efficient production of vincamine by *V. minor*.

References

Bhadra R, Vani S, Shanks JV (1993) Production of indole alkaloids by selected hairy root lines of *Catharanthus roseus*. Biotechnol Bioeng 37:673–680

Garnier F, Label P, Hallard D, Chenieux J-C (1996a) Transgenic periwinkle tissue overproducing cytokinins do not accumulate enhanced levels of indole alkaloids. Plant Cell Tissue Organ Cult 45:223–230

Garnier F, Carpin S, Label P, Rideau M, Hamdi S (1996b) Effect of cytokinin on alkaloid accumulation in periwinkle callus cultures transformed with a light-inducible ipt gene. Plant Sci 120:47–55

Garnier J, Kunesch N, Siou E, Poisson J, Kunesch G, Koch M (1975) Etude des cultures de tissue de *Vinca minor* isolement d'un lignane, le lirioresinol B. Phytochemistry 14:1385–1387

Hirata K, Yamanaka A, Kurano N, Miyamoto K, Miura Y (1987) Production of indole alkaloids in multiple-shoot culture of *Catharanthus roseus* (L). G. Don. Agric Biol Chem 51:1311–1317

Jefferson RA, Kavanagh TA, Bevan MW (1987) GUS fusions: β-glucuronidase as a sensitive and versatile gene fusion marker in higher plants. EMBO J 6:3901–3907

Murashige T, Skoog F (1962) A revised medium for rapid growth and bioassays with tobacco tissue cultures. Physiol Plant 15:473–497

Oniscu C, Mocoveanu M, Horoba E, Cojocaru M, Potorac E (1985) The study of the extraction process of vincamine. Rev Roum Chem 30:807–815

Parr AJ, Peerless ACJ, Hamill JD, Walton NJ, Robins RJ, Rhodes MJC (1988) Alkaloid production by transformed root cultures of *Catharanthus roseus*. Plant Cell Rep 7:309–312

Payne GF, Payne NN, Shuler MI, Asada M (1988) In situ adsorption for enhanced alkaloid production by *Catharanthus roseus*. Biotechnol Lett 10:187–192

Petiard V, Demarly Y (1972) Mise en evidence de glucosides et d'alcaloides dans les cultures de tissue vegetaux. Ann Amelior Plantes 22:361–374

Scragg AH, Cresswell RC, Ashton S, York A, Bond PA, Fowler MW (1989) Growth and alkaloid production in bioreactors by a selected *Catharanthus roseus* cell line. Enzyme Microb Technol 11:329–333

Stapfer RE, Heuser CW (1985) In vitro propagation of periwinkle. Hortscience 20:141–142

Tanaka N, Matsumoto T (1993) Characterization of *Ajuga* plant regenerated from hairy roots. Plant Tissue Cult Lett 10:78–83

Tanaka N, Oka A (1994) Restriction endonuclease map of the root-inducing plasmid (pRi1724) of *Agrobacterium rhizogenes* strain MAFF03-01724. Biosci Biotechnol Biochem 58:297–299

Tanaka N, Takao M, Matsumoto T, Machida Y (1993) Transformation of *Agrobacterium rhizogenes* strain MAFF03-01724 by electroporation. Ann Phytopath Soc Jpn 59:587–593

Tanaka N, Takao M, Matsumoto T (1994) *Agrobacterium rhizogenes*-mediated transformation and regeneration of *Vinca minor* L. Plant Tissue Cult Lett 11:191–198

Tanaka N, Takao M, Matsumoto T (1995) Vincamine production in multiple-shoot culture derived from hairy roots of *Vinca minor*. Plant Cell Tissue Organ Cult 41:61–64

Toivonen L, Balsevich J, Kurz WGW (1989) Indole alkaloid production by hairy root cultures of *Catharanthus roseus*. Plant Cell Tissue Organ Cult 18:79–93

van der Fits L, Memelink J (1997) Comparison of the activities of CaMV 35S and FMV 34S promoter derivatives in *Catharanthus roseus* cells transiently and stably transformed by particle bombardment. Plant Mol Biol 33:943–946

Subject Index